Organic Reactions

γ-Lactones and Unsaturated Acids. When the product of the Stobbe condensation with a ketone RCOR' is heated with a mixture of halogen acid, water, and acetic acid, the half-ester is hydrolyzed and the unsaturated dicarboxylic acid loses carbon dioxide to produce a γ-lactone according to the scheme shown in Chart 3. In some reactions an isomeric unsaturated acid is also produced, and in four of these reactions it has been demonstrated that the two products are interconvertible, thus rendering the method useful for the synthesis of either the lactones or unsaturated acids. This interconvertibility represents a true "lactoenoic" tautomerism, and the proportion of products produced in the decarboxylation generally represents the equilibrium mixture.[81]

Apparently it is necessary that at least one of the R groups be aryl in order to realize decarboxylation by this method. The Stobbe condensation product from cyclohexanone, for example, fails to lose carbon dioxide even on prolonged heating with the hydrobromic-acetic acid mixture, and gives exclusively γ,γ-pentamethyleneparaconic acid (98%).[27] Decarboxylation of the paraconic acid, however, can be effected by pyrolysis.[82]

Another γ-lactone synthesis, quite unrelated to those described above, involves the aluminum-amalgam reduction of substituted succinic anhydrides which may be prepared by hydrogenation and cyclodehydration of products of the Stobbe condensation. Dialkylidenesuccinic acids have been employed in this manner by the scheme indicated in the accompanying formulas.[19, 83, 84, 85] When Ar = 3,4-methylenedioxyphenyl the product is dl-hinokinin and on resolution gives material identical with the natural product.[19, 83] The related lactone, matairesinol, was synthesized in a similar manner (Ar = 3-methoxy-4-hydroxyphenyl).[85]

$$\begin{array}{c} ArCH=CCO_2H \\ | \\ ArCH=CCO_2H \end{array} \rightarrow \begin{array}{c} ArCH_2CHCO_2H \\ | \\ ArCH_2CHCO_2H \end{array} \rightarrow \begin{array}{c} ArCH_2CHCO \\ | \quad \rangle O \\ ArCH_2CHCO \end{array} \rightarrow \begin{array}{c} ArCH_2CHCH_2 \\ | \quad \rangle O \\ ArCH_2CHCO \end{array}$$

The Naphthol Synthesis. Alkylidenesuccinic acids or half-esters having the appropriate stereochemical configuration, viz., an aryl group cis to the CH_2CO_2H group, may undergo cyclodehydration and enolization to give a substituted 1-naphthol-3-carboxylic acid as represented

---

[81] For a discussion of the mechanism see Johnson and Heinz, *J. Am. Chem. Soc.*, **71**, 2913 (1949).
[82] Johnson and Hunt, *J. Am. Chem. Soc.*, **72**, 935 (1950).
[83] Keimatsu, Ishiguro, and Nakamura, *J. Pharm. Soc. Japan*, **55**, 775 (1935), [*C. A.*, **29**, 7961 (1935)].
[84] Haworth and Woodcock, *J. Chem. Soc.*, **1939**, 154.
[85] Haworth and Slinger, *J. Chem. Soc.*, **1940**, 1098.

$$\text{LVI} + \text{CH}_2\text{C}(\text{CH}_3)=\text{CHCO}_2\text{CH}_3 \rightarrow \text{LVII}$$

(Structure LVI: 4,4-dimethylcyclohexanone with CO$_2$CH$_3$ at α-position; Structure LVII: 4,4-dimethylcyclohexene with CHC(CH$_3$)=CHCO$_2$H and CO$_2$CH$_3$ substituents)

The condensation of the esters of o-benzoylbenzoic acid with t-butyl acetate (p. 4) is also related to the Stobbe condensation.

APPLICATIONS

Besides the obvious general use for preparing many varieties of unsaturated and (by hydrogenation) saturated substituted succinic acids, the Stobbe condensation has found wide application in the synthesis of other types of substances, including substituted lactones, naphthols, indones, tetrahydroindanones, and tetralones. These applications have led to the synthesis of such substances as hinokinin, matairesinol, 2-methylazulene, cadalene; structures related to the steroids, including equilenin and bisdehydrodoisynolic acid; and polycyclic aromatic compounds in the benzanthracene, naphthacene, and 3,4-benzphenanthrene series. The general synthetic methods and their applications are considered below.

Lactonic Acids. Alkylidenesuccinic acids (or half-esters) on treatment with bromine give substituted bromoparaconic acids (or esters) according to the first step of the equation in Chart 1, p. 25. When these bromo lactonic acids are treated with boiling water they lose hydrogen bromide, generally giving α,β-unsaturated lactonic acids ("aconic acids") according to the second step of the same equation. The bromo lactonic acids and unsaturated lactones that have been prepared in this manner are summarized in Chart 1.

The action of bromine on alkenylsuccinic acids takes a somewhat different course (see p. 15 for discussion). Bromo lactonic acids are formed, which on heating with water or dilute alkali usually yield dilactones, as well as isomeric unsaturated lactonic acids as depicted in Chart 2.

Saturated lactonic acids (substituted paraconic acids) have been prepared both by reduction of the bromo or unsaturated lactonic acids and by direct lactonization of the alkylidene- or alkenyl-succinic acids, but as yet these reactions have not received extensive application.[4, 27, 80]

[80] Linstead and Mann, *J. Chem. Soc.*, **1930**, 2064.

with a highly reactive α-methylene group. The condensation of ethyl β-veratroylpropionate with benzaldehyde and sodium methoxide has also been described as giving the β-benzylidene derivative.[73]

Certain aspects of the use of diethyl glutarate in a Stobbe type of condensation have been considered above (p. 5). This ester fails to condense to any appreciable extent with benzophenone under the same conditions that promote condensation with diethyl succinate in 90% yield.[74] When di-t-butyl glutarate was employed instead of the diethyl ester with the hope of inhibiting the competing self-condensation of the ester, the half-ester, $(C_6H_5)_2C{=}C(CO_2C_4H_9\text{-}t)CH_2CH_2CO_2H$, was obtained in poor yield.[3] With a ketone containing a reactive methylene group, diethyl glutarate reacts preferentially by the acetoacetic ester type of condensation. Thus cyclohexanone, which is highly reactive in the Stobbe condensation,[27] gives only the diketo ester,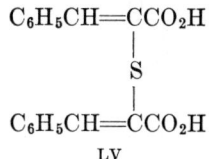

$COCH_2CH_2CH_2CO_2C_2H_5$.[5] 1-Tetralone gives an analogous product with diethyl glutarate,[75] even though this ketone undergoes the Stobbe condensation readily.[76]

Diethyl thiodiglycolate, the sulfur analog of diethyl glutarate, appears to be reactive in the Stobbe type of condensation as indicated by the condensation with benzaldehyde to give dibenzylidenethiodiglycolic acid (LV) in 62–74% yield.[77] The sulfur atom probably assists the reaction by exerting a proton-releasing effect on the α-carbon atoms.[78]

$$C_6H_5CH{=}CCO_2H$$
$$|$$
$$S$$
$$|$$
$$C_6H_5CH{=}CCO_2H$$
$$\text{LV}$$

The condensation of dimethyl β-methylglutaconate (LVI) with 3,3-dimethylcyclohexanone in the presence of potassium t-butoxide [79] appears to proceed by a Stobbe type of mechanism. The product is an oily mixture of half-esters which is produced in good yield and presumably contains LVII.

[73] Borsche, Hofmann, and Kühn, *Ann.*, **554**, 23 (1943).
[74] Johnson, unpublished observation.
[75] Johnson, Johnson, and Petersen, *J. Am. Chem. Soc.*, **68**, 1926 (1946).
[76] Johnson, Johnson, and Petersen, *J. Am. Chem. Soc.*, **67**, 1360 (1945).
[77] Stobbe, Ljungren, and Freyberg, *Ber.*, **59B**, 265 (1926).
[78] *Cf.* Woodward and Eastman, *J. Am. Chem. Soc.*, **68**, 2229 (1946).
[79] Bischof, Jeger, and Ruzicka, *Helv. Chim. Acta*, **32**, 1911 (1949).

ester LII. α,α-Disubstituted succinic esters would be expected to react normally in the Stobbe condensation, but no record of such an experiment has been found.

```
                                    C6H5
                                    |
    RCHCO2C2H5              C6H5CH——CCO2C2H5
    |                           |    |
    CH2CO2C2H5                  O    CH2
                                 \  /
                                  CO

        LI                          LII
```

Related Condensations

Reaction of benzaldehyde with triethyl carballylate and sodium ethoxide gives an acidic ester mixture which after saponification, acidification, and steam distillation affords in unspecified yield what is probably β-benzylideneglutaric anhydride LIII along with other unidentified products.[71] The formation of LIII is reasonable on the basis of the decarboxylation of an intermediary lactonic di-acid LIV as in the reaction of diethyl phenylsuccinate with benzaldehyde described in the preceding section.

```
                                              CH2CO2H
                                              |
              CH2—CO              C6H5CH——CCO2H
             /      \                 |    |
   C6H5CH=C          O                O    CH2
             \      /                  \  /
              CH2—CO                    CO

            LIII                        LIV
```

The condensation of benzaldehyde with ethyl β-benzoylpropionate and sodium ethoxide to give β-benzylidene-β-benzoylpropionic acid, $C_6H_5CH=C(COC_6H_5)CH_2CO_2H$, in 90% yield [72] resembles the Stobbe condensation in that the course of the attack (at the β- rather than the α-methylene group) may be explained by the formation of an intermediary keto lactone $C_6H_5\overline{CHCH(COC_6H_5)CH_2COO}$. The Perkin condensation of β-benzoylpropionic acid with benzaldehyde gives, in contrast, exclusively α-benzylidene-β-benzoylpropionic acid, $C_6H_5CH=C(CO_2H)CH_2COC_6H_5$, and this behavior may be rationalized by assuming the preliminary formation of the enol lactone, $C_6H_5\overline{C=CHCH_2COO}$,

---

[71] Müller, *Ber.*, **39**, 3590 (1906).
[72] Borsche, *Ber.*, **47**, 1108 (1914).

idenesuccinic acids XLIX have been named "fulgenic acids;" their anhydrides L, which are highly colored substances, are called "fulgides." More than 50 different fulgenic acids and fulgides varying both in the nature of the substituent groups and in configuration have been prepared.[61] The alkylidenesuccinic esters appear to condense somewhat more readily than diethyl succinate, possibly owing to an additional activating influence of the olefinic bond on the methylene group. The nitrobenzaldehydes, for example, gave only resinous substances with diethyl succinate, but with both diethyl isopropylidene- and benzhydryl-idene-succinate good yields of the dialkylidenesuccinic acids were realized.[62, 63, 64]

$$\begin{array}{cc}
\mathrm{R} \diagdown \\
\mathrm{C}\!\!=\!\!\mathrm{CCO_2H} \\
\mathrm{R'} \diagup \\
 | \\
\mathrm{R''} \diagdown \\
\mathrm{C}\!\!=\!\!\mathrm{CCO_2H} \\
\mathrm{R'''} \diagup \\
\mathrm{XLIX}
\end{array}
\qquad
\begin{array}{c}
\mathrm{R}\diagdown \\
\mathrm{C}\!\!=\!\!\mathrm{C}\!\!-\!\!\mathrm{CO}\diagdown \\
\mathrm{R'}\diagup \\
 \mathrm{O} \\
\mathrm{R''}\diagdown \diagup \\
\mathrm{C}\!\!=\!\!\mathrm{C}\!\!-\!\!\mathrm{CO} \\
\mathrm{R'''}\diagup \\
\mathrm{L}
\end{array}$$

The Stobbe condensation has also been carried out with saturated substituted succinic esters of the type represented by formula LI. The reaction proceeds normally when R = CH_3,[65, 66] $C_6H_5CH_2$,[67] $C_6H_5CH_2CH_2$,[68] and 3,4-$(CH_3O)_2C_6H_3$,[69] the aldehyde or ketone condensing at the methylene group. The condensation of benzaldehyde with diethyl phenylsuccinate, LI (R = C_6H_5), and sodium ethoxide, however, seems to involve the α-carbon holding (and thus activated by) the phenyl group; since the product isolated appeared to be β,γ-diphenylvinylacetic acid, $C_6H_5CH\!\!=\!\!C(C_6H_5)CH_2CO_2H$,[70] which could reasonably arise from hydrolysis and decarboxylation of an intermediary paraconic

---

[61] Many of these substances are described in an extensive work by Stobbe and his collaborators, *Ann.*, **380**, 1–129 (1911).

[62] Stobbe and Leuner, *Ber.*, **39**, 292 (1906).

[63] Stobbe and Küllenberg, *Ber.*, **38**, 4081 (1905).

[64] Bachman and Hoaglin, *J. Org. Chem.*, **8**, 300 (1943).

[65] Stobbe and Noetzel, *Ber.*, **39**, 1070 (1906).

[66] Stobbe and Gollücke, *Ber.*, **39**, 1066 (1906).

[67] Weizmann, *J. Org. Chem.*, **8**, 285 (1943).

[68] Bougault, *Bull. soc. chim. France*, **41**, 663 (1927).

[69] Richardson, Robinson, and Seijo, *J. Chem. Soc.*, **1937**, 835.

[70] Fichter and Latzko, *J. prakt. Chem.*, [2], **74**, 327 (1906).

XLVII (along with the corresponding lactone XLVIII), respectively, thus proving the configurations.[11] Another effective reagent is zinc chloride in a mixture of acetic acid and acetic anhydride, which promotes cyclization of the half-esters as well as the dibasic acids. The behavior with this reagent cannot be used to determine the position of the ester group, since the ring closure may be attended by ester exchange.[52, 58] With sodium acetate and acetic anhydride, however, no ester exchange occurs, and the half ethyl ester corresponding to XXXVI (β-$C_{10}H_7$ group instead of C_6H_5) cyclizes to the ester XLVI (R = $COCH_3$, R' = C_2H_5). This reagent has been used similarly to prove the configuration of unsymmetrical diarylmethylenesuccinic acids.[59, 60, 20]

Substituted Succinic Esters

An excellent method for obtaining dialkylidenesuccinic acids (XLIX) is the Stobbe condensation of a ketone or aldehyde with the di-ester of an alkylidenesuccinic acid (prepared by esterification of the product of a normal Stobbe condensation with an aldehyde or ketone). In certain cases lactonic acids are isolated; they are easily converted to the dibasic acids by heating with alcoholic metal alkoxide. These dialkyl-

[58] Johnson, Stromberg, and Petersen, *J. Am. Chem. Soc.*, **71**, 1384 (1949).
[59] Borsche and Leditschke, *Ann.*, **529**, 108 (1937).
[60] Borsche, Kettner, Gillies, Kuhn, and Manteuffel, *Ann.*, **526**, 1 (1936).

An alkylidenesuccinic acid (XXXVI or XXXVII, for example) generally reacts similarly to give a bromolactone (XL), which, however, on heating in water loses hydrogen bromide to give an unsaturated lactonic acid XLI. The reaction with the acids XXXVI and XXXVII is stereo-selective, giving different diastereoisomeric forms of XL, each of which gives the same lactonic acid XLI on dehydrobromination. In the benzophenone series,[53] treatment of the bromolactone XL (C_6H_5 in place of CH_3) with water effects decarboxylation as well as dehydrobromination to give the unsaturated lactone XLII.

$$\begin{array}{ccc}
\text{Br} & & \\
| & & \\
HO_2CC\!-\!-\!-\!CH_2 & HO_2CC\!=\!=\!=\!CH & HC\!=\!=\!=\!CH \\
| \quad\quad\quad\ | & | \quad\quad\quad\ | & | \quad\quad\quad\ | \\
H_5C_6C \quad\quad CO & H_5C_6C \quad\quad CO & (H_5C_6)_2C \quad\quad CO \\
| \ \diagdown \ \diagup & | \ \diagdown \ \diagup & \diagdown \ \diagup \\
H_3C \quad O & H_3C \quad O & O \\
\text{XL} & \text{XLI} & \text{XLII}
\end{array}$$

The behavior of the half-esters toward bromine has also afforded evidence for the position of the ester group. For example, the crystalline half-ester of XXXVII which could be isolated from the acetophenone condensation gave the ethyl ester of the bromo lactone XL. This result is consistent with the formulation of the ester grouping at that carboxyl of XXXVII which is attached to the doubly bonded carbon.[25, 31] Similar experiments have been performed in the benzophenone series.[25, 53]

The configurations of a few alkylidenesuccinic acids have been shown by cyclization experiments. With concentrated sulfuric acid the acid XXXVI was converted to the anhydride XLIII, and the acid XXXVII was cyclized to a mixture of the yellow indoneacetic acid XLIV and the corresponding colorless lactone XLV.[54] This treatment has been used in other related series, [55, 20] and in some instances the development of a deep color with sulfuric acid has been interpreted as an indication of the production of an indone derivative, suggesting that the aryl group is *cis* to CO_2H as in formula XXXVII.[56, 57]

The sulfuric acid method fails to give the expected products in the 2-acetylnaphthalene series, probably because of sulfonation of the naphthalene nucleus. Hydrogen fluoride, however, cyclizes the acids corresponding to XXXVI and XXXVII (β-$C_{10}H_7$ group instead of C_6H_5) to the phenanthrol XLVI (R = R' = H) and the benzindone

[53] Stobbe, *Ann.*, **308**, 89 (1899).
[54] Stobbe, *Ber.*, **37**, 1619 (1904).
[55] Stobbe and Horn, *Ber.*, **41**, 3983 (1908).
[56] Stobbe, Gademann, and Rose, *Ann.*, **380**, 87 (1911).
[57] Stobbe and Gademann, *Ann.*, **380**, 39 (1911).

drides which could be separated by fractional crystallization from carbon disulfide.[31] Hydrolysis of the anhydrides gave the pure stereoisomeric acids XXXVI and XXXVII.

$$\begin{array}{ccc}
\underset{\text{XXXV}}{\begin{array}{c}CH_2CO_2H\\|\\CHCO_2H\\|\\C\\\diagup\;\diagdown\\C_6H_5\quad CH_2\end{array}} &
\underset{\text{XXXVI}}{\begin{array}{c}HO_2CCH_2\diagdown\;\diagup CO_2H\\C\\\|\\C\\\diagup\;\diagdown\\C_6H_5\quad CH_3\end{array}} &
\underset{\text{XXXVII}}{\begin{array}{c}HO_2C\diagdown\;\diagup CH_2CO_2H\\C\\\|\\C\\\diagup\;\diagdown\\C_6H_5\quad CH_3\end{array}}
\end{array}$$

Evidence for the position of the double bond may be provided by oxidation of the sodium salts of the dibasic acids in aqueous solution with cold dilute potassium permanganate. The acid XXXV, for example, thus gave β-benzoylpropionic acid, $C_6H_5COCH_2CH_2CO_2H$, and the stereoisomers XXXVI and XXXVII both gave acetophenone.[31] An alkylidenesuccinic half-ester on ozonization has been observed to form the expected ketone and in addition ethyl pyruvate (from decarboxylation of $HO_2CCH_2COCO_2C_2H_5$), proving that the carbethoxyl group was located on a carbon attached to the double bond.[52] The alkylidene- but not the alkenyl-succinic acids are generally reduced by sodium amalgam. In this manner the acid XXXVI was converted to the saturated acid, $C_6H_5CH(CH_3)CH(CO_2H)CH_2CO_2H$, in almost quantitative yield with 4% sodium amalgam, while XXXV was unaffected.[31] The alkenylsuccinic acid XIII (R = R′ = C_6H_5) derived from desoxybenzoin is exceptional in that it is reduced under these conditions.[21]

Further evidence for the structures of these acids may be obtained from their behavior with bromine. An alkenylsuccinic acid (XXXV, for example) reacts rapidly to give a bromolactone (possibly XXXVIII) from which a second molecule of hydrogen bromide is eliminated on heating with water, giving a dilactone XXXIX.[31]

[52] Johnson and Goldman, *J. Am. Chem. Soc.*, **67**, 430 (1945).

cleavage is also observed, but is evidently slower, with the α-aryl cyano ketones XXVIII. α-Cyanosuberone has also been reported not to undergo the Stobbe condensation,[51] but this might be expected on account of the presence of the strongly enolizable α-hydrogen atom.

XXXIII   XXXIV

Methods of Isolation of Products and Proof of Structures and Configurations

The isolation of the Stobbe condensation product from a ketone that gives a single substance offers no unusual problems. If the half-ester cannot be obtained crystalline, the dibasic acid produced by saponification usually can. For the saponification of the half-ester aqueous barium or sodium hydroxide is used. The former reagent is preferred for substances that are sensitive to alkali, because the barium salts of the dibasic acids are generally insoluble and are thus essentially removed from the reaction medium as they are formed.

For the isolation of Stobbe condensation products from ketones which, like acetophenone, give rise to mixtures certain general techniques have proved useful. Sometimes a portion of one of the half-esters crystallizes,[31, 11] but frequently the crude half-ester mixture is obtained as an oil. Though this product is generally satisfactory for synthetic purposes, separation is necessary for the fundamental study of the reaction. Such a mixture of half-esters can be saponified with barium hydroxide to yield a mixture of solid dibasic acids which can be separated by crystallization or as described below.

Many alkylidenesuccinic acids have been separated from the alkenylsuccinic acids successfully by taking advantage of the difference in ease with which they undergo anhydride formation. Although there are exceptions,[27] it is generally true that alkylidenesuccinic acids form anhydrides more readily than do the alkenyl isomers. For example, the crude mixture of dibasic acids from acetophenone was treated for twelve hours at room temperature with acetyl chloride and the product washed with sodium bicarbonate solution. This extracted the alkenylsuccinic acid XXXV, leaving the neutral mixture of alkylidenesuccinic anhy-

[51] Cook, Philip, and Somerville, *J. Chem. Soc.*, **1948**, 164.

Cyano Ketones

The condensation of the α-cyano ketones XXVIII (R = R' = H),[46] XXVIII (R = OCH$_3$, R' = H),[46] and XXVIII (R = H, R' = OCH$_3$)[47] with dimethyl succinate and potassium t-butoxide does not give the normal Stobbe condensation product. Instead the cyano group is involved in the reaction, which produces a cyclic product XXIX via an intramolecular Thorpe type of reaction (possibly upon the expected intermediary paraconic ester XXX). The ring closure XXX → XXXI probably precedes the cleavage of the lactone ring to the half-ester salt (XXXI → XXXII), since the methylene group of the latter would probably be too unreactive to condense with the cyano group. Hydrolysis and decarboxylation of the intermediary imino keto acid XXXII to give XXIX apparently occurs during the isolation of the product. By using a large excess of succinate and t-butoxide, it is possible to obtain XXIX (R = R' = H) in 75–83% yields,[48] XXIX (R = OCH$_3$, R' = H) in 83% yield,[46] and XXIX (R = H, R' = OCH$_3$) in 73–78% yields.[47]

The presence of an aromatic nucleus in conjugation with the carbonyl group appears to be necessary for successful condensation of α-cyano ketones with succinates. Thus both 2-cyano-2-methylcyclohexanone[49] and the related tricyclic cyano ketone XXXIII[50] fail to condense, apparently because of the more rapid competing cleavage of the cyano ketone giving ring opening (to XXXIV in the latter instance). This

---

[47] Hirschmann and Johnson, unpublished observation.
[48] Johnson and Sharpe, unpublished observation.
[49] Johnson and Bumpus, unpublished observation.
[50] Johnson and Shelberg, unpublished observation.

Diketones

Only one diketone, benzil, has been employed in the Stobbe condensation.[44] With diethyl α-(1-phenethylidene)succinate and sodium ethoxide the product of mono condensation XXIV was isolated in unspecified yield. Benzilic and benzoic acids were also produced.

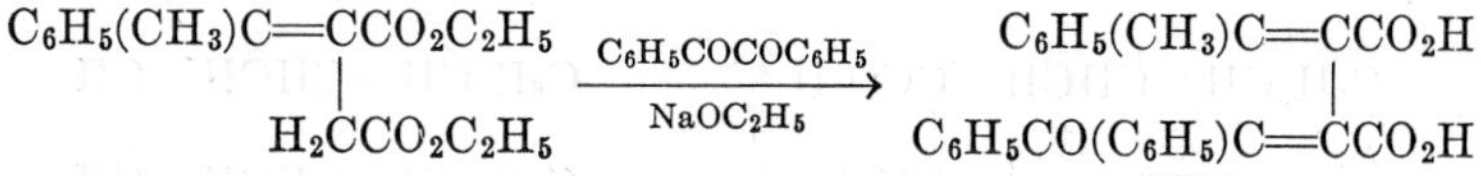

$$\begin{array}{c} C_6H_5(CH_3)C{=}CCO_2C_2H_5 \\ | \\ H_2CCO_2C_2H_5 \end{array} \xrightarrow[NaOC_2H_5]{C_6H_5COCOC_6H_5} \begin{array}{c} C_6H_5(CH_3)C{=}CCO_2H \\ | \\ C_6H_5CO(C_6H_5)C{=}CCO_2H \end{array}$$
$$\text{XXIV}$$

Keto Esters

The condensation of ethyl γ-anisoylbutyrate (XXV) with diethyl succinate and potassium *t*-butoxide fails under normal conditions (refluxing in *t*-butyl alcohol), only γ-anisoylbutyric acid being isolated from the reaction mixture. At room temperature, however, the condensation proceeds excellently to give an oily mixture of acid esters (formula XXVI representing one of the probable structures) in 98% yield.[45]

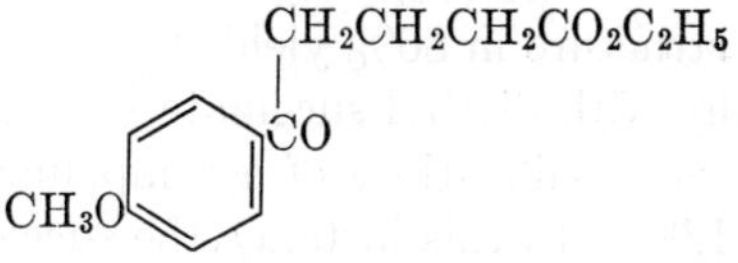

XXV

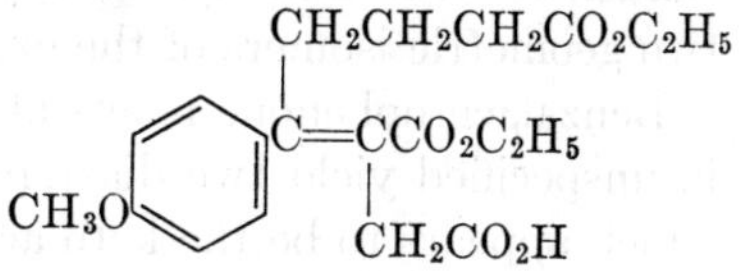

XXVI

Preliminary attempts to effect a Stobbe condensation between the β-keto ester XXVII and diethyl succinate, however, failed. The only product which could be isolated was 2-methyl-1-keto-1,2,3,4-tetrahydrophenanthrene (formula XXVII, H in place of CO_2CH_3), resulting from ketonic cleavage of the keto ester.[46]

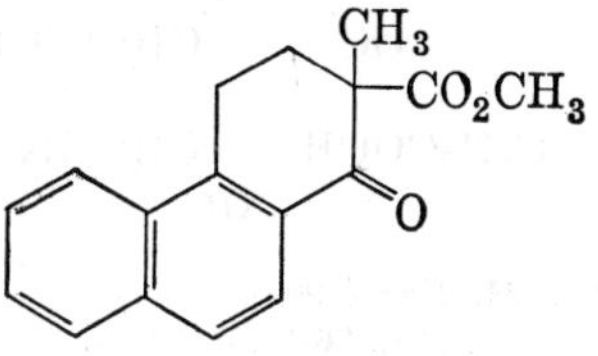

XXVII

---

[44] Stobbe, *Ber.*, **30**, 94 (1897).
[45] Johnson, Jones, and Schneider, *J. Am. Chem. Soc.*, **72**, 2395 (1950).
[46] Johnson, Petersen, and Gutsche, *J. Am. Chem. Soc.*, **69**, 2942 (1947).

α,β-Unsaturated Aldehydes and Ketones

Cinnamaldehyde behaves normally in the Stobbe condensation with diethyl succinate to give cinnamylidenesuccinic acid (XVIII).[35, 36] Distillation of the crude product yields also a hydrocarbon, probably 1,8-diphenyloctatetraene (XIX), arising from decarboxylation of dicinnamylidenesuccinic acid (XX).

$$\underset{\text{XVIII}}{\begin{array}{c} C_6H_5CH\!=\!CHCH\!=\!CCO_2H \\ | \\ CH_2CO_2H \end{array}} \qquad \underset{\text{XIX}}{\begin{array}{c} C_6H_5CH\!=\!CHCH\!=\!CH \\ | \\ C_6H_5CH\!=\!CHCH\!=\!CH \end{array}}$$

$$\underset{\text{XX}}{\begin{array}{c} C_6H_5CH\!=\!CHCH\!=\!CCO_2H \\ | \\ C_6H_5CH\!=\!CHCH\!=\!CCO_2H \end{array}}$$

The condensation of cinnamaldehyde with diethyl benzylidene-, isopropylidene-, and benzhydrylidene-succinate proceeds as expected, giving the dialkylidenesuccinic acids.[37] Also β-phenylcinnamaldehyde condenses with dimethyl benzhydrylidenesuccinate, giving a mixture of two geometric isomers of the expected structure in 80% yield.[38]

Benzalacetophenone reacts abnormally with diethyl succinate, giving in unspecified yield two diastereoisomeric modifications of a compound which appears to be the keto acid XXI.[39, 40] In this instance the succinate evidently reacts at the β-carbon atom by a Michael type of addition. Cyclization of the dimethyl ester of XXI gives the diketo ester XXII which is cleaved to the original acid XXI on alkaline hydrolysis.[41] It is noteworthy that ethyl cinnamate reacts similarly with diethyl succinate to give the addition product XXIII in 26–28% yield.[42, 43]

$$\underset{\text{XXI}}{\begin{array}{c} HO_2CCH_2CHCO_2H \\ | \\ C_6H_5COCH_2CHC_6H_5 \end{array}} \quad \underset{\text{XXII}}{\begin{array}{c} CH_2 \\ \diagup \quad \diagdown \\ OC \qquad CHCO_2CH_3 \\ | \qquad\quad | \\ C_6H_5COCH\!\text{———}\!CHC_6H_5 \end{array}} \quad \underset{\text{XXIII}}{\begin{array}{c} HO_2CCHCH_2CO_2H \\ | \\ C_6H_5CHCH_2CO_2H \end{array}}$$

---

[35] Fichter and Hirsch, *Ber.*, **34**, 2188 (1901).
[36] Alder, Pascher, and Schmitz, *Ber.*, **76B**, 27 (1943).
[37] Stobbe, Benary, and Seydel, *Ann.*, **380**, 113 (1911).
[38] Koelsch and Richter, *J. Org. Chem.*, **3**, 473 (1938).
[39] Stobbe, *Ann.*, **314**, 111 (1901).
[40] Stobbe and Russwurm, *Ann.*, **314**, 125 (1901).
[41] Stobbe and Fischer, *Ann.*, **314**, 142 (1901).
[42] Stobbe, *Ann.*, **315**, 219 (1901).
[43] Stobbe and Fischer, *Ann.*, **315**, 232 (1901).

alkenyl:alkylidene ratio is approximately 1 to 9,[31] but when R = CH$_3$ (propiophenone) this ratio is reversed.[29,32] With desoxybenzoin (R = C$_6$H$_5$) [25,33] it is expected that the alkenyl form XIII (R = R' = C$_6$H$_5$) would be favored,[24] and indeed this is the only product isolated.

Although all three isomeric condensation products are undoubtedly formed in many Stobbe condensations, their presence has been demonstrated infrequently, probably because of the experimental difficulties involved in separation. This problem is considered below (p. 14). It is noteworthy that formation of such a mixture usually does not interfere with the usefulness of the Stobbe condensation in many types of synthesis which eliminate the isomerism at a subsequent step (see p. 21).

Hindered Ketones. There are surprisingly few reports of failure of ketones to react at least to some extent with diethyl succinate in the Stobbe condensation. Perhaps this is a result of the lack of any special effort to study the limitations. The only comparative data available are in the desoxybenzoin series.[34] Desoxybenzoin (XVI, R = R' = R'' = R''' = H), itself undergoes satisfactory condensation with diethyl succinate and potassium *t*-butoxide to give, after saponification, exclusively the alkenylsuccinic acid XVII (R = H). Under similar conditions α-methyldesoxybenzoin (XVI, R = CH$_3$, R' = R'' = R''' = H) gives the acid XVII (R = CH$_3$) in 42% yield, but the second α-methyl substituent in α,α-dimethyldesoxybenzoin (XVI, R = R' = CH$_3$, R'' = R''' = H) prevents reaction completely. Even a single methyl group in the *ortho* position of the benzene nucleus inhibits the condensation of the ketone XVI (R = R' = H, R'' = R''' = CH$_3$). Similarly the ketones XVI (R = R'' = CH$_3$, R' = R''' = H) and XVI (R = R' = R'' = CH$_3$, R''' = H) fail to react. Thus in a variety of structures two α-methyl groups or a single *ortho* methyl group is sufficient to prevent condensation.

$$\underset{\text{XVI}}{\text{C}_6\text{H}_5\overset{\overset{\text{R}}{|}}{\underset{\underset{\text{R}'}{|}}{\text{C}}}\text{CO}\underset{\underset{\text{R}'''}{|}}{\overset{\overset{\text{R}''}{|}}{\bigcirc}}} \qquad \underset{\text{XVII}}{\text{C}_6\text{H}_5\overset{\overset{\text{R}}{|}}{\text{C}}=\overset{\overset{\text{CH}_2\text{CO}_2\text{H}}{|}}{\underset{\underset{\text{C}_6\text{H}_5}{|}}{\text{C}}}\text{CHCO}_2\text{H}}$$

[31] Stobbe, *Ann.*, **308**, 114 (1899).
[32] Stobbe and Niedenzu, *Ann.*, **321**, 94 (1902).
[33] Stobbe and Russwurm, *Ann.*, **308**, 156 (1899).
[34] Newman and Linsk, *J. Am. Chem. Soc.*, **71**, 936 (1949).

bond may be quite mobile, as illustrated by the behavior with cyclohexanone. Condensation with diethyl succinate and potassium *t*-butoxide yields a half-ester which is principally the cyclohexenyl compound (half-ester corresponding to XIV, $n = 3$).[27] Saponification,

$$\begin{array}{cc} \text{CH}_2\text{CO}_2\text{H} & \text{CH}_2\text{CO}_2\text{H} \\ | & | \\ \text{CHCO}_2\text{H} & \text{CCO}_2\text{H} \\ | & \| \\ \text{C} & \text{C} \\ \diagup\diagdown & \diagup\diagdown \\ \text{CH}_2 \quad \text{CH} & \text{CH}_2 \quad \text{CH}_2 \\ \diagdown\diagup & \diagdown\diagup \\ (\text{CH}_2)_n & (\text{CH}_2)_n \\ \text{XIV} & \text{XV} \end{array}$$

however, yields a mixture of cyclohexenyl-, XIV ($n = 3$), and cyclohexylidene-succinic acid, XV ($n = 3$).[27] This behavior shows that the ratio of products isolated does not necessarily correspond to the proportion produced in the reaction. Cycloheptanone affords a striking example of this phenomenon, since the half-ester from the Stobbe condensation appears to exist entirely in the cycloheptenyl form, XIV ($n = 4$), while the dibasic acid obtained on saponification is exclusively XV ($n = 4$).[28]

Unsymmetrical ketones having one or more α-hydrogen atoms may give rise to as many as three condensation products, two stereoisomeric alkylidenesuccinic acids (XI and XII) and an alkenylsuccinic acid (XIII). There is a possibility of the formation of two alkenyl acids, depending on whether the bond of XI or XII shifts toward R or R', but there is no report of the isolation of both these forms, probably because the appropriate structures have not been studied. Nor have geometrical isomers of the alkenylsuccinic acids been found (see above). As observed generally in cases of 3-carbon tautomerism,[24] the bond moves toward that γ-carbon which is most highly substituted or carries an aryl group. Thus the Stobbe condensation with methyl ethyl ketone affords in addition to both XI and XII (R = CH_3, R' = C_2H_5), the alkenylsuccinic acid XIII (R = R' = CH_3). No substance corresponding to XIII (R = C_2H_5, R' = H) has been found.[4, 29, 30]

The effect of an α substituent in the ketone on the alkenyl:alkylidene ratio in the products of the Stobbe condensation is demonstrated with the ketone type $C_6H_5COCH_2R$. When R = H (acetophenone) the

[27] Johnson, Davis, Hunt, and Stork, *J. Am. Chem. Soc.*, **70**, 3021 (1948).
[28] Plattner and Büchi, *Helv. Chim. Acta*, **29**, 1608 (1946).
[29] Stobbe, *Ann.*, **321**, 83 (1902).
[30] Stobbe, *Ann.*, **321**, 105 (1902).

R′ = 2-furyl).[20] The configuration of these isomers may be established by cyclization experiments (see p. 16). From some ketones, e.g., 2-benzoylnaphthalene, only one of the two possible alkylidenesuccinic acids is isolated.[21]

$$\underset{\text{XI}}{\underset{R'}{\overset{R}{>}}C=C\underset{CH_2CO_2H}{\overset{(C_2H_5)\;CO_2H}{<}}} \qquad \underset{\text{XII}}{\underset{R'}{\overset{R}{>}}C=C\underset{CO_2H\;(C_2H_5)}{\overset{CH_2CO_2H}{<}}} \qquad \underset{\text{XIII}}{\underset{R'CH}{\overset{R}{>}}\overset{(C_2H_5)\;CO_2H}{\underset{\|}{C}}CHCH_2CO_2H}$$

Symmetrical ketones having one or more α-hydrogen atoms can give only one alkylidenesuccinic acid ester, but the alkenylsuccinic acid ester XIII with the double bond β,γ to the carbethoxyl group may also be produced, presumably by rearrangement of the bond from the α,β position (3-carbon tautomerism).[22] Thus acetone and diethyl succinate in a molecular ratio of 2 to 1 condense to give (after saponification) predominantly isopropylidenesuccinic acid accompanied by a trace of the isopropenyl isomer, (XIII, R = CH$_3$, R′ = H).[1,4] With a 1:1 ratio of ketone to ester, the isopropenyl isomer is the predominant product.[4] In addition the product formed by condensation of two molecules of acetone with one of diethyl succinate has been isolated.[23] This type of behavior is discussed above under the section on aldehydes.

$$(CH_3)_2C=CCO_2H$$
$$|$$
$$(CH_3)_2C=CCO_2H$$

An α-phenyl group in the ketone favors the formation of an alkenylsuccinic acid structure XIII (R′ = C$_6$H$_5$) in which the double bond is conjugated with the phenyl group.[24] Dibenzyl ketone, for example, gives the alkenylsuccinic acid XIII (R = C$_6$H$_5$CH$_2$, R′ = C$_6$H$_5$) as the exclusive product.[25, 26] Although two geometrical isomers of the alkenyl structure are theoretically possible only one has been found. The double

---

[20] Knott, *J. Chem. Soc.*, **1945**, 189.
[21] Hewett, *J. Chem. Soc.*, **1942**, 585.
[22] Bond rearrangement to the α,α′ position is also possible; *cf.* the itaconic-citraconic-mesaconic acid tautomerism, Coulson and Kon, *J. Chem. Soc.*, **1932**, 2568. The presence of such isomers, however, has never been demonstrated.
[23] Stollé, *J. prakt. Chem.*, [2], **67**, 197 (1903).
[24] *Cf.* the tendency of γ substituents, particularly aryl groups, to favor the β,γ form in simple 3-carbon systems, Linstead, *J. Chem. Soc.*, **1929**, 2498.
[25] Stobbe, *Ann.*, **308**, 67 (1899).
[26] Stobbe, Russwurm, and Schulz, *Ann.*, **308**, 175 (1899).

(−15 to 0°) in ether solution, dipiperonylidenesuccinic acid (VII, Ar = C₆H₃OCH₂O) is formed in 36% yield.[19] The lactone acid VIII (Ar = C₆H₃OCH₂O) could be produced in yields as high as 30% by use of short reaction periods and low temperatures,[6] and it has been clearly demonstrated that the proportion of dibasic acid VII to lactone acid VIII is greater with longer reaction periods.

The behavior described above suggests that the condensation of a second molecule of aldehyde occurs with an intermediary paraconic ester IX, which would have a longer existence at lower temperatures (higher temperatures promoting conversion to the half-ester salt which would not be expected to condense further because of the less reactive methylene group). It is possible that a dilactone like X has a transient existence in this scheme.

```
ArCH——CHCO₂C₂H₅          ArCH——CH    CO
  |       |                 |     |   /  \
  O      CH₂                O    CH    O
   \    /                    \  /      |
    CO                        CO    CHAr
    IX                         X
```

The bis(arylmethylene)succinic acids (VII), called "fulgenic" acids, are also prepared by the Stobbe condensation with the diester of an alkylidenesuccinic acid (see p. 17). The anhydrides of these acids are called "fulgides" and are of interest because of their intense color.

Ketones

The Stobbe condensation of a ketone RCOR′ with diethyl succinate may give rise to one or more isomeric half-esters, depending largely on the nature of R and R′.

Symmetrical ketones having no α-hydrogen atoms can give only one product, the alkylidenesuccinic acid ester XI (R = R′). This class is exemplified by the diaryl ketones like benzophenone which generally give excellent yields of homogeneous products.

Unsymmetrical ketones having no α-hydrogen atoms generally give both stereoisomeric alkylidenesuccinic acids XI and XII. This class is typified by the unsymmetrical ketones like 2-benzoylfuran which affords two crystalline *cis* and *trans* isomeric half-esters XI and XII (R = C₆H₅,

---

[19] Haworth and Woodcock, *J. Chem. Soc.*, **1938**, 1985.

panied by some of the isomeric lactonic acid. With potassium *t*-butoxide, however, the yield of condensation product is 85%.[16] Other aldehydes which have been condensed by this last method are decanal (40%), dodecanal (58%), and heptanal (59%).[16]

A wide variety of aromatic aldehydes has been used in the Stobbe condensation. The products are the expected arylmethylenesuccinic acid (V), occasionally the isomeric arylparaconic acid (VI), the bis-(arylmethylene)succinic acid (VII) arising from the condensation of two molecules of aldehyde with one of ester, and the corresponding lactonic acid VIII.

$$\begin{array}{cc}
\text{ArCH=CCO}_2\text{H} & \text{ArCH——CHCO}_2\text{H} \\
| & |\quad\quad\quad| \\
\text{CH}_2\text{CO}_2\text{H} & \text{O}\quad\text{CH}_2 \\
& \diagdown\diagup \\
& \text{CO} \\
\text{V} & \text{VI}
\end{array}$$

$$\begin{array}{cc}
\text{ArCH=CCO}_2\text{H} & \text{ArCH——CHCO}_2\text{H} \\
| & |\quad\quad\quad| \\
\text{ArCH=CCO}_2\text{H} & \text{O}\quad\text{C=CHAr} \\
& \diagdown\diagup \\
& \text{CO} \\
\text{VII} & \text{VIII}
\end{array}$$

The proportion of mono- to di-substituted products depends to a considerable extent upon the conditions of reaction, low temperatures favoring the formation of the latter. Benzaldehyde, for example, condenses with diethyl succinate and sodium ethoxide in refluxing ether to give mainly the benzylidenesuccinic acid, V (Ar = C_6H_5) in 35% yield.[9] This product consists of a mixture of the stereoisomers C_6H_5/CO_2H *trans* and C_6H_5/CO_2H *cis* in the ratio 9:1, which is the same proportion obtained when the pure *trans* form is heated in sodium hydroxide solution. The configurations were established by cyclization with sulfuric acid as described below (p. 16). When the condensation is carried out at $-10°$, the main product (35–40%) is the dibenzylidenesuccinic acid VII (Ar = C_6H_5), only a small proportion of benzylidenesuccinic acid being formed.[17] Even more striking is the behavior of piperonal, which condenses with diethyl succinate and sodium ethoxide in refluxing ethanol to give piperonylidenesuccinic acid (V, Ar = $C_6H_3OCH_2O$) in 90% yield.[18] In contrast, at low temperatures

---

[16] Overberger and Roberts, *J. Am. Chem. Soc.*, **71**, 3618 (1949).
[17] Stobbe and Naoum, *Ber.*, **37**, 2240 (1904).
[18] Cornforth, Hughes, and Lions, *J. Proc. Roy. Soc., N. S. Wales*, **72**, 228 (1939) [*C. A.*, **33**, 6816 (1939)].

benzoic acid which condenses readily with *t*-butyl acetate to give the half-ester IV.³ Since the condensation fails without the CO_2R group (i.e., *t*-butyl acetate does not condense with benzophenone), the participation of an intermediary lactone ester III is suggested.

$$\underset{\text{II}}{\text{Ar-CO-Ar'-CO}_2\text{R}} \xrightarrow[\text{KOC(CH}_3)_3]{\text{CH}_2\text{CO}_2\text{C(CH}_3)_3} \left[\underset{\text{III}}{\text{lactone}}\right] \rightarrow \underset{\text{IV}}{\text{Ar-C(CHCO}_2\text{C(CH}_3)_3)\text{-Ar'-CO}_2\text{H}}$$

Diethyl glutarate might be expected to react like its lower homolog in a Stobbe type of condensation, since it too should give rise to an aldol capable of lactonization. The main difference would be that the glutarate would form a δ- rather than a γ-lactonic ring. Surprisingly, however, this ester is relatively unreactive in the Stobbe condensation (see below, under Related Condensations), which may be partly attributable to a lower susceptibility to formation of the δ- in comparison with the γ-lactone ring so that competing reactions such as self-condensation of the ester take precedence.³

SCOPE AND LIMITATIONS

The carbonyl compounds that undergo the Stobbe condensation include at least one member, and in some cases many members, of the following classes of substances: aliphatic, aromatic, and α,β-unsaturated aldehydes; aliphatic, alicyclic, and aromatic ketones; diketones; keto esters; and cyano ketones. The succinic esters that have been employed are diethyl, dimethyl, and di-*t*-butyl succinate, and also α-substituted aryl-, aralkyl-, alkyl-, and alkylidene-succinic esters. A variety of condensing agents has been used, including sodium ethoxide, potassium *t*-butoxide, and sodium hydride. Sodium methoxide, metallic sodium, potassium ethoxide, and sodium triphenylmethyl have also had limited application.

Aldehydes

Isobutyraldehyde has been employed in the Stobbe condensation with diethyl succinate. When sodium metal [14] or sodium ethoxide [15] is used as the condensing agent the expected isobutylidenesuccinic acid, $(CH_3)_2CHCH=C(CO_2H)CH_2CO_2H$, is obtained in low yield accom-

[14] Fittig and Thron, *Ann.*, **304**, 288 (1899).
[15] Stobbe and Leuner, *Ber.*, **38**, 3682 (1905).

rationalization of the course of the Stobbe condensation, the irreversibility of the second step driving the reaction to completion.

(1) $(C_6H_5)_2\underset{\parallel}{C} + (-)\overset{CO_2C_2H_5}{\overset{|}{C}}HCH_2CO_2C_2H_5 \rightleftharpoons (C_6H_5)_2\overset{CO_2C_2H_5}{\underset{(-)O}{\overset{|}{C}}}CHCH_2\overset{}{C}OC_2H_5$
$\quad\quad\quad O \quad\quad\quad\quad\quad\quad\quad\quad\quad\quad\quad\quad\quad\quad\quad\quad\quad\quad\quad\quad\quad\quad\quad\quad\quad\quad\quad\quad\quad\quad\quad\quad\quad\quad\quad\quad\quad\quad\quad\quad\quad\quad\quad\quad\quad\quad\quad\quad\quad\quad\quad\quad\quad O$

$\updownarrow$

$C_2H_5O^{(-)} + (C_6H_5)_2\underset{\underset{I}{O\text{——}}}{\overset{CO_2C_2H_5}{\overset{|}{C}}CHCH_2\overset{}{C}}{\diagdown O} \rightleftharpoons (C_6H_5)_2\underset{O\text{——}}{\overset{CO_2C_2H_5}{\overset{|}{C}}CHCH_2\overset{}{C}OC_2H_5}{O_{(-)}}$

(2) $(C_6H_5)_2\underset{\underset{I}{O\text{——}}}{\overset{CO_2C_2H_5}{\overset{|}{C}}CHCH_2\overset{}{C}O} \underset{}{\overset{-OC_2H_5}{\rightleftharpoons}} (C_6H_5)_2\underset{\underset{(-)}{O}}{\overset{CO_2C_2H_5}{\overset{|}{C}}CCH_2CO} \rightarrow (C_6H_5)_2C=\overset{CO_2C_2H_5}{\overset{|}{C}}CH_2CO$
$\quad\quad\quad\quad\quad\quad\quad\quad\quad\quad\quad\quad\quad\quad\quad\quad\quad\quad\quad\quad\quad\quad\quad\quad\quad\quad\quad\quad\quad\quad\quad\quad\quad\quad\quad\quad\quad\quad\quad\quad\quad\quad\quad\quad\quad\quad\quad\quad\quad\quad\quad\quad\quad\quad\quad\quad\quad\quad\quad\quad\quad\quad\quad\quad\quad\quad\quad\quad\quad\quad\quad\quad (-)O$

The more obvious mechanism, in which the ketone first condenses with the succinate eliminating water which then reacts with the alkoxide to form hydroxide ion which in turn effects partial saponification of the di-ester, is not tenable in view of: (a) the failure to isolate the postulated intermediary di-ester, even when a large excess of diethyl succinate was employed in the condensation, thus affording a highly competitive source of ester groups to react with the limited amount of hydroxide ion;[8] (b) the failure of other esters with comparably reactive methylene groups to condense readily (see above); (c) the failure of the appropriate unsaturated di-ester to give a good yield of half-ester on partial saponification;[9-12] (d) the fact that isomers of the citraconic and mesaconic acid type, which would be expected tautomers of certain alkylidenesuccinic di-esters,[13] have never been found as products of the Stobbe condensation.

The importance of an appropriately situated carbalkoxyl group is strikingly illustrated by an experiment with an ester (II) of o-benzoyl-

[8] Johnson and Miller, *J. Am. Chem. Soc.*, **72**, 511 (1950).
[9] Stobbe, *Ber.*, **41**, 4350 (1908).
[10] W. S. Johnson and H. C. E. Johnson, unpublished observation.
[11] Johnson and Goldman, *J. Am. Chem. Soc.*, **66**, 1030 (1944).
[12] Johnson and Graber, *J. Am. Chem. Soc.*, **72**, 925 (1950).
[13] Coulson and Kon, *J. Chem. Soc.*, **1932**, 2568.

teraconic acid, $(CH_3)_2C{=}C(CO_2H)CH_2CO_2H$, formed by an aldol type of condensation between the carbonyl group of the ketone and an α-methylene group of the ester. This reaction was indeed surprising in view of the numerous precedents from the work of Claisen for the former type of behavior. Stobbe and his collaborators, therefore, undertook an extensive study which revealed that both aldehydes and ketones generally condense with succinic esters in this special manner, the stoichiometry of the reaction being expressed by the equation above. The liberation of the acidic material from the salt fraction affords the alkylidenesuccinic acid, or a tautomer, in the form of either the half-ester or the dibasic acid produced by hydrolysis.

$$\underset{R_2C=CCH_2CO_2Na}{\overset{CO_2C_2H_5}{|}} + HCl \rightarrow \underset{R_2C=CCH_2CO_2H}{\overset{\overset{(H)}{|}{CO_2C_2H_5}}{|}} + NaCl$$

It is striking that this facile aldol type of condensation of esters with ketones is limited to succinic and substituted succinic esters, with few exceptions. Benzophenone condenses with diethyl succinate to give pure β-carbethoxy-γ,γ-diphenylvinylacetic acid, $(C_6H_5)_2C{=}C(CO_2C_2H_5)$-$CH_2CO_2H$, in 90% yield;[2] under the same conditions this ketone, in contrast, fails altogether to react with ethyl or t-butyl acetate.[3] The success of the Stobbe condensation is not attributable solely to a high reactivity of the α-methylene groups of succinic esters, as shown by the failure of diethyl malonate, which has a more reactive α-methylene group, to condense to any appreciable extent with benzophenone.[3] The specificity of succinic esters in this reaction may be associated with the juxtaposition of a carbethoxyl group for ring formation as indicated in reaction sequence 1, below. The postulation of an intermediary paraconic ester (I) [1,4] is reasonable in view of the fact that such substances are isolable,[5] particularly when shorter reaction periods are employed,[6] and that they are cleaved by alkoxides in excellent yield to salts of the unsaturated half-esters.[7] This cleavage may be represented by reaction sequence 2. The combined steps 1 and 2 thus constitute a satisfactory

[2] Johnson, Petersen, and Schneider, *J. Am. Chem. Soc.*, **69**, 74 (1947).
[3] Johnson, McCloskey, and Dunnigan, *J. Am. Chem. Soc.*, **72**, 514 (1950).
[4] Stobbe, *Ann.*, **282**, 280 (1894).
[5] Robinson and Seijo, *J. Chem. Soc.*, **1941**, 582.
[6] Stobbe, Vieweg, Eckert, and Reddelien, *Ann.*, **380**, 78 (1911).
[7] Roser, *Ann.*, **220**, 258 (1883); Fittig, *Ann.*, **256**, 50 (1890); Fittig, *Ber.*, **27**, 2681 (1894).

	PAGE
EXPERIMENTAL PROCEDURES	41
Sodium Ethoxide Method	42
Sodium Methoxide Method	42
β-Carbethoxy-γ,γ-diphenylvinylacetic Acid. (Use of Potassium *t*-Butoxide and Diethyl Succinate)	42
β-Carbomethoxy-β-(2-methyl-1,2,3,4-tetrahydro-1-phenanthrylidene)propionic Acid. (Use of Potassium *t*-Butoxide and Dimethyl Succinate with Unreactive Ketones.)	44
The Condensation of Acetophenone and Diethyl Succinate (Sodium Hydride Method)	46
TABULAR SURVEY OF THE STOBBE CONDENSATION	48
Structural Key for Generic Names in Tables I and II	48
Table I. Aldehydes	50
Table II. Ketones	58

INTRODUCTION

The reaction of aldehydes or ketones with an ester of succinic acid to form alkylidenesuccinic acids (substituted itaconic acids), or isomers formed by a tautomeric shift of hydrogen, is known as the Stobbe condensation.[1] One mole of a metal alkoxide is required per mole of carbonyl compound and ester, and the primary product is the salt of the half-ester, i.e.,

$$R_2C=O + \overset{\overset{\displaystyle CO_2C_2H_5}{|}}{CH_2}CH_2CO_2C_2H_5 + NaOR' \rightarrow$$

$$R_2C=\overset{\overset{\displaystyle CO_2C_2H_5}{|}}{C}CH_2CO_2Na + C_2H_5OH + R'OH$$

GENERAL CHARACTER AND MECHANISM

In 1893 Hans Stobbe[1] demonstrated that when a mixture of acetone and diethyl succinate was treated with sodium ethoxide the expected acetoacetic ester type of condensation to give a β-diketo compound, $CH_3COCH_2COCH_2CH_2CO_2C_2H_5$ or $CH_3COCH_2COCH_2CH_2COCH_2COCH_3$, did not take place; but that the main reaction product was

---

[1] Stobbe, *Ber.*, **26**, 2312 (1893). A review article dealing, in part, with the Stobbe condensation has been published by Mlle. D. Billet, *Bull. soc. chim. France*, [5], **16**, D297–321 (1949).

CHAPTER 1

THE STOBBE CONDENSATION

WILLIAM S. JOHNSON and GUIDO H. DAUB

University of Wisconsin

CONTENTS

	PAGE
INTRODUCTION	2
GENERAL CHARACTER AND MECHANISM	2
SCOPE AND LIMITATIONS	5
Aldehydes	5
Ketones	7
α,β-Unsaturated Aldehydes and Ketones	11
Diketones	12
Keto Esters	12
Cyano Ketones	13
Methods of Isolation of Products and Proof of Structures and Configurations	14
Substituted Succinic Esters	17
Related Condensations	19
APPLICATIONS	21
Lactonic Acids	21
γ-Lactones and Unsaturated Acids	22
The Naphthol Synthesis	22
The Indone Synthesis	23
The Tetrahydroindanone Synthesis	23
The Tetralonecarboxylic Acid Synthesis	24
The Tetralone Synthesis	24
The Naphthalic Anhydride Synthesis	25
The Equilenone Synthesis	25
Charts for Syntheses	25
EXPERIMENTAL CONDITIONS AND SIDE REACTIONS	36
The Sodium Ethoxide and Methoxide Methods. The Oxidation-Reduction Side Reaction	36
Potassium t-Butoxide	38
Reducing Action and Self-Condensation of Succinates	38
Di-t-butyl Succinate	39
Sodium Hydride	39
Succinoylation	40
Other Side Reactions	41

SUBJECTS OF PREVIOUS VOLUMES

	VOLUME
Acetoacetic Ester Condensation and Related Reactions	I
Acetylenes	V
Acyloins	IV
Aliphatic Fluorine Compounds	II
Alkylation of Aromatic Compounds by the Friedel-Crafts Method	III
Amination of Heterocyclic Bases by Alkali Amides	I
Arndt-Eistert Reaction	I
Aromatic Arsonic and Arsinic Acids	II
Aromatic Fluorine Compounds	V
Azlactones	III
Benzoins	IV
Biaryls	II
Bucherer Reaction	I
Cannizzaro Reaction	II
Chloromethylation of Aromatic Compounds	I
Claisen Rearrangement	II
Clemmensen Reduction	I
Curtius Reaction	III
Cyanoethylation	V
Cyclic Ketones	II
Darzens Glycidic Ester Condensation	V
Diels-Alder Reaction with Cyclenones	V
Diels-Alder Reaction: Ethylenic and Acetylenic Dienophiles	IV
Diels-Alder Reaction with Maleic Anhydride	IV
Direct Sulfonation of Aromatic Hydrocarbons and Their Halogen Derivatives	III
Elbs Reaction	I
Friedel-Crafts Reaction with Aliphatic Dibasic Acid Anhydrides	V
Fries Reaction	I
Gattermann-Koch Reaction	V
Hoesch Synthesis	V
Hofmann Reaction	III
Jacobsen Reaction	I
Leuckart Reaction	V
Mannich Reaction	I
Periodic Acid Oxidation	II
Perkin Reaction and Related Reactions	I
Preparation of Amines by Reductive Alkylation	IV
Preparation of Benzoquinones by Oxidation	IV
Preparation of Ketenes and Ketene Dimers	III
Reduction with Aluminum Alkoxides	II
Reformatsky Reaction	I
Replacement of Aromatic Primary Amino Group by Hydrogen	II
Resolution of Alcohols	II
Rosenmund Reduction	IV
Schmidt Reaction	III
Selenium Dioxide Oxidation	V
Substitution and Addition Reactions of Thiocyanogen	III
Willgerodt Reaction	III
Wolff-Kishner Reduction	IV

CONTENTS

CHAPTER	PAGE
1. The Stobbe Condensation—*William S. Johnson and Guido H. Daub*	1
2. The Preparation of 3,4-Dihydroisoquinolines and Related Compounds by the Bischler-Napieralski Reaction—*Wilson M. Whaley and Tuticorin R. Govindachari*	74
3. The Pictet-Spengler Synthesis of Tetrahydroisoquinolines and Related Compounds—*Wilson M. Whaley and Tuticorin R. Govindachari*	151
4. The Synthesis of Isoquinolines by the Pomeranz-Fritsch Reaction—*Walter J. Gensler*	191
5. The Oppenauer Oxidation—*Carl Djerassi*	207
6. The Synthesis of Phosphonic and Phosphinic Acids—*Gennady M. Kosolapoff*	273
7. The Halogen-Metal Interconversion Reaction with Organolithium Compounds—*Reuben G. Jones and Henry Gilman*	339
8. The Preparation of Thiazoles—*Richard H. Wiley, D. C. England, and Lyell C. Behr*	367
9. The Preparation of Thiophenes and Tetrahydrothiophenes—*Donald E. Wolf and Karl Folkers*	410
10. Reductions by Lithium Aluminum Hydride—*Weldon G. Brown*	469
Index	511

theless, the investigator will be able to use the tables and their accompanying bibliographies in place of most or all of the literature search so often required.

Because of the systematic arrangement of the material in the chapters and the entries in the tables, users of the books will be able to find information desired by reference to the table of contents of the appropriate chapter. In the interest of economy the entries in the indices have been kept to a minimum, and, in particular, the compounds listed in the tables are not repeated in the indices.

The success of this publication, which will appear periodically in volumes of about ten chapters, depends upon the cooperation of organic chemists and their willingness to devote time and effort to the preparation of the chapters. They have manifested their interest already by the almost unanimous acceptance of invitations to contribute to the work. The editors will welcome their continued interest and their suggestions for improvements in *Organic Reactions*.

PREFACE TO THE SERIES

In the course of nearly every program of research in organic chemistry the investigator finds it necessary to use several of the better-known synthetic reactions. To discover the optimum conditions for the application of even the most familiar one to a compound not previously subjected to the reaction often requires an extensive search of the literature; even then a series of experiments may be necessary. When the results of the investigation are published, the synthesis, which may have required months of work, is usually described without comment. The background of knowledge and experience gained in the literature search and experimentation is thus lost to those who subsequently have occasion to apply the general method. The student of preparative organic chemistry faces similar difficulties. The textbooks and laboratory manuals furnish numerous examples of the application of various syntheses, but only rarely do they convey an accurate conception of the scope and usefulness of the processes.

For many years American organic chemists have discussed these problems. The plan of compiling critical discussions of the more important reactions thus was evolved. The volumes of *Organic Reactions* are collections of about ten chapters, each devoted to a single reaction, or a definite phase of a reaction, of wide applicability. The authors have had experience with the processes surveyed. The subjects are presented from the preparative viewpoint, and particular attention is given to limitations, interfering influences, effects of structure, and the selection of experimental techniques. Each chapter includes several detailed procedures illustrating the significant modifications of the method. Most of these procedures have been found satisfactory by the author or one of the editors, but unlike those in *Organic Syntheses* they have not been subjected to careful testing in two or more laboratories. When all known examples of the reaction are not mentioned in the text, tables are given to list compounds which have been prepared by or subjected to the reaction. Every effort has been made to include in the tables all such compounds and references; however, because of the very nature of the reactions discussed and their frequent use as one of the several steps of syntheses in which not all of the intermediates have been isolated, some instances may well have been missed. Never-

COPYRIGHT, 1951
BY
ROGER ADAMS

All Rights Reserved

This book or any part thereof must not be reproduced in any form without the written permission of the publisher.

FIFTH PRINTING, FEBRUARY, 1967

PRINTED IN THE UNITED STATES OF AMERICA

Organic Reactions

VOLUME VI

EDITORIAL BOARD

ROGER ADAMS, *Editor-in-Chief*

HOMER ADKINS (Deceased)   FRANK C. MCGREW
A. H. BLATT   CARL NIEMANN
ARTHUR C. COPE   HAROLD R. SNYDER

ADVISORY BOARD

W. E. BACHMANN (Deceased)   LOUIS F. FIESER
JOHN R. JOHNSON

ASSOCIATE EDITORS

LYELL C. BEHR   HENRY GILMAN
WELDON G. BROWN   TUTICORIN R. GOVINDACHARI
GUIDO H. DAUB   WILLIAM S. JOHNSON
CARL DJERASSI   REUBEN G. JONES
D. C. ENGLAND   GENNADY M. KOSOLAPOFF
KARL FOLKERS   WILSON M. WHALEY
WALTER J. GENSLER   RICHARD H. WILEY
DONALD E. WOLF

NEW YORK
JOHN WILEY & SONS, INC.
LONDON · SYDNEY

by the equation in Chart 4. Probably the best way of effecting the cyclization is to use sodium acetate and acetic anhydride, which gives the acetate of the phenol.[60] Zinc chloride in acetic acid-acetic anhydride as well as anhydrous hydrogen fluoride has also been used for the ring closure.[11]

The Indone Synthesis. Alkylidenesuccinic acids having an aryl group *cis* to the carboxyl group may undergo cyclodehydration to form a substituted indoneacetic acid, usually accompanied by some of the isomeric lactone as indicated in Chart 5. A variety of reagents has been employed for this ring closure, e.g., sulfuric acid,[20, 21, 54, 55, 66, 86] hydrogen fluoride,[11] zinc chloride-acetic acid-acetic anhydride,[52] sodium acetate-acetic acid-acetic anhydride,[52] and aluminum chloride (on the alkylidenesuccinic anhydride).[60, 87, 88] The indoneacetic acids are usually colored, and the isomeric lactones colorless. Longer reaction periods favor the formation of the lactones.

The alkylidenesuccinic half-esters with $Ar/CO_2C_2H_5$ *cis* will undergo simultaneous cyclization and intramolecular ester-exchange with the zinc chloride-acetic acid-acetic anhydride reagent to give the indoneacetic esters.[52] For example, by this treatment the half-ester derived from benzophenone gives mainly ethyl 3-phenyl-1-indone-2-acetate (79%); but, with sodium acetate in place of zinc chloride, only the naphtholacetate cyclization occurs as indicated in the accompanying flow sheet.

The Tetrahydroindanone Synthesis. When a cyclic ketone is employed in the Stobbe condensation, the resulting half-ester may be decarboxylated according to (and with the limitations of) the method described above for the preparation of γ-lactones and unsaturated acids

---

[86] Stobbe and Vieweg, *Ber.*, **35**, 1727 (1902).
[87] Haworth and Sheldrick, *J. Chem. Soc.*, **1935**, 636.
[88] Koelsch and Richter, *J. Org. Chem.*, **3**, 465 (1938).

(Chart 3). Either the lactone or the unsaturated acid thus produced may be cyclized with zinc chloride in acetic acid and acetic anhydride to give a fused cyclopentenone nucleus as represented in Chart 6, Scheme A. An alternative approach, Scheme B, is to treat the half-ester with the zinc chloride reagent, which effects cyclization to a keto ester which then can be easily hydrolyzed and decarboxylated by hydrochloric-acetic acid. Scheme B is generally preferred, since the two steps may be carried out in a single operation; i.e., after the cyclization reaction is completed, water is added to decompose the acetic anhydride, then hydrochloric acid is introduced and the heating continued. The portion of the product that is neutral after saponification consists largely of the desired ketone. In the 1-keto-2-methyl-1,2,3,4-tetrahydrophenanthrene series, however, it was necessary to employ Scheme A, because the half-ester underwent the indone cyclization with ester exchange, the ring closing into the aromatic nucleus (see Chart 5). Examples of the tetrahydroindanone synthesis are tabulated in Chart 6.

The Tetralonecarboxylic Acid Synthesis. Catalytic or chemical reduction of the Stobbe condensation product from an aromatic aldehyde or diaryl or aryl alkyl ketone yields an arylmethylsuccinic acid which on cyclization generally gives a substituted 1-tetralone-3-carboxylic acid as represented in Chart 7. The ring closure is usually effected by the action of aluminum chloride on the arylmethylsuccinic anhydride,[21, 87, 89, 12] and the six-membered ring is generally formed in preference to the five.[90] In certain cases, as with desoxybenzoin, it is possible to realize double cyclization.[21, 91, 92]

The Tetralone Synthesis. γ-Arylbutyrolactones produced via the Stobbe condensation according to the γ-lactone synthesis (Chart 3) may be reduced to substituted γ-arylbutyric acids which on cyclodehydration yield substituted 1-tetralones according to the scheme outlined in Chart 8. The reduction of the lactones may be effected by the Clemmensen method,[93, 94] with phosphorus and hydriodic acid,[93] phosphorus and iodine,[94] or probably best by catalytic hydrogenation.[93, 94, 95] The cyclization may be carried out by one of the conventional methods.[90]

The isomeric unsaturated acids (Chart 3) can be employed as well as the γ-lactones as, for example, in the synthesis of tetrahydroperinaphthanone from 1-tetralone. Decarboxylation of the Stobbe condensation

---

[89] Hewett, *J. Chem. Soc.*, **1936**, 596.
[90] Johnson in Adams, *Organic Reactions*, Vol. 2, John Wiley & Sons, New York, 1944.
[91] Borsche and Sinn, *Ann.*, **555**, 70 (1945).
[92] Newman and Hart, *J. Am. Chem. Soc.*, **69**, 298 (1947).
[93] Johnson and Jones, *J. Am. Chem. Soc.*, **69**, 792 (1947).
[94] Riegel and Burr, *J. Am. Chem. Soc.*, **70**, 1070 (1948).
[95] Johnson, Goldman, and Schneider, *J. Am. Chem. Soc.*, **67**, 1357 (1945).

product gives predominantly the unsaturated acid, which is reduced readily by catalytic hydrogenation and cyclized with hydrogen fluoride.[76]

The Naphthalic Anhydride Synthesis. This synthesis is typified by the cyclodehydrogenation of dibenzylidenesuccinic anhydride, obtained via the Stobbe condensation with benzaldehyde, to give 1-phenyl-2,3-naphthalenedicarboxylic anhydride. The ring closure may be effected by the action of sunlight on a benzene or chloroform solution of the anhydride containing a trace of iodine [96, 97] or by heating at 200–280°.[97] The naphthalic anhydrides prepared by this method are tabulated in Chart 9.

The Equilenone Synthesis. The Stobbe condensation with a β-keto nitrile has been discussed in some detail on p. 13. The resulting keto ester may be hydrolyzed and decarboxylated, giving an unsaturated ketone which on catalytic hydrogenation yields an equilenone. The 7-methoxy keto nitrile has thus been employed in the synthesis of the natural hormone equilenin.

CHART 1

BROMO AND UNSATURATED LACTONIC ACIDS FROM ALKYLIDENESUCCINIC ACIDS [25]

R	R'	R''	REFERENCE
CH_3	C_2H_5	H	29, 30
CH_3	C_6H_5	H	25, 31
C_6H_5	C_6H_5	H [a]	25, 53
C_6H_5	C_6H_5	CH_3	65
$-CH_2CH_2CH_2CH_2-$		H	98
$-CH_2CH(CH_3)CH_2CH_2CH_2-$		H	99

[a] The bromo lactonic acid loses carbon dioxide and hydrogen bromide on boiling with water to give an α,β-unsaturated γ-lactone.

[96] Stobbe, *Ber.*, **40**, 3372 (1907).
[97] Baddar, El-Assal, and Gindy, *J. Chem. Soc.*, **1948**, 1270.
[98] Stobbe, *J. prakt. Chem.*, [2], **89**, 329 (1914).
[99] Stobbe, *J. prakt. Chem.*, [2], **89**, 341 (1914).

CHART 2
Dilactones and Lactonic Acids from Alkenylsuccinic Acids [25]

$$RCH=CR' \atop HO_2CCHCH_2CO_2H \xrightarrow{Br_2} \underset{CO}{\underset{|}{RCH}}\!\!-\!\!\underset{|}{\overset{Br}{CR'}}\;\underset{}{\overset{}{CHCH_2CO_2H}} \text{ or } \underset{HO_2CCH}{\underset{|}{RCH}}\!\!-\!\!\underset{CH_2}{\overset{Br\ R'}{\underset{|}{C}}}\!\!-\!\!O \atop CO$$

↓

$$\underset{CO}{\underset{|}{RCH}}\!\!-\!\!\underset{CH}{\overset{R'}{\underset{|}{CH}}}\!\!-\!\!\underset{CH_2}{\overset{}{CO}} \quad \text{and an isomeric unsaturated lactonic acid}$$

R	R'	Reference
CH₃	CH₃	29, 30, 80
H	C₆H₅ [a]	25, 31
C₆H₅	C₆H₅	25, 33
C₆H₅	C₆H₅CH₂	25, 26

[a] The product is the dilactone; none of the isomeric unsaturated lactonic acid is produced.

CHART 3
γ-Lactone Synthesis [2]

$$\underset{R'}{\overset{R}{\diagdown}}\!\!CO \xrightarrow[\text{condensation}]{\text{Stobbe}} \underset{R'}{\overset{R}{\diagdown}}\!\!C\!\!=\!\!CCH_2CO_2H \atop CO_2C_2H_5 \atop \text{(H)} \atop \text{or/and isomers} \xrightarrow[H_2O]{HX-CH_3CO_2H} \underset{R'}{\overset{R}{\diagdown}}\!\!\overset{O\rule{1.2cm}{0.4pt}}{\underset{|}{C}}CH_2CH_2CO \atop \text{or/and unsaturated acid}$$

Ketone	Product	Reference
Acetophenone	$C_6H_5\underset{CH_3}{\overset{O\rule{1cm}{0.4pt}}{\underset{|}{C}}}CH_2CH_2CO$	2
Propiophenone	$C_6H_5\underset{C_2H_5}{\overset{O\rule{1cm}{0.4pt}}{\underset{|}{C}}}CH_2CH_2CO$	2
p-CH₃C₆H₄COCH₃	$p\text{-}CH_3C_6H_4\underset{CH_3}{\overset{O\rule{1cm}{0.4pt}}{\underset{|}{C}}}CH_2CH_2CO$	93

CHART 3—Continued
γ-LACTONE SYNTHESIS—Continued

KETONE	PRODUCT		REFERENCE
2-Acetylnaphthalene	naphthyl-C(CH₃)(O-)CH₂CH₂CO (γ-lactone)		95
3-Acetylphenanthrene	phenanthryl-C(CH₃)(O-)CH₂CH₂CO (γ-lactone)		94
Benzophenone	$(C_6H_5)_2\overset{O-}{C}CH_2CH_2CO \rightleftharpoons$ $(C_6H_5)_2C{=}CHCH_2CO_2H$		2
$(p\text{-}CH_3OC_6H_4)_2CO$	$(p\text{-}CH_3OC_6H_4)_2\overset{O-}{C}CH_2CH_2CO \rightleftharpoons$ $(p\text{-}CH_3OC_6H_4)_2C{=}CHCH_2CO_2H$		8
(α-tetralone)	(tetralin-lactone) ⇌ (dihydronaphthyl)-CH₂CH₂CO₂H		76
(phenanthrenone, partially reduced)	(lactone) ⇌ (dihydrophenanthryl)-CH₂CH₂CO₂H		100
(methyl-substituted phenanthrenone with R)	(product with CH₃ and CH₂CH₂CO₂H) and an oily lactone	R=H 58 R=OCH₃ 101	
(R-substituted phenanthrenone)	(R-substituted lactone) R=CH₃, C₂H₅, CH(CH₃)₂		102

[100] Johnson and Petersen, *J. Am. Chem. Soc.*, **67**, 1366 (1945).
[101] Johnson and Stromberg, *J. Am. Chem. Soc.*, **72**, 505 (1950).
[102] Riegel, Siegel, and Kritchevsky, *J. Am. Chem. Soc.*, **70**, 2950 (1948).

CHART 4
The Naphthol Synthesis [59, 60]

PhCOR → [HO₂C-CH₂, CCO₂H(C₂H₅), R substituted benzene intermediate] → [naphthol with OH (COCH₃), CO₂H (C₂H₅), R]

KETONE	PRODUCT		REFERENCE
C₆H₅COC₆H₅	naphthalene with OCOCH₃(H), CO₂C₂H₅(H), C₆H₅		60
furan-CO-C₆H₄R	benzofuran with OCOCH₃(H), CO₂C₂H₅(H), C₆H₄R + naphthalene with OCOCH₃(H), C₂H₅O₂C, R, furyl	R=H 20 R=CH₃ 59
furan-CO-C₆H₃(R)(R′)	benzofuran with OH, CO₂H, C₆H₃(R)(R′)	R=OCH₃, R′=H R=R′=OCH₃	59 59
C₆H₅-CO-thienyl	naphthalene with OH, CO₂H, 2-thienyl (Isolated as a diazo coupled product)		59
methylenedioxyphenyl-COC₆H₅	methylenedioxy-naphthalene with OH, CO₂H, C₆H₅		60

CHART 4—Continued

The Naphthol Synthesis—Continued

Ketone	Product	Reference
2-acetylnaphthalene (COCH$_3$)	phenanthrene with CH$_3$COO, CO$_2$C$_2$H$_5$, CH$_3$, (H), (H)	11, 52
6-methoxy-2-propionylnaphthalene (CH$_3$O, COC$_2$H$_5$)	phenanthrene with CH$_3$COO, CO$_2$C$_2$H$_5$, C$_2$H$_5$, CH$_3$O	12

CHART 5

The Indone Synthesis

PhCOR → Ph(HO$_2$C)C=C(R)CH$_2$CO$_2$H (C$_2$H$_5$) → indanone-CH$_2$CO$_2$H (C$_2$H$_5$), R + lactone (indanone fused with CH$_2$-CO-O, R)

Aldehyde or Ketone	Products	Reference
Benzaldehyde	indanone-CH$_2$CO$_2$H + lactone	55
Benzaldehyde (with benzhydrylidenesuccinate)	indanone with CO$_2$H, C=CHC$_6$H$_5$, C$_6$H$_5$	88
Acetophenone	indanone with CH$_2$CO$_2$H, CH$_3$ + lactone	54
Benzophenone	indanone with R, CHCO$_2$H (C$_2$H$_5$), C$_6$H$_5$ + lactone; R=H; R=CH$_3$	86, 60; 66
Benzophenone (with benzhydrylidenesuccinate)	indanone with CO$_2$H, C=C(C$_6$H$_5$)$_2$, C$_6$H$_5$	88

CHART 5—Continued

The Indone Synthesis—Continued

Aldehyde or Ketone	Products	Reference
Benzophenone (with β-phenylcinnamylidenesuccinate)	[indone with C=CH—CH=C(C_6H_5)$_2$, CO_2H, C_6H_5]	88
CH_3O, CH_3O—C$_6H_3$—CO—C$_6H_3$(OCH$_3$)OCH$_3$	[indone with CH_2CO_2H, dimethoxyphenyl, CH_3O, CH_3O]	87
C_6H_5—CO—(furyl)	[indone with CH_2CO_2H, furyl]	20
tetrahydronaphthyl—COC_6H_5	[indone with CH_2CO_2H, C_6H_5] or [isomer]	21
naphthyl—COCH$_3$	[benz-indone with CH_2CO_2H, (C_2H_5), CH_3]	11, 52
CH_3O-naphthyl—COC$_2H_5$	[benz-indone with $CH_2CO_2C_2H_5$, C_2H_5, CH_3O]	12
[2-methyl-dihydrophenanthrone with CH$_3$]	[cyclopenta-phenanthrene with CH$_3$, (C_2H_5), $CH_2CO_2CH_3$]	58

CHART 6

THE TETRAHYDROINDANONE SYNTHESIS [76]

KETONE	SCHEME	PRODUCTS		REFERENCE
(cyclohexanone with R)	B	(tetrahydroindanone with R)	R=H	27
			R=CH$_3$	103
Cycloheptanone	B	(bicyclic ketone) (→ methylazulene)		28, 51
α-Tetralone	A and B	(tricyclic ketone)		76
(benzosuberone)	B	(tricyclic ketone)		51
(dihydrophenanthrone with R)	A and B	(tetracyclic ketone with R)	R=H	100
			R=CH$_3$	102
			R=C$_2$H$_5$	102
			R=CH(CH$_3$)$_2$	102

[103] Cook and Phillip, *J. Chem. Soc.*, **1948**, 162.

CHART 6—*Continued*

THE TETRAHYDROINDANONE SYNTHESIS—*Continued*

KETONE	SCHEME	PRODUCTS		REFERENCE
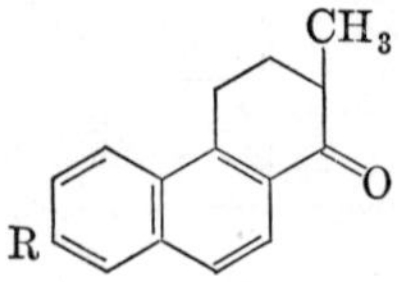 (structure with CH₃, R)	A	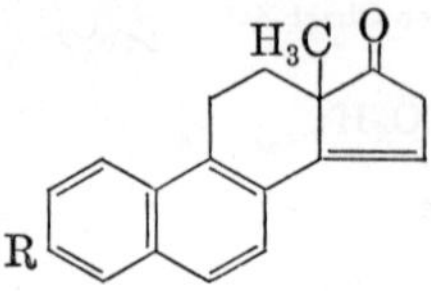 (structure with H₃C, R)	R=H R=OCH₃	58 101

CHART 7

THE TETRALONECARBOXYLIC ACID SYNTHESIS [57]

(reaction scheme: PhCOR → phenyl-substituted diacid intermediate with CO₂H, CH₂, CCO₂H, R; or/and isomers → HO₂C, CH₂, CHCO₂H intermediate → tetralone with CO₂H and R)

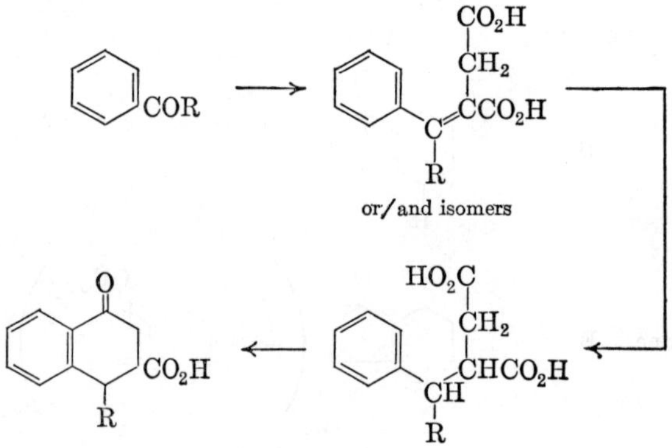

ALDEHYDE OR KETONE	PRODUCTS		REFERENCES
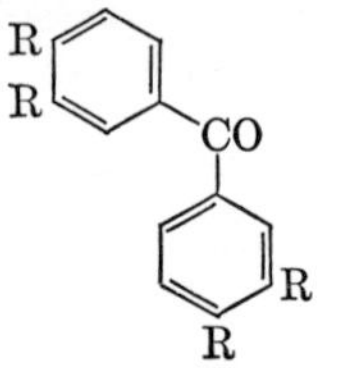 (diaryl ketone with R substituents)	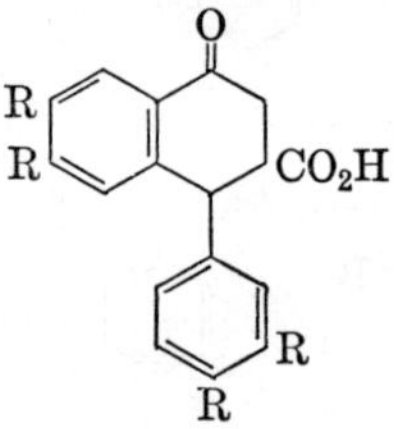 (tetralone with R substituents, CO₂H, and phenyl-R group)	R=H R=OCH₃	89 87

(Converted to 3,4-benzphenanthrene, R=H)

| 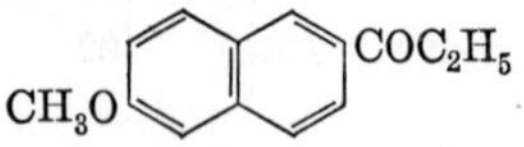 CH₃O-naphthyl-COC₂H₅ | 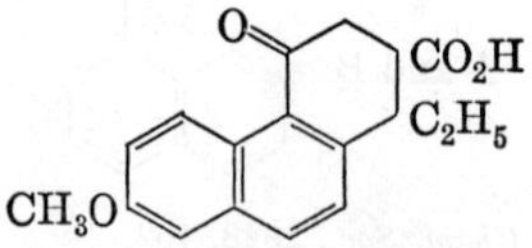 CH₃O-phenanthrenone with CO₂H, C₂H₅ | 12 |

(Converted to bisdehydrodoisynolic acid)

CHART 7—*Continued*

The Tetralonecarboxylic Acid Synthesis—*Continued*

Aldehyde or Ketone	Products	References
naphthyl–COC_6H_5	phenanthrenone with CO_2H, C_6H_5	21
phenanthryl–$COCH_3$	tetracyclic with CH_3, CO_2H, ketone (Converted to substituted 1,2-benzanthracenes)	104
$(C_6H_5CH_2)_2CO$	tetralone with phenyl, CH_2CO_2H + tetracyclic diketone (Converted to substituted 1,2-benzanthracenes)[92]	21, 91, 92
Benzaldehyde	$\left[\begin{array}{c}CH_2\text{—}CO_2H\\ \mid\\ CH\text{—}C_6H_5\\ \mid\\ HO_2C\text{—}CH\\ \mid\\ CH_2\text{—}C_6H_5\end{array}\right] \rightarrow$ pentacyclic diketone; Unsaturated analogs	67; 105, 106
CH_3O–C$_6H_4$–CO–$CH_2CH_2CH_2CO_2C_2H_5$	CH_3O–tetralone with $CH_2CH_2CH_2CO_2H$ and CO_2H	45

[104] Cook and Robinson, *J. Chem. Soc.*, **1938**, 505.
[105] Bergmann and Weizmann, *Compt. rend.*, **209**, 539 (1939).
[106] Dufraisse and Houpillart, *Compt. rend.*, **206**, 756 (1938).

CHART 8

The Tetralone Synthesis [95]

Ketone	Products	Reference
p-$CH_3C_6H_4COCH_3$	(methyl-tetralone with CH$_3$) (Converted to cadalene)	93
α-Tetralone	(tricyclic ketone)	76
(2-naphthyl)COCH$_3$	(Converted to substituted phenanthrenes)	95
(phenanthrenyl)COCH$_3$	(Converted to substituted 1,2-benzanthracenes)	94

THE STOBBE CONDENSATION

CHART 9
THE NAPHTHALIC ANHYDRIDE SYNTHESIS [97]

$C_6H_5CHO \longrightarrow$ [intermediate] $\xrightarrow{\text{Heat or sunlight}}$ [phenylnaphthalic anhydride]

ALDEHYDE	PRODUCTS	REFERENCE
Benzaldehyde	4-phenyl-2,3-naphthalic anhydride (C_6H_5)	96
$o\text{-}CH_3OC_6H_4CHO$	4-(o-methoxyphenyl)-methoxy-naphthalic anhydride (OCH_3, CH_3O)	97
Anisaldehyde	methoxy-4-(p-methoxyphenyl)-naphthalic anhydride (CH_3O, OCH_3)	97
$p\text{-}CH_3C_6H_4CHO$	methyl-4-(p-tolyl)-naphthalic anhydride (H_3C, CH_3)	97

CHART 10

The Equilenone Synthesis [46]

or a tautomer

Keto Nitrile	Product	Reference
		46, 48
	(Converted to equilenin)	46
		47

EXPERIMENTAL CONDITIONS AND SIDE REACTIONS

In the following discussion an attempt is made to evaluate various experimental methods and to compare the usefulness of particular reagents and conditions. Such treatment necessarily involves a consideration of many of the side reactions of the Stobbe condensation.

The Sodium Ethoxide and Methoxide Methods. The Oxidation-Reduction Side Reaction

According to the classical procedure for the Stobbe condensation, a mixture of the ketone, diethyl succinate, and sodium ethoxide in ether is allowed to stand in the cold for several days to weeks. It is then heated for a short period, and treated with water. The half-ester or hydrolysis product is recovered from the aqueous layer by acidification. Satisfactory yields can sometimes be realized if ethanol is substituted for ether; moreover the reaction period may be shortened considerably by heating the mixture at the start. In general, however, the ether-sodium ethoxide method gives better results. In either reaction medium there is almost always a significant amount of reduction of the ketone to the corresponding carbinol. For example, methylphenylcarbinol [31] and benzhydrol [53] were thus obtained in the condensation with acetophenone and benzophenone, respectively. This reduction is evidently effected by the ethoxide, which is converted to acetaldehyde, which in turn is largely responsible for the formation of resinous material and darkening usually observed with this procedure. Evidence for the presence of acetaldehyde has been provided by Koelsch and Richter [38] in a careful reexamination of the condensation of benzophenone with diethyl benzhydrylidenesuccinate by the classical procedure.[107] The crude acidic fraction was hydrolyzed and treated with acetyl chloride to convert the dibasic acids to anhydrides, which were separated into the expected dibenzhydrylidenesuccinic anhydride (LVIII) in 40% yield, and two stereochemical forms of 1,1,6,6-tetraphenylhexatriene-1,2-dicarboxylic anhydride (LIX) in 10% yield each. The formation of LIX could be explained only by the participation of acetaldehyde in the condensation. Considerable benzhydrol was found in the neutral fraction. The structure of LIX was confirmed by direct synthesis from β-phenylcinnamaldehyde and dimethyl benzhydrylidenesuccinate. The formation of this by-product in the original condensation with benzophenone was avoided altogether by the use of the dimethyl ester and sodium methoxide. This behavior is in accord with the demonstration that metal methoxides are weaker reducing agents than ethoxides.[108] Although sodium methoxide largely eliminates the oxidation-reduction complication, it is a weaker condensing agent than sodium ethoxide and has therefore not found general use.

$$(C_6H_5)_2C=CCO \atop (C_6H_5)_2C=CCO \Big\rangle O$$
LVIII

$$(C_6H_5)_2C=CCO \atop (C_6H_5)_2C=CHCH=CCO \Big\rangle O$$
LIX

[107] Stobbe, *Ber.*, **38**, 3673 (1905).
[108] Adkins, Elofson, Rossow, and Robinson, *J. Am. Chem. Soc.*, **71**, 3622 (1949).

Potassium *t*-Butoxide

The oxidation-reduction reaction can also be inhibited by the use of potassium *t*-butoxide in *t*-butyl alcohol. Potassium *t*-butoxide is a considerably stronger condensing agent than sodium ethoxide and in general affords better yields of pure products in much shorter reaction periods. For example, the condensation of 1-tetralone with diethyl succinate [76] by the ether-sodium ethoxide method gave after two to three days (optimum time) a red, gummy, semi-solid acidic product in 83% yield from which pure half-ester was obtained in less than 50% yield based on the original ketone. By the potassium *t*-butoxide method, however, a pale yellow crystalline product was produced in 89–94% yield after a reaction period of only forty-five minutes, and a single crystallization gave practically pure colorless half-ester with a recovery of 90%. With cyclohexanone the *t*-butoxide method afforded distilled half-ester in 84% yield after a reaction period of only ten minutes,[27] whereas the best yields that have been reported by other methods do not exceed 40%.[36]

Reducing Action and Self-Condensation of Succinates

Even potassium *t*-butoxide and diethyl succinate cause some reduction of the ketone. With these reactants the reduction is effected by the alcohol formed as a by-product in the Stobbe condensation and as a product of the self-condensation of the succinate to produce diethyl cyclohexane-1,4-dione-2,5-dicarboxylate, LX (R = C_2H_5). A significant amount of reduction occurs only with ketones that react slowly in the Stobbe condensation, thus allowing a considerable concentration of ethoxide to build up by the competing self-condensation reaction. This was the case with the cyano ketone XXVIII, R = OCH_3, R' = H (p. 13), which under these conditions was partly reduced to the cyano carbinol LXI.[46] This reduction could be almost completely eliminated by the use of dimethyl instead of diethyl succinate, a result that is in accord with the comparative reducing properties of methoxide and ethoxide considered above. Dimethyl succinate is therefore useful in

LX LXI

conjunction with *t*-butoxide for condensation with slowly reacting ketones. This ester, however, is more susceptible to self-condensation to form the cyclic keto ester LX (R = CH$_3$) than the diethyl compound; therefore it is usually necessary to employ a larger excess of dimethyl succinate and potassium *t*-butoxide, added gradually, in order to obtain good yields. By such a procedure 2-methyl-1-keto-1,2,3,4-tetrahydrophenanthrene undergoes condensation in 93% yield as compared with 75% by the usual *t*-butoxide procedure.[58, 101]

Di-*t*-butyl Succinate. Another solution to the problem of the competing self-condensation of the esters lies in the use of an ester like di-*t*-butyl succinate, which reacts in this way relatively slowly. The unreactive ketone *p,p'*-dimethoxybenzophenone undergoes condensation with diethyl succinate by the classical sodium ethoxide procedure only to a slight extent,[109] while by the conventional *t*-butoxide method the yield is about 47%.[8] The yield can be raised to 83% by use of a large excess of reagents as described above, or to 90% by employing di-*t*-butyl succinate in slight excess.[8] No reduction of the ketone is possible, and the self-condensation of the ester is virtually eliminated. The Stobbe condensation itself, however, is considerably slower with di-*t*-butyl than with dimethyl or diethyl succinate, presumably owing to steric resistance of the carbo-*t*-butoxy group to participation in the lactonization step. Longer periods of heating, therefore, are required, and such treatment may be undesirable in condensations involving aldehydes or ketones which are themselves sensitive to the alkaline conditions. Thus the cyano ketone XXVIII (R = OCH$_3$, R' = H), p. 13, gave the expected *t*-butyl keto ester in only 13% yield, probably because the competing ring-opening reaction to produce the compound corresponding to XXXIV (ring B aromatic), p. 14, took precedence.[46] Another limitation of di-*t*-butyl succinate is that currently it is considerably more difficult to prepare than the dimethyl and diethyl esters.

Sodium Hydride

Although this reagent has not been studied extensively, it promises to be particularly effective in the Stobbe condensation.[110] It has the advantage of being inexpensive and especially easy to use as a condensing agent. For example, a mixture of benzophenone, diethyl succinate, and sodium hydride is stirred for about five hours at room temperature, ether (or benzene) being added as a diluent. The mixture is acidified and the product extracted with bicarbonate solution, acidification of

[109] Johnson and Goldman, unpublished observation.
[110] Daub and Johnson, *J. Am. Chem. Soc.*, **70**, 418 (1948); **72**, 501 (1950).

which gives essentially pure crystalline half-ester in 97% yield. A trace of ethanol is usually required to initiate the reaction. The alcohol reacts rapidly with the sodium hydride to produce sodium ethoxide, which may be the true condensing agent. As the reaction proceeds, more alcohol is formed as a by-product. This reacts rapidly with the sodium hydride, producing additional sodium ethoxide; and, as the concentration of the latter gradually increases, there is a corresponding increase in the rate of condensation as evidenced by the rate of evolution of hydrogen. The essential difference between this and the classical sodium ethoxide method is that there is no accumulation of alcohol as the reaction progresses, even if there is considerable self-condensation of the ester.

When di-*t*-butyl succinate is used with sodium hydride the self-condensation reaction is essentially eliminated so that the progress of the Stobbe condensation can be observed conveniently by measuring the volume of evolved hydrogen, two moles of gas being produced for each of half-ester salt formed.[110] With enolizable ketones, like desoxybenzoin, a competing reaction to form the sodio derivative may be involved, with the production of hydrogen in a mole-to-mole ratio.

Succinoylation. With some ketones having reactive α-methyl or methylene groups the sodium hydride method tends to promote a small amount of succinoylation of the ketone by the diethyl succinate. This acetoacetic ester type of condensation is the reaction which was originally expected (see Scope and Limitations) but was never definitely observed until the recent study with sodium hydride.[110] In the condensation of 1-keto-1,2,3,4-tetrahydrophenanthrene with diethyl succinate, in addition to the expected half-ester (yield 86%) a product shown to be the succinoyl derivative LXII (R = C_2H_5) was isolated in 3% yield. With dimethyl succinate the yield of the corresponding by-product LXII (R = CH_3) was 9%. No succinoylation was observed with acetophenone, diethyl succinate, and sodium hydride, the mixture of half-esters being produced in 93% yield. Surprisingly, with di-*t*-butyl succinate the succinoylation product, $C_6H_5COCH_2COCH_2CH_2COCH_2COC_6H_5$, was isolated in 33% yield, the Stobbe condensation proceeding in only 57% yield.

LXII

Other Side Reactions

With aldehydes in the Stobbe condensation, expected side reactions involving the aldehyde alone have been observed in the presence of alkoxide. Among these are the Cannizzaro reaction [17, 111, 112] and the aldol condensation.[14, 15]

The failure of certain ketones containing highly active α-methyl or α-methylene groups to give good yields in the Stobbe condensation may be due in part to a tendency for these ketones to enolize. The anion produced may be relatively stable, as with desoxybenzoin and dibenzyl ketone, in which event the ketone is recovered unchanged. On the other hand the anion may compete with the ester anion in reaction with free ketone, in which event self-condensation of the ketone is effected. 1-Hydrindone falls into this latter category, since considerable hydrindylidenehydrindone is produced in the Stobbe condensation with t-butoxide.[113] Cyclopentanone also fails to react well in the Stobbe condensation. In addition to considerable cyclopentylidenecyclopentanone,[113] a lactonic acid apparently produced by condensation of two moles of ketone and one of ester has been obtained.[98] This substance is either a lactone of a dibasic acid like LXIII, produced by the condensation of succinic ester with cyclopentylidenecyclopentanone, or of a dibasic acid like LXIV. The former structure is perhaps preferred because no comparable product was found in the condensation with cyclohexanone, which has an even more reactive carbonyl group (thus favoring the formation of a product like LXIV) and a less reactive methylene group for self-condensation. However, a product of "dicondensation" was isolated in poor yield in the 3-methylcyclohexanone series.[99]

LXIII

LXIV

EXPERIMENTAL PROCEDURES

The following procedures represent typical examples of different methods for effecting the Stobbe condensation. The selection and adaptation of these procedures for application to other ketones may be

---

[111] Fichter and Scheuermann, *Ber.*, **34**, 1626 (1901).
[112] Stobbe, *Ann.*, **380**, 49 (1911).
[113] Johnson and Davis, unpublished observation.

facilitated by a consideration of the preceding section on experimental conditions and side reactions.

Sodium Ethoxide Method. Since this method generally gives poorer results than those outlined below, it is not described in detail. A fairly complete procedure for the ether method is given elsewhere for the condensation of α-tetralone with diethyl succinate.[76] Details for the use of sodium ethoxide in ethanol are described for the condensation of 2-acetylnaphthalene with diethyl succinate.[11]

Sodium Methoxide Method. Successful uses of this reagent are described in the literature for the condensation of benzophenone with dimethyl benzhydrylidenesuccinate [38] and for the condensation of desoxybenzoin with dimethyl succinate.[92]

β-Carbethoxy-γ,γ-diphenylvinylacetic Acid. (Use of Potassium t-Butoxide and Diethyl Succinate.) The following directions for the condensation of benzophenone with diethyl succinate represent a modification of a procedure previously reported.[2] This method is applicable to many ketones, although for best yields it may be necessary to vary the reaction period from ten minutes (for cyclohexanone) to forty-five minutes (for α-tetralone). In some reactions it may prove effective to increase the concentration of alkoxide by reducing the volume of solvent.[102]

The following procedure is recommended for the safe handling of potassium. The metal may be cut conveniently under xylene, which has been dried over sodium wire, contained in a mortar. A beaker or crystallizing dish should not be used as it is too fragile. Each scrap obtained in cutting off the outer oxide-coated surface of the metal should be immediately transferred with tweezers to a second deep mortar containing dry xylene where the accumulated residues are decomposed as described below as soon as the cutting operation is complete. In order to weigh the freshly cut metal it may be removed with tweezers, blotted rapidly with a piece of filter paper, and introduced into a tared beaker containing dry xylene. The weighed potassium is then introduced into the reaction mixture, the proper precautions, such as judicious rate of addition, exclusion of air and moisture, etc., being taken depending on the nature of the reaction involved.

Caution: It is the small scraps of metal which adhere to the knife or float on top of the xylene that are most likely to start a fire.

Danger: Potassium residues have been known to explode even under a protective liquid. It is therefore important that all such residues be decomposed immediately; under no circumstances should they be stored. The mortar containing the scraps is moved to the rear of the hood, and t-butyl (not methyl or ethyl) alcohol is added in small portions from a medicine dropper or beaker at such a rate that the reaction does not become too vigorous. A square sheet of asbestos large enough to cover the mortar should be at hand. If the liquid should catch fire it may be extinguished easily by covering the mortar with the asbestos sheet. There should be no other inflammable material or flames in the hood during this treatment. Sufficient t-butyl alcohol must be employed

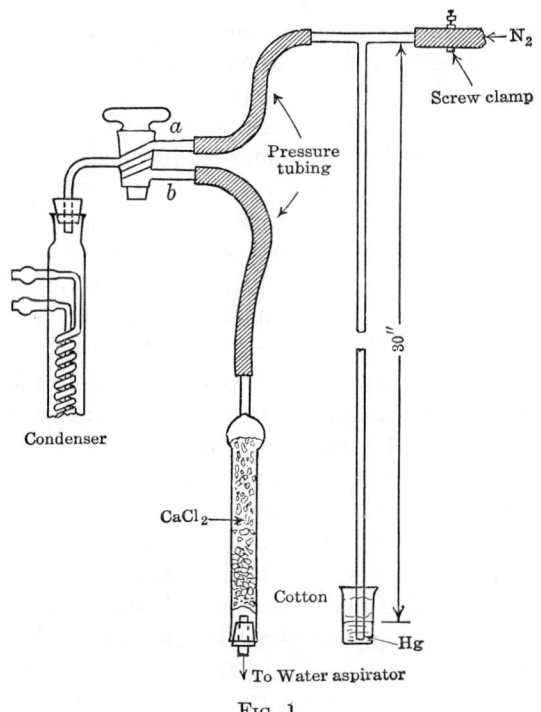

Fig. 1.

to ensure complete decomposition of all the potassium, or a serious fire may result when the reactants are washed down the drain. Small specks of potassium usually remain in the first mortar used for the cutting operation and should be decomposed in the hood by cautious addition of small amounts of t-butyl alcohol as described above.

The reaction is conducted in a 500-ml. round-bottomed flask attached by a ground-glass joint to a reflux condenser, the top of which is connected to a three-way stopcock leading to a, a source of nitrogen and a mercury trap, and b, a water aspirator (Fig. 1). The flask and condenser are dried by warming with a free flame while the system is under reduced pressure (cock turned to b to engage aspirator). Tank nitrogen, dried by passage through concentrated sulfuric acid and soda-lime, is then admitted to the apparatus by turning the cock slowly to the position indicated in Fig. 1 while nitrogen is bubbling through the mercury. The cooled flask is quickly charged with 45 ml. of dry *t*-butyl alcohol and 2.15 g. of potassium and then reconnected to the apparatus. It is particularly important that the *t*-butyl alcohol be thoroughly anhydrous.* The flow of nitrogen is stopped, the screw clamp is closed, and the mixture is boiled under reflux until the potassium is dissolved, hydrogen being liberated through the mercury trap. The complete dissolution of the potassium will require more than four hours if the *t*-butyl alcohol and apparatus have been properly dried. The solution is then cooled to room temperature, nitrogen being admitted to equalize the pressure. The flask is quickly disconnected just long enough to add 9.11 g. of dry, distilled benzophenone and 13.05 g. of freshly distilled diethyl succinate. The system is then evacuated (until the alcohol begins to boil) and filled with nitrogen. With the stopcock in the position shown in Fig. 1 and the screw clamp closed, the mixture is refluxed gently for thirty minutes. The potassium salt of the half-ester may precipitate during this period.

The mixture is then chilled, acidified with about 10 ml. of cold 1:1 hydrochloric acid, and distilled under reduced pressure (water aspirator) until most of the alcohol is removed. Water is added to the residue, which is extracted thoroughly with ether, and the combined extracts are washed with successive portions of 1 N aqueous ammonia until a test portion gives no precipitate on acidification. The combined alkaline solutions are washed once with a fresh portion of ether and then are added slowly with stirring to an excess of cold dilute hydrochloric acid. When the addition is complete the mixture should still be acidic to Congo red. The pale tan, crystalline half-ester is separated on a suction funnel, washed well with water, and dried. The yield is 14.0–14.5 g. (90–94%), m.p. 120–124°. If a purer material is desired the crude product can be recrystallized by dissolving in about 50 ml. of warm benzene, filtering, and adding an equal volume of petroleum ether (b.p. 40–60°). Upon cooling 13.0–13.4 g. of almost colorless half-ester, m.p. 123–124.5°, crystallizes.

β-Carbomethoxy-β-(2-methyl-1,2,3,4,-tetrahydro-1-phenanthrylidene)propionic Acid. (Use of Potassium *t*-Butoxide and Dimethyl Succinate with Unreactive Ketones.) If the procedure described above

---

* Anhydrous *t*-butyl alcohol can be prepared by refluxing the commercial product with sodium (about 3 g. of sodium per 100 ml. of alcohol) until about two-thirds of the metal has dissolved and then distilling the *t*-butyl alcohol. It may be necessary to add fresh sodium in order to have free metal present throughout the distillation. A highly effective and convenient method of drying *t*-butyl alcohol is with calcium hydride, which can be obtained from Metal Hydrides Inc.. Beverly, Mass.

fails to give good yields and unchanged ketone is recovered, the following procedure for the condensation of 2-methyl-1-keto-1,2,3,4-tetrahydrophenanthrene with excess dimethyl succinate [58, 101] may be useful. This represents a reaction that gives rise to a mixture of isomeric half-esters. Dimethyl succinate is used instead of the diethyl ester to avoid reduction of the ketone. Dimethyl succinate is conveniently prepared in 85–90% yields on a scale as large as 2 kg. by the general procedure of Clinton and Laskowski [114] utilizing ethylene dichloride as the solvent. The product is purified by a single distillation through a short Vigreux column; b.p. 192–195°/750 mm., n_D^{25} 1.4173–1.4174.

A 500-ml. three-necked flask with ground-glass joints is fitted with a Hershberg dropping funnel [115] and a Hershberg wire stirrer passing through a glass bearing capped with a silicone-lubricated rubber sleeve. The third neck of the flask is connected with pressure tubing to a T-tube leading to the top of the dropping funnel and to the arm of a three-way stopcock which leads to a source of nitrogen and reduced pressure as shown in Fig. 1 (T-tube replacing condenser). The apparatus is flame-dried, and dry nitrogen is admitted as described on p. 43. The dropping funnel is charged with a mixture prepared by adding 12.9 g. of dimethyl succinate to a solution of 3.02 g. of potassium in 63 ml. of dry t-butyl alcohol. (See the procedures for handling potassium and drying t-butyl alcohol on pp. 43–44.) Two and one-half grams of 2-methyl-1-keto-1,2,3,4-tetrahydrophenanthrene (m.p. 72–73°) is placed in the flask, and the system is then evacuated and filled with nitrogen as described above.

With the stopcock in the position a indicated in Fig. 1 the screw clamp is closed and about 15 ml. of the solution is added from the dropping funnel. The stirrer is started and the flask heated with an oil bath maintained at 50–55° while the remainder of the mixture is dropped in over a period of about four hours. After an additional hour at 50°, the mixture is cooled, acidified with excess 1:1 hydrochloric acid, and most of the alcohol removed under reduced pressure. Water is added, and the semi-solid organic residue is taken up in ether, washed with water, and extracted with successive portions of 1 N aqueous ammonia. Acidification of the combined alkaline solutions gives 3.57 g. (93%) of a yellow oily mixture of half-esters which solidifies on standing, m.p. 119–143°. The predominant isomer can be separated in 61% yield by crystallization of the crude product from dilute methanol, giving 2.37 g. of colorless needles, m.p. 156–159°.

[114] Clinton and Laskowski, *J. Am. Chem. Soc.*, **70**, 3135 (1948).
[115] *Organic Syntheses*, Coll. Vol. 2, 129 (1943); see also Fieser, *Experiments in Organic Chemistry*, 2nd ed., D. C. Heath and Co., Boston, Mass., 1941, p. 312.

When applied to the even less reactive 7-methoxy ketone, the above procedure gives only a 61% yield. However, by conducting the reaction at the boiling point instead of 50° and by adding the reagents over a period of six instead of four hours, practically quantitative condensation is realized.[101]

The Condensation of Acetophenone and Diethyl Succinate (Sodium Hydride Method). The following description of the condensation of acetophenone with diethyl succinate is a modification of a reported procedure [110] and is typical for a reactive ketone that gives a mixture of half-esters.

The following procedure is recommended for handling sodium hydride.[116, 117] Sodium hydride is a grayish white, crystalline, free-flowing powder; it must be kept in air-tight containers for protection against atmospheric moisture and oxygen. The hermetically sealed tin containers in which it is supplied may be opened without hazard in ordinary air. Because sodium hydride is a free-flowing powder, it may be quickly transferred from a container to another vessel, e.g., to a reaction vessel, in the open air. If exposed to the air unduly, traces of sodium hydroxide formed on the surface render the material hygroscopic. Rapid absorption of atmospheric moisture may then take place, and the heat generated by reaction with water may be great enough to cause the solid to ignite. The fire is not violent, however, and may be extinguished readily by excluding air. Dry sodium carbonate or an asbestos blanket may be used to extinguish a sodium hydride fire. Carbon tetrachloride and carbon dioxide are not safe materials for extinguishing sodium hydride fires since some metallic sodium may be liberated which will react with these two fire-extinguishing agents. In using sodium hydride in the presence of inflammables (low-flash-point combustible liquids, vapors, or gases), the safest procedure is to measure or weigh the hydride away from the immediate vicinity of the inflammable liquid and bring the two together either in an inert medium or blanketed by an inert gas such as nitrogen. Keeping the above precautions in mind, it is general practice to weigh sodium hydride on an ordinary analytical balance in the open air, if it is done fairly rapidly. The period of time during which sodium hydride can be exposed to air without difficulty increases as the relative humidity of the air decreases. Adequate protection, such as goggles, gloves, and face shield, should be worn by the operator just as in the handling of metallic sodium.

The reaction is conducted in a 125-ml. round-bottomed three-necked

---

[116] *New Products Bulletin*, No. 25, E. I. du Pont de Nemours and Co., Electrochemical Department, Wilmington 98, Delaware, March 18, 1947.

[117] Hansley and Carlisle, *Chem. Eng. News*, **23**, 1332 (1945).

(ground-glass joints) flask equipped with a Hershberg wire stirrer passing through a glass bearing capped with a silicone-lubricated rubber sleeve, and a condenser the top of which leads to a source of nitrogen and reduced pressure as shown in Fig. 1. The third neck of the flask carries a ground-glass stopper, which is removed for the addition of reagents. For larger runs the stopper may be replaced by a special addition tube for the introduction of sodium hydride.[116] The apparatus is evacuated, flame-dried, and filled with nitrogen as described above. With nitrogen flowing, the stopper is removed and 3.6 g. (0.15 mole) of sodium hydride is washed into the flask with the aid of about 25 ml. of dry benzene, followed by 6.0 g. (0.05 mole) of freshly distilled acetophenone and 26.13 g. (0.15 mole) of freshly distilled diethyl succinate, which are washed into the flask with an additional 25 ml. of dry benzene. A little ethanol (0.73 ml.) is then added, the stopper is replaced, and the flow of nitrogen is stopped, the pinch-clamp (Fig. 1) being closed. The stirrer is started, and hydrogen gas is evolved through the mercury bubbler trap, slowly at first and then more rapidly as the reaction progresses. The flask is cooled as needed with a cold-water bath to maintain the temperature below 40°. At the end of about one hour the evolution of gas has usually almost subsided and the reaction is essentially over.

The mixture is cooled with an ice bath, and 10.5 ml. of glacial acetic acid is added dropwise (to avoid excessive foaming). Water and ether are then added, and the aqueous layer is separated and washed once with ether. The combined ethereal solutions are extracted repeatedly with 5% sodium carbonate solution until a test portion shows no appreciable cloudiness on acidification. The combined alkaline solutions are acidified, and the precipitated oil is collected by ether extraction. The ethereal solution is dried over anhydrous sodium sulfate and evaporated in vacuum, leaving 11.4–11.6 g. (92–93%) of a pale yellow semi-solid mixture of isomeric half-esters. This crude product has a neutral equivalent of about 261 (calculated 248) and may be employed directly in synthetic operations. Such a product, for example, when heated with a mixture of hydrobromic acid, water, and acetic acid, is hydrolyzed and decarboxylated giving γ-phenylvalerolactone in about 85% yield.[2] However, if it is desired, the crude product may be crystallized from petroleum ether (60–68°), and thus about one-third of the material may be rendered crystalline (m.p. 111–112° after recrystallization). This product is the half-ester, $C_6H_5C(CH_3)$=$C(CO_2C_2H_5)CH_2CO_2H$, in which the phenyl and carbethoxyl groups are *cis*.

When the above procedure is applied to benzophenone, β-carbethoxy-γ,γ-diphenylvinylacetic acid, m.p.124.5–125.5°, is obtained in 97% yield.

TABULAR SURVEY OF THE STOBBE CONDENSATION

Tables I and II include examples of the Stobbe condensation with aldehydes and with ketones, which are arranged in a conventional order according to molecular formula. The following information is provided: ester employed, solvent, condensing agent, time, temperature, products isolated, and reference. The molecular ratio of reactants and the yields of products are also given where available. The literature has been reviewed through 1949.

Generic names are given in the tables for the products isolated. A key for the structures of the products is shown on p. 49. The substances listed on the right are the isomeric lactonic forms of the unsaturated dibasic acids in the left-hand column.

Structural Key for Generic Names in Tables I and II

Dibasic Acids

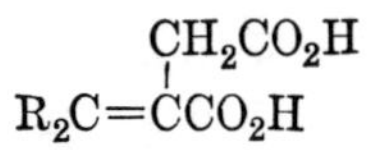

$$\begin{array}{c} \text{CH}_2\text{CO}_2\text{H} \\ | \\ \text{R}_2\text{C}=\text{CCO}_2\text{H} \end{array}$$

Alkylidenesuccinic acid

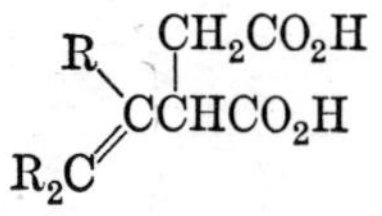

$$\begin{array}{c} \text{R} \quad \text{CH}_2\text{CO}_2\text{H} \\ \diagdown \quad | \\ \text{CCHCO}_2\text{H} \\ \diagup \\ \text{R}_2\text{C} \end{array}$$

Alkenylsuccinic acid

$$\begin{array}{c} \text{HO}_2\text{CC}=\text{CR}_2 \\ | \\ \text{R}_2\text{C}=\text{CCO}_2\text{H} \end{array}$$

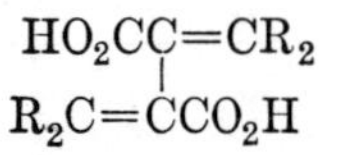

Dialkylidenesuccinic acid

Lactonic Acids

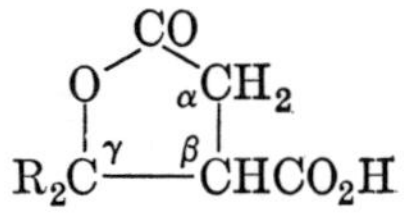

$$\begin{array}{c} \quad\quad \text{CO} \\ \quad\diagup \quad\diagdown \\ \text{O} \quad\quad {}_\alpha\text{CH}_2 \\ | \quad\quad\quad | \\ \text{R}_2\text{C}\underset{\gamma}{\rule{1cm}{0.4pt}}\underset{\beta}{\text{CHCO}_2\text{H}} \end{array}$$

Paraconic acid

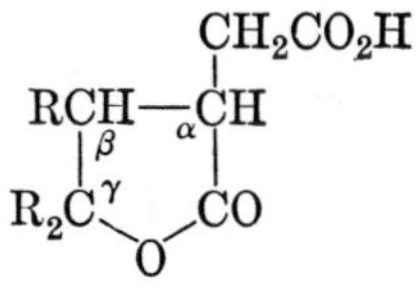

$$\begin{array}{c} \quad\quad\quad\quad \text{CH}_2\text{CO}_2\text{H} \\ \quad\quad\quad\quad | \\ \text{RCH}\underset{\beta}{\rule{0.5cm}{0.4pt}}{}_\alpha\text{CH} \\ | \quad\quad\quad | \\ \text{R}_2\text{C}_\gamma \quad\quad \text{CO} \\ \quad \diagdown\text{O}\diagup \end{array}$$

Isoparaconic acid

$$\begin{array}{c} \quad\quad \text{CO} \\ \quad\diagup \quad\diagdown \\ \text{O} \quad\quad {}_\alpha\text{C}=\text{CR}_2 \\ | \quad\quad\quad | \\ \text{R}_2\text{C}\underset{\gamma}{\rule{1cm}{0.4pt}}\underset{\beta}{\text{CCO}_2\text{H}} \\ \quad\quad\quad\quad \text{H} \end{array}$$

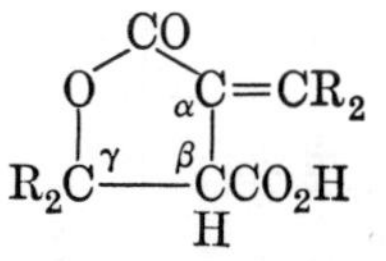

Alkylideneparaconic acid

TABLE I

THE STOBBE CONDENSATION WITH ALDEHYDES

Aldehyde			Ester (moles per mole of aldehyde)	Condensing Agent (moles per mole of aldehyde), Solvent	Time, Temp.	Products Isolated (% yield)	Reference [*]
Formula	Name or Structure						
C_4H_8O	Isobutyraldehyde		Diethyl succinate (1)	Na (1), ether	14 d, cold	Alkylidenesuccinic acid (low); alkylisoparaconic acid (low)	14
			Diethyl succinate (0.5)	$NaOC_2H_5$ (1), ether	Varied conditions	Alkylidenesuccinic acid (poor); alkylisoparaconic acid (poor)	15
			Diethyl succinate (1.25)	$KOC(CH_3)_3$ (1.13), $(CH_3)_3COH$	1.5 hr., reflux	Unsaturated diethyl ester [a] (85)	16
			Diethyl isopropylidenesuccinate (1)	$NaOC_2H_5$ (2), ether	3 d., $-15°$	Dialkylidenesuccinic acid (low)	15
$C_5H_4O_2$	Furfural		Diethyl succinate (2)	$NaOC_2H_5$ (1), ether	Several days	Alkylidenesuccinic acid; dialkylidenesuccinic acid	111
			Diethyl succinate (0.33)	$NaOC_2H_5$ (0.66), ether	Several days, $-10°$	Dialkylidenesuccinic acid (15)	118
			Diethyl isopropylidenesuccinate (0.67)	$NaOC_2H_5$ (1.3), ethanol	5 d., cold to reflux	Dialkylidenesuccinic acid (25)	118
			Diethyl benzhydrylidenesuccinate (0.67)	$NaOC_2H_5$ (1.3), ethanol	2 d., cold to reflux	Dialkylidenesuccinic acid	118
C_7H_6O	Benzaldehyde		Diethyl succinate (0.5)	$NaOC_2H_5$ (1), ether	Several days, $23°$	Alkylidenesuccinic acid; dialkylidenesuccinic acid	119

Diethyl succinate	Na or NaOC$_2$H$_5$, ether	—	Alkylidenesuccinic acid (7)	120
Diethyl succinate (1)	NaOC$_2$H$_5$ (2.5), ethanol	—, cold to reflux	Alkylidenesuccinic acid (29)	121
Diethyl succinate (0.5)	NaOC$_2$H$_5$ (1), ether	Several days, −10°	Dialkylidenesuccinic acid (35–40); alkylidenesuccinic acid (trace)	17
Diethyl succinate (1)	Na (0.5), ether	17 d., −10° to 8°	Dialkylidenesuccinic acid (18); alkylidenesuccinic acid (6)	77
Diethyl succinate (1)	NaOC$_2$H$_5$ (1), ether	7 d., −10°	Dialkylidenesuccinic acid (15); alkylidenesuccinic acid (7.5)	77
Diethyl succinate (1)	NaOC$_2$H$_5$ (2), ether	3 d., −10° to 0°	Dialkylidenesuccinic acid (35); alkylidenesuccinic acid (trace)	77
Diethyl succinate (1)	NaOC$_2$H$_5$ (2.5), ether	3 hr, reflux	Alkylidenesuccinic acid (35); dialkylidenesuccinic acid (trace)	9, 77
Diethyl succinate (1)	NaOC$_2$H$_5$ (2.5), ethanol	—, 10° to reflux	Alkylidenesuccinic acid (21–26); dialkylidenesuccinic acid (trace)	77
Diethyl succinate	NaOC$_2$H$_5$, ether	—	Alkylidenesuccinic acid; dialkylidenesuccinic acid	122
Diethyl succinate (0.5)	NaOC$_2$H$_5$ (1), ether	3.5 d., −14° to 0°	Dialkylidenesuccinic acid (15–20)	97
Dimethyl succinate (1)	Na (1.1), ether	27 hr., reflux to 23°	Dialkylidenesuccinic acid (20–25); alkylidenesuccinic acid (10–12)	123

* References 118–147 are on p. 73.
[a] This ester was obtained by direct esterification of the crude half-ester.

TABLE I—Continued
The Stobbe Condensation with Aldehydes

Formula	Aldehyde Name or Structure	Ester (moles per mole of aldehyde)	Condensing Agent (moles per mole of aldehyde), Solvent	Time, Temp.	Products Isolated (% yield)	Reference *
C_7H_6O	Benzaldehyde (Cont'd)	Succinic anhydride	$(C_6H_5)_3CNa$, benzene	3 wk, 23°	Dialkylidenesuccinic acid	124
		Diethyl isopropylidenesuccinate (1) HCC_6H_5 (1)	$NaOC_2H_5$ (2), ethanol	1 d., cold to reflux	Dialkylidenesuccinic acid	125
		$C_2H_5O_2CCCH_2CO_2C_2H_5$ $\|\|$ $H_3CCC_6H_5$ (1)	$NaOC_2H_5$ (2), ethanol	Several days, $-15°$ to reflux	Dialkylidenesuccinic acid (excellent)	77, 17
		$C_2H_5O_2CCCH_2CO_2C_2H_5$ Diethyl benzhydrylidenesuccinate (1)	$NaOC_2H_5$ (2), ethanol	4 hr., $-15°$	Dialkylidenesuccinic acid (70)	112
		Diethyl cinnamylidenesuccinate (1)	$NaOC_2H_5$ (2), ethanol	6 hr., $-15°$ to reflux	Dialkylidenesuccinic acid (good)	44, 126
		Diethyl phenylsuccinate	$NaOC_2H_5$ (2), ethanol	—	Dialkylidenesuccinic acids	37
			$NaOC_2H_5$, ether	—	Mixture β,γ-diphenylvinylacetic acids	70
		Dimethyl benzylsuccinate (1)	Na (1), ether	3 hr., reflux	Alkylalkylidenesuccinic acid (39)	67, 123, 127, 105
		Dimethyl (β-phenethyl)-succinate (0.8)	Na (3), ether	1 d., cold to reflux	Alkylalkylidenesuccinic acid	68, 128
		Diethyl thiodiglycolate [b] (2.6)	$NaOCH_3$ (1.2), methanol	—	Dialkylidenethiodiglycolic acid (satisfactory)	129

THE STOBBE CONDENSATION

	Ester	Catalyst, Solvent	Time, Temp.	Product	Ref.
	Diethyl thiodiglycolate [b] (0.5)	NaOC$_2$H$_5$ (1), ethanol	2 hr, −10° to reflux	Dialkylidenethiodiglycolic acid (74)	77
	Diethyl thiodiglycolate [b] (0.5)	NaOC$_2$H$_5$ (1), ether	—	Dialkylidenethiodiglycolic acid (62)	77
	Triethyl carballylate [b]	NaOC$_2$H$_5$, ether	30 d, 23°	Mixture containing some β-benzylideneglutaric anhydride [c]	71
	Ethyl β-benzoylpropionate [b] (1)	NaOC$_2$H$_5$ (1), 96% ethanol	1 d, 23°	β-Benzylidene-β-benzoylpropionic acid (90)	72
	Methyl β-veratroylpropionate (1)	NaOCH$_3$ (1), methanol	1 d, 23°	β-Benzylidene-β-veratroylpropionic acid (50–60)	73
CH$_3$(CH$_2$)$_5$CHO	Diethyl succinate (1.25)	KOC(CH$_3$)$_3$ (1.13), (CH$_3$)$_3$COH	14 hr, reflux	Unsaturated diethylester [a] (59)	16
p-ClC$_6$H$_4$CHO	Diethyl isopropylidenesuccinate (1)	NaOC$_2$H$_5$ (2), ether	Several days, −15°	α-Alkylidene-γ-aryl- or α-alkylidene-γ,γ-dialkylparaconic acid	130
	Diethyl benzhydrylidenesuccinate (0.9)	NaOC$_2$H$_5$ (2.4), ethanol	2 d., −15° to reflux	Dialkylidenesuccinic acid	131
	Diethyl succinate (0.5)	NaOC$_2$H$_5$ (1), ether	—	Resinous products	112
o-O$_2$NC$_6$H$_4$CHO	Diethyl isopropylidenesuccinate (1)	NaOC$_2$H$_5$ (2), ethanol	1 d, 0° to reflux	Dialkylidenesuccinic acid (75–85)	62
	Diethyl benzhydrylidenesuccinate (1)	NaOC$_2$H$_5$ (2), ethanol	3 d, —	Dialkylidenesuccinic acid (60–70)	63, 64
m-O$_2$NC$_6$H$_4$CHO	Diethyl succinate (1)	NaOC$_2$H$_5$ (1), ether	Several days, cold	Resinous products	112
	Diethyl isopropylidenesuccinate (1)	NaOC$_2$H$_5$ (2), ethanol	1 d, 0° to reflux	Dialkylidenesuccinic acid (85)	62

* References 118–147 are on p. 73.
[b] This reaction is related to but is not a Stobbe condensation.
[c] This is the probable structure of the products.

TABLE I—*Continued*

THE STOBBE CONDENSATION WITH ALDEHYDES

Aldehyde		Ester (moles per mole of aldehyde)	Condensing Agent (moles per mole of aldehyde), Solvent	Time, Temp.	Products Isolated (% yield)	Reference *
Formula	Name or Structure					
$C_7H_5O_3N$	m-$O_2NC_6H_4CHO$ (*Cont'd*)	Diethyl benzhydrylidenesuccinate (1)	$NaOC_2H_5$ (2), ethanol	3 d.—	Dialkylidenesuccinic acid (85)	63
	p-$O_2NC_6H_4CHO$	Diethyl succinate (0.5)	$NaOC_2H_5$ (1), ether	Varied conditions	Resinous products	112
		Diethyl isopropylidenesuccinate (1)	$NaOC_2H_5$ (2), ethanol	1 d., 0° to reflux	Dialkylidenesuccinic acid (80)	62
		Diethyl benzhydrylidenesuccinate (1)	$NaOC_2H_5$ (2), ethanol	3 d.—	Dialkylidenesuccinic acid (65)	63
$C_8H_6O_3$	Piperonal	Diethyl succinate (0.5)	$NaOC_2H_5$ (1), ether	8 d., −15°	Dialkylidenesuccinic acid [d]	6
		Diethyl succinate (0.5)	$NaOC_2H_5$ (1), ether	21 hr., −10° to 0°	Dialkylidenesuccinic acid (48)	83
		Diethyl succinate (0.5)	$NaOC_2H_5$ (1), ether	7 d., cold	Dialkylidenesuccinic acid (35)	19
		Diethyl succinate (1)	$NaOC_2H_5$ (2.5), ethanol	2 hr., reflux	Alkylidenesuccinic acid (90)	18
		Diethyl isopropylidenesuccinate (1)	$NaOC_2H_5$ (2), ethanol	Several hr., 23° to reflux	Dialkylidenesuccinic acid (75)	130
		HCC_6H_5 (1) $\|\|$ $C_2H_5O_2CCCH_2CO_2C_2H_5$	$NaOC_2H_5$ (2), ethanol	2 d., cold to reflux	Dialkylidenesuccinic acid (70)	112

		$C_6H_5CCH_3$ (1)	$NaOC_2H_5$ (1), ether	5 d., $-15°$	Dialkylidenesuccinic acid (good)	56
		$C_2H_5O_2CCCH_2CO_2C_2H_5$ $CH_3CC_6H_5$ (1)	$NaOC_2H_5$ (2), ethanol	Several days, cold to reflux	Dialkylidenesuccinic acid (good)	56
		$C_2H_5O_2CCCH_2CO_2C_2H_5$				
		Diethyl benzhydrylidene-succinate (0.9)	$NaOC_2H_5$ (2), ethanol	Several hr., $-15°$ to reflux	Dialkylidenesuccinic acid	131
		β-$C_{10}H_7C$==$CCO_2C_2H_5$ (1) CH_3 $CH_2CO_2C_2H_5$	$NaOC_2H_5$ (2), ether	Several d., $-15°$ to $23°$	Dialkylidenesuccinic acid	132
		Dimethyl veratrylsuccinate (1)	$NaOC_2H_5$ (2), ether	Several hr., cold	Alkylalkylidenesuccinic anhydride (poor)	69
C_8H_8O	p-Tolualdehyde	Diethyl succinate (0.5)	$NaOC_2H_5$ (1), ether	3.5 d., $-18°$ to 35°	Dialkylidenesuccinic acid (20)	97
		Diethyl isopropylidene-succinate (1)	$NaOC_2H_5$ (2), ether	Several days, $-15°$	Dialkylidenesuccinic acid	125
		Diethyl benzhydrylidene-succinate (1)	$NaOC_2H_5$ (2.2), ethanol	2 d., $-15°$ to reflux	Dialkylidenesuccinic acid	126
$C_8H_8O_2$	o-Methoxybenzalde-hyde	Diethyl succinate (0.5)	$NaOC_2H_5$ (1), ether	3.5 d., $-18°$ to 0°	Dialkylidenesuccinic acid (20)	97
		Diethyl isopropylidene-succinate (1)	$NaOC_2H_5$ (2), ethanol	—, $-15°$ to reflux	Dialkylidenesuccinic acid	133
		Diethyl benzhydrylidene-succinate (1)	$NaOC_2H_5$ (2), ethanol	2 d., cold to reflux	Dialkylidenesuccinic acid (60)	133
$C_8H_8O_2$	Anisaldehyde	Diethyl succinate (0.5)	$NaOC_2H_5$ (1), ether	4 d., $-15°$ to 23°	Dialkylidenesuccinic acid; α-alkylidene-γ-arylparaconic acid	134

* References 118–147 are on p. 73.
^{d} Shorter reaction periods gave also the α-alkylidene-γ-arylparaconic acid in yields as high as 30%.

TABLE I—*Continued*

The Stobbe Condensation with Aldehydes

Aldehyde		Ester (moles per mole of aldehyde)	Condensing Agent (moles per mole of aldehyde), Solvent	Time, Temp.	Products Isolated (% yield)	Reference *
Formula	Name or Structure					
$C_8H_8O_2$	Anisaldehyde (*Cont'd*)	Diethyl succinate (0.5)	$NaOC_2H_5$ (1), ether	3.5 d., $-18°$ to $35°$	Dialkylidenesuccinic acid	97
		Diethyl isopropylidenesuccinate (1) HCC_6H_5 (1)	$NaOC_2H_5$ (2), ethanol	1 d., $-15°$ to reflux	Dialkylidenesuccinic acid	133
		$C_2H_5O_2CCCH_2CO_2C_2H_5$	$NaOC_2H_5$ (2), ethanol	Several days, $-15°$ to reflux	Dialkylidenesuccinic acid (80)	133
		Diethyl benzhydrylidenesuccinate (1)	$NaOC_2H_5$ (2), ethanol	2 d., cold	Dialkylidenesuccinic acid	133
		Dimethyl benzylsuccinate (1.1)	Na (1.1), ether	24 hr., 23°	Alkylalkylidenesuccinic acid (26)	123
		Dimethyl (β-phenethyl)-succinate (0.9)	Na (1.1), ether	24 hr., cold	Alkylalkylidenesuccinic acid (28)	123, 135
C_9H_8O	Cinnamaldehyde	Diethyl succinate (1)	$NaOC_2H_5$, —	—	Alkylidenesuccinic acid	35, 36
		Diethyl isopropylidenesuccinate (1) HCC_6H_5	$NaOC_2H_5$ (2), ethanol	—	Dialkylidenesuccinic acids	37
		$C_2H_5O_2CCCH_2CO_2C_2H_5$	$NaOC_2H_5$ (2), ethanol	—	Dialkylidenesuccinic acids	37
		Diethyl benzhydrylidenesuccinate (1)	$NaOC_2H_5$ (2), ethanol	Several days, cold to reflux	Dialkylidenesuccinic acid (90)	37

THE STOBBE CONDENSATION

Formula	Aldehyde/Ketone	Succinic Ester	Base, Solvent	Conditions	Products	Ref.
$C_9H_{10}O_2$	$o\text{-}C_2H_5OC_6H_4CHO$	Diethyl benzhydrylidenesuccinate (1)	$NaOC_2H_5$ (2.2), ethanol	7 hr., $-15°$ to reflux	Dialkylidenesuccinic acid (>56)	133
$C_9H_{10}O_3$	$3,4\text{-}(CH_3O)_2C_6H_3CHO$	Diethyl succinate (0.5)	$NaOC_2H_5$ (1), ether	8 d., $-15°$ to reflux	Dialkylidenesuccinic acid; alkylidenesuccinic acid	136
		Diethyl succinate (0.5)	$NaOC_2H_5$ (1), ether	4 hr., —	Dialkylsuccinic acid [e] (17)	84
		Diethyl isopropylidenesuccinate (1)	$NaOC_2H_5$ (2), ethanol	5 d., $-15°$ to reflux	Dialkylidenesuccinic acid	130
		Diethyl benzhydrylidenesuccinate (1)	$NaOC_2H_5$ (2.2), ethanol	3 d., $-15°$ to reflux	Dialkylidenesuccinic acid	131
$C_{10}H_{12}O$	$p\text{-}(CH_3)_2CHC_6H_4CHO$	Diethyl succinate (0.5)	$NaOC_2H_5$ (1), ether	Several days, $-15°$	Dialkylidenesuccinic acids; alkylidenesuccinic acid; γ-arylparaconic acid	137
		Diethyl isopropylidenesuccinate (1)	$NaOC_2H_5$ (2), ethanol	8 d., $-10°$ to reflux	Dialkylidenesuccinic acids; α-alkylidene-γ-aryl- or α-alkylidene-γ,γ-dialkylparaconic acid	138
		Diethyl benzhydrylidenesuccinate (1)	$NaOC_2H_5$ (2.2), ethanol	2 d., $-15°$ to reflux	Dialkylidenesuccinic acid	126
$C_{10}H_{20}O$	n-Decanal	Diethyl succinate (1.25)	$KOC(CH_3)_3$ (1.13), $(CH_3)_3COH$	2 hr., reflux	Unsaturated diethyl ester [a] (40)	16
$C_{12}H_{24}O$	Lauraldehyde	Diethyl succinate (1.25)	$KOC(CH_3)_3$ (1.13), $(CH_3)_3COH$	8 hr., reflux	Unsaturated diethyl ester [a] (58)	16
$C_{15}H_{12}O$	$(C_6H_5)_2C{=}CHCHO$	Dimethyl benzhydrylidenesuccinate (1)	$NaOCH_3$ (2.2), methanol	3 hr., reflux	Dialkylidenesuccinic acids; α (27); β (53)	38
$C_{15}H_{14}O_3$	CHO–C_6H_3(OCH_3)($OCH_2C_6H_5$)	Diethyl succinate (0.5)	$NaOC_2H_5$ (1), ether	1 d., cold	Dialkylsuccinic acid [e] (25)	85

* References 118–147 are on p. 73.
[e] The crude dialkylidenesuccinic acid was reduced directly with 4% sodium amalgam.

TABLE II
THE STOBBE CONDENSATION WITH KETONES

Ketone Formula	Ketone Name or Structure	Ester (moles per mole of ketone)	Condensing Agent (moles per mole of ketone), Solvent	Time, Temp.	Products Isolated (% yield)	Reference *
C_3H_6O	Acetone	Diethyl succinate (0.5)	$NaOC_2H_5$ (1), ether	Several days, $-15°$ to $23°$	Alkylidenesuccinic acid (55); alkenylsuccinic acid (trace)	1
		Diethyl succinate (1)	$NaOC_2H_5$ (2), ether	2 wk., $-15°$ to $23°$	Alkenylsuccinic acid (47); alkylidenesuccinic acid (trace)	4
		Diethyl succinate (0.5)	$NaOC_2H_5$ (1), ether	5–6 d., $-17°$ to $23°$	Alkylidenesuccinic acid (54)	139
		Diethyl succinate (0.5)	$NaOC_2H_5$ (1), ether	Several days, $-15°$ to $23°$	Unsaturated diethyl esters [a] (41)	140
		Diethyl succinate (1.25)	$KOC(CH_3)_3$ (1.13), $(CH_3)_3COH$	0.5 hr, reflux	Unsaturated diethyl ester [a] (92)	16
		Diethyl cinnamylidenesuccinate (1)	$NaOC_2H_5$ (2), ethanol		Amorphous acidic mixture	37
		Diethyl isopropylidenesuccinate (0.67)	$NaOC_2H_5$ (1.3), ether	Several days, $-15°$	Dialkylidenesuccinic acid (40); ethyl α-alkylidene-γ,γ-dialkylparaconate (low)	107
		$CH_3CC_6H_5$ $\|\|$ $C_2H_5O_2CCCH_2CO_2C_2H_5$ (1)	$NaOC_2H_5$ (2), ethanol	2 hr., $-15°$ to reflux	cis-Dialkylidenesuccinic acid [b]	44

Ketone	Ester (ratio)	Reagent (equiv.), Solvent	Conditions	Products (%)	Ref.
	$CH_3CC_6H_5$ (0.77) / $C_2H_5O_2CCH_2CO_2C_2H_5$	$NaOC_2H_5$ (1.5), ethanol	Several days, $-17°$ to reflux	cis- and trans-Dialkylidenesuccinic acids [b]	107
	$CH_3CC_6H_5$ (1) / $C_2H_5O_2CCH_2CO_2C_2H_5$	$NaOC_2H_5$ (2), ethanol	8 d, $-15°$	cis- and trans-Dialkylidenesuccinic acids [b]	57
	$CH_3CC_6H_5$ (1) / $C_2H_5O_2CCH_2CO_2C_2H_5$	$NaOC_2H_5$ (2), ether	8 d, $-15°$	cis- and trans-Dialkylidenesuccinic acids [b]	57
	$C_6H_5CCH_3$ (1) / $C_2H_5O_2CCH_2CO_2C_2H_5$	$NaOC_2H_5$ (1.6), ether	6 d, $-15°$	cis-Dialkylidenesuccinic acid [b] (excellent)	57
	Diethyl benzhydrylidenesuccinate (1)	$NaOC_2H_5$ (2), ether	Several days, $-15°$ to $23°$	Dialkylidenesuccinic acid	112
C_4H_8O Methyl ethyl ketone	Diethyl succinate (0.5)	$NaOC_2H_5$ (1), ether	2 wk., $-15°$ to $23°$	Alkylidene- and alkenylsuccinic acids	4
	Diethyl succinate (1)	$NaOC_2H_5$ (2), ether	2 wk., $-15°$ to $23°$	Alkylidene- and alkenylsuccinic acids [c]	30
	Diethyl succinate (1)	$NaOC_2H_5$ (2), ether	7–10 d, $-15°$ to $23°$	Alkylidenesuccinic acid (6); alkenylsuccinic acid (37)	80
	Diethyl succinate (1.25)	$KOC(CH_3)_3$ (1.13), $(CH_3)_3COH$	0.5 hr, reflux	Unsaturated diethyl esters [a] (83)	16
C_5H_8O Cyclopentanone	Diethyl succinate (1)	$NaOC_2H_5$ (2), ether	1 wk., $-15°$ to $23°$	Alkylidenesuccinic acid; lactonic acid, $C_{14}H_{18}O_4$	141

* References 118–147 are on p. 73.

[a] The product was obtained by direct esterification of crude half-ester.

[b] cis-Acid: C_6H_5/CO_2H cis. trans-Acid: C_6H_5/CO_2H trans.

[c] The ratio of alkylidene- to alkenyl-succinic acid was 1:12.

TABLE II—*Continued*

THE STOBBE CONDENSATION WITH KETONES

Ketone		Ester (moles per mole of ketone)	Condensing Agent (moles per mole of ketone), Solvent	Time, Temp.	Products Isolated (% yield)	Reference *
Formula	Name or Structure					
C_5H_8O	Cyclopentanone (*Cont'd*)	Diethyl succinate (1)	$NaOC_2H_5$ (2), ether	10 d., $-15°$ to $23°$	Alkylidenesuccinic acid (low); alkenylsuccinic acid (low); lactonic acid $C_{14}H_{18}O_4$ (trace)	98
$C_5H_{10}O$	Methyl isopropyl ketone	Diethyl succinate (1.25)	$KOC(CH_3)_3$ (1.13), $(CH_3)_3COH$	7 hr., reflux	Unsaturated diethyl ester ^{a} (78)	16
$C_6H_{10}O$	Cyclohexanone	Dimethyl succinate (1)	$NaOCH_3$ (1), methanol	2 d., 0°	Methyl γ,γ-pentamethyleneparaconate	5
		Diethyl succinate (1)	$NaOC_2H_5$ (2.5), ethanol	2 hr., reflux	Alkenylsuccinic acid (40)	36
		Diethyl succinate (0.7)	$NaOC_2H_5$ (1.4), ether	2 wk., $-15°$ to $23°$	Alkenylsuccinic acid (37)	142
		Diethyl succinate (1.4)	$KOC(CH_3)_3$ (1.2), $(CH_3)_3COH$	10 min., reflux	Alkenylsuccinic half-ester (84)	27
		Diethyl succinate (1.25)	$KOC(CH_3)_3$ (1.13), $(CH_3)_3COH$	20 min., reflux	Unsaturated diethyl ester ^{a} (72)	16
$C_7H_{12}O$	2-Methylcyclohexanone	Diethyl succinate (1.5)	$KOC(CH_3)_3$ (1.1), $(CH_3)_3COH$	280 min., reflux to 23°	Oily half-esters ^{d}	103

THE STOBBE CONDENSATION

Formula	Ketone	Ester	Base	Conditions	Product	Ref.
$C_7H_{14}O$	3-Methylcyclohexanone	Diethyl succinate (0.67)	$NaOC_2H_5$ (1.3), ether	10 d., $-15°$ to 23°	Alkylidenesuccinic acid (20–25); alkenylsuccinic acid (10–13); lactonic acids, $C_{18}H_{26}O_4$	99
$C_7H_{12}O$	Cycloheptanone	Diethyl succinate (1.5)	$KOC(CH_3)_3$ (1.1), $(CH_3)_3COH$	45 min., reflux	Alkenylsuccinic half-ester (80–87)	28, 51
C_8H_8O	Acetophenone	Diethyl succinate (0.5)	$NaOC_2H_5$ (1), ether	2 wk., $-15°$ to 23°	Oily half-esters	4
		Diethyl succinate (1)	$NaOC_2H_5$ (2), ethanol	9 d., $-15°$ to reflux	Alkenyl- and isomeric alkylidene-succinic acids (60–75) [e]	31
		Diethyl succinate (1.5)	$KOC(CH_3)_3$ (1.1), $(CH_3)_3COH$	40 min., reflux	γ-Methyl-γ-phenylbutyrolactone (85) [f]	2
		Diethyl succinate (3)	NaH (2), benzene	3.7 hr., 23°	Oily half-esters (93)	110
		Di-t-butyl succinate (1.25)	NaH (2.75), benzene	8 hr., 50°	Oily half-esters (57) [g]	110
		Diethyl isopropylidenesuccinate (0.77)	$NaOC_2H_5$ (1.5), ethanol	—, $-15°$ to reflux	cis- and trans-Dialkylidenesuccinic acids [b]	57
$C_8H_{14}O$	3,3-Dimethylcyclohexanone	Dimethyl β-methylglutaconate (1.5)	$KOC(CH_3)_3$ (1.2), $(CH_3)_3COH$	50 min., reflux	Oily half-ester	79
$C_8H_{11}ON$	2-Cyanocycloheptanone	Diethyl succinate	$KOC(CH_3)_3$, $(CH_3)_3COH$	—	Failed	51
$C_9H_{10}O$	p-Methylacetophenone	Diethyl succinate (1.5)	$KOC(CH_3)_3$ (1.1), $(CH_3)_3COH$	45 min., reflux	γ-Methyl-γ-p-tolylbutyrolactone (76) [f]	93

* References 118–147 are on p. 73.
[d] One isomer was isolated in a pure crystalline form.
[e] The ratio of alkylidene- to alkenyl-succinic acid was 9:1.
[f] This represents the over-all yield of pure lactone (from ketone) obtained by hydrolysis and decarboxylation of the crude Stobbe condensation product with a boiling mixture of hydrobromic (or hydrochloric) and acetic acids.
[g] A by-product $C_{20}H_{18}O_4$ which gave a deep red color with alcoholic ferric chloride, possibly $(C_6H_5COCH_2COCH_2)_2$, was isolated in about 33% yield.

TABLE II—*Continued*

The Stobbe Condensation with Ketones

Ketone			Ester (moles per mole of ketone)	Condensing Agent (moles per mole of ketone), Solvent	Time, Temp.	Products Isolated (% yield)	Reference*
Formula	Name or Structure						
$C_9H_{10}O$	Propiophenone		Diethyl succinate (1)	$NaOC_2H_5$ (2), ether	5 d., cold	Alkenylsuccinic acid (80); isomeric alkylidenesuccinic acids (9)	32
			Diethyl succinate (1.5)	$KOC(CH_3)_3$ (1.1), $(CH_3)_3COH$	40 min., reflux	γ-Ethyl-γ-phenyl-butyrolactone (82) *f*	2
$C_{10}H_{10}O$	α-Tetralone		Diethyl succinate (1)	$NaOC_2H_5$ (2), ether	52 hr., 23°	Crude half-ester (83)	76
			Diethyl succinate (1)	$NaOC_2H_5$ (2), ethanol	6 hr., reflux	Alkenylsuccinic half-ester (79)	76
			Diethyl succinate (1.5)	$KOC(CH_3)_3$ (1.1), $(CH_3)_3COH$	40 min., reflux	Alkenylsuccinic half-ester (89–94) *h*	76
			Diethyl succinate (3)	NaH (3.3), benzene	3 to 3.5 hr., 23°	Alkenylsuccinic half-ester (70–73)	110
			Di-t-butyl succinate (1.25)	NaH (2.75), benzene	8.5 hr., 50°	Alkenylsuccinic half-ester (72)	110
$C_{11}H_8O_2$	2-Benzoylfuran		Diethyl succinate (1)	$NaOC_2H_5$ (2), ethanol	30 min., reflux	Oily half-esters (66) *i*	20
			Diethyl succinate (1)	$NaOC_2H_5$ (2), ethanol	12 hr., 23° to reflux	Mixture of half-esters (63) *j*	20

		Diethyl succinate (1.5)	KOC(CH$_3$)$_3$ (1.2), (CH$_3$)$_3$COH	45 min., reflux	Oily half-ester (58)	51
C$_{11}$H$_8$OS	2-Benzoylthiophene	Diethyl succinate (1)	NaOC$_2$H$_5$ (2), ether	1.5 hr, 23° to reflux	Oily half-ester (85)	59
C$_{12}$H$_{10}$O	1-Acetylnaphthalene	Diethyl succinate (1)	NaOC$_2$H$_5$ (1.8), ether	3 d., −15°	Alkylidenesuccinic half-ester	132
C$_{12}$H$_{10}$O	2-Acetylnaphthalene	Diethyl succinate (1)	NaOC$_2$H$_5$ (1.8), ether	3 d., −15°	Alkylidenesuccinic half-ester	132
		Diethyl succinate (1)	NaOC$_2$H$_5$ (1.1), ethanol	15 hr, reflux	Mixture of half-esters (65) $^{d, k}$	11
		Diethyl succinate (1)	NaOC$_2$H$_5$ (2), ether	5 d., cold	γ-Methyl-γ-(2-naphthyl)butyrolactone (69) ^{f}	95
		Diethyl succinate (1)	NaOC$_2$H$_5$ (1.1), ethanol	19 hr, reflux	γ-Methyl-γ-(2-naphthyl)butyrolactone (57) ^{f}	95
		Diethyl succinate (1.5)	KOC(CH$_3$)$_3$ (1.1), (CH$_3$)$_3$COH	40 min., reflux	γ-Methyl-γ-(2-naphthyl)butyrolactone (82) ^{f}	95
C$_{12}$H$_{10}$O$_2$	⌬—CO—⌬—CH$_3$	Diethyl succinate (1)	NaOC$_2$H$_5$ (2), ether	—	Resinous half-esters (83)	59
C$_{12}$H$_{10}$O$_3$	⌬—CO—⌬—OCH$_3$	Diethyl succinate (1)	NaOC$_2$H$_5$ (2), ether	—	Oily half-esters (81)	59

* References 118–147 are on p. 73.

^{h} With equimolar amounts of ester and ketone, the best yields of half-ester after one hour of refluxing (optimum time) were only 54–58%.

^{i} The *cis* isomer (C$_6$H$_5$/CO$_2$C$_2$H$_5$ *cis*) was isolated crystalline in 33% yield.

^{j} The *cis* isomer (C$_6$H$_5$/CO$_2$C$_2$H$_5$ *cis*) was isolated crystalline in 6% yield; the *trans* isomer (C$_6$H$_5$/CO$_2$C$_2$H$_5$ *trans*) in 40% yield.

^{k} The mixture was shown to consist principally of about equal quantities of the *cis*- and *trans*-alkylidenesuccinic acid derivatives.

TABLE II—*Continued*

THE STOBBE CONDENSATION WITH KETONES

Formula	Ketone		Ester (moles per mole of ketone)	Condensing Agent (moles per mole of ketone), Solvent	Time, Temp.	Products Isolated (% yield)	Reference *
	Name or Structure						
$C_{12}H_{20}O$	(cyclohexyl-cyclohexanone structure)		Diethyl succinate (1.5)	$KOC(CH_3)_3$ (1.1), $(CH_3)_3COH$	40 min, reflux	Alkylsuccinic acid (30) ^{l}	143
$C_{13}H_8O$	Fluorenone		Diethyl methylsuccinate (1)	$NaOC_2H_5$ (2.1), ether	4 d., $-15°$ to 23°	Alkylalkylidenesuccinic acid (low)	66
			Diethyl isopropylidene-succinate (1)	$NaOC_2H_5$ (2), ether	Several days, $-15°$ to 23°	Dialkylidenesuccinic acid	144
			HCC_6H_5 (1) $\parallel$ $C_2H_5O_2CCCH_2CO_2C_2H_5$	$NaOC_2H_5$ (2), ether	Several days, $-15°$	Dialkylidenesuccinic acid; isomeric lactonic acid	144
			Diethyl benzhydryl-idenesuccinate (1)	$NaOC_2H_5$ (2), none	—, 100°	Dialkylidenesuccinic acid	144
$C_{13}H_{10}O$	Benzophenone		Diethyl succinate (0.5)	$NaOC_2H_5$ (1), ether	2 wk., $-15°$ to 23°	Alkylidenesuccinic half-ester	4
			Diethyl succinate (1)	$NaOC_2H_5$ (2), ether	Several days, $-15°$ to 23°	Alkylidenesuccinic half-ester (58–62)	53
			Diethyl succinate (1)	$NaOC_2H_5$ (2), none	—, 100°	Alkylidenesuccinic acid (90)	53

THE STOBBE CONDENSATION

Diethyl succinate (1)	NaOC$_2$H$_5$ (2), ethanol	6 d., −15° to reflux	Alkylidenesuccinic half-ester (50) ^{m}	53
Diethyl succinate (1)	NaOC$_2$H$_5$ (2), ether	Several days, 0° to 23°	Alkylidenesuccinic acid (90)	64
Diethyl succinate (1.5)	KOC(CH$_3$)$_3$ (1.1), (CH$_3$)$_3$COH	30 min., reflux	Alkylidenesuccinic half-ester (90) ^{n}	2
Diethyl succinate (3)	NaH (2), ether or benzene	8 hr., 23°	Alkylidenesuccinic half-ester (97)	110
Dimethyl succinate (3)	BF$_3$, CS$_2$	2 hr., 0° to 23°	Failed	145
Di-t-butyl succinate (1.6)	KOC(CH$_3$)$_3$ (1.2), (CH$_3$)$_3$COH	1 hr., reflux	Alkylidenesuccinic half-ester (80) ^{o}	2
Di-t-butyl succinate (1.25)	NaH (2.75), benzene	3.5 hr., 50°	Alkylidenesuccinic half-ester (98)	110
Di-t-butyl succinate (2)	(C$_6$H$_5$)$_3$CNa (3), ether	3 d., 23°	Alkylidenesuccinic half-ester (87) ^{p}	145
Diethyl methylsuccinate (1)	NaOC$_2$H$_5$ (2), ether	Several days, −15° to 23°	Alkylalkylidenesuccinic half-ester	65
Diethyl isopropylidenesuccinate (1)	NaOC$_2$H$_5$ (2), ether	Several days, −15° to 23°	Dialkylidenesuccinic half-ester (67)	107
HCC$_6$H$_5$ (1.2) ‖ C$_2$H$_5$O$_2$CCCH$_2$CO$_2$C$_2$H$_5$	NaOC$_2$H$_5$ (1.9), ether	—	α-Alkylidene-γ-aryl- or α-alkylidene-γ,γ-diarylparaconic acid	146

* References 118–147 are on p. 73.
^{l} The unsaturated half-ester was hydrogenated and then saponified.
^{m} The condensation failed when carried out for six days at −10° or "several hours" at reflux.
^{n} With equimolar amounts of ketone and ester an 80% yield was realized after heating for twelve hours. A trace of benzhydrol was identified as a by-product.
^{o} When the heating period was reduced to one-half hour, the yield was 63%. After twelve hours of heating, the yield was 77%.
^{p} After two days at 23° the yield was 79%.

TABLE II—Continued

The Stobbe Condensation with Ketones

Formula	Ketone Name or Structure	Ester (moles per mole of ketone)	Condensing Agent (moles per mole of ketone), Solvent	Time, Temp.	Products Isolated (% yield)	Reference *
$C_{13}H_{10}O$	Benzophenone (Cont'd)	Diethyl benzhydryli-denesuccinate (1)	$NaOC_2H_5$ (2), none	—, 80°	Dialkylidenesuccinic acid	107
		Diethyl benzhydryli-denesuccinate (1)	$NaOC_2H_5$ (2), none	15 min., 80°	Dialkylidenesuccinic anhydride (40) [q, r]	38
		Dimethyl benzhydryl-idenesuccinate (1)	$NaOCH_3$ (2), none	30 min., 80°	Dibenzhydrylidenesuccinic anhydride (20); [q] $(C_6H_5)_2C{=}CCO_2H$ [q] $(C_6H_5)_2C{-}CHCO_2H$ $\quad\quad\quad OCOCH_3$ (39)	38
$C_{13}H_{12}O_4$	[structure: furan ring with $-CO-$ attached to 2,5-dimethoxyphenyl group with OCH_3 groups]	Di-t-butyl glutarate (1.3)	$KOC(CH_3)_3$ (1.6), $(CH_3)_3COH$	1.5 hr., reflux	α-Alkylideneglutaric half-ester (10)	3
		Diethyl succinate (1)	$NaOC_2H_5$ (2), ether	—	Resinous half-esters (44)	59
$C_{14}H_{10}O_2$	Benzil	$CH_3CC_6H_5$ (0.8) $\Vert$ $C_2H_5O_2CCCH_2CO_2C_2H_5$	$NaOC_2H_5$ (1.7), ethanol	2 hr., cold to reflux	Dialkylidenesuccinic acid	44

$C_{14}H_{10}O_3$ (H₂C structure with COC₆H₅)		Diethyl succinate (1)	$NaOC_2H_5$ (2), ether	—	Oily half-esters [d]	60
$C_{14}H_{12}O$ $C_6H_5COCH_2C_6H_5$		Diethyl succinate (1)	$NaOC_2H_5$ (2), none	—, 35° to 100°	Alkenylsuccinic acid (50)	33, 21
		Diethyl succinate (1)	$NaOC_2H_5$ (2), ether	8 wk., −15° to 23°	Alkenylsuccinic acid (16)	33
		Diethyl succinate (1)	$NaOC_2H_5$ (2), ether	—, reflux	Diethyl alkenylsuccinate (51–54) [q]	91
		Diethyl succinate (3)	NaH (3.3), benzene	5 hr, 23°	Oily half-ester (19)	110
		Dimethyl succinate (1)	$NaOCH_3$ (2), none	3 hr, hot to 23°	Alkenylsuccinic acid (60)	92
		Diethyl succinate (1.5)	$KOC(CH_3)_3$ (1.1), $(CH_3)_3COH$	50 min., reflux	Alkenylsuccinic half-ester (88)	100
		Diethyl succinate (3)	NaH (2.25), benzene	1 hr, 23°	Alkenylsuccinic half-ester (88) [s]	110
		Dimethyl succinate (3)	NaH (2.25), benzene	2.25 hr, 23°	Alkenylsuccinic half-ester (81) [t]	110
		Di-t-butyl succinate (1.25)	NaH (2.75), benzene	5.5 hr, 50°	Alkenylsuccinic half-ester (92)	110
$C_{14}H_{12}O_2$ p-Methoxybenzophenone		Diethyl succinate (1)	$NaOC_2H_5$ (2), ether	—	Alkylidenesuccinic half-esters (75)	60
$C_{14}H_{14}O_2$ (naphthalene with COC₂H₅ and CH₃O)		Diethyl succinate (1.5)	$KOC(CH_3)_3$ (1.1), $(CH_3)_3COH$	40 min., reflux	Oily half-esters (96)	12
		Diethyl succinate (3)	NaH (2.25), benzene	2 hr, 23°	Oily half-ester (91)	110

* References 118–147 are on p. 73.
[q] This product was obtained by treatment of the saponified condensation product with acetyl chloride.
[r] The compounds 1,1,6,6-tetraphenylhexatriene-2,3-dicarboxylic anhydride (two stereoisomeric forms each obtained in 10% yield) and 1,1,6,6-tetraphenylhexadiene-1,2-dicarboxylic anhydride (1% yield) were isolated as by-products resulting from the participation of the by-product acetaldehyde in the condensation.
[s] After one and one-half hours at 0° the yield of half-ester was 86%. A by-product, 2-succinoyl-1-keto-1,2,3,4-tetrahydrophenanthrene, was isolated in both cases in 3–5% yield.
[t] 2-Succinoyl-1-keto-1,2,3,4-tetrahydrophenanthrene was isolated in 9% yield.

68 ORGANIC REACTIONS

TABLE II—*Continued*

THE STOBBE CONDENSATION WITH KETONES

Formula	Ketone Name or Structure	Ester (moles per mole of ketone)	Condensing Agent (moles per mole of ketone), Solvent	Time, Temp.	Products Isolated (% yield)	Reference *
$C_{14}H_{18}O_4$	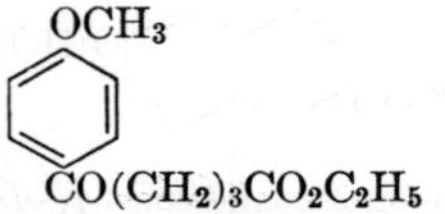 OCH$_3$ — C$_6$H$_4$ — CO(CH$_2$)$_3$CO$_2$C$_2$H$_5$	Diethyl succinate (1.9)	KOC(CH$_3$)$_3$ (1.4), (CH$_3$)$_3$COH	12 hr, 23°	Acidic ester (98)	45
$C_{15}H_{12}O$	$C_6H_5CH=CHCOC_6H_5$	Diethyl succinate (1)	NaOC$_2$H$_5$ (2), ether	Several days, 23°	Abnormal product: $C_6H_5CHCH_2COC_6H_5$ $HO_2CCHCH_2CO_2H$ (64)	40
$C_{15}H_{14}O$	$C_6H_5CH_2COCH_2C_6H_5$	Diethyl succinate (1)	NaOC$_2$H$_5$ (2–4), ether	Several wk., −10° to 23°	Alkenylsuccinic acid (44–50)	26
	$C_6H_5CH(CH_3)COC_6H_5$	Diethyl succinate (1.8)	KOC(CH$_3$)$_2$ (1.1), (CH$_3$)$_3$COH	1.7 hr, reflux	Alkenylsuccinic acid (42)	34
	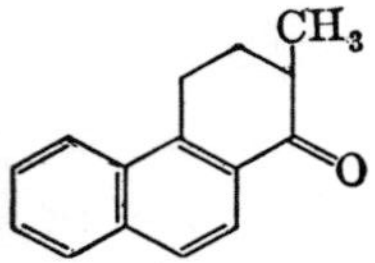 (structure: methyl-dihydrophenanthrenone)	Dimethyl succinate (7.4)	KOC(CH$_3$)$_3$ (6.5), (CH$_3$)$_3$COH	5 hr, 50°	Half-esters (93) ^{a}	58
		Dimethyl succinate (3)	NaH (2.25), benzene	11.5 hr, 23°	Half-esters (41)	110
		Di-*t*-butyl succinate (1.25)	NaH (2.75), benzene	20.5 hr, 50°	Half-esters (86) ^{d}	110
	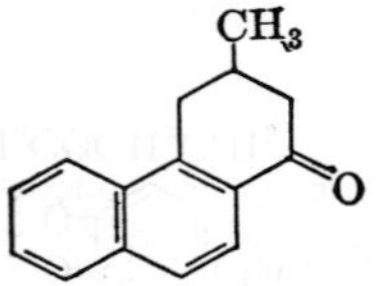 (structure: methyl-dihydrophenanthrenone)	Diethyl succinate (1.5)	KOC(CH$_3$)$_3$ (1.45), (CH$_3$)$_3$COH	—, reflux	Oily half-ester (55–60)	102

	Ketone	Ester (moles)	Condensing agent (moles)	Conditions	Product (% yield)	Reference
$C_{15}H_{14}O_3$	p,p'-Dimethoxybenzophenone	Diethyl succinate (3)	$KOC(CH_3)_3$ (2.2), $(CH_3)_3COH$	1 hr., reflux	Alkylidenesuccinic acid (83) ^{v}	8
		Diethyl succinate (3)	NaH (4), benzene	22.5 hr., 23°	Oily half-ester (64) ^{w}	110
		Di-t-butyl succinate (1.4)	$KOC(CH_3)_3$, $(CH_3)_3COH$	3 hr., reflux	Alkylidenesuccinic half-ester (89)	8
		Di-t-butyl succinate (1.25)	NaH (2.75), benzene	11 hr., 50°	Alkylidenesuccinic half-ester (91)	110
$C_{16}H_{12}O$	3-Acetylphenanthrene	Diethyl succinate (1)	$NaOC_2H_5$ (2), ether	—, warm	Alkylidenesuccinic acid (46.5)	104
		Diethyl succinate (2)	$KOC(CH_3)_3$ (1.3), $(CH_3)_3COH$	5 hr., reflux	Oily half-ester (72) ^{x}	94
		Diethyl succinate (3)	NaH (3.3), benzene	1.5 hr., 23°	Oily half-ester (89)	110
$C_{16}H_{16}O$	$C_6H_5CH(CH_3)COC_6H_5$	y	$KOC(CH_3)_3$, $(CH_3)_3COH$	y	Failed	34
	$C_6H_5C(CH_3)_2COC_6H_5$	y	$KOC(CH_3)_3$, $(CH_3)_3COH$	y	Failed	34
	$C_6H_5CH_2C$(...)CH_3 ring	y	$KOC(CH_3)_3$, $(CH_3)_3COH$	y	Failed	34

* References 118–147 are on p. 73.

^{u} One isomer was isolated pure in 65% yield.

^{v} With one-half the amount of ester and condensing agent the yield of dibasic acid was 47% after one-half hour of refluxing.

^{w} After seven hours at room temperature the yield of half-ester was only 12%; after four hours at room temperature followed by five and one-half hours at 50°, the yield of half-ester was 64%.

^{x} More dilute solutions of potassium t-butoxide gave lower yields (42–45%) of half-ester, correspondingly more ketone being recovered.

^{y} The condensation was tried with diethyl succinate, dimethyl succinate, succinic anhydride, and N-methylsuccinimide under various conditions.

TABLE II—*Continued*

THE STOBBE CONDENSATION WITH KETONES

Formula	Ketone Name or Structure	Ester (moles per mole of ketone)	Condensing Agent (moles per mole of ketone), Solvent	Time, Temp.	Products Isolated (% yield)	Reference*
$C_{16}H_{16}O$	(ketone with C_2H_5 substituent)	Diethyl succinate (1.5)	$KOC(CH_3)_3$ (1.45), $(CH_3)_3COH$	—, reflux	Oily half-ester (45–50)	102
		Diethyl succinate	NaH, benzene	—	Oily half-ester (90)	147
$C_{16}H_{16}O_2$	(ketone with CH_3 and CH_3O substituents)	Dimethyl succinate (7.6)	$KOC(CH_3)_3$ (6.5), $(CH_3)_3COH$	6 hr, reflux	Dibasic acids (98)	101
$C_{16}H_{13}ON$	(ketone with CH_3 and CN substituents)	Diethyl succinate (4.6)	$KOC(CH_3)_3$ (2.2), $(CH_3)_3COH$	7 hr, 23°	LXV, R = C_2H_5 (60) LXV	46
		Diethyl succinate (10)	NaH (24), benzene	23 hr, 23°	LXV, R = C_2H_5 (low)	110

Formula	Structure	Ester (equiv.)	Condensing agent (equiv.), solvent	Time, temp.	Product (% yield)	Ref.
	(naphthyl-COC$_6$H$_5$)	Dimethyl succinate (9.4)	KOC(CH$_3$)$_3$ (7.1), (CH$_3$)$_3$COH	5 hr., 23°	LXV, R = CH$_3$ (75–83)	48
		Dimethyl succinate (10)	NaH (24), benzene	10.5 hr., 50°	LXV, R = CH$_3$ (45)	110
		Di-t-butyl succinate (2)	NaH (4), benzene	8.5 hr., 80°	LXV, R = t-C$_4$H$_9$ (24)	110
		Di-t-butyl succinate (2)	(C$_6$H$_5$)$_3$CNa (4), ether	2 d., 23°	LXV, R = t-C$_4$H$_9$ (low)	145
C$_{17}$H$_{12}$O	(naphthyl-COC$_6$H$_5$)	Diethyl succinate (1.3)	NaOC$_2$H$_5$ (2), ether	1 hr., reflux	Alkylidenesuccinic acid (65)	21
C$_{17}$H$_{16}$O	(tetrahydronaphthyl-COC$_6$H$_5$)	Diethyl succinate (1.35)	NaOC$_2$H$_5$ (2.05), ether	2.5 hr., reflux	Alkylidenesuccinic acids: α (23), β (34)	21
C$_{17}$H$_{16}$O$_3$	(structure with CH$_3$, CO$_2$CH$_3$)	Diethyl succinate	KOC(CH$_3$)$_3$, (CH$_3$)$_3$COH	—	Failed	46
C$_{17}$H$_{18}$O	C$_6$H$_5$C(CH$_3$)$_2$– (with CH$_3$ aryl)	y		y	Failed	34
C$_{17}$H$_{18}$O	(CH(CH$_3$)$_2$ substituted tetrahydrophenanthrenone)	Diethyl succinate (1.5)	KOC(CH$_3$)$_3$ (1.45), (CH$_3$)$_3$COH	—, reflux	Oily half-ester (55–60)	102
		Diethyl succinate	NaH, benzene	—	Oily half-ester (94)	147
C$_{17}$H$_{18}$O$_5$	CH$_3$O, CH$_3$O, CO, OCH$_3$, OCH$_3$	Diethyl succinate (1)	KOC$_2$H$_5$ (1.9), benzene	12 hr., reflux	Alkylidenesuccinic acid (62)	87

* References 118–147 are on p. 73.
y The condensation was tried with diethyl succinate, dimethyl succinate, succinic anhydride, and N-methylsuccinimide under various conditions.

TABLE II—*Continued*

THE STOBBE CONDENSATION WITH KETONES

Ketone		Ester (moles per mole of ketone)	Condensing Agent (moles per mole of ketone), Solvent	Time, Temp.	Products Isolated (% yield)	Reference
Formula	Name or Structure					
$C_{17}H_{15}O_2N$	(ketone with CH₃, CN, and CH₃O substituents)	Diethyl succinate (2.5)	$KOC(CH_3)_3$ (1.7), $(CH_3)_3COH$	6.5 hr., 55° to 57°	LXVI, R = C_2H_5 (50)	46
		Dimethyl succinate (7.6)	$KOC(CH_3)_3$ (6.75), $(CH_3)_3COH$	6 hr., 53° to 55°	LXVI, R = CH_3 (77–83)	46
		Di-*t*-butyl succinate (3.2)	$KOC(CH_3)_3$ (2.2), $(CH_3)_3COH$	1.75 hr., reflux	LXVI, R = *t*-C_4H_9 (13)	46
	(ketone with CH₃, CN, and OCH₃ substituents)	Dimethyl succinate (7.8)	$KOC(CH_3)_3$ (6.8), $(CH_3)_3COH$	6 hr., 53° to 55°	(73–78)	47

REFERENCES FOR TABLES

[118] Stobbe and Eckert, *Ber.*, **38**, 4075 (1905).
[119] Stobbe and Kloeppel, *Ber.*, **27**, 2405 (1894).
[120] Fittig, *Ann.*, **305**, 50 (1899).
[121] Hecht, *Monatsh.*, **24**, 367 (1903).
[122] Cordier, *Compt. rend.*, **192**, 361 (1931).
[123] Cordier, *Ann. chim.*, (10), **15**, 228 (1931).
[124] Müller, Gawlick, and Kreutzmann, *Ann.*, **515**, 97 (1935).
[125] Stobbe, *Ber.*, **38**, 3893 (1905).
[126] Stobbe, *Ber.*, **37**, 2656 (1904).
[127] Cordier, *Compt. rend.*, **190**, 1191 (1930).
[128] Bougault, *Compt. rend.*, **181**, 247 (1925).
[129] Hinsberg, *J. prakt. Chem.*, [2], **84**, 192 (1911).
[130] Stobbe, *Ann.*, **380**, 26 (1911).
[131] Stobbe, *Ann.*, **380**, 99 (1911).
[132] Stobbe and Lenzner, *Ann.*, **380**, 93 (1911).
[133] Stobbe, *Ber.*, **39**, 761 (1906).
[134] Stobbe and Benary, *Ann.*, **380**, 71 (1911).
[135] Cordier, *Compt. rend.*, **189**, 538 (1929).
[136] Stobbe and Leuner, *Ann.*, **380**, 75 (1911).
[137] Stobbe and Härtel, *Ann.*, **380**, 59 (1911).
[138] Stobbe and Leuner, *Ber.*, **38**, 3897 (1905).
[139] Petkow, *Ber.*, **35**, 4322 (1902).
[140] Wojcik and Adkins, *J. Am. Chem. Soc.*, **56**, 2424 (1934).
[141] Stobbe, *Ber.*, **32**, 3354 (1899).
[142] Swain, Todd, and Waring, *J. Chem. Soc.*, **1944**, 548.
[143] Fieser, Leffler, et al., *J. Am. Chem. Soc.*, **70**, 3194 (1948).
[144] Stobbe, Badenhausen, Hennicke, and Wahl, *Ann.*, **380**, 120 (1911).
[145] Daub and Johnson, unpublished observation.
[146] Stobbe and Badenhausen, *Ber.*, **39**, 769 (1906).
[147] Riegel and Kritchevsky, private communication.

CHAPTER 2

THE PREPARATION OF 3,4-DIHYDROISOQUINOLINES AND RELATED COMPOUNDS BY THE BISCHLER-NAPIERALSKI REACTION

Wilson M. Whaley * and Tuticorin R. Govindachari †

University of Illinois

CONTENTS

	PAGE
Introduction	75
The Course of the Reaction	80
Direction of Ring Closure	80
Position of the Double Bond Formed	83
Side Reactions	85
Factors Affecting the Ease of Cyclization	90
Reactivity of the Aromatic Nucleus	90
Table I. Preparation of Substituted Phenanthridines	91
Table II. 3,4-Dihydro-2-carbolines	94
Substituents in the Ethylamine Side Chain	94
Table III. 3,4-Dihydroisoquinolines	95
Table IV. Isoquinolines	95
Nature of the Acyl Residue	96
Table V. 3,4-Dihydroisoquinolines	97
Table VI. Amides that Could Not Be Cyclized	97
Experimental Conditions and Condensing Agents	98
Phosphorus Oxychloride	98
Phosphorus Pentoxide	99
Phosphorus Pentachloride	99
Other Agents	99
Experimental Procedures	100
1-Methyl-3,4-dihydroisoquinoline	100
1-(2,3-Dimethoxybenzyl)-6,7-dimethoxy-3,4-dihydroisoquinoline	100
1-Homoveratryl-6,7-dimethoxy-3,4-dihydroisoquinoline	100
1-(o-Nitrobenzyl)-6,7-dimethoxy-3,4-dihydroisoquinoline	101
1-Phenylisoquinoline	101
1,3-Dimethyl-6,7-dimethoxyisoquinoline	101
9-Ethylphenanthridine	102
7-Nitro-9-phenylphenanthridine	102

* Present address: University of Tennessee, Knoxville, Tennessee.
† Present address: 25, Thanikachalam Chetty Road, T. Nagar, Madras, India.

	PAGE
3,11-Dimethoxy-5,6-dihydro-8H-dibenzo[a,g]quinolizine	102
2,3-Methylenedioxy-11,12-dimethoxy-5,6,8,9-tetrahydrodibenzo[a,h]quinolizinium Iodide	102
1-Benzyl-3,4-dihydro-2-carboline	103
TABULAR SURVEY OF THE BISCHLER-NAPIERALSKI REACTION	103
Table VII. 3,4-Dihydroisoquinolines	104
Supplement to Table VII	120
Table VIII. Isoquinolines	124
Supplement to Table VIII	130
Table IX. Benzisoquinolines	131
Supplement to Table IX	135
Table X. Naphthisoquinolines	136
Supplement to Table X	136
Table XI. Benzoquinolizines	137
Table XII. Dibenzoquinolizines	138
Supplement to Table XII	141
Table XIII. 2-Carbolines	142
Supplement to Table XIII	144
Table XIV. Miscellaneous Compounds	145
Supplement to Table XIV	150

INTRODUCTION

The frequent occurrence of the isoquinoline nucleus in alkaloids has led to considerable interest in the synthesis of isoquinoline derivatives. Many methods have been developed, but only three have enjoyed much popularity: the Bischler-Napieralski reaction discussed in this chapter, the Pictet-Spengler reaction treated in Chapter 3, and the Pomeranz-Fritsch synthesis which is the subject of Chapter 4. It will be of value to the reader to recall that the isoquinoline ring is numbered as shown in the following formula.

The Bischler-Napieralski reaction consists in the cyclodehydration of β-phenethylamides to 3,4-dihydroisoquinolines (I) by heating to high temperatures with phosphorus pentoxide or anhydrous zinc chloride.[1] No yields were given by the discoverers of the reaction, but

[1] Bischler and Napieralski, *Ber.*, **26**, 1903 (1893).

later workers have shown that the yields are very poor under the conditions originally described for the reaction.[2,3,4] Modifications using lower temperatures and milder condensing agents have improved the reaction, and it has become the most frequently used method of preparing isoquinoline derivatives.

The most important variation in the reaction is that introduced by Pictet and Gams,[4,5] which yields the isoquinoline directly from a β-hydroxy-β-phenethylamide and eliminates the dehydrogenation necessary when the original Bischler-Napieralski reaction is used for preparing isoquinolines. The classical synthesis of papaverine (II) by Pictet and Gams is given here as an example of their variation.[4] Removal of water

from the ethylamine side chain to create a double bond has been found to precede cyclization, the intermediate vinylamide (III) being easily isolable in certain reactions,[6,7,8] The isoquinolines produced in this stepwise manner had no substituents in the 5,6,7,8 positions, but ultra-

[2] Pictet and Kay, *Ber.*, **42**, 1973 (1909).
[3] Pictet and Finkelstein, *Compt. rend.*, **148**, 925 (1909).
[4] Pictet and Gams, *Ber.*, **42**, 2943 (1909).
[5] Pictet and Gams, *Ber.*, **43**, 2384 (1910).
[6] Krabbe, *Ber.*, **69**, 1569 (1936).
[7] Krabbe, Böhlk, and Schmidt, *Ber.*, **71**, 64 (1938).
[8] Krabbe, Eisenlohr, and Schöne, *Ber.*, **73**, 656 (1940).

violet absorption studies indicate that the same sequence of steps is involved in the cyclization of hydroxyamides having activating groups on the benzene ring.[9,10] Experiments on the cyclization of various stereoisomeric N-acyl-β-phenyl-β-hydroxyisopropylamines failed to reveal any significant differences in the ease of ring closure between diastereoisomers.[11]

A further extension of the Pictet-Gams modification, utilizing a methoxyethylamine (IV) rather than a hydroxyethylamine, has been found equally useful, and the starting materials are available through

[Structure IV: phenyl-CH(OCH$_3$)-CH$_2$-NH-CO-R, with $-CH_3OH$, $-H_2O$ → isoquinoline-R]

several efficient syntheses.[12,13,14] The choice between the two modifications is probably best made according to the availability of the respective intermediates.

An oxime capable of undergoing a Beckmann rearrangement[15] to an N-acyl-β-phenethylamine (V) or an N-acylstyrylamine (VI) may be

[Structure V: phenyl-CH$_2$-CH$_2$-C(R)=NOH → POCl$_3$ → phenyl-CH$_2$-CH$_2$-NH-CO-R → $-H_2O$ → isoquinoline-like with C-R]

[Structure VI: phenyl-CH=CH-C(R)=NOH → POCl$_3$ → phenyl-CH=CH-NH-CO-R → $-H_2O$ → isoquinoline-R]

[9] Gerendás and Varga, *J. prakt. Chem.*, **149**, 175 (1937).
[10] Varga and Fodor, *J. prakt. Chem.*, **150**, 94 (1938).
[11] Bruckner, Fodor, Kiss, and Kovács, *J. Chem. Soc.*, **1948**, 885.
[12] Mannich and Walther, *Arch. Pharm.*, **265**, 1 (1927).
[13] Rosenmund, Nothnagel, and Riesenfeldt, *Ber.*, **60**, 392 (1927).
[14] Mannich and Falber, *Arch. Pharm.*, **267**, 601 (1929).
[15] Komatsu, *Mem. Coll. Sci. Kyoto Imp. Univ.*, **7**, 147 (1924) [*C. A.*, **18**, 2126 (1924)].

used as the initial reactant of the Bischler-Napieralski reaction.[16,17] It is not necessary to isolate the amide, and the product is either an isoquinoline or a dihydroisoquinoline, depending on the oxime used.[18] No condensing agent is needed if the benzenesulfonyl ester of the oxime (VII) is used, only gentle heating being required to effect the transformation.[19,20] Very few isoquinolines have been prepared by the rearrangement and cyclization of oximes; consequently the synthetic value of the method is undetermined.

A less significant variation is the use of an amidine instead of the corresponding amide. Amidines have been converted in good yields to substituted phenanthridines.[20a,20b]

Isoquinoline derivatives having a hydroxyl or an amino function in the 1 position may be obtained by replacing the starting amide with a substituted urethan[21] or urea. The urethan VIII has been converted to 1-hydroxy-6,7-methylenedioxy-3,4-dihydroisoquinoline (IX) in 42% yield,[22] but the yields in this type of reaction are generally lower. Similarly, 1-hydroxy-6,7-dimethoxy-3,4-dihydroisoquinoline was pre-

[16] Bamberger and Goldschmidt, *Ber.*, **27**, 1954 (1894).
[17] Burstin, *Monatsh.*, **34**, 1443 (1913).
[18] Kaufmann and Radosević, *Ber.*, **49**, 675 (1916).
[19] Scheuing and Walach, Ger. pat. 576,532 [*Frdl.*, **20**, 719 (1933)].
[20] Scheuing and Walach, Ger. pat. 579,227 [*Frdl.*, **20**, 722 (1933)].
[20a] Barber, Holt, and Wragg, Brit. pat. 631, 651 [*C. A.*, **44**, 5401 (1950)].
[20b] Cymerman and Short, *J. Chem. Soc.*, **1949**, 703. The compounds are not listed in the tables.
[21] Späth and Dobrowsky, *Ber.*, **58**, 1274 (1925).
[22] Dey and Parikshit, *Proc. Natl. Inst. Sci. India*, **11**, 37 (1945).

pared in poor yield from homoveratryl isocyanate.[23] Phenanthridone has been prepared in excellent yield from *o*-xenyl isocyanate.[24] The substituted urea X was cyclized in 70% yield to 1-(*m*-toluino)-6,7-dimethoxy-3,4-dihydroisoquinoline (XI) in a similar manner.[23] N-Homoveratryl-N′-phenylthiourea could not be cyclized by lead oxide at 80° according to a method used for preparing carbodiimides.[23a]

The Bischler-Napieralski reaction is applicable to the synthesis of ring systems other than isoquinoline, such as phenanthridine, benzoquinolizine, and 2-carboline. The fundamental reaction is the same, however, and the syntheses will be discussed as a group, with occasional notation of exceptions to the usual behavior. Although many examples of the Bischler-Napieralski reaction have been recorded, they are not of sufficient variety to allow precise definition of the effects of various substituents upon the course of the reaction. The reaction has been seldom studied in itself but has been employed mainly as a convenient route to various classes of alkaloids and their synthetic analogs.

One novel use of the Bischler-Napieralski reaction is in the synthesis of phthalazines by dehydration of benzaldehyde acylhydrazones.[24a, 24b] Veratraldehyde benzoylhydrazone was dehydrated to 1-phenyl-6,7-dimethoxyphthalazine in 50% yield when heated with hydrogen chloride in amyl alcohol.

Numerous less important methods of synthesizing isoquinoline derivatives will not be mentioned because they have been described in available review articles [25, 26] and standard treatises.[27, 28] One new method,

---

[23] Mohunta and Rây, *J. Chem. Soc.*, **1934**, 1263.
[23a] Whaley and White, unpublished results.
[24] Butler, *J. Am. Chem. Soc.*, **71**, 2578 (1949).
[24a] Aggarwal, Darbari, and Rây, *J. Chem. Soc.*, **1929**, 1941.
[24b] Aggarwal, Khera, and Rây, *J. Chem. Soc.*, **1930**, 2354.
[25] Bergstrom, *Chem. Revs.*, **35**, 217 (1944).
[26] Manske, *Chem. Revs.*, **30**, 145 (1942).
[27] Hollins, *The Synthesis of Nitrogen Ring Compounds*, Benn, London, 1924, pp. 308–331.
[28] Morton, *The Chemistry of Heterocyclic Compounds*, McGraw-Hill, 1946, p. 301.

similar in principle to the aminoacetal synthesis, has appeared recently.[29] It is discussed in Chapter 4.

THE COURSE OF THE REACTION

Direction of Ring Closure. Cyclization of a *m*-methoxy-β-phenethylamide (XII) may be expected to lead to either a 6-methoxy- or an 8-methoxy-3,4-dihydroisoquinoline, depending upon the direction of ring closure. When the position *para* to the methoxyl group is free it is

invariably the point of closure leading to a 6-methoxyisoquinoline derivative. This fact is the logical result of an electrophilic attack upon an aromatic ring by a carbonium ion (XIII), that ion being necessarily involved in an acid-catalyzed reaction.* The reported [30] preferential

cyclization of the amide XIV to the 7,8-dimethoxyisoquinoline XV rather than the expected isomer XVI has been shown to be erroneous by oxidative degradation of the product to *m*-hemipinic acid (XVII).[31]

* See the discussion of the mechanism of the reaction in ref. 101, below.
[29] Schlittler and Müller, *Helv. Chim. Acta*, **31**, 914 (1948).
[30] Pfeiffer, Breitbach, and Scholl, *J. prakt. Chem.*, **154**, 157 (1940).
[31] Bruckner, Fodor, Kovács, and Kiss, *J. Am. Chem. Soc.*, **70**, 2697 (1948).

Para orientation is not so pronounced with activation due to a carbethoxyamino group. 2-(*p*-Nitrobenzamido)-3'-carbethoxyaminobiphenyl yielded a mixture of the 6- and 8-carbethoxyaminophenanthridines.[32] Cyclization may proceed *ortho* to the *m*-alkoxyl group of a β-phenethylamide if the *para* position is blocked. N-Acetyl-2,5-dimethoxy-β-phenethylamine (XVIII) may thus be readily converted to 1-methyl-5,8-dimethoxy-3,4-dihydroisoquinoline.[33] If both available positions are activated to a similar degree a mixture of products is obtained, as

in the cyclization of N-phenylacetylhomomyristicylamine to the 6,7-methylenedioxy-8-methoxy- (XIX) and 6-methoxy-7,8-methylenedioxy-3,4-dihydroisoquinolines (XX).[34] In an attempted synthesis of berberine, the formamide XXI was heated with phosphorus oxychloride, yielding the bromine-free compound XXII rather than the expected bromodihydroberberine (XXIII). This result is remarkable as an

[32] Caldwell and Walls, *J. Chem. Soc.*, **1948**, 188.
[33] Sugasawa and Shigehara, *Ber.*, **74**, 459 (1941).
[34] Salway, *J. Chem. Soc.*, **97**, 1208 (1910).

[Structures XIX, XX, XXI, XXII, XXIII shown]

instance of the preferred direction of ring closure, a bromine atom being ejected to allow cyclization to proceed *para* to the electron-releasing group.[35]

Cyclization of amides to benz-, dibenz-, and naphth-isoquinolines can usually proceed in more than one direction. For example, 2-(β-acetamidoethyl)naphthalene (XXIV) may be expected to yield either a 6,7-benz- (XXV) or a 7,8-benz-3,4-dihydroisoquinoline (XXVI). Though the structures of none of the compounds of these three classes have been established experimentally, the workers [36] prefer structure XXV. By analogy with results [37] of the Pictet-Spengler reaction, it is more probable that the correct structure for the product is that shown

[35] Haworth and Perkin, *J. Chem. Soc.*, **127**, 1448 (1925).
[36] Kindler and Peschke, Ger. pat. 704,762 [*C. A.*, **36**, 1956 (1942)].
[37] Mayer and Schnecko, *Ber.*, **56**, 1408 (1923).

in formula XXVI. Amides derived from 5-indanylethylamine and 6-tetrahydronaphthylethylamine cyclize so as to place the polymethylene ring in the 6,7-positions of the products, direction of ring closure being proved by oxidation of the products to pyromellitic acid.[38]

Position of the Double Bond Formed. Most Bischler-Napieralski reactions yield 3,4-dihydroisoquinolines; i.e., the double bond is formed between the carbonyl carbon atom and the nitrogen atom in the cyclodehydration. If the acyl derivative of a secondary amine is cyclized, the double bond may also appear in the 1,2 position even though this involves the formation of an ammonium salt. Thus, the amide XXVII may be cyclized in the usual way to 2-piperonyl-6,7-methylenedioxy-3,4-dihydroisoquinolinium chloride (XXVIII).[39]

The presence of an active methylene group in the 1 position in compounds analogous to XXVIII allows the double bond to become exo-

---

[38] Schultz and Arnold, *J. Am. Chem. Soc.*, **71**, 1911 (1949).
[39] Malan and Robinson, *J. Chem. Soc.*, **1927**, 2653.

cyclic in the free base, as in 1-benzal-2-methyl-1,2,3,4-tetrahydroisoquinoline (XXIX), which is yellow because of the extended conjugation. The colorless salt of the base has been shown to be quaternary.[40] Even

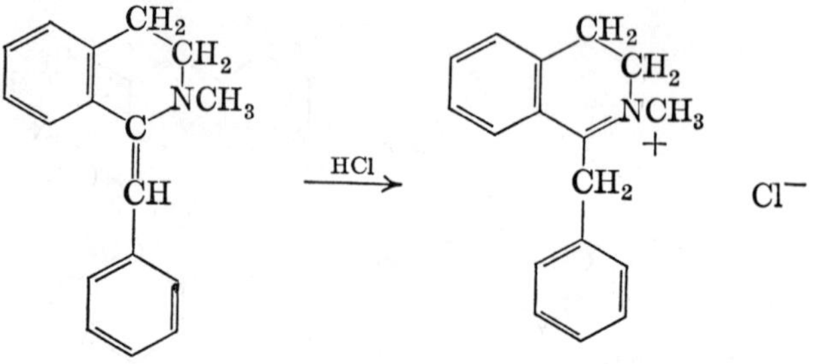

XXIX

without a substituent on the nitrogen atom, there is evidence for the existence of an exocyclic double bond in equilibrium with the normal endocyclic form (XXX).[41] Ultraviolet absorption studies indicate that 1-(α-picolyl)-6,7-methylenedioxy-3,4-dihydroisoquinoline exists entirely

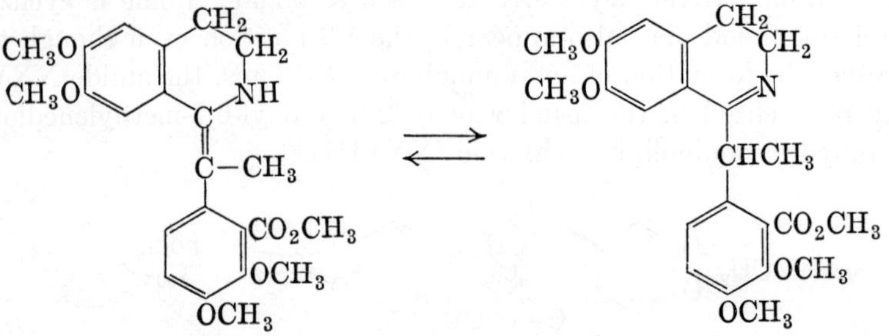

XXX

in the form with an exocyclic double bond, though its hydrochloride has the normal structure.[42]

Another instance of the shift of a double bond into conjugation between two aromatic rings is found in the synthesis of dibenzoquinolizines from N-formyl-1-benzyl-1,2,3,4-tetrahydroisoquinolines. In these compounds the double bond appears in the 3,4 position of the isoquinoline ring. Thus, the formamide XXXI yields 5,6-dihydro-8H-dibenzo[a,g]-quinolizine (XXXII).[43]

[40] Hamilton and Robinson, *J. Chem. Soc.*, **109**, 1029 (1916).
[41] Koepfli and Perkin, *J. Chem. Soc.*, **1928**, 2989.
[42] Bills and Noller, *J. Am. Chem. Soc.*, **70**, 957 (1948).
[43] Chakravarti, Haworth, and Perkin, *J. Chem. Soc.*, **1927**, 2275.

[Structures XXXI → XXXII with POCl₃, 110°]

Some investigators have preferred to express the structures of such compounds as pseudobases, the hydrated form which was used to depict compound XXII (p. 82).

Side Reactions. The Bischler-Napieralski reaction usually runs its course unhindered by specific side reactions, though the use of drastic cyclizing conditions may result in production of tars from amides which are not easily cyclized. Competing reactions which have been recorded apply to exceptional amides and have never been suggested as general side reactions.

Treatment of N-formyl-β-phenethylamine with phosphorus pentoxide yielded a small amount of 3,4-dihydroisoquinoline but mostly the aminomalondiamide XXXIII.[44] No instance of a similar by-product is known.

[Reaction scheme with P₂O₅ giving dihydroisoquinoline + XXXIII]

In the cyclization of *m*- and *p*-nitrobenzoyl derivatives of unactivated β-phenethylamines, it was found that considerable proportions of the corresponding nitrobenzonitriles were formed as by-products.[45] The formation of such substances may be attributed to the resistance of the amides to cyclodehydration and has recently been encountered with other unactivated amides.[45a]

Unactivated 2-nitrohomoveratroyl-β-phenethylamines have been found to undergo dehydration without cyclization, affording products which have been formulated as vinylideneamines (XXXIV and XXXV) and as an acetylene derivative (XXXVI).

[44] Decker, Kropp, Hoyer, and Becker, *Ann.*, **395**, 299 (1913).
[45] McCoubrey and Mathieson, *J. Chem. Soc.*, **1949**, 696.
[45a] Hill and Holliday, American Chemical Society Meeting, Chicago, September, 1950.

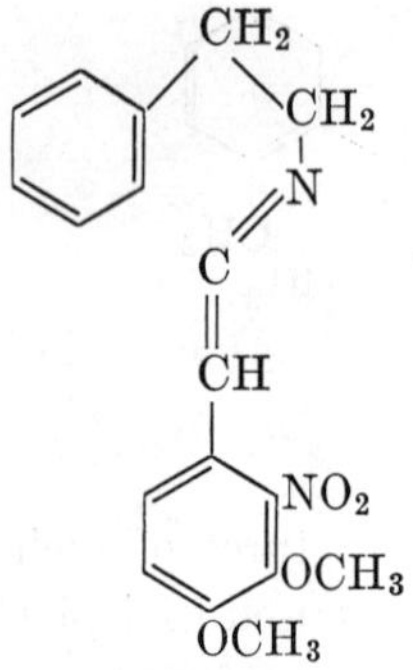

XXXIV [46,47] 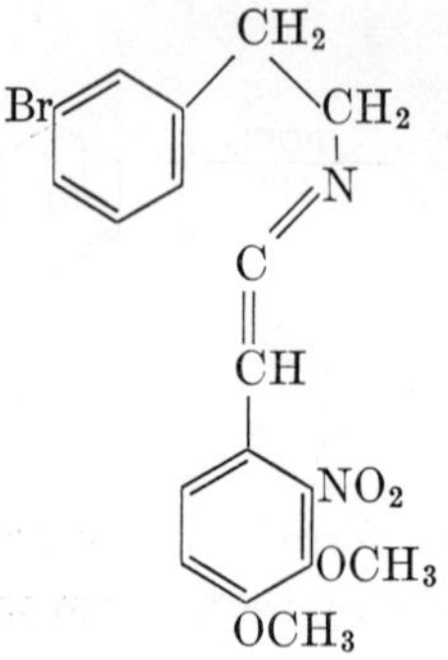   XXXV [48] 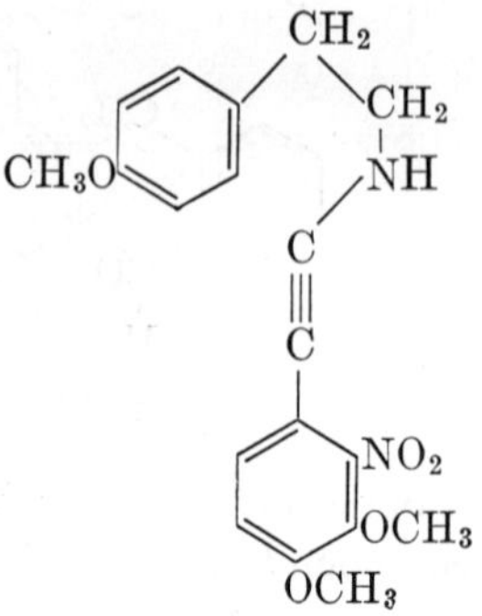   XXXVI [49]

All attempts to cyclize N-acylphenacylamines to the corresponding 4(3H)-isoquinolones (XXXVII) have failed, the products obtained being oxazoles (XXXVIII). Certain investigators [50–54] thought the products of this reaction to be the desired isoquinoline derivatives, but their nature was correctly interpreted by Robinson.[55]

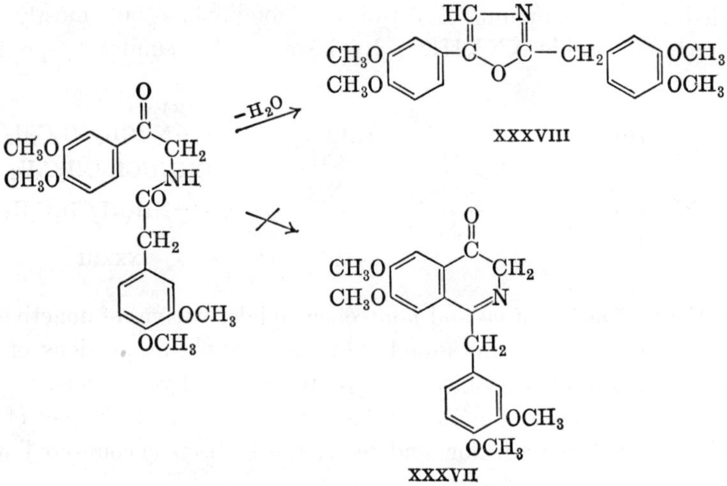

XXXVIII

XXXVII

[46] Kay and Pictet, *J. Chem. Soc.*, **103**, 947 (1913).
[47] Späth and Hromatka, *Ber.*, **62**, 325 (1929).
[48] Kondo and Ishiwata, *Ber.*, **64**, 1533 (1931).
[49] Callow, Gulland, and Haworth, *J. Chem. Soc.*, **1929**, 1444.
[50] Buck, *J. Am. Chem. Soc.*, **52**, 3610 (1930).
[51] Buck, *J. Chem. Soc.*, **1933**, 740.
[52] Dey and Rajagopalan, *Arch. Pharm.*, **277**, 359 (1939).
[53] Dey and Rajagopalan, *Arch. Pharm.*, **277**, 377 (1939).
[54] Dey and Rajagopalan, *Current Sci.*, **13**, 204 (1944).
[55] Young and Robinson, *J. Chem. Soc.*, **1933**, 275.

The Pictet-Gams modification does not always run a smooth course if the hydroxyphenethylamine is not activated by a *meta* alkoxyl group. The side reaction encountered is similar to that just discussed and results in formation of an oxazoline (XXXIX) instead of the intermediate vinylamide.[8] It has been found desirable in such cases to carry out the first step with a Grignard reagent, which does not promote oxazoline formation.

In the cyclization of N-(o-carbomethoxyphenylacetyl)homopiperonylamine there was obtained 2,3-methylenedioxy-5,6-dihydro-8-oxo-8H-dibenzo[a,h]quinolizine (XL) as well as the expected 2,3-methylenedioxy-5,6-dihydro-8-oxo-8H-dibenzo[a,g]quinolizine (XLI).[56] It is probable that the starting material was a mixture of the two isomeric amides obtainable by cleaving the parent homophthalimide.

[56] Haworth, Perkin, and Pink, *J. Chem. Soc.*, **127**, 1709 (1925).

A number of secondary reactions have been encountered in which the isoquinoline ring first formed was immediately modified by further reaction of the 1 substituent. Typical secondary reactions involve γ-chloropropyl (XLII),[57] benzamidomethyl (XLIII),[57] and o-carbomethoxy-

benzyl (XLIV)[56] groups. The last reaction does not always occur spontaneously, and the expected o-carbomethoxybenzyl derivative is then isolated.[41]

Substituted 1-benzyl-3,4-dihydroisoquinolines have a characteristic tendency to undergo air oxidation to 1-benzoyl-3,4-dihydroisoquinolines (XLV) when in neutral or alkaline solution. The change does not occur when dilute acidic solutions are exposed to air.[58] It takes place rapidly in the presence of alkali, and occasionally the oxidized product has been the only one isolated from a cyclization.[59,60] It is surprising that more examples of the oxidation have not been reported. A more remarkable

[57] Child and Pyman, *J. Chem. Soc.*, **1931**, 36.
[58] Buck, Haworth, and Perkin, *J. Chem. Soc.*, **125**, 2176 (1924).
[59] Lindenmann, *Helv. Chim. Acta*, **32**, 69 (1949).
[60] Livshits, Bazilevskaya, Bainova, Dobrovinskaya, and Preobrazhenskiĭ, *J. Gen Chem U.S.S.R.*, **17**, 1671 (1947) [*C. A.*, **42**, 2606 (1948)].

SYNTHESIS OF ISOQUINOLINES 1

[Structure diagrams showing conversion to XLV via O₂/Ethanol]

instance of oxidation by atmospheric oxygen is the simultaneous oxidation and dehydrogenation of 1-(o-methylbenzyl)-3,4-dihydro-2-carboline (XLVI) to yobyrone (XLVII) upon slow evaporation of an ethereal solution.[61] These changes may be effected more rapidly by boiling the

XLVI → (O₂) → XLVII

dihydro compound with strong methanolic potassium hydroxide,[62] but fission of the molecule may also result from alkaline treatment at elevated temperatures.[63,64]

A somewhat similar reaction has been encountered in the cyclization of amides derived from phenylalanine and tryptophan, in which cyclodehydration was accompanied by decarboxylation and dehydrogenation.[65] N-Formyltryptophan, when heated at 125° with phosphorus oxychloride and polyphosphoric acid, yielded 36% of the theoretically possible quantity of norharman (XLVIII). The reaction could not be

[Structure: N-formyltryptophan] → POCl₃, Polyphosphoric acid → XLVIII

[61] Julian, Karpel, Magnani, and Meyer, *J. Am. Chem. Soc.*, **70**, 180 (1948).
[62] Späth, Riedl, and Kubiczek, *Monatsh.*, **79**, 72 (1948) [*C. A.*, **42**, 6821 (1948)].
[63] Clemo and Swan, *J. Chem. Soc.*, **1949**, 487.
[64] Huntress and Shaw, *J. Org. Chem.*, **13**, 674 (1948).
[65] Snyder and Werber, *J. Am. Chem. Soc.*, **72**, 2962 (1950).

effected by other condensing agents and apparently did not depend upon the presence of atmospheric oxygen. In an analogous reaction N-(β-phenethyl)cyanoacetamide was cyclized, hydrolyzed, and decarboxylated by polyphosphoric acid at 170°, forming 1-methyl-3,4-dihydroisoquinoline.[65a]

A further side reaction encountered with 3,4-dihydroisoquinolines is disproportionation at distillation temperatures to the corresponding isoquinolines and tetrahydroisoquinolines.[65b]

FACTORS AFFECTING THE EASE OF CYCLIZATION

Reactivity of the Aromatic Nucleus. The Bischler-Napieralski reaction embodies an electrophilic attack upon the benzenoid ring of the β-phenethylamine and is dependent upon increased electron density at the position of ring closure. It is readily apparent that acyl derivatives of β-phenethylamine would not be so easily cyclized as compounds in which there is a *meta* alkoxyl group, and that an electron-attracting group such as nitro would inhibit the reaction. Preparation of the 3,4-dihydroisoquinoline XLIX [45] in 13% yield illustrates that the presence of an electron-attracting group does not prevent the reaction altogether.

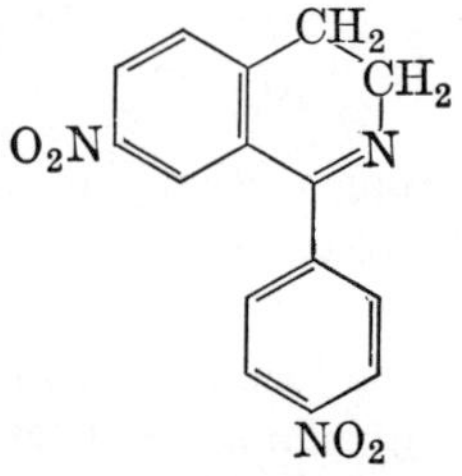

XLIX

Cyclization in the phenanthridine series also affords compounds (L) containing a nitro group on the reacting ring. Inspection of Table I reveals the effects of various substituents upon the formation of phenan-

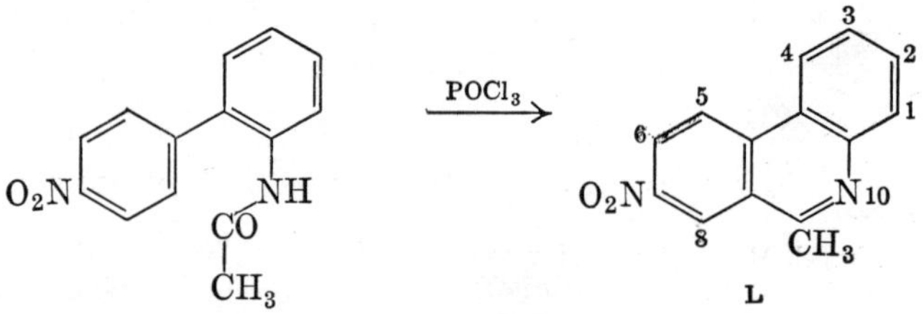

L

[65a] Leonard and Boyer, *J. Am. Chem. Soc.*, **72**, 2980 (1950).
[65b] Broderick and Short, *J. Chem. Soc.*, **1949**, 2587.

thridines. As would be expected, 7-nitro derivatives may be obtained with much less ease than the 3-nitro compounds, and amides with electron-releasing groups are readily cyclized. The 7-nitro derivatives may be prepared in excellent yield by using a higher reaction temperature.

TABLE I

PREPARATION OF SUBSTITUTED PHENANTHRIDINES

(Phosphorus oxychloride was used as the condensing agent.)

Substituents	Temperature °C.	Yield %	Reference
9-Methyl-	110	70	66
7-Nitro-9-methyl-	110	4	67
7-Carbethoxyamino-9-methyl-	110	85	68
2,3,6,7-Tetramethoxy-9-methyl-	110	85	69
9-(p-Nitrophenyl)-	110	65	66
3-Nitro-9-(p-nitrophenyl)-	110	61	70
7-Nitro-9-(p-nitrophenyl)-	110	30	70
3,7-Dinitro-9-(p-nitrophenyl)-	110	0	71

The effect of electron-releasing groups is even more obvious in the synthesis of 3,4-dihydroisoquinolines. Under identical conditions, the yield of 1-methyl-3,4-dihydroisoquinoline (LI) [72] is only a fraction of that of 1-methyl-6,7-methylenedioxy-3,4-dihydroisoquinoline (LII).[73]

LI   LII

Very little is known of the activating influence of groups other than alkoxyl, though 1-methyl-6-benzamido-3,4-dihydroisoquinoline (LIII) [74] and 1-phenyl-6-(β-benzamidoethyl)-3,4-dihydroisoquinoline (LIV) [75] have been prepared in good yield.

[66] Morgan and Walls, *J. Chem. Soc.*, **1931**, 2447.
[67] Petrow, *J. Chem. Soc.*, **1945**, 18.
[68] Walls, *J. Chem. Soc.*, **1947**, 67.
[69] Ritchie, *J. Proc. Roy. Soc. N. S. Wales*, **78**, 134 (1945) [*C. A.*, **40**, 876 (1946)].
[70] Morgan and Walls, *J. Chem. Soc.*, **1938**, 389.
[71] Morgan and Walls, Brit. pat. 520,273 [*C. A.*, **36**, 495 (1942)].
[72] Dey and Ramanathan, *Proc. Natl. Inst. Sci. India*, **9**, 193 (1943).
[73] Dey and Govindachari, *Proc. Natl. Inst. Sci. India*, **6**, 219 (1940).
[74] Fries and Bestian, *Ann.*, **533**, 72 (1937).
[75] Leupin and Dahn, *Helv. Chim. Acta*, **30**, 1945 (1947).

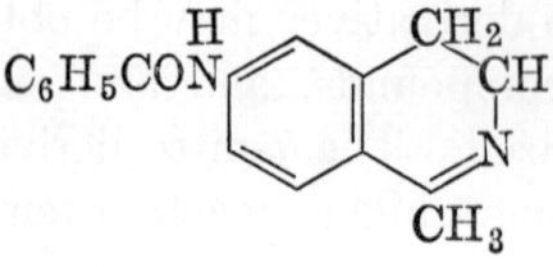

LIII 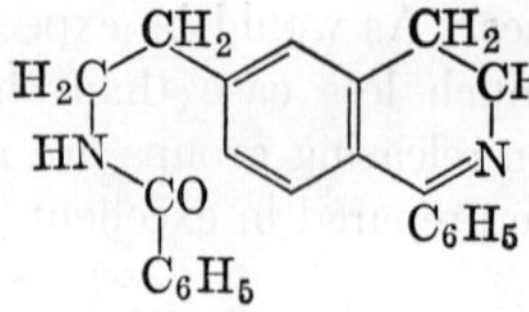 LIV

N-Formyl-β-(9-phenanthryl)ethylamine and similar amides (LV) could not be cyclized to dibenzisoquinolines.[76] Amides of β-(3-phenanthryl)ethylamine could not be cyclized either,[76] but N-acetyl-β-(9,10-dihydro-2-phenanthryl)ethylamine (LVI) was efficiently condensed to the corresponding tetrahydronaphthisoquinoline.[77]

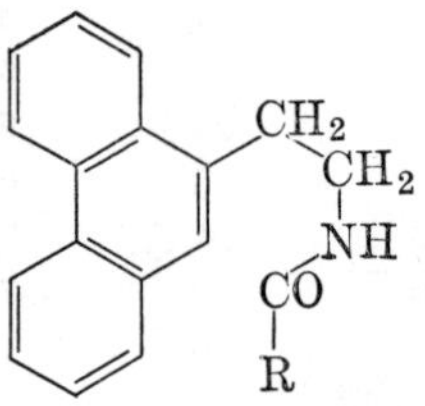

LV 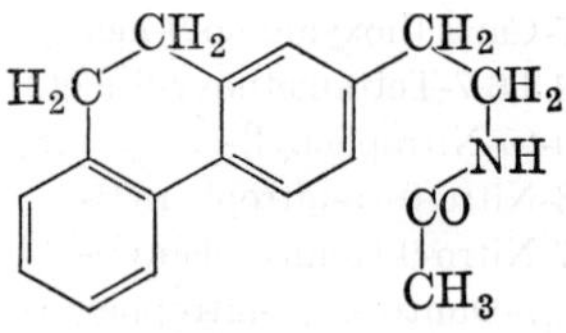 LVI

Peri ring closure of α-acetamidomethyl-β-methoxynaphthalene (LVII) to the desired 4,5-benzisoquinoline could not be effected.[53] However, the analogous 4-formamidophenanthrene (LVIII) was converted in fair yield to 4-azapyrene under the same conditions.[78]

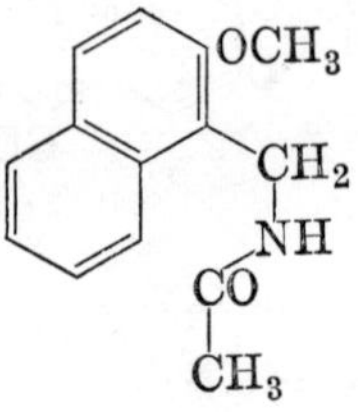

LVII 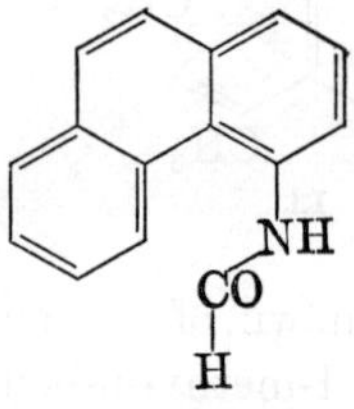 LVIII

Several attempts to obtain double ring closure of 2-phenyl-1,3-diamidopropanes (LIX), which would involve closure at a *peri* position, resulted only in the formation of one ring.[79] A second closure not involving a *peri* position failed also in attempted cyclodehydration of 6- or 7-benzamidoethylisoquinoline.[75] Successful double cyclization of a

[76] Mosettig and May, *J. Am. Chem. Soc.*, **60**, 2962 (1938).
[77] Stuart and Mosettig, *J. Am. Chem. Soc.*, **62**, 1110 (1940).
[78] Cook and Thomson, *J. Chem. Soc.*, **1945**, 395.
[79] Jackson and Kenner, *J. Chem. Soc.*, **1928**, 1657.

diamide with participation of two benzene nuclei takes place in the formation of 5,10-di(*o*-carboxyphenyl)pyrido[2,3,4,5-*l,m,n*]phenanthridine (LX).[80]

Cyclization of β-indolylethylamines to 2-carbolines generally proceeds with greater ease than cyclization of β-phenethylamines. Treatment of N-phenylacetyl-β-(3-indolyl)ethylamine (LXI) with phosphorus oxychloride afforded 90% of 1-benzyl-3,4-dihydro-2-carboline,[81] whereas the corresponding 1-benzyl-3,4-dihydroisoquinoline has been prepared in 9% yield under comparable conditions.[72] 3,4-Benzo-2-carboline was obtained in 76% yield by treatment of the appropriate formamide with

[80] Křepelka and Štefec, *Collection Czechoslov. Chem. Commun.*, **9**, 29 (1937) [*C. A.*, **31**, 3909 (1937)].
[81] Hahn and Ludewig, *Ber.*, **67**, 2031 (1934).

phosphorus oxychloride at 110°,[82] but it has not been found possible to prepare phenanthridine under such mild conditions.

The preparation of 3,4-dihydro-2-carbolines may be facilitated in some measure by the presence of electron-releasing groups in the 6 position of the indole nucleus, as seen in Table II. The mechanism of activation

TABLE II

3,4-Dihydro-2-carbolines

(All reactions were run at 140° with phosphorus pentoxide as condensing agent.)

Substituents	Yield %	Reference
1-Methyl-	56	83
1-Methyl-6-methoxy-	58	84
1-Methyl-7-methoxy-	78	83
1-Methyl-8-methoxy-	32	84

by a 6-alkoxyl group is illustrated by the accompanying figure, in which the path of electron shift is shown (LXII).

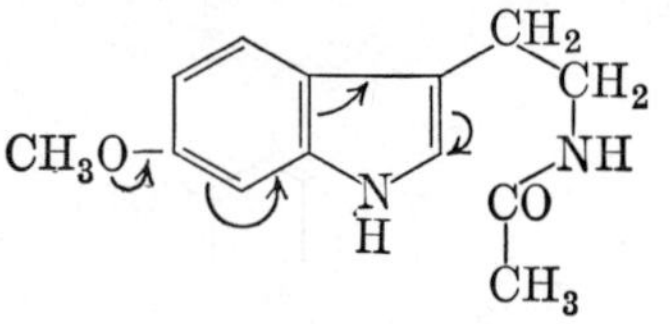

LXII 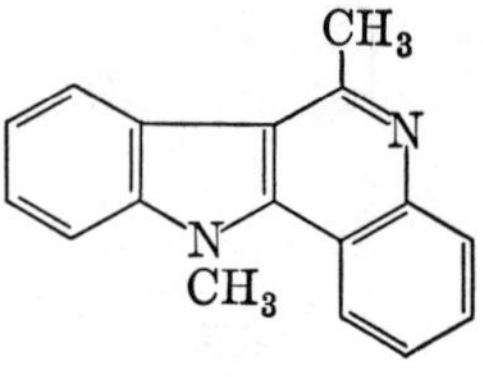   LXIII

Ring closure to the 3 position of indole has been obtained in the synthesis of 4,9-dimethyl-1,2-benzo-3-carboline (LXIII).[85]

Substituents in the Ethylamine Side Chain. The nature of the side chain of a β-phenethylamine has a profound influence on the ease of cyclization of its acyl derivatives. In Tables III and IV are listed isoquinolines and dihydroisoquinolines which are unsubstituted in the isocyclic ring and which were for the most part prepared under similar conditions. All the compounds listed in the two tables lack alkoxyl groups in the isocyclic ring and their formation is susceptive to adverse influences. The isoquinolines and dihydroisoquinolines having alkyl, aryl, or aralkyl groups in the 3 position have generally been obtained in lower yield than the derivatives unsubstituted in that position. The yield among the isoquinolines was progressively less as the 3-alkyl group

[82] Kermack and Slater, *J. Chem. Soc.*, **1928**, 32.
[83] Späth and Lederer, *Ber.*, **63**, 120 (1930).
[84] Späth and Lederer, *Ber.*, **63**, 2102 (1930).
[85] Kermack and Smith, *J. Chem. Soc.*, **1930**, 1999.

TABLE III
3,4-DIHYDROISOQUINOLINES

$$\text{PhCH}_2\text{CH}_2\text{NHCOC}_6\text{H}_5 \xrightarrow{\text{POCl}_3, 110°} \text{1-phenyl-3,4-dihydroisoquinoline}$$

Substituents	Yield %	Reference
1-Phenyl-	26	72
1-Phenyl-3-methyl-	35	72
1,3-Diphenyl-	0	72
1-Phenyl-3-benzyl-	11	72
1-Phenyl-4-methyl-	45	72
1,4-Diphenyl-	53	72

TABLE IV
ISOQUINOLINES

$$\text{PhCH(OH)CH}_2\text{NHCOC}_6\text{H}_5 \xrightarrow{-2\text{H}_2\text{O}} \text{1-phenylisoquinoline}$$

Substituents	Condensing Agent	Temperature °C.	Yield %	Reference
1-Phenyl-	$P_2O_5 + POCl_3$	140	91	86
1-Phenyl-3-methyl-	$P_2O_5 + POCl_3$	140	50	86
1-Phenyl-3-ethyl-	$P_2O_5 + POCl_3$	140	26	86
1-Phenyl-3-propyl-	$P_2O_5 + POCl_3$	140	20	86
1-Phenyl-3-butyl-	$P_2O_5 + POCl_3$	140	1	86
1-Phenyl-3-hexyl-	$P_2O_5 + POCl_3$	140	0	86
1,3-Diphenyl-	P_2O_5	140	20	86
1-Phenyl-4-methyl- *	P_2O_5	110	93	8
1-Phenyl-4-ethyl-	P_2O_5	140	10	86
1,4-Diphenyl-	P_2O_5	110	80	87
1-Phenyl-3,4-dimethyl-	$P_2O_5 + POCl_3$	140	0	86
1,3,4-Triphenyl-	P_2O_5	110	21	7

* From the styrylamide.

[86] Whaley and Hartung, *J. Org. Chem.*, **14**, 650 (1949).
[87] Krabbe, Ger. pat. 652,041 [*Frdl.*, **24**, 378 (1937)].

increased in length, but an aryl group was not so inhibitive as an alkyl group of comparable size.

Compounds having substituents in the 4 position have been prepared in generally better yield than those with corresponding groups in the 3 position, but the data at hand are too meager to allow this statement to serve as a reliable basis for prediction.

Activating alkoxyl groups on the ring of the phenethylamine counteract to a considerable degree the inhibition arising from alkylation of the side chain. 1,3-Diphenyl-3,4-dihydroisoquinoline was not obtained by heating the corresponding amide with phosphorus oxychloride, but 1,3-diphenyl-6,7-methylenedioxy-3,4-dihydroisoquinoline was formed in 28% yield under the same mild conditions.[72]

1-Phenyl-3-ethyl-3,4-dihydro-2-carboline (LXIV) has been prepared in 80% yield,[88] whereas 1-phenyl-3,4-dihydro-2-carboline (LXV) was formed in only 36% yield under identical conditions,[84] constituting a reversal of the effects noted among isoquinolines. Side reactions that

occur when the β-phenethylamine has a β-hydroxyl or β-ketonic function have already been discussed (pp. 86–87).

Nature of the Acyl Residue. The influence of the acyl residue on the ease of cyclization is usually of a minor order; consequently the 1 substituent has been varied to a great extent. Nearly all aryl and aralkyl groups in the acid moiety permit the reaction to proceed in excellent yield, but the yields tend to be somewhat less with alkyl groups under similar conditions (Table V). Under special conditions the 1-alkyl-3,4-dihydroisoquinolines are available in good yields.[89, 86]

The synthesis of 1-(o-nitrobenzyl)-3,4-dihydroisoquinolines presents a special difficulty. In the absence of nuclear activation the o-nitrophenylacetamides have usually failed to yield 3,4-dihydroisoquino-

[88] Snyder and Katz, *J. Am. Chem. Soc.*, **69**, 3140 (1947).
[89] Späth, Berger, and Kuntara, *Ber.*, **63**, 134 (1930).

TABLE V

3,4-DIHYDROISOQUINOLINES

(Phosphorus pentoxide was used as the condensing agent.)

Substituents	Temperature °C.	Yield %	Reference
1-Methyl-	110	35	2
1-Phenyl-	140	75	2
1-Phenyl-	110	83	86
1-Benzyl-	140	75	2
1-(o-Nitrophenyl)-	140	73	90
1-(o-Nitrobenzyl)-	140	0	91, 46
1-(2-Nitroveratryl)-	110	21	47

lines,[49,92,46,93] although a single successful instance has been recorded.[47] A large number of activated amides have been cyclized. Some workers believe that the cyclization can be effected only by phosphorus pentachloride in chloroform at room temperature;[94] others state that phosphorus pentoxide is a more general reagent and succeeds when phosphorus pentachloride does not.[95,96]

In Table VI are listed various amides that could not be cyclized

TABLE VI

AMIDES THAT COULD NOT BE CYCLIZED

N-(Substituted-β-phenethyl)-	Reference	N-(o-Xenyl)-	Reference
Phthalimidoacetamide	97	Dichloroacetamide	100
Aminoacetamide	97	Trichloroacetamide	100
o-Benzamidophenylglyoxamide	92	β-Carboxypropionamide	101
Succindiamide	98	Glutardiamide	101
Triazoacetamide	97	Acetoacetamide	101
β-Furylacrylamide	99	Crotonamide	101
β-Chloropropionamide	57	Oxamic Acid	100
2-Aminohomoveratramide	48		

[90] Rodionov and Yavorskaya, *J. Gen. Chem. U.S.S.R.*, **13**, 491 (1943) [*C. A.*, **38**, 3285 (1944)].
[91] Gadamer, Oberlin, and Schoeler, *Arch. Pharm.*, **263**, 81 (1925).
[92] Gulland, Haworth, Virden, and Callow, *J. Chem. Soc.*, **1929**, 1666.
[93] Kondo, *J. Pharm. Soc. Japan*, **519**, 429 (1925) [*C. A.*, **20**, 604 (1926)].
[94] Gulland and Haworth, *J. Chem. Soc.*, **1928**, 581.
[95] Barger and Schlittler, *Helv. Chim. Acta*, **15**, 381 (1932).
[96] Späth and Hromatka, *Ber.*, **61**, 1692 (1928).
[97] Harwood and Johnson, *J. Am. Chem. Soc.*, **55**, 4178 (1933).
[98] Child and Pyman, *J. Chem. Soc.*, **1929**, 2010.
[99] Harwood and Johnson, *J. Am. Chem. Soc.*, **55**, 2555 (1933).
[100] Walls, *J. Chem. Soc.*, **1934**, 104.
[101] Ritchie, *J. Proc. Roy. Soc. N. S. Wales*, **78**, 147 (1945) [*C. A.*, **40**, 877 (1946)].

to the corresponding dihydroisoquinolines or phenanthridines. No reason has been suggested for most of the failures. Many other amides have not been cyclized, but they do not differ greatly from those that have been listed here or discussed elsewhere in the chapter.

An interesting application of the Bischler-Napieralski reaction utilizes lactams, pyridones, and other cyclic amides to produce substituted quinolizines. The reaction is generally effected with phosphorus oxychloride and frequently affords excellent yields. 2-Homopiperonyl-6,7-dimethoxy-3,4-dihydroisocarbostyril (LXVI) in this fashion yields the dibenzoquinolizinium chloride LXVII.[102]

LXVI → (POCl₃, 80%) → LXVII Cl⁻

EXPERIMENTAL CONDITIONS AND CONDENSING AGENTS

The Bischler-Napieralski reaction is usually conducted by heating the appropriate amide with a dehydrating agent in the presence of a solvent. The solvents must be inert and anhydrous, and they may be used to establish a moderate refluxing temperature or to provide a high reaction temperature. Solvents frequently encountered are chloroform, benzene, toluene, xylene, nitrobenzene, and tetralin, selection being based on the refluxing temperature desired. Cyclizations conducted with phosphorus oxychloride often do not require additional solvent if an excess of the condensing agent is used. Phosphorus oxychloride has been the most commonly employed dehydrating agent, but phosphorus pentoxide has specific uses and various other agents have found occasional use.

Phosphorus Oxychloride. Phosphorus oxychloride is a relatively mild dehydrating agent when employed at or near its own boiling point (107°). It is very useful for those cyclizations which proceed with ease owing to inherent or induced reactivity of the aromatic nucleus. Phosphorus oxychloride has been employed almost exclusively in the synthesis of phenanthridines; when drastic dehydrating conditions were required nitrobenzene has been used as a solvent to provide reaction

[102] Kakemi, *J. Pharm. Soc. Japan*, **60**, 6 (1940) [*C. A.*, **34**, 3747 (1940)].

temperatures of approximately 180°. This modification has not been extended to other phases of the Bischler-Napieralski reaction. A low temperature of cyclization has been obtained by using phosphorus oxychloride in refluxing chloroform.

Duration of the reaction is usually one-half to three hours, though longer periods are often employed in the preparation of phenanthridines.

Phosphorus Pentoxide. Phosphorus pentoxide has been used for many cyclizations which phosphorus oxychloride could not be expected to effect. It is a stronger dehydrating agent and is required for difficultly cyclized amides. However, it is less convenient than phosphorus oxychloride because of difficulties in handling the reagent and stirring the reaction mixture. Hence, several small runs may be more convenient and efficient than one large-scale dehydration. Toluene (110°) and xylene (140°) have usually been the solvents; the combination of phosphorus pentoxide with boiling tetralin (205°) has provided the most drastic dehydrating conditions that have been found practicable. Cyclizations requiring phosphorus pentoxide may be facilitated by the addition of phosphorus oxychloride to the mixture. The reaction is usually complete in one-half to three hours.

Phosphorus Pentachloride. Phosphorus pentachloride has found particular application to the synthesis of 3,4-dihydroisoquinolines having a 1-(o-nitrobenzyl) substituent. Difficulties have been encountered in obtaining such derivatives by the action of phosphorus oxychloride or pentoxide, though some investigators have had success with those agents. The relative merits of phosphorus pentachloride and phosphorus pentoxide in the synthesis of 1-(o-nitrobenzyl)-3,4-dihydroisoquinolines are still debatable; both agents enjoy considerable support.[95, 103, 92, 104] Phosphorus pentachloride in chloroform at 25° has also been used in a few cyclizations of activated amides having no nitro group in the acyl residue, the reaction requiring from one day to a week for completion. Unpublished reports [105, 106] indicate that this technique is of considerable general value and is frequently the method of choice.

Other Agents. Dehydrating agents that have been tried but not generally used include aluminum chloride,[107, 108] thionyl chloride,[109, 110]

[103] Faltis, Wagner, and Adler, *Ber.*, **77**, 686 (1944).
[104] Schlittler, *Helv. Chim. Acta*, **15**, 394 (1932).
[105] M. B. Moore, A. W. Weston, A. H. Sommers, H. B. Wright, M. R. Vernsten, R. J. Michaels, R. W. DeNet, M. Freifelder, and E. J. Matson, private communication.
[106] C. Schöpf, private communication.
[107] Decker and Kropp, *Ber.*, **42**, 2075 (1909).
[108] Ebel, Ger. pat. 614,196 [*Frdl.*, **22**, 1126 (1935)].
[109] Avenarius and Pschorr, *Ber.*, **62**, 321 (1929).
[110] Gulland and Virden, *J. Chem. Soc.*, **1929**, 1791.

zinc chloride-acetic anhydride,[111, 112] zinc chloride,[113] alumina,[13] phosphorus oxybromide,[114] and silicon tetrachloride.[115, 116] There seems to be little to recommend these less common reagents. Phosphorus oxybromide is useful for cyclizing bromoamides with which phosphorus oxychloride can effect halogen interchange.[57, 114] Polyphosphoric acid has also been found to effect the Bischler-Napieralski reaction. It was an essential agent in the simultaneous cyclization, decarboxylation, and dehydrogenation of N-acyl-β-aryl-α-amino acids discussed on pp. 89–90.[65]

EXPERIMENTAL PROCEDURES

1-Methyl-3,4-dihydroisoquinoline.[89] (The use of phosphorus pentoxide and tetralin to cyclize an unactivated amide.) A solution of 0.5 g. of N-acetyl-β-phenethylamine in 25 ml. of dry tetralin was boiled fifteen minutes with 3.8 g. of phosphorus pentoxide. After another 3.8 g. of phosphorus pentoxide was added the mixture was refluxed fifteen minutes longer. The tetralin was decanted; the residue was treated with water and steam-distilled to remove traces of tetralin. The cooled solution was made strongly alkaline and steam-distilled; the distillate was made alkaline and extracted with ether. Evaporation of the extract yielded 0.37 g. (83%) of base boiling at 130°/10 mm. The picrate melted at 188–190°.

1-(2,3-Dimethoxybenzyl)-6,7-dimethoxy-3,4-dihydroisoquinoline.[117] (The use of phosphorus pentoxide and toluene to cyclize an activated amide.) A solution of 4.2 g. of N-(2,3-dimethoxyphenylacetyl)homoveratrylamine in 100 ml. of refluxing, dry toluene was treated with 16 g. of phosphorus pentoxide in small portions during thirty minutes. After the mixture had refluxed another thirty minutes the toluene was decanted and the sticky residue was dissolved in water and washed with ether. The aqueous solution was made alkaline and extracted with ether, evaporation of which yielded 3.55 g. (89%) of amorphous product. The picrate melted at 172–174°.

1-Homoveratryl-6,7-dimethoxy-3,4-dihydroisoquinoline.[118] (a) (The cyclization of an activated amide by phosphorus oxychloride and

---

[111] Kermack, Perkin, and Robinson, *J. Chem. Soc.*, **119**, 1602 (1921).
[112] Leonard and Elderfield, *J. Org. Chem.*, **7**, 556 (1942).
[113] Pictet and Hubert, *Ber.*, **29**, 1182 (1896).
[114] Rajagopalan, *Proc. Indian Acad Sci.*, **14A**, 126 (1941).
[115] Asta Akt.-Ges. Chem. Fabrik, Ger. pat. 614,703 [*Frdl.*, **21**, 688 (1934)].
[116] Koschara, Fr. pat. 760,825 [*C. A.*, **28**, 4178 (1934)]; Brit. pat. 424,348 [*C. A.*, **29**, 4524 (1935)].
[117] Späth and Mosettig, *Ann.*, **433**, 138 (1923).
[118] Sugasawa and Yoshikawa, *J. Chem. Soc.*, **1933**, 1583.

toluene.) A mixture of 15 g. of N-(3,4-dimethoxyhydrocinnamoyl)-homoveratrylamine, 80 ml. of dry toluene, and 60 g. of phosphorus oxychloride was refluxed for two hours. The amide soon disappeared, and after some time a yellow, crystalline substance separated. It was collected, washed with petroleum ether, and dissolved in water. The filtered solution was made alkaline and extracted with ether. Evaporation of the ether yielded 13 g. (91%) of colorless needles, m.p. 96–97°.

(b) (Preparation by the rearrangement and cyclization of an oxime.) A solution of 5 g. of bis(homoveratryl) ketoxime, 25 ml. of dry toluene, and 20 g. of phosphorus oxychloride was refluxed until hydrogen chloride was no longer evolved (two hours). Sufficient petroleum ether was added to produce a thick, brown precipitate, which was purified by the method recorded in the previous paragraph to yield 4 g. (85%) of the pure base.

1-(o-Nitrobenzyl)-6,7-dimethoxy-3,4-dihydroisoquinoline.[119] (The use of phosphorus pentachloride to prepare an o-nitrobenzyl derivative.) A mixture of 4 g. of N-(o-nitrophenylacetyl)homoveratrylamine, 5 g. of phosphorus pentachloride, and 30 ml. of chloroform was allowed to stand for twenty-four hours at room temperature. The solvent was evaporated under reduced pressure from the crystalline material which had separated; the latter was extracted with boiling water and filtered from traces of tar. The crude base was precipitated by addition of ammonia. It was recrystallized from methanol as large prisms weighing 3.5 g. (92%) and melting at 132°.

1-Phenylisoquinoline.[86] (The cyclization of an unactivated hydroxyamide by a mixture of phosphorus pentoxide and phosphorus oxychloride.) One gram of N-benzoyl-β-hydroxy-β-phenethylamine, 5 g. of phosphorus pentoxide, 10 g. of phosphorus oxychloride, and 25 ml. of dry xylene were refluxed for three hours. The excess condensing agents were cautiously decomposed with ice, the layers were separated, and the aqueous layer was made strongly alkaline with 20% sodium hydroxide. The benzene extract of the precipitated oil was dried over magnesium sulfate and treated with hydrogen chloride to yield 0.91 g. (91%) of crystalline hydrochloride melting at 233–236°.

1,3-Dimethyl-6,7-dimethoxyisoquinoline.[120] (The cyclization of an activated hydroxyamide by phosphorus oxychloride in chloroform.) A solution of 2.5 g. of N-acetyl-β-hydroxy-β-(3,4-dimethoxyphenyl)isopropylamine, 3 ml. of phosphorus oxychloride, and 20 ml. of chloroform was refluxed for three hours, then poured into hot water and made alka-

[119] Gulland and Haworth, *J. Chem. Soc.*, **1928**, 581.
[120] Bruckner, *Ann.*, **518**, 226 (1935).

line with 10% aqueous sodium hydroxide. The yellow base that precipitated was collected and recrystallized from ligroin to yield 1.65 g. (77%) of colorless needles melting at 121.5°.

9-Ethylphenanthridine.[66] (The use of phosphorus oxychloride to cyclize an o-xenylamide.) Five grams of N-propionyl-o-xenylamine and 10 g. of phosphorus oxychloride were heated gently in a dry atmosphere for one hour. Excess of phosphorus oxychloride was removed by distillation under reduced pressure, and the residual gum was warmed with dilute hydrochloric acid. The acid solution was filtered and made alkaline with aqueous ammonia. An ethereal extract of the liberated oil was dried over sodium sulfate and evaporated. The residual base was crystallized from petroleum ether to yield 3.6 g. (80%) of colorless plates melting at 56.5°.

7-Nitro-9-phenylphenanthridine.[121] (The use of phosphorus oxychloride in nitrobenzene to cyclize a p'-nitro-o-xenylamide.) A mixture of 15 g. of 2-benzamido-4'-nitrobiphenyl, 30 g. of phosphorus oxychloride, and 45 g. of nitrobenzene was refluxed at 180° for twelve hours. The product was carefully stirred into water, and a salt of the desired base separated. When the salt was heated with aqueous alkali 14 g. (99%) of the base was liberated. After crystallization from pyridine the yellow needles melted at 237°.

3,11-Dimethoxy-5,6-dihydro-8H-dibenzo[a,g]quinolizine.[122] A solution of 13 g. of N-formyl-1-(m-methoxybenzyl)-6-methoxy-1,2,3,4-tetrahydroisoquinoline, 30 ml. of phosphorus oxychloride, and 50 ml. of dry toluene was boiled for one and a half hours. Dilution with petroleum ether yielded a brown oil which was dissolved in ethanol, made alkaline with sodium hydroxide, and diluted with water. The free base separated as a yellow powder (8 g., 66%); it was recrystallized from ethanol as yellow prisms melting at 130°.

2,3-Methylenedioxy-11,12-dimethoxy-5,6,8,9-tetrahydrodibenzo[a,h]-quinolizinium Iodide.[123] A solution of 2.2 g. of unpurified 2-homopiperonyl-6,7-dimethoxy-3,4-dihydroisocarbostyril in 30 ml. of benzene was treated with 8 ml. of freshly distilled phosphorus oxychloride and heated for one hour on the steam bath. A large volume of petroleum ether was added, and after a while the supernatant liquid was decanted. An aqueous solution of the residue was decolorized with charcoal and treated with 5 g. of sodium iodide. After a few hours the precipitated quinolizinium iodide was collected (2.3 g., 80%) and recrystallized from ethanol as yellow needles melting at 188–189°.

[121] Walls, *J. Chem. Soc.*, **1945**, 294.
[122] Chakravarti, Haworth, and Perkin, *J. Chem. Soc.*, **1927**, 2265.
[123] Sugasawa and Kakemi, *Ber.*, **72**, 980 (1939).

1-Benzyl-3,4-dihydro-2-carboline.[81] A mixture of 2.5 g. of N-phenylacetyltryptamine, 5 ml. of phosphorus oxychloride, and 100 ml. of pure benzene was refluxed for one hour. The benzene was removed in vacuum; the residue was dissolved in dilute acetic acid, filtered, and treated with aqueous ammonia. The orange base that precipitated weighed 2.1 g. (90%). It was sensitive to atmospheric oxidation and was handled under carbon dioxide. The picrate melted at 225°.

TABULAR SURVEY OF THE BISCHLER-NAPIERALSKI REACTION

The following tables are based on a literature survey embracing all available reports published before July, 1949. The compounds in the tables are listed in order of increasing substitution in the basic nucleus. Among compounds having the same number of substituents, precedence has been given those having a substituent at the point of ring closure (position 1 for isoquinolines and 2-carbolines, position 9 for phenanthridines). Compounds with a substituent at the point of cyclization have been arranged in order of increasing complexity of that substituent (alkyl, aryl, aralkyl, heterocyclic).

Parenthetical notes, such as "(from oxime)," following the name of a compound refer to the starting material in a particular case, and *not* to the following lines. Data for more than one preparation of a single compound are listed in order of increasing yield.

Nearly all patents were consulted in the original, though secondary references are given for the convenience of the reader.

Note to Table VII. Compounds arising from cyclization of formamides of secondary amines have been considered as 1-hydroxy derivatives (LXVIII) by some investigators. These pseudobases are listed in the

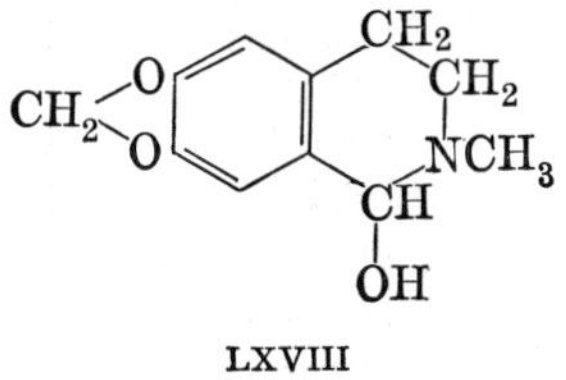

LXVIII

table as 1,2-dihydro derivatives of the parent 3,4-dihydroisoquinoline. Such compounds are frequently isolated as quaternary ammonium salts (LXIX).

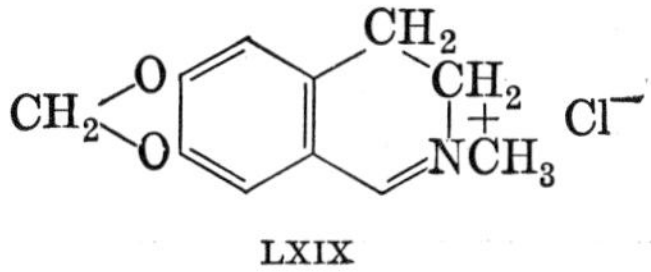

LXIX

TABLE VII
3,4-Dihydroisoquinolines

Substituents	Condensing Agent	Temperature °C.	Yield %	Reference
A. *Unsubstituted and monosubstituted*				
None	P₂O₅	—	0	124
	—	—	Poor	44
	P₂O₅	205	18	89
	Polyphosphoric acid	145	31	65
1-Methyl-	P₂O₅	—	—	1
	ZnCl₂	240	—	1
	P₂O₅	110	11	86
	POCl₃	110	22	72
	Polyphosphoric acid	160	23	65
	P₂O₅	110	35	2
	P₂O₅ + POCl₃	140	70	86
	P₂O₅	205	83	89
(from the cyanoacetamide)	Polyphosphoric acid	170	20	65a
1-Chloromethyl-	P₂O₅	140	38	99
1-Ethoxymethyl-	P₂O₅	140	45	125
1-Ethyl-	P₂O₅	205	50	89
1-*n*-Propyl-	P₂O₅	205	80	89
1-*n*-Butyl-	P₂O₅	205	70	89
1-(β-Phenethyl)carbamyl-	P₂O₅	110	—	126, 44
1-Cyclohexyl-	POCl₃	140	—	127, 128, 129
1-Phenyl-	P₂O₅	250	—	1
	Various	—	—	44
	Al₂O₃	195	5	86
	POCl₃	110	26	72, 130
	P₂O₅	205	67	89
	P₂O₅	140	75	2, 65b
	PCl₃ + AlCl₃	—	Good	107
	P₂O₅	110	83	86
	P₂O₅ + POCl₃	140	100	86
1-(*o*-Hydroxyphenyl)-	POCl₃	—	—	130
1-(*p*-Methoxyphenyl)-	P₂O₅	110	Poor	105
1-(*o*-Nitrophenyl)-	P₂O₅	140	73	90
1-(*m*-Nitrophenyl)-	P₂O₅	140	64	90
	P₂O₅	205	42	45
1-(*p*-Nitrophenyl)-	P₂O₅	140	73	131
	P₂O₅	205	74	45

TABLE VII—Continued

3,4-Dihydroisoquinolines

Substituents	Condensing Agent	Temperature °C.	Yield %	Reference
1-(o-Carboxyphenyl)-	NaCl + AlCl₃	150	—	108, 132, 133, 134
1-(2-Carboxy-4-chlorophenyl)-	NaCl + AlCl₃	160	—	108, 134, 132, 133
1-Benzyl-	POCl₃	110	0	86
	POCl₃	110	—	135
	P₂O₅	140	—	43
	POCl₃	110	9	72
	P₂O₅	205	54	136
	P₂O₅	205	65	89
	P₂O₅	205	Good	137
	P₂O₅	140	75	2, 65b
	P₂O₅	140	80	86
	P₂O₅	205	84	64
	PCl₅ + AlCl₃	—	86	44
1-(p-Methoxybenzyl)-	POCl₃	110	51	48
1-Veratryl-	POCl₃	140	—	93
1-(2-Aminoveratryl)-	—	—	Trace	48
1-(2-Acetamidoveratryl)-	—	—	Trace	48
1-(2-Nitroveratryl)-	Various	—	0	93
	P₂O₅	110	0	46
	PCl₅	—	0	92
	P₂O₅	140	11	47
	P₂O₅	110	21	47
4-Methyl-	P₂O₅	—	37	89
5-Methyl-	P₂O₅	—	34	89
6-Methoxy-	POCl₃	100	—	110
6-Ethoxy-	POCl₃	110	—	138
B. *Disubstituted*				
1-Hydroxy-2-methyl-1,2-dihydro-	SOCl₂	110	0	110
	SOCl₂	110	—	109
1,3-Dimethyl-	P₂O₅	110	—	139
	POCl₃	110	48	72
1-Methyl-3-benzyl-	POCl₃	110	38	72
1,4-Dimethyl-	POCl₃	110	31	72
	—	—	81	89
1-Methyl-4-aminomethyl-	P₂O₅	110	40	79
1-Methyl-4-phenyl-	POCl₃	110	30	72
1,5-Dimethyl-	—	—	59	89
1-Methyl-6-methoxy-	POCl₃	110	51	105

TABLE VII—*Continued*

3,4-DIHYDROISOQUINOLINES

Substituents	Condensing Agent	Temperature °C.	Yield %	Reference
1-Methyl-6-benzyloxy-	PCl$_5$	45	Poor	105
1-Methyl-6-benzamido-	P$_2$O$_5$	130	75	74
1-Methyl-7-isopropyl-	P$_2$O$_5$	110	16	105
1-Methyl-7-methoxy-	P$_2$O$_5$	207	0	105
	P$_2$O$_5$	140	2	105
	POCl$_3$	110	Poor	105
1-Chloromethyl-6-methoxy-	POCl$_3$	110	—	140
	POCl$_3$	—	41	57
1-Phenyl-3-methyl-	P$_2$O$_5$	110	12–19	86
	P$_2$O$_5$ + POCl$_3$	140	24	86
	POCl$_3$	110	35	72, 130
1-Phenyl-3-benzyl-	POCl$_3$	110	11	72
1-Phenyl-4-methyl-	POCl$_3$	110	45	72
	P$_2$O$_5$	110	92	86
1-Phenyl-4-benzamidomethyl-	P$_2$O$_5$	110	74	79
1,4-Diphenyl-	POCl$_3$	110	53	72
	P$_2$O$_5$	110	69	6
1-Phenyl-6-(β-benzamidoethyl)-	POCl$_3$	125	62	75
1-Phenyl-7-(β-benzamidoethyl)-	POCl$_3$	120	82	75
1-Phenyl-7-nitro-	P$_2$O$_5$	210	1.9	45
1-(*p*-Methoxyphenyl)-3-methyl-	POCl$_3$	—	—	130
1-(*p*-Chlorophenyl)-5-chloro-	P$_2$O$_5$	205	30	45, 141
1-(*m*-Nitrophenyl)-7-nitro-	P$_2$O$_5$	210	2.1	45
1-(*p*-Nitrophenyl)-7-nitro-	P$_2$O$_5$	210	13	45, 141
	POCl$_3$ + AlCl$_3$	210	6.3	45
1-(*o*-Carboxyphenyl)-7-chloro-	NaCl + AlCl$_3$	180	—	108, 132, 133, 134
1-(3,4,5-Trimethoxyphenyl)-6-propoxy-	PCl$_5$	25	—	142
1-(3,4,5-Trimethoxyphenyl)-6-isopropoxy-	PCl$_5$	25	—	142
1-Benzyl-2-methyl- (quaternary phosphate)	P$_2$O$_5$	140	—	40
1-Benzyl-3-methyl-	POCl$_3$	110	45	72
1,3-Dibenzyl-	POCl$_3$	110	22	72
1-Benzyl-4-methyl-	POCl$_3$	110	38	72
1-Benzyl-4-phenyl-	POCl$_3$	110	9	72
1-(*m*-Methoxybenzyl)-6-methoxy-	POCl$_3$	100	80	122
1-(*p*-Methoxybenzyl)-6-methoxy-	POCl$_3$	100	—	143
	POCl$_3$	100	80	144

TABLE VII—Continued
3,4-Dihydroisoquinolines

Substituents	Condensing Agent	Temperature °C.	Yield %	Reference
1-(p-Methoxybenzyl)-6-benzyloxy-	PCl₅	25	—	48
1-(3-Benzyloxy-4-methoxy-benzyl)-6-benzyloxy-	PCl₅	25	88	145
1-(2-Nitroveratryl)-6-methoxy-	PCl₅	25	—	146
1-(2-Nitroveratryl)-6-benzyloxy-	PCl₅	—	100	48
1-(p-Methoxybenzoyl)-6-methoxy- *	POCl₃	100	74	60
3,4-Dimethyl-	P₂O₅	205	—	147
5,6-Dimethoxy-	POCl₃	110	55	148
6,7-Methylenedioxy-		—	—	149, 44
	PCl₅	100	—	150, 126
	POCl₃	100	Good	150, 126
	P₂O₅	110	61	151
	POCl₃	110	66	73
6,7-Dimethoxy-	POCl₃	—	—	152
	P₂O₅	110	69	153
	P₂O₅	110	72	151
6-Methoxy-7-ethoxy-	P₂O₅	110	52–59	154, 21
	POCl₃	110	87	155
6-Ethoxy-7-methoxy-	P₂O₅	140	42–74	154, 21
6-Methoxy-7-benzyloxy-	P₂O₅	110	—	154
(also some of the 7-hydroxy compound)	POCl₃	110	24	248
C. Trisubstituted				
1-Hydroxy-6,7-methylenedioxy- (from urethan)	POCl₃	140	3	21
	POCl₃	100	42	22
1-Hydroxy-5,6-dimethoxy- (from urethan)	P₂O₅	110	17	156
1-Hydroxy-6,7-dimethoxy- (from isocyanate)	POCl₃	110	Poor	23
	POCl₃	140	—	21
1-Hydroxy-6-methoxy-7-ethoxy- (from urethan)	POCl₃ + P₂O₅	140	14	21
1-Hydroxy-6-ethoxy-7-methoxy- (from urethan)	POCl₃ + P₂O₅	—	26	21
1-Anilino-6,7-dimethoxy- (from urea)	POCl₃	110	—	23

* This product was apparently formed by an oxidation of the expected benzyl derivative during the alkaline phase of the isolation.

TABLE VII—Continued

3,4-Dihydroisoquinolines

Substituents	Condensing Agent	Temperature °C.	Yield %	Reference
1-p-Phenetidino-6,7-dimethoxy- (from urea)	POCl$_3$	110	Good	23
1-o-Toluino-6,7-dimethoxy- (from urea)	POCl$_3$	110	Poor	23
1-m-Toluino-6,7-dimethoxy- (from urea)	POCl$_3$	110	70	23
1-p-Toluino-6,7-dimethoxy- (from urea)	POCl$_3$	110	—	23
1-(N-Methylanilino)-6,7-dimethoxy- (from urea)	POCl$_3$	110	—	23
1-Methyl-5,6-dimethoxy-	POCl$_3$	110	81	157
1-Methyl-5-butoxy-6-methoxy-	PCl$_5$	45	25	105
1-Methyl-5,8-dimethoxy-	POCl$_3$	—	82	33
1-Methyl-6-hydroxy-7-methoxy- *	P$_2$O$_5$	—	>10	158
1-Methyl-6-methoxy-7-hydroxy- *	P$_2$O$_5$	—	>30	158
1-Methyl-6,7-methylenedioxy-	P$_2$O$_5$	110	—	159
	—	—	—	44
(from oxime)	P$_2$O$_5$	110	—	18
	POCl$_3$	—	>39	160
(from oxime)	POCl$_3$	110	60	161
	POCl$_3$	110	78	42
	P$_2$O$_5$	110	86	151
	POCl$_3$	110	92	73
1-Methyl-6,7-dimethoxy- (from oxime)	P$_2$O$_5$	110	—	18
	POCl$_3$	110	72	162
	P$_2$O$_5$	110	89	163, 164, 165, 151
1-Methyl-6-methoxy-7-butoxy-	PCl$_5$	50	57	105
1-Methyl-6-methoxy-7-hexyloxy-	PCl$_5$	45	64	105
1-Methyl-6-methoxy-7-benzyloxy-	PCl$_5$	40	46	105
	POCl$_3$	60	50	20
(from the benzenesulfonyl derivative of the oxime)	—	140	70	20
1-Methyl-6,7-diethoxy-	PCl$_5$	50	74	105
1-Methyl-6-ethoxy-7-butoxy-	PCl$_5$	45	77	105
1-Methyl-6-ethoxy-7-benzyloxy-	PCl$_5$	47	75	105
1-Methyl-6,7-dipropoxy-	PCl$_5$	50	53	105
1-Methyl-6,7-diisopropoxy-	PCl$_5$	45	74	105
1-Methyl-6,7-dibutoxy-	PCl$_5$	45	4.6	105
1-Methyl-6-butoxy-7-methoxy-	PCl$_5$	45	82	105

* The starting amide was the corresponding O-benzyl ether.

TABLE VII—Continued

3,4-Dihydroisoquinolines

Substituents	Condensing Agent	Temperature °C.	Yield %	Reference
1-Methyl-6-benzyloxy-7-methoxy-	PCl$_5$	45	75	105
1-Methyl-6-benzyloxy-7-ethoxy-	PCl$_5$	45	49	105
1-Chloromethyl-6,7-methylenedioxy-	POCl$_3$	—	82	140, 57
	POCl$_3$	100	89	160
1-Chloromethyl-6,7-dimethoxy- (also from α-bromoacetamide)	POCl$_3$	110	80–94	140, 57
1-Bromomethyl-6,7-dimethoxy-	P$_2$O$_5$	140	70	140, 57
1-Cyanomethyl-6,7-dimethoxy-	POCl$_3$	110	40	140, 57
1-Ethyl-6,7-methylenedioxy-	POCl$_3$	—	64	160
	P$_2$O$_5$	110	Good	151
1-Ethyl-6,7-dimethoxy-	POCl$_3$	—	66	160
	P$_2$O$_5$	110	94	151
1-(α-Chloroethyl)-6,7-methylenedioxy- (from the lactamide)	POCl$_3$	—	29	160
1-(α-Chloroethyl)-6,7-dimethoxy- (from the lactamide)	POCl$_3$	—	29	160
1-(β-Bromoethyl)-6,7-dimethoxy-	POBr$_3$	25	18	114
1-Propyl-6,7-methylenedioxy-	P$_2$O$_5$	110	—	151
1-Propyl-6,7-dimethoxy-	P$_2$O$_5$	110	93	151
1-(γ-Chloropropyl)-6,7-dimethoxy-	POCl$_3$	110	Poor	57
1-(δ-Chlorobutyl)-6,7-dimethoxy- (from the bromoamide)	POCl$_3$	110	—	57
	POCl$_3$	110	93	140, 57
1-Pentadecyl-6,7-dimethoxy-	POCl$_3$	110	78	166
1-(1,2,2-Trimethyl-3-carboxycyclopentyl)-6,7-dimethoxy-	POCl$_3$	110	40	167
1-Cyclohexyl-6-methoxy-7-hydroxy-	POCl$_3$	60	—	127, 128, 129
1-Cyclohexyl-6,7-methylenedioxy-	POCl$_3$	110	77	168, 169
1-Cyclohexyl-6,7-dimethoxy-	POCl$_3$	60	—	127, 128, 129
1-Cyclohexyl-6,7-ethylenedioxy-	POCl$_3$	60	—	127, 128, 129
1-Cyclohexylmethyl-6,7-methylenedioxy-	POCl$_3$	110	80	170

TABLE VII—*Continued*

3,4-Dihydroisoquinolines

Substituents	Condensing Agent	Temperature °C.	Yield %	Reference
1-Phenyl-6,7-methylenedioxy-	P_2O_5	110	—	159
	—	—	—	44
	$POCl_3$	140	32	171
	$POCl_3$	100	75	73
1-Phenyl-6,7-dimethoxy-	$POCl_3$	100	—	172
	$POCl_3$	—	Good	173
	$POCl_3$	140	70	171
	$POCl_3$	110	85	174
1-(*p*-Methoxyphenyl)-6,7-dimethoxy-	$POCl_3$	110	65	175, 171
1-(*o*-Nitrophenyl)-6,7-methylenedioxy-	$POCl_3$	110	82	176
1-(*o*-Nitrophenyl)-6,7-dimethoxy-	$POCl_3$	110	99	114
1-(*m*-Nitrophenyl)-6,7-methylenedioxy-	$POCl_3$	110	99	176
1-(*p*-Nitrophenyl)-3,4-dimethyl-	P_2O_5	110	32	45
1-(*p*-Nitrophenyl)-6,7-methylenedioxy-	$POCl_3$	110	90	176
1-(*p*-Nitrophenyl)-6,7-dimethoxy-	$POCl_3$	110	95	177
1-(*o*-Carboxyphenyl)-6,7-methylenedioxy-	PCl_5	60	18	167
1-(3,4-Methylenedioxyphenyl)-6,7-methylenedioxy-	—	—	—	178
1-(3,4-Dimethoxyphenyl)-6,7-dimethoxy-	—	—	—	178
1-(3,4-Diethoxyphenyl)-6,7-diethoxy-	—	—	—	178
1-(3,4,5-Trimethoxyphenyl)-6,7-methylenedioxy-	$POCl_3$	110	90	178a, 178b
1-(3,4,5-Trimethoxyphenyl)-6,7-dimethoxy-	—	—	—	178
1-(3,4,5-Trimethoxyphenyl)-6,7-diethoxy-	$POCl_3$	110	42	174
	$POCl_3$	80	87	178
1-(3,4,5-Trimethoxyphenyl)-6-propoxy-7-methoxy-	PCl_5	25	33	142, 179
1-(3,4,5-Triethoxyphenyl)-6,7-diethoxy-	—	—	—	178
1-(α-Naphthyl)-6,7-methylenedioxy-	$POCl_3$	110	Good	180

TABLE VII—*Continued*

3,4-Dihydroisoquinolines

Substituents	Condensing Agent	Temperature °C.	Yield %	Reference
1-(β-Naphthyl)-6,7-methylenedioxy-	POCl₃	110	Good	180
1-(9-Phenanthryl)-6,7-methylenedioxy-	POCl₃	110	—	181
1-Benzyl-6,7-methylenedioxy-	P₂O₅	110	—	159
	POCl₃	110	—	135
	—	—	—	44, 182
	POCl₃	110	—	56
	POCl₃	110	70	42
	POCl₃	110	82	183
1-Benzyl-6,8-dimethoxy-	P₂O₅	140	85	184
1-(p-Methoxybenzyl)-6,7-methylenedioxy-	POCl₃	110	94	183
1-(p-Methoxybenzyl)-6,7-dimethoxy-	POCl₃	110	—	185, 186, 182
1-(o-Nitrobenzyl)-6,7-methylenedioxy-	PCl₅	25	39	187
	POCl₃	25	51	176
	POCl₃	25	79	188
1-(o-Nitrobenzyl)-6,7-dimethoxy-	P₂O₅ or POCl₃	—	0	94
	PCl₅	25	92	94
1-(m-Nitrobenzyl)-6,7-methylenedioxy-	POCl₃	100	93	176
1-(p-Nitrobenzyl)-6,7-methylenedioxy-	POCl₃	110	85	176
1-(2,3-Dimethoxybenzyl)-5,6-dimethoxy-	POCl₃	140	>80	62
1-(2,3-Dimethoxybenzyl)-6,7-methylenedioxy-	POCl₃	100	80	189
1-(2,3-Dimethoxybenzyl)-6,7-dimethoxy-	POCl₃	100	—	189
	P₂O₅	110	89	117
1-Piperonyl-6,7-methylenedioxy-	POCl₃	110	—	190
	POCl₃	110	—	191
	—	—	—	44, 192
	POCl₃	110	>76	193
1-Piperonyl-6,7-dimethoxy-	POCl₃	110	Good	194
	P₂O₅	140	30	195
	POCl₃	110	>80	195
1-Piperonyl-6-methoxy-7-ethoxy-	POCl₃	110	Good	196
1-Piperonyl-6-ethoxy-7-methoxy-	POCl₃	110	—	197

TABLE VII—*Continued*

3,4-Dihydroisoquinolines

Substituents	Condensing Agent	Temperature °C.	Yield %	Reference
1-Piperonyl-6,7-dibenzyloxy-	PCl_5	25	70	198
1-Veratryl-5,6-dimethoxy-	PCl_5	25	—	59
1-Veratryl-6,7-methylenedioxy-	P_2O_5	—	—	199
	—	—	—	93
	P_2O_5	140	Poor	200
	P_2O_5	140	35	201, 202
	$POCl_3$	110	75	200
	$POCl_3$	80	>80	136
	—	—	92	203
1-Veratryl-6,7-dimethoxy-	$POCl_3$	110	—	58
	P_2O_5	140	—	204, 3
	P_2O_5	110	56	205
	P_2O_5	140	65	205
	$POCl_3$	110	95	206
	$POCl_3$	80	100	119
1-(3,4-Diethoxybenzyl)-6,7-diethoxy-	$POCl_3$	80	97	206a
1-(3-Benzyloxy-4-methoxybenzyl)-6-benzyloxy-7-methoxy-	PCl_5	25	85	207
1-(3-Benzyloxy-4-methoxybenzyl)-6,7-dibenzyloxy-	PCl_5	—	68	208
6-Methoxy-3,4′-bis[(6,7-dimethoxy-3,4-dihydroisoquinolyl-1)methyl]diphenyl ether (full name)	PCl_5	—	68	209

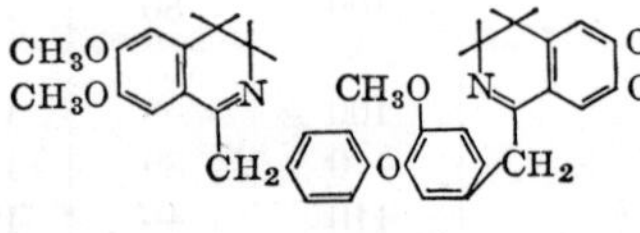				
1-(2-Nitro-3-methoxybenzyl)-6,7-methylenedioxy-	PCl_5	25	65	95
1-(2-Nitro-4-methoxybenzyl)-6,7-methylenedioxy-	P_2O_5	110	55	103
	P_2O_5	110	59	104
	PCl_5	25	68–75	103

TABLE VII—*Continued*

3,4-Dihydroisoquinolines

Substituents	Condensing Agent	Temperature °C.	Yield %	Reference
1-(2-Nitro-4-methoxybenzyl)-6,7-dimethoxy-	P_2O_5	110	85	210
1-(2-Nitro-4-ethoxybenzyl)-6,7-dimethoxy-	P_2O_5	110	—	211
1-(2-Nitro-5-methoxybenzyl)-6,7-methylenedioxy-	P_2O_5	140	34	103
	PCl_5	25	63	212
	PCl_5	45	66	103
	PCl_5	25	70	213
1-(2-Nitro-5-benzyloxybenzyl)-6,7-methylenedioxy-	PCl_5	25	80	213
1-(3-Methoxy-4-carbethoxybenzyl)-6,7-methylenedioxy-	$POCl_3$	110	—	135
1-(3-Carbethoxyoxy-4-methoxybenzyl)-6,7-dimethoxy-	P_2O_5	110	55	214
1-(3,4,5-Trimethoxybenzyl)-6,7-dimethoxy-	P_2O_5	110	Good	215
	P_2O_5	110	75	216
1-(6-Bromopiperonyl)-5,6-dimethoxy-	$POCl_3$	110	—	148
1-(6-Bromoveratryl)-6,7-methylenedioxy-	$POCl_3$	110	—	35
1-(2-Nitroveratryl)-5,6-dimethoxy-	PCl_5	25	58	217
1-(2-Nitroveratryl)-6,7-methylenedioxy-	P_2O_5	110	34–60	218
	PCl_5	25	80	219
1-(2-Nitroveratryl)-6,7-dimethoxy-	P_2O_5	110	72–93	96
	PCl_5	25	82	220
1-(2-Nitro-3-methoxy-4-benzyloxybenzyl)-6,7-dimethoxy-	PCl_5	25	65	221
1-(6-Nitroveratryl)-5,6-dimethoxy-	PCl_5	25	66	217
1-(3-Methoxy-4-ethoxy-6-nitrobenzyl)-6,7-dimethoxy-	P_2O_5	110	46	222
1-(3-Ethoxy-4-methoxy-6-nitrobenzyl)-6,7-dimethoxy-	—	—	47	222
1-(3-Ethoxy-4-methoxy-6-nitrobenzyl)-6-methoxy-7-ethoxy-	—	—	78	223

TABLE VII—Continued

3,4-Dihydroisoquinolines

Substituents	Condensing Agent	Temperature °C.	Yield %	Reference
1-(3-Ethoxy-4-methoxy-6-nitrobenzyl)-6-ethoxy-7-methoxy-	—	—	76	223
1-(3-Methoxy-4-benzyloxy-6-nitrobenzyl)-6,7-dimethoxy-	PCl$_5$	25	—	224
1-(2-Nitro-3-methoxy-4-carbethoxyoxybenzyl)-6,7-dimethoxy-	PCl$_5$	25	—	225
1-[2-Nitro-3,4-bis(carbethoxyoxy)benzyl]-6,7-dimethoxy-	PCl$_5$	25	—	225
1-Veratroyl-5,6-dimethoxy- *	POCl$_3$	110	88	59
1-(β-Phenethyl)-6,7-dimethoxy-	POCl$_3$	110	83	162
1-Homopiperonyl-6,7-methylenedioxy-	PCl$_5$	25	33	167
	POCl$_3$	110	73	167
1-Homopiperonyl-6,7-dimethoxy-	PCl$_5$	25	42	167
1-Homoveratryl-6,7-methylenedioxy-	POCl$_3$	110	53	167
1-Homoveratryl-6,7-dimethoxy-	—	—	—	167
(from oxime)	POCl$_3$	110	85	118
	POCl$_3$	110	91	226, 118
1-(3,4-Dimethoxystyryl)-6,7-dimethoxy- (from benzenesulfonyl ester of oxime)	—	140	—	19
	P$_2$O$_5$	140	—	19
1-(2-Carbomethoxy-3,4-dimethoxy-α-methylbenzyl)-6,7-dimethoxy-	POCl$_3$	110	58	41
1-Benzohydryl-6,7-methylenedioxy-	POCl$_3$	110	93	114
1-(5-Indanylmethyl)-6,7-methylenedioxy-	P$_2$O$_5$	110	37	38
1-(α-Naphthylmethyl)-6,7-methylenedioxy-	POCl$_3$	110	Poor	180
1-(2-Furyl)-6,7-methylenedioxy-	POCl$_3$	—	51	181
1-(2-Furylvinyl)-6,7-dimethoxy-	POCl$_3$	110	80	99
1-(7-Methoxy-2-coumaronyl)-6,7-methylenedioxy-	POCl$_3$	110	79	181

* This product was apparently formed by air oxidation of the expected benzyl derivative during the alkaline phase of the isolation.

TABLE VII—Continued

3,4-Dihydroisoquinolines

Substituents	Condensing Agent	Temperature °C.	Yield %	Reference
1-Opianyl-6,7-methylenedioxy-	POCl$_3$	100	>14	227, 228
1-(1-Oxo-4-methoxy-5,6-methylenedioxy-7-isobenzofuranylmethyl)-6,7-dimethoxy- (?)	POCl$_3$	—	—	229
1-(1-Oxo-4,5,6-trimethoxy-7-isobenzofuranylmethyl)-6,7-dimethoxy-	—	—	—	230
1-(3-Coumarinyl)-6,7-methylenedioxy-	POCl$_3$	110	74	231
1-(3-Coumarinyl)-6,7-dimethoxy-	POCl$_3$	110	63	231
1-(3-Coumarinylmethyl)-6,7-methylenedioxy-	POCl$_3$	110	—	231
1-(3-Coumarinylmethyl)-6,7-dimethoxy-	POCl$_3$	110	—	231
1-(7-Methyl-4-coumarinylmethyl)-6,7-methylenedioxy-	POCl$_3$	110	69	231
1-(7-Methyl-4-coumarinylmethyl)-6,7-dimethoxy-	POCl$_3$	110	55	231
α,ω-Bis(6,7-methylenedioxy-3,4-dihydroisoquinolyl-1)-butane (full name)	POCl$_3$	—	87	98
α,ω-Bis(6,7-dimethoxy-3,4-dihydroisoquinolyl-1)butane (full name)	POCl$_3$	110	80	166
	POCl$_3$	110	89	98
α,ω-Bis(6,7-dimethoxy-3,4-dihydroisoquinolyl-1)pentane (full name)	POCl$_3$	—	83	98
	POCl$_3$	110	95	166
α,ω-Bis(6,7-dimethoxy-3,4-dihydroisoquinolyl-1)hexane (full name)	POCl$_3$	110	94	166
α,ω-Bis(6,7-dimethoxy-3,4-dihydroisoquinolyl-1)heptane (full name)	POCl$_3$	110	100	166
α,ω-Bis(6,7-dimethoxy-3,4-dihydro-1-isoquinolyl)-4-heptanone (full name)	POCl$_3$	110	36	232

TABLE VII—*Continued*

3,4-DIHYDROISOQUINOLINES

Substituents	Condensing Agent	Temperature °C.	Yield %	Reference
α,ω-Bis(6,7-dimethoxy-3,4-dihydroisoquinolyl-1)octane (full name)	POCl₃	110	43	166
	POCl₃	—	85	98
1-Phthalimidomethyl-6,7-dimethoxy-	POCl₃	110	87	97
1-(N-Methylpiperidyl-2)-6,7-dimethoxy-	POCl₃	110	—	233
1-(2-Pyridyl)-6,7-dimethoxy-	POCl₃	110	42	234
1-(2-Picolyl)-6,7-methylenedioxy-	POCl₃	110	53	235
	POCl₃	110	62	42
1-(2-Quinolyl)-6,7-methylenedioxy-	POCl₃	110	—	233, 179
	POCl₃	110	74	236, 237
1-(4-Quinolyl)-6,7-methylenedioxy-	POCl₃	110	93	238
1-(6-Quinolyl)-6,7-methylenedioxy-	POCl₃	110	99	239
1-(8-Quinolyl)-6,7-methylenedioxy-	POCl₃	110	95	239
1-(2-Methyl-4-quinolyl)-6,7-methylenedioxy-	POCl₃	110	85	236, 237
1-(2-Phenyl-4-quinolyl)-6,7-methylenedioxy-	POCl₃	110	85	236, 237
1-(6-Methoxy-4-quinolyl)-6,7-methylenedioxy-	POCl₃	110	90	238
2-Methyl-6,7-methylenedioxy- (chloride)	SOCl₂	80	100	240, 241
2-Methyl-6,7-dimethoxy- (chloride)	SOCl₂	80	92–100	240, 241
2-Ethyl-6,7-methylenedioxy- (chloride)	SOCl₂	80	—	241
2-Piperonyl-6,7-methylenedioxy- (chloride)	POCl₃	100	88	39
3-Methyl-6,7-methylenedioxy-	P₂O₅	140	—	242
	POCl₃	110	Good	243
3-Methyl-6,7-dimethoxy-	POCl₃	110	Good	243
3-Methyl-6-methoxy-7-benzyloxy-	PCl₅	60	—	105
3-Methyl-6-methoxy-7-benzoyloxy-	POCl₃	60	22	244
3-Veratryl-6,7-dimethoxy-	POCl₃	140	Good	245
	POCl₃	100	95	246

TABLE VII—*Continued*

3,4-Dihydroisoquinolines

Substituents	Condensing Agent	Temperature °C.	Yield %	Reference
6,7-Methylenedioxy-8-methoxy-	POCl$_3$	110	—	126, 247
6,7-Dimethoxy-8-bromo-	—	—	—	248
D. *Tetra- and penta-substituted*				
1-Hydroxy-2-methyl-6,7-methylenedioxy-1,2-dihydro-	P$_2$O$_5$	80	—	249, 247, 250, 251
1-Hydroxy-2-ethyl-6,7-methylenedioxy-1,2-dihydro-	P$_2$O$_5$	80	—	249
	POCl$_3$	80	—	247
1-Hydroxy-6,7,8-trimethoxy-				
(from urethan)	P$_2$O$_5$ + POCl$_3$	140	10	252
(from isocyanate)	P$_2$O$_5$ + POCl$_3$	140	19	252
1,3-Dimethyl-6-methoxy-7-acetoxy-	POCl$_3$	110	56	253
1,3-Dimethyl-6-methoxy-7-butoxy-	PCl$_5$	50	62	105
1,3-Dimethyl-6-methoxy-7-benzyloxy-	PCl$_5$	50	58	105
1,3-Dimethyl-6-benzyloxy-7-methoxy-	PCl$_5$	50	59	105
1-Methyl-3-phenyl-6,7-methylenedioxy-	POCl$_3$	110	60	72
1,4-Dimethyl-6-methoxy-7-benzyloxy-	PCl$_5$	50	57	105
1-Methyl-6-methoxy-7,8-methylenedioxy-	P$_2$O$_5$	110	82	254
1-Methyl-6,7,8-trimethoxy-	P$_2$O$_5$	110	65	255
1-Methyl-6-acetoxy-7,8-dimethoxy- (?)	P$_2$O$_5$	110	43	256
1-Phenoxymethyl-3-methyl-6-methoxy-7-benzyloxy-	PCl$_5$	50	47	105
1-Ethyl-3-methyl-6-methoxy-7-benzyloxy-	PCl$_5$	70	19	105
1-Ethyl-4-methyl-6-methoxy-7-benzyloxy-	PCl$_5$	50	55	105
1-Propyl-3-methyl-6-methoxy-7-benzyloxy-	PCl$_5$	50	58	105
1-Isopropyl-3-methyl-6-methoxy-7-benzyloxy-	PCl$_5$	50	26	105
1-Isobutyl-3-methyl-6-methoxy-7-benzyloxy-	PCl$_5$	50	18	105

TABLE VII—*Continued*

3,4-Dihydroisoquinolines

Substituents	Condensing Agent	Temperature °C.	Yield %	Reference
1-Phenyl-2-methyl-6,7-methylenedioxy- (chloride)	$POCl_3$	110	—	249
1-Phenyl-3-methyl-6,7-methylenedioxy-	$POCl_3$	110	—	257, 258
1-Phenyl-3-methyl-6-methoxy-7-benzoyloxy-	$POCl_3$	110	95	244
1-Phenyl-3-carbomethoxy-6,7-methylenedioxy-	$POCl_3$	110	Good	259, 260, 261, 173
1-Phenyl-3-carbomethoxy-6,7-dimethoxy-	$POCl_3$	130	—	262, 260, 261
	P_2O_5	140	69	172
1-Phenyl-3-carbethoxy-6,7-dimethoxy-	P_2O_5	140	79	172
1,3-Diphenyl-6,7-methylenedioxy-	$POCl_3$	110	28	72
1-(3,4-Methylenedioxyphenyl)-3-methyl-6,7-methylenedioxy-	$POCl_3$	110	90	258, 263, 257
1-(3,4-Methylenedioxyphenyl)-3-carbomethoxy-6,7-methylenedioxy-	$POCl_3$	110	Good	264, 260, 261, 173
1-(3,4-Dimethoxyphenyl)-4-methyl-6,7-dimethoxy-	$POCl_3$	130	95 (Crude)	265, 179
1-(3,4,5-Trimethoxyphenyl)-6,7,8-trimethoxy-	—	—	—	178
1-Benzyl-3-methyl-6,7-dimethoxy-	P_2O_5	110	>60	266, 267, 263
1-Benzyl-3-methyl-6-methoxy-7-phenylacetoxy-	$POCl_3$	60	56	244
1-Benzyl-3-phenyl-6,7-methylenedioxy-	$POCl_3$	110	46	72
1-Benzyl-6,7-methylenedioxy-8-methoxy- and 1-Benzyl-6-methoxy-7,8-methylenedioxy-	P_2O_5	140	65	34
1-Piperonyl-3-methyl-6,7-methylenedioxy-	$POCl_3$	100	90	266, 267
1-Piperonyl-3-methyl-6,7-dimethoxy-	$POCl_3$	125	>60	266, 267, 263

TABLE VII—Continued

3,4-Dihydroisoquinolines

Substituents	Condensing Agent	Temperature °C.	Yield %	Reference
1-Veratryl-3-methyl-6,7-methylenedioxy-	P_2O_5	100	>60	267, 263
	$POCl_3$	140	>60	266, 263
1-Veratryl-3-methyl-6,7-dimethoxy-	$POCl_3$	100	>60	266, 267
1-Veratryl-3-methyl-6-methoxy-7-(3,4-dimethoxyphenylacetoxy)-	$POCl_3$	60	49	244
1,3-Bis(veratryl)-6,7-dimethoxy-	$POCl_3$	140	—	206, 246
1-Veratryl-4-methyl-6,7-dimethoxy-	$POCl_3$	110	85	265, 179
1-(β-Phenethyl)-3-carbomethoxy-6,7-methylenedioxy-	$POCl_3$	130	Good	268, 260, 261, 173
1-(α-Ethylbenzyl)-3-methyl-6,7-methylenedioxy-	$POCl_3$	110	>70	266, 267, 263
1-(3-Pyridyl)-3-methyl-6,7-methylenedioxy-	$POCl_3$	110	—	258, 257
1-(2-Quinolyl)-3-methyl-6,7-methylenedioxy-	$POCl_3$	110	80	233
2,3-Dimethyl-6,7-methylenedioxy- (phosphate)	P_2O_5	140	—	242
2-Methyl-6,7-methylenedioxy-8-methoxy- (chloride)	$SOCl_2$	80	—	240
1,3,4-Trimethyl-6-methoxy-7-benzyloxy-	PCl_5	50	11	105
1-(3,4-Dimethoxyphenyl)-3,4-dimethyl-6,7-dimethoxy-	$POCl_3$	110	88	265, 179

Supplement to Table VII
Amides That Could Not Be Cyclized

Name	Condensing Agent	Temperature °C.	Reference
N-Formyl-1,3-diphenylisopropylamine	—	—	269
N-Acetyl-β-(3-acetylaminoethylphenyl)ethylamine	—	—	75
N-Acetyl-α,β-diphenylethylamine	POCl$_3$	110	72
N-Phthalimidoacetyl-β-phenethylamine	—	—	97
N-Triazoacetyl-β-(3,4-dimethoxyphenyl)ethylamine	—	—	97
N-Glycyl-β-(3,4-dimethoxyphenyl)ethylamine	—	—	97
N-(β-Chloropropionyl)-β-(3,4-dimethoxyphenyl)-ethylamine	POCl$_3$	110	57
N-Benzoyl-α,β-diphenylethylamine	POCl$_3$	110	72
N-Benzoyl-α-aminoacetophenone	POCl$_3$	110	53
N-Benzoyl-α-aminopropiophenone	P$_2$O$_5$ + POCl$_3$	140	86
N-Benzoyl-α-amino-3,4-diethoxypropiophenone	POCl$_3$	110	11
N-(p-Nitrobenzoyl)-β-(m-benzamidophenyl)ethylamine	—	—	45
N-Phenylacetyl-α,β-diphenylethylamine	POCl$_3$	110	72
N-Phenylacetyl-α-aminoacetophenone	H$_2$SO$_4$	—	270
N-(2-Nitrophenylacetyl)-β-phenethylamine	P$_2$O$_5$	140	91, 46
N-Homoveratroyl-α-aminoacetoveratrone	POCl$_3$	110	50, 51, 55
N-(3-Benzyloxy-4-methoxyphenylacetyl)-2-(3-benzyloxy-4-methoxyphenyl)ethylamine	—	—	271
N-(2-Nitro-3,4-dimethoxyphenylacetyl)-β-[3-(2-nitro-3,4-dimethoxyphenylacetamido)phenyl]ethylamine	PCl$_5$	25	92
N-(2-Nitro-3,4-dimethoxyphenylacetyl)-β-(4-methoxyphenyl)ethylamine	Various	—	49
N-(2-Nitro-3,4-dimethoxyphenylacetyl)-β-[3-(2-nitro-3,4-dimethoxyphenylacetamido)-4-methoxyphenyl]ethylamine	Various	—	49
N-(2-Nitro-3,4-dimethoxyphenylacetyl)-β-(3-bromophenyl)ethylamine	POCl$_3$	110	48
N-(2-Nitro-3,4-dimethoxyphenylacetyl)-β-(3-bromo-4-methoxyphenyl)ethylamine	POCl$_3$	110	48
N-(2-Benzamidophenylglyoxalyl)-β-phenethylamine	PCl$_5$	25	92
N-(2-Furylpropionyl)-β-(3,4-methylenedioxyphenyl)-ethylamine	—	—	181
N-[2-(5-Phenylfuryl)propionyl]-β-(3,4-methylenedioxyphenyl)ethylamine	—	—	181
N-(2-Furylacrylyl)-β-phenethylamine	Various	—	99
C$_6$H$_5$CH$_2$CH$_2$NHCO(CH$_2$)$_{3-10}$CONHCH$_2$CH$_2$C$_6$H$_5$	Various	—	98
3,4-(CH$_3$O)$_2$C$_6$H$_3$CH$_2$CH$_2$NHCO(CH$_2$)$_{2-3}$CONHCH$_2$CH$_2$C$_6$H$_3$(OCH$_3$)$_2$-3,4	—	—	98

REFERENCES TO TABLE VII

[124] Decker and Becker, *Ann.*, **382**, 369 (1911).
[125] Craig and Tarbell, *J. Am. Chem. Soc.*, **71**, 462 (1949).
[126] Decker, Ger. pat. 245,095 [*Frdl.*, **10**, 1187 (1910–1912)].
[127] Bockmühl and Hermann, Ger. pat. 670,683 [*C. A.*, **33**, 6527 (1939)].
[128] Bockmühl and Hermann, U. S. pat. 2,168,929 [*C. A.*, **33**, 9552 (1939)].
[129] I.G. Farbenind., Brit. pat. 488,423 [*C. A.*, **33**, 180 (1939)].
[130] Keil and Dobke, Ger. pat. 739,866 [*Chem. Zentr.*, **I**, 36 (1944)].
[131] Rodionov and Yavorskaya, *J. Gen. Chem. U.S.S.R.*, **11**, 446 (1941) [*C. A.*, **35**, 6592, (1941)].
[132] Ebel, U. S. pat. 2,069,473 [*C. A.*, **31**, 1823 (1937)].
[133] I.G. Farbenind., Brit. pat. 431,790 [*C. A.*, **29**, 8004 (1935)].
[134] I.G. Farbenind., Fr. pat. 781,562 [*C. A.*, **29**, 6249 (1935)].
[135] Kitasato, *Acta Phytochim.*, **3**, 215 (1927) [*C. A.*, **22**, 1779 (1928)].
[136] Craig and Tarbell, *J. Am. Chem. Soc.*, **70**, 2783 (1948).
[137] Leithe, *Ber.*, **63**, 1498 (1930).
[138] Hjort, deBeer, Buck, and Randall, *J. Pharmacol.*, **76**, 64 (1942) [*C. A.*, **36**, 7133 (1942)].
[139] Hey, *J. Chem. Soc.*, **1930**, 18.
[140] Child and Pyman, Brit. pat. 344,166 [*C. A.*, **26**, 154 (1932)].
[141] Mathieson and McCoubrey, *Nature*, **162**, 73 (1948).
[142] Sugasawa and Kakemi, *J. Pharm. Soc. Japan*, **55**, 1283 (1935) [*C. A.*, **30**, 2572 (1936)].
[143] Chakravarti, Vaidyanathan, and Venkatasubban, *J. Annamalai Univ.*, **1**, 190 (1932) [*C. A.*, **27**, 1351 (1933)].
[144] Chakravarti, Vaidyanathan, and Venkatasubban, *J. Indian Chem. Soc.*, **9**, 573 (1932).
[145] Schöpf, Perrey, and Jäckh, *Ann.*, **497**, 47 (1932).
[146] Gulland and Haworth, *J. Chem. Soc.*, **1928**, 2083.
[147] Witkop, *J. Am. Chem. Soc.*, **70**, 1424 (1948).
[148] Haworth, *J. Chem. Soc.*, **1927**, 2281.
[149] Decker, U. S. pat. 1,010,598 [*C. A.*, **6**, 413 (1912)].
[150] Decker, Ger. pat. 234,850 [*Frdl.*, **10**, 1186 (1910–1912)].
[151] Späth and Polgar, *Monatsh.*, **51**, 190 (1929).
[152] Buck and Ide, *J. Am. Chem. Soc.*, **60**, 2101 (1938).
[153] Späth and Epstein, *Ber.*, **59**, 2791 (1926).
[154] Manske, *Can. J. Research*, **15B**, 159 (1937).
[155] Kondo and Tanaka, *J. Pharm. Soc. Japan*, **49**, 4 (1929) (English).
[156] Späth and Strauhal, *Ber.*, **61**, 2395 (1928).
[157] Rajagopalan, *Proc. Indian Acad. Sci.*, **13A**, 566 (1941).
[158] Späth, Orechoff, and Kuffner, *Ber.*, **67**, 1214 (1934).
[159] Bayer and Co., Ger. pat. 235,358 [*Frdl.*, **10**, 1189 (1910–1912)].
[160] Dey and Govindachari, *Arch. Pharm.*, **277**, 177 (1939).
[161] Gaind, Kapoor, and Rây, *J. Indian Chem. Soc.*, **18**, 213 (1941).
[162] Adams and Whaley, unpublished work.
[163] Schöpf and Bayerle, *Ann.*, **513**, 190 (1934).
[164] Späth, *Ber.*, **62**, 1021 (1929).
[165] Späth and Dengel, *Ber.*, **71**, 113 (1938).
[166] Hahn and Gudjons, *Ber.*, **71**, 2183 (1938).
[167] Narang, Rây, and Silooja, *J. Chem. Soc.*, **1932**, 2510.
[168] Dey and Govindachari, *Current Sci.*, **13**, 203 (1944).
[169] Govindachari, Ph.D. thesis, Univ. of Madras, 1946.
[170] Dey and Venkataraman, *Proc. Natl. Inst. Sci. India*, **6**, 209 (1940).
[171] Weinbach and Hartung, unpublished work.
[172] Harwood and Johnson, *J. Am. Chem. Soc.*, **56**, 468 (1934).
[173] Soc. Anon. pour l'Ind. Chim. à Bâle, Brit. pat. 191,233 [*C. A.*, **17**, 3073 (1923)].
[174] Sugasawa, *J. Pharm. Soc. Japan*, **55**, 224 (1935) [*C. A.*, **29**, 5116 (1935)].
[175] Ahluwalia, Narang, and Rây, *J. Chem. Soc.*, **1931**, 2057.

[176] Dey and Parikshit, *Proc. Natl. Inst. Sci. India*, **11**, 30 (1945).
[177] Rajagopalan and Ganapathi, *Proc. Indian Acad. Sci.*, **15A**, 432 (1942).
[178] Slotta and Haberland, *Angew. Chem.*, **46**, 766 (1933).
[178a] Reeve and Eareckson, *J. Am. Chem. Soc.*, **72**, 5195 (1950).
[178b] Gensler and Samour, *J. Am. Chem. Soc.*, **72**, 3318 (1950).
[179] S. Sugasawa, private communication.
[180] Dey and Rajagopalan, *Current Sci.*, **13**, 202 (1944).
[181] Raman, *J. Indian Chem. Soc.*, **17**, 715 (1940).
[182] Tomita and Satomi, *J. Pharm. Soc. Japan*, **58**, 165 (1938) [*Chem. Zentr.*, II, 3396 (1938)].
[183] Tomita and Satomi, *J. Pharm. Soc. Japan*, **58**, 617 (1938) [*C. A.*, **32**, 8424 (1938)].
[184] Salway, *J. Chem. Soc.*, **99**, 1320 (1911).
[185] Kondo and Kondo, *J. Pharm. Soc. Japan*, **48**, 324 (1928) [*C. A.*, **22**, 3414 (1928)].
[186] Kondo, *J. Pharm. Soc. Japan*, **48**, 56 [*Chem. Zentr.*, II, 55 (1928)].
[187] Marion and Grassie, *J. Am. Chem. Soc.*, **66**, 1290 (1944).
[188] Barger and Weitnauer, *Helv. Chim. Acta*, **22**, 1036 (1939).
[189] Chakravarti and Swaminathan, *J. Indian Chem. Soc.*, **11**, 107 (1934).
[190] Kitasato, *Acta Phytochim.*, **3**, 203 (1927) [*C. A.*, **22**, 1779 (1928)].
[191] Buck, Perkin, and Stevens, *J. Chem. Soc.*, **127**, 1462 (1925).
[192] Kropp and Decker, *Ber.*, **42**, 1184 (1909).
[193] Späth, Kuffner, and Kesztler, *Ber.*, **69**, 378 (1936).
[194] Kitasato, *Acta Phytochim.*, **3**, 195 (1927) [*C. A.*, **22**, 1779 (1928)].
[195] Buck and Perkin, *J. Chem. Soc.*, **125**, 1675 (1924).
[196] Shishido, *Bull. Chem. Soc. Japan*, **12**, 150 (1937) [*C. A.*, **31**, 5802 (1937)].
[197] Shishido, *Bull. Chem. Soc. Japan*, **12**, 419 (1937) [*C. A.*, **32**, 944 (1938)].
[198] Schöpf and Salzer, *Ann.*, **544**, 1 (1940).
[199] Buck and Davis, *J. Am. Chem. Soc.*, **52**, 660 (1930).
[200] Haworth, Perkin, and Rankin, *J. Chem. Soc.*, **125**, 1686 (1924).
[201] Pictet and Gams, *Ber.*, **44**, 2480 (1911).
[202] Pictet and Gams, *Compt. rend.*, **153**, 386 (1911).
[203] Haworth, Perkin, and Rankin, *J. Chem. Soc.*, **127**, 2019 (1925).
[204] Pictet and Finkelstein, *Ber.*, **42**, 1979 (1909); *Arch. sci. phys. nat.*, [4], **29**, 245 [*C. A.*, **4**, 2281 (1910)].
[205] Späth and Burger, *Ber.*, **60**, 704 (1927).
[206] Kakemi, *J. Pharm. Soc. Japan*, **60**, 11 (1940) [*C. A.*, **34**, 3748 (1940)].
[206a] Weijlard, Swanezy, and Tashjian, *J. Am. Chem. Soc.*, **71**, 1889 (1949).
[207] Robinson and Sugasawa, *J. Chem. Soc.*, **1933**, 280.
[208] Schöpf, Jäckh, and Perrey, *Ann.*, **497**, 59 (1932).
[209] Kondo, Narita, and Uyeo, *Ber.*, **68**, 519 (1935).
[210] Goto, Inaba, and Nozaki, *Ann.*, **530**, 142 (1937).
[211] Goto and Shishido, *Ann.*, **539**, 262 (1939).
[212] Marion, *J. Am. Chem. Soc.*, **66**, 1125 (1944).
[213] Govindachari, *Current Sci.*, **10**, 76 (1941).
[214] Späth and Lang, *Monatsh.*, **42**, 273 (1921).
[215] Späth and Meinhard, *Ber.*, **75**, 400 (1942).
[216] Späth and Böhm, *Ber.*, **55**, 2985 (1922).
[217] Callow, Gulland, and Haworth, *J. Chem. Soc.*, **1929**, 658.
[218] Späth and Hromatka, *Ber.*, **61**, 1334 (1928).
[219] Gulland and Haworth, *J. Chem. Soc.*, **1928**, 1132.
[220] Gulland and Haworth, *J. Chem. Soc.*, **1928**, 1834.
[221] Gulland, Ross, and Smellie, *J. Chem. Soc.*, **1931**, 2885.
[222] Barger, Eisenbrand, Eisenbrand, and Schlittler, *Ber.*, **66**, 450 (1933).
[223] Schlittler, *Ber.*, **66**, 988 (1933).
[224] Douglas and Gulland, *J. Chem. Soc.*, **1931**, 2893.
[225] Gulland, Ross, and Virden, *J. Chem. Soc.*, **1931**, 2881.
[226] Sugasawa and Yoshikawa, *J. Pharm. Soc. Japan*, **54**, 305 (1934) [*C. A.*, **29**, 169 (1935)].

[227] Perkin, Rây, and Robinson, *J. Chem. Soc.*, **127**, 740 (1925); Späth and Quietensky, *Ber.*, **58**, 2267 (1925).
[228] Freundler, *Bull. soc. chim. France*, [4], **15**, 465 (1914).
[229] Paul, *Science and Culture*, **1**, 781 (1936) [*C. A.*, **30**, 5990 (1936)].
[230] Paul, *Science and Culture*, **1**, 659 (1936) (*Chem. Zentr.*, **1936**, II, 87).
[231] Dey and Sankaran, *Proc. Natl. Inst. Sci. India*, **6**, 173 (1940).
[232] King and Robinson, *J. Chem. Soc.*, **1938**, 2119.
[233] Sugasawa, Sakurai, Huzisawa, and Sugimoto, *J. Pharm. Soc. Japan*, **60**, 140 (1940) [*C. A.*, **34**, 5086 (1940)].
[234] Sugasawa and Kuriyagawa, *Ber.*, **69**, 2068 (1936).
[235] Clemo, McIlwain, and Morgan, *J. Chem. Soc.*, **1936**, 610.
[236] Alamela and Dey, *Proc. Natl. Inst. Sci. India*, **6**, 195 (1940).
[237] Dey and Alamela, *Arch. Pharm.*, **280**, 245 (1942) [*Chem. Zentr.*, I, 952 (1943)].
[238] Alamela and Dey, *Proc. Natl. Inst. Sci. India*, **7**, 207 (1941).
[239] Alamela and Dey, *Proc. Natl. Inst. Sci. India*, **7**, 215 (1941).
[240] Kindler and Peschke, *Arch. Pharm.*, **270**, 353 (1932).
[241] Kindler and Peschke, Ger. pat. 579,819 [*Frdl.*, **20**, 723 (1933)].
[242] E. Merck, Ger. pat. 279,194 [*Frdl.*, **12**, 758 (1914–1916)].
[243] Ide and Buck, *J. Am. Chem. Soc.*, **62**, 425 (1940).
[244] Clemo and Turnbull, *J. Chem. Soc.*, **1946**, 701.
[245] Sugasawa, Kakemi, and Kazumi, *Proc. Imp. Acad. Tokyo*, **15**, 223 (1939) [*C. A.*, **33**, 8617 (1939)].
[246] Sugasawa, Kakemi, and Kazumi, *Ber.*, **73**, 782 (1940).
[247] Decker and Becker, *Ann.*, **395**, 328 (1913).
[248] Tomita and Watanabe, *J. Pharm. Soc. Japan*, **58**, 783 (1938) [*C. A.*, **33**, 2524 (1939)].
[249] Decker, Ger. pat. 267,699 [*Frdl.*, **11**, 999 (1912–1914)].
[250] Kindler and Giese, *Ann.*, **431**, 228 (1923).
[251] Kindler, *Arch. Pharm.*, **265**, 411 (1927).
[252] Manske and Holmes, *J. Am. Chem. Soc.*, **67**, 95 (1945).
[253] Clemo and Turnbull, *J. Chem. Soc.*, **1945**, 533.
[254] Späth and Gangl, *Monatsh.*, **44**, 103 (1923).
[255] Späth, *Monatsh.*, **42**, 97 (1921).
[256] Späth, *Monatsh.*, **43**, 477 (1923).
[257] Wolfes and Dobrowsky, Ger. pat. 549,967 [*Frdl.*, **19**, 1113 (1932)].
[258] E. Merck, Brit. pat. 374,627 [*C. A.*, **27**, 3947 (1933)].
[259] Ges. Chem. Ind. Basel, Swiss pat. 92,610 [*Chem. Zentr.*, II, 574 (1923)].
[260] Ges. Chem. Ind. Basel, Ger. pat. 399,805 [*Frdl.*, **14**, 1313 (1925–1926)].
[261] Hartmann and Kägi, U. S. pat. 1,437,802 [*C. A.*, **17**, 854 (1923)].
[262] Ges. Chem. Ind. Basel, Swiss pat. 92,004 [*Chem. Zentr.*, II, 574 (1923)].
[263] Wolfes, U. S. pat. 1,941,647 [*C. A.*, **28**, 1717 (1934)].
[264] Ges. Chem. Ind. Basel, Swiss pat. 92,611 (1920) [*Chem. Zentr.*, II, 574 (1923)].
[265] Sugasawa and Sugimoto, *J. Pharm. Soc. Japan*, **61**, 62 (1941) [*C. A.*, **36**, 92 (1942)].
[266] E. Merck, Brit. pat. 348,956 [*C. A.*, **26**, 1944 (1932)].
[267] Wolfes, Ger. pat. 550,122 [*Frdl.*, **17**, 2310 (1930)].
[268] Ges. Chem. Ind. Basel, Swiss pat. 92,612 [*Chem. Zentr.*, II, 574 (1923)].
[269] Chakravarti and Ganapati, *J. Annamalai Univ.*, **3**, 208 (1934) [*C. A.*, **29**, 1094 (1935)].
[270] Robinson, *J. Chem. Soc.*, **95**, 2167 (1909).
[271] Robinson and Sugasawa, *J. Chem. Soc.*, **1931**, 3163.

TABLE VIII
Isoquinolines

Substituents	Condensing Agent	Temperature °C.	Yield %	Reference
A. *From β-hydroxyethylamides*				
None	P_2O_5	110	21	5
1-Methyl-	P_2O_5	140	—	272, 5
1-Phenyl-	P_2O_5	140	60	5
	P_2O_5	110	81	86
	$P_2O_5 + POCl_3$	140	91	86
1-Benzyl-	P_2O_5	140	4	273
	P_2O_5	80	Poor	270
	P_2O_5	140	40	5
4-Phenyl-	P_2O_5	110	—	87
	P_2O_5	140	35	7
1,3-Dimethyl-	P_2O_5	110	37	86
1-Methyl-4-phenyl-	P_2O_5	140	80	87
	P_2O_5	110	82	7
1-*n*-Propyl-3-methyl-	$P_2O_5 + POCl_3$	140	35	86
1-Phenyl-3-methyl-	P_2O_5	205	35	86
	$POCl_3$	140	45	86
	$P_2O_5 + POCl_3$	140	50	86
1-Phenyl-3-ethyl-	$P_2O_5 + POCl_3$	140	26	86
1-Phenyl-3-*n*-propyl-	$P_2O_5 + POCl_3$	140	20	86
1-Phenyl-3-butyl-	$P_2O_5 + POCl_3$	140	1	86
1,3-Diphenyl-	P_2O_5	140	20	86
1-Phenyl-4-ethyl-	P_2O_5	140	5–10	86
1,4-Diphenyl-	$P_2O_5 + POCl_3$	140	0	86
	P_2O_5	110	80	87
1-Benzyl-3-methyl-	P_2O_5	110	10	86
	$P_2O_5 + POCl_3$	140	16	86
	P_2O_5	205	20	86
1-Methyl-3,4-diphenyl-	P_2O_5	110	—	87
	P_2O_5	110	58	7
1,3,4-Triphenyl-	P_2O_5	110	21	7
1-Phenyl-3-methyl-7-methoxy-	—	—	—	274
1-(*p*-Methoxyphenyl)-3-methyl-7-methoxy-	—	—	—	274
1-(*p*-Ethoxyphenyl)-3-methyl-7-methoxy-	—	—	—	274
1-(3,4-Methylenedioxyphenyl)-3-methyl-7-methoxy-	—	—	—	274

TABLE VIII—Continued

Isoquinolines

Substituents	Condensing Agent	Temperature °C.	Yield %	Reference
1-(3,4-Dimethoxyphenyl)-3-methyl-7-methoxy-	—	—	—	274
1-Phenyl-6,7-dimethoxy-	$POCl_3$	110	—	275
1-(3,4-Dimethoxyphenyl)-6,7-methylenedioxy-	$POCl_3$	—	—	93
1-(2,3-Dimethoxybenzyl)-5,6-dimethoxy-	$POCl_3$	140	77	62
1-Veratryl-6,7-dimethoxy-	P_2O_5	140	—	276
	P_2O_5	140	30	4
	$POCl_3$	80	60–65	277
	$POCl_3$	60	70–75	277
	$POCl_3$	60–80	75	278, 279, 280
1-Veratryl-6,7-diethoxy-	$POCl_3$	60	—	281, 282
1-(3,4-Diethoxybenzyl)-6,7-dimethoxy-	$POCl_3$	80	—	281
	PCl_5	80	—	280
1-(3,4-Diethoxybenzyl)-6,7-diethoxy-	$POCl_3$	60–80	—	280
1-(3,5-Diethoxybenzyl)-6,8-dimethoxy-	PCl_5	80	75	278
1-(3,5-Diethoxybenzyl)-6,8-diethoxy-	$POCl_3$	60–80	—	278
1-(Diethoxybenzyl)dimethoxy-	PCl_5	80	75	279
1-(Diethoxybenzyl)diethoxy-	$POCl_3$	60–80	75	279
1,3-Dimethyl-6,7-methylenedioxy-	$POCl_3$	60	—	120
1,3-Dimethyl-6,7-dimethoxy-	$POCl_3$	60	77	120
1,3-Dimethyl-6-methoxy-7-ethoxy-	$POCl_3$	110	75	283
1,3-Dimethyl-6-methoxy-7-benzyloxy-	$POCl_3$	110	57	284
	$POCl_3$	60	69	284
1,3-Dimethyl-6,7-diethoxy-	$POCl_3$	60	74–92	11
1,3-Dimethyl-6,7-dibenzyloxy-	$POCl_3$	110	50	285
1-Cyclohexylmethyl-3-methyl-6,7-ethylenedioxy-	$POCl_3$	80	—	127, 129
1-Δ^1-Cyclohexenyl-3-methyl-6,7-dimethoxy-	$POCl_3$	60	—	127, 129, 128
1-Δ^1-Cyclohexenyl-3-cyclohexyl-6,7-dimethoxy-	$POCl_3$	80	—	127, 128, 129

TABLE VIII—Continued

Isoquinolines

Substituents	Condensing Agent	Temperature °C.	Yield %	Reference
1-Phenyl-3-methyl-6,7-methylenedioxy-	POCl$_3$	110	—	286, 287
1-Phenyl-3-methyl-6,7-dimethoxy-	POCl$_3$	110	—	288
1-Phenyl-3-methyl-6,7-diethoxy-	POCl$_3$	110	72–82	11
1-(p-Methoxyphenyl)-3-methyl-6,7-methylenedioxy-	POCl$_3$	140	—	288
1-(p-Methoxyphenyl)-3-methyl-6,7-dimethoxy-	POCl$_3$	120	—	288
1-(3,4-Methylenedioxyphenyl)-3-methyl-6,7-methylenedioxy-	POCl$_3$	120	9	286, 287
1-(3,4-Methylenedioxyphenyl)-3-methyl-6,7-dimethoxy-	POCl$_3$	120	—	288, 289
1-(2,4-Dimethoxyphenyl)-3-methyl-7,8-dimethoxy-	POCl$_3$	110	76	30
1-(3,4-Dimethoxyphenyl)-3-methyl-6,7-methylenedioxy-	POCl$_3$	110	—	288
1-(3,4-Dimethoxyphenyl)-3-methyl-6,7-dimethoxy-	POCl$_3$	110	—	288, 289
1-(3,4-Dimethoxyphenyl)-3-methyl-6,7-diethoxy-	POCl$_3$	110	65–80	11
1-(3,4-Diethoxyphenyl)-3-methyl-6,7-diethoxy-	POCl$_3$	110	50–63	11
1-(2-Benzyloxy-4-methoxyphenyl)-3-methyl-6,7-dimethoxy-	POCl$_3$	110	75	31
1-(2-Carbethoxyoxy-4-methoxyphenyl)-3-methyl-7,8-dimethoxy-	POCl$_3$	110	>1.5	30
1-(2,3,4-Trimethoxyphenyl)-3-methyl-7,8-dimethoxy-	POCl$_3$	110	94	30
1-(2,4,5-Trimethoxyphenyl)-3-methyl-6,7-dimethoxy-	—	—	—	289
1-(3,4,5-Trimethoxyphenyl)-3-methyl-6,7-methylenedioxy-	POCl$_3$	115	—	288
1-(3,4,5-Trimethoxyphenyl)-3-methyl-6,7-dimethoxy-	POCl$_3$	110	—	288, 289
1-(3,4,5-Triethoxyphenyl)-3-methyl-6,7-dimethoxy-	POCl$_3$	115	—	288, 289
1-Benzyl-3-methyl-6,7-methylenedioxy-	POCl$_3$	110	62	286, 287

TABLE VIII—Continued

ISOQUINOLINES

Substituents	Condensing Agent	Temperature °C.	Yield %	Reference
1-Benzyl-3-methyl-6,7-dimethoxy-	POCl₃	110	—	288, 289, 275
	POCl₃	140	—	290
1-Benzyl-3-methyl-6,7-diethoxy-	POCl₃	110	73	11
1-Piperonyl-3-methyl-6,7-methylenedioxy-	POCl₃	110	39	286, 287
1-Piperonyl-3-methyl-6,7-dimethoxy-	POCl₃	110	—	288, 289
1-Veratryl-3-methyl-6,7-methylenedioxy-	POCl₃	110	46	286
1-Veratryl-3-methyl-6,7-dimethoxy-	POCl₃	120	—	288, 289
1-(3,4-Diethoxybenzyl)-3-methyl-6,7-diethoxy-	POCl₃	110	80–90	11
B. *From β-methoxyethylamides*				
1-Phenyl-3-methyl-	P₂O₅	205	—	291
1-Cyclohexyl-6,7-dimethoxy-	POCl₃	60	—	127, 128, 129
1-Cyclohexyl-6,7-ethylenedioxy-	POCl₃	80	—	127, 128, 129
1-Cyclohexyl-6,7-diethoxy-	POCl₃	80	—	127, 128, 129
1-Cyclohexylmethyl-6,7-dimethoxy-	POCl₃	140	—	127, 128, 129
1-Δ¹-Cyclohexenyl-6,7-dimethoxy-	POCl₃	60	—	127, 128, 129
1-Phenyl-6,7-methylenedioxy-	POCl₃	140	30–40	12
1-Phenyl-6,7-dimethoxy-	P₂O₅	110	—	13
1-(3,4,5-Trimethoxyphenyl)-6,7-dimethoxy-	SiCl₄	110	50–80	115, 116
1-(3,4,5-Triethoxyphenyl)-6,7-methylenedioxy-	PCl₅	130	50–80	115, 116
1-(3,4,5-Triethoxyphenyl)-6,7-dimethoxy-	POCl₃	80	50–80	115, 116
1-Benzyl-6,7-methylenedioxy-	POCl₃	80	50	12
1-Benzyl-6,7-dimethoxy-	POCl₃	140	60	12
1-Piperonyl-6,7-methylenedioxy-	POCl₃	140	30	12

TABLE VIII—Continued

Isoquinolines

Substituents	Condensing Agent	Temperature °C.	Yield %	Reference
1-Piperonyl-6,7-dimethoxy-	$POCl_3$	140	40	12
1-Veratryl-6,7-methylenedioxy-	$POCl_3$	140	—	12
1-Veratryl-6,7-dimethoxy-	P_2O_5	110	7	13
	$POCl_3$	140	40	12
1-(3-Benzyloxy-4-methoxybenzyl)-6,7-dimethoxy-	$POCl_3$	110	—	271
1-Phenyl-3-methyl-6,7-methylenedioxy-	—	—	75	292, 293
1-Phenyl-3-methyl-6,7-dimethoxy-	$POCl_3$	140	60	294
1-Phenyl-3-methyl-6-methoxy-7-benzyloxy-	$POCl_3$	110	65	295
1-(3,4-Methylenedioxyphenyl)-3-methyl-6,7-methylenedioxy-	$POCl_3$	110	85	293, 179
1-Piperonyl-3-methyl-6,7-methylenedioxy-	$POCl_3$	110	70	291
	—	—	Good	292
1-(Piperidinomethyl)-3-methyl-6,7-methylenedioxy-	$POCl_3$	80	85	233
1-Phenyl-3-methyl-6,7,8-trimethoxy-	$POCl_3$	140	80	294, 179
1-(3,4,5-Trimethoxyphenyl)-3-methyl-6,7,8-trimethoxy-	$POCl_3$	140	90	294, 179
1-Benzyl-5,8-dimethoxy-6,7-methylenedioxy-	$POCl_3$	80	13	14
1-Piperonyl-5,8-dimethoxy-6,7-methylenedioxy-	$POCl_3$	80	17	14
1-Veratryl-5,8-dimethoxy-6,7-methylenedioxy-	$POCl_3$	80	18	14
C. *From styrylamides*				
None (from oxime)	P_2O_5	—	2	16, 17
(from oxime)	P_2O_5	100	10	296, 297
1-Phenyl-	Al_2O_3	195	—	13
1,4-Dimethyl-	P_2O_5	110	>50	298
1,3-Diphenyl-	P_2O_5	110	51	298
1-Phenyl-4-methyl-	P_2O_5	110	93	8
1,4-Diphenyl-	P_2O_5	110	86	6
1-Benzyl-4-methyl-	P_2O_5	110	27	298

TABLE VIII—*Continued*

ISOQUINOLINES

Substituents	Condensing Agent	Temperature °C.	Yield %	Reference

D. From N-acylphenylalanine

$$\underset{\underset{\underset{CH_3}{|}}{CO}}{\overset{CH_2}{\underset{NH}{\overset{|}{C}H-CO_2H}}} \xrightarrow[-CO_2\ -H_2]{-H_2O} \underset{CH_3}{\bigodot{N}}$$

Substituents	Condensing Agent	Temperature °C.	Yield %	Reference
1-Methyl-	Polyphosphoric acid + POCl$_3$	125	2	65

[272] Mills and Smith, *J. Chem. Soc.*, **121**, 2724 (1922).
[273] Forsyth, Kelly, and Pyman, *J. Chem. Soc.*, **127**, 1659 (1925).
[274] Bruckner and Bodnár, *Magyar Biol. Kutatóintézet Munkái*, **15**, 404 (1943) [*C. A.*, **42**, 172 (1948)].
[275] Vinkler and Bruckner, *Magyar Chem. Folyóirat*, **45**, 147 (1939) [*C. A.*, **34**, 3747 (1940)].
[276] Pictet and Gams, *Compt. rend.*, **149**, 210 (1909).
[277] Kereszty and Wolf, Ger. pat. 613,830 [*Frdl.*, **20**, 812 (1933)].
[278] Wolf, Brit. pat. 380,874 [*C. A.*, **27**, 3948 (1933)].
[279] Kereszty and Wolf, Fr. pat. 719,638 [*C. A.*, **26**, 3806 (1932)].
[280] Wolf, U. S. pat. 1,962,224 [*C. A.*, **28**, 4841 (1934)].
[281] Chemische-Pharmazeutische A.-G. Bad Homburg, Ger. pat. 574,656 [*Frdl.*, **19**, 1116 (1932)].
[282] Wolf, Hung. pat. 108,865 [*Chem. Zentr.*, I, 1900 (1935)].
[283] Bruckner, Kovács, and Kovács, *Ber.*, **77**, 610 (1944).
[284] Fodor, *Ber.*, **76**, 1216 (1943).
[285] Bruckner and Fodor, *Ber.*, **76**, 466 (1943).
[286] Bruckner and Krámli, *J. prakt. Chem.*, **145**, 291 (1936).
[287] Bruckner and Krámli, *Magyar Chem. Folyóirat*, **43**, 23 (1937) [*C. A.*, **31**, 6238 (1937)].
[288] Bruckner and Fodor, *Ber.*, **71**, 541 (1938).
[289] Fodor, *Acta Lit. Sci. Regiae Univ. Hung. Francisco-Josephinae, Sect. Chem., Mineral. Phys.*, **6**, 1 (1937) [*C. A.*, **32**, 2124 (1938)].
[290] Vinkler and Bruckner, *J. prakt. Chem.*, **151**, 17 (1938).
[291] Wolfes and Dobrowsky, Ger. pat. 556,709 [*Frdl.*, **19**, 1114 (1932)].
[292] Keimatsu, *J. Pharm. Soc. Japan*, **53**, 1070 (1933) [*C. A.*, **29**, 7989 (1935)].
[293] Sugasawa and Sakurai, *J. Pharm. Soc. Japan*, **56**, 563 (1936) [*C. A.*, **33**, 9307 (1939)].
[294] Sugasawa and Kakemi, *J. Pharm. Soc. Japan*, **57**, 172 (1937) [*C. A.*, **33**, 9307 (1939)].
[295] Sugasawa, *J. Pharm. Soc. Japan*, **58**, 265 (1938) [*C. A.*, **32**, 5402 (1938)].
[296] Goldschmidt, *Ber.*, **27**, 2795 (1894).
[297] Goldschmidt, *Ber.*, **28**, 818 (1895).
[298] Krabbe, Schmidt, and Eisenlohr, *Ber.*, **74**, 1905 (1941).

Supplement to Table VIII

Amides That Could Not Be Cyclized

Name	Condensing Agent	Temperature °C.	Reference
N-Phenylacetyl-2-phenyl-2-methoxyethylamine	$POCl_3$	140	12
N-(3,4-Methylenedioxyphenylacetyl)-2-phenyl-2-methoxyethylamine	$POCl_3$	140	12
N-(α-Pyridylacetyl)-2-hydroxy-2-phenethylamine	P_2O_5	205	42
N-Benzoyl-1-hydroxy-1-phenyl-2-aminoöctane	P_2O_5	205	86
N-Benzoyl-2-hydroxy-2-phenyl-3-aminobutane	$POCl_3 + P_2O_5$	140	86
N-(2-Carbomethoxybenzoyl)-2-phenyl-2-methoxyethylamine	$POCl_3$	140	299
N-(2-Carbomethoxybenzoyl)-2-(3,4-methylenedioxyphenyl)-2-methoxyethylamine	$POCl_3$	140	299
N-Phenylacetyl-2-(2,4-dimethoxyphenyl)-2-methoxyethylamine	Various	—	14
N-(3,4-Methylenedioxyphenylacetyl)-2-(2,4-dimethoxyphenyl)-2-methoxyethylamine	Various	—	14
N-(3,4-Dimethoxyphenylacetyl)-2-(2,4-dimethoxyphenyl)-2-methoxyethylamine	Various	—	14
N-(α-Pyridylacetyl)-2-(3,4-dimethoxyphenyl)-2-hydroxyethylamine	P_2O_5 or $POCl_3$	—	42
Benzylideneacetophenone oxime	P_2O_5	—	297

[299] Mannich and Walther, *Arch. Pharm.*, **265**, 11 (1927).

TABLE IX
BENZISOQUINOLINES

Substituents	Condensing Agent	Temperature °C.	Yield %	Reference

A. *Phenanthridines*

Substituents	Condensing Agent	Temperature °C.	Yield %	Reference
None	ZnCl$_2$	Melt	—	113
	POCl$_3$	—	0	66
	ZnCl$_2$	220–280	42	300
9-Hydroxy-	ZnCl$_2$	Melt	—	113
	ZnCl$_2$	250	29	300
(from isocyanate)	AlCl$_3$	80	78	24
9-Methyl-	POCl$_3$	110	—	301
	ZnCl$_2$	250–300	—	66
	ZnCl$_2$	Melt	—	113
	POCl$_3$	110	70	66
9-Chloromethyl-	POCl$_3$	110	80	66, 301
9-Phenoxymethyl-	POCl$_3$	110	65	101
9-Ethyl-	ZnCl$_2$	Melt	—	113
	POCl$_3$	110	80	66, 301
9-(γ-Carbethoxypropyl)-	POCl$_3$	110	64	101
9-(δ-Carbethoxybutyl)-	POCl$_3$	110	68	101
9-Carbethoxy-	POCl$_3$	110	20	100
9-Phenyl-	ZnCl$_2$	Melt	—	113
	POCl$_3$	110	75	66, 301
9-(2,4,6-Trimethylphenyl)-	POCl$_3$	110	—	101
9-(o-Nitrophenyl)-	POCl$_3$	110	74	66, 301
9-(m-Nitrophenyl)-	POCl$_3$	110	61	66, 301
9-(p-Nitrophenyl)-	POCl$_3$	110	65	66, 301
9-(3,5-Dinitrophenyl)-	POCl$_3$	180	Good	302, 121
9-(o-Carboxyphenyl)-	NaCl + AlCl$_3$	180	—	132, 133, 134
	ZnCl$_2$	275	77	303
9-(α-Naphthyl)-	POCl$_3$	110	—	101
9-Benzyl-	POCl$_3$	110	20	101
9-(p-Nitrobenzyl)-	POCl$_3$	110	67	32
9-Phenethyl-	POCl$_3$	110	70	101

TABLE IX—Continued

BENZISOQUINOLINES

Substituents	Condensing Agent	Temperature °C.	Yield %	Reference
9-Styryl-	POCl$_3$	110	12	101
9-(3-Pyridyl)-	POCl$_3$	110	0	304
	POCl$_3$	180	72	304
9,9′-Tetramethylene-bis-	POCl$_3$	110	3	101
1-Nitro-9-methyl-	POCl$_3$	140	60	304a
2,9-Dimethyl-	POCl$_3$	—	Good	67
	POCl$_3$	110	79	305
2-Carbethoxyamino-9-methyl-	POCl$_3$	110	81	32
3-Bromo-9-methyl-	POCl$_3$	110	89	306
3-Cyano-9-methyl-	POCl$_3$	110	31	307
3-Nitro-9-methyl-	POCl$_3$	110	80	301, 300
6-Carbethoxyamino-9-methyl- and 8-Carbethoxyamino-9-methyl-	POCl$_3$	110	70 10	32
7-Carbethoxyamino-9-methyl-	POCl$_3$	110	85	68
	POCl$_3$	110	96	308, 309
7-Benzamido-9-methyl-	POCl$_3$	—	62	67
7-Nitro-9-methyl-	POCl$_3$	—	Poor	301, 300
	POCl$_3$	—	4	67
2-Methyl-9-phenyl-	POCl$_3$	110	98	305
7-Nitro-9-phenyl-	POCl$_3$	110	23	71
	POCl$_3$	110	24	70
	POCl$_3$	180	45	71
	POCl$_3$	180	99	121
3-Bromo-9-(p-bromophenyl)-	POCl$_3$	180	100	307
3-Nitro-9-(o-nitrophenyl)-	POCl$_3$	—	Poor	300
7-Carbethoxyamino-9-(o-nitrophenyl)-	POCl$_3$	100	>60	309, 308, 68
3-Nitro-9-(m-nitrophenyl)-	POCl$_3$	180	Good	121, 310
7-Nitro-9-(m-nitrophenyl)-	POCl$_3$	180	82–87	121, 310
2-Carbethoxyamino-9-(p-nitrophenyl)-	POCl$_3$	110	59	32
3-Nitro-9-(p-nitrophenyl)-	POCl$_3$	110	61	70, 302
6-Carbethoxyamino-9-(p-nitrophenyl)- and 8-Carbethoxyamino-9-(p-nitrophenyl)-	POCl$_3$	130	33 50	32
7-Carbethoxyamino-9-(p-nitrophenyl)-	POCl$_3$	150	62	308, 309, 68
	POCl$_3$	110	ca. 100	32
7-Nitro-9-(p-nitrophenyl)-	POCl$_3$	110	30	70, 302
	POCl$_3$	180	95	302, 71
7-Nitro-9-(3,5-dinitrophenyl)-	POCl$_3$	180	—	302, 71

TABLE IX—Continued

BENZISOQUINOLINES

Substituents	Condensing Agent	Temperature °C.	Yield %	Reference
3-Nitro-9-(3-pyridyl)-	POCl$_3$	110	0	304
	POCl$_3$	180	94	304
7-Nitro-9-(3-pyridyl)-	POCl$_3$	110	0	304
	POCl$_3$	180	37	304
2,7-Dicarbethoxyamino-9-methyl-	POCl$_3$	110	>70	309, 308, 68
2,7-Dibromo-9-methyl-	POCl$_3$	110	—	311
3,7-Dinitro-9-(t-butyl)-	POCl$_3$	180	32	71
2,7-Dicarbethoxyamino-9-phenyl-	POCl$_3$	110	69	309, 312, 68
2,7-Dibromo-9-phenyl-	POCl$_3$	210	98	311
2,7-Dinitro-9-phenyl-	POCl$_3$	180	50	121, 312
	POCl$_3$	180	66	313
3,7-Dinitro-9-phenyl-	POCl$_3$	110	Poor	70
	POCl$_3$	180	58	302, 71, 121
4,5-Dimethyl-9-phenyl-	POCl$_3$	110	—	314
3-Bromo-7-nitro-9-(p-nitrophenyl)-	POCl$_3$	180	Good	121
	POCl$_3$	110	Good	302
3,7-Dinitro-9-(p-nitrophenyl)-	POCl$_3$	110	0	71
	POCl$_3$	180	>30	71
4,5-Dimethyl-9-(p-nitrophenyl)-	POCl$_3$	180	74	314
2,3-Dimethyl-6,7-methylenedioxy-1,4,11,12-tetrahydro-	POCl$_3$	140	70	315
2,3,6,7-Tetramethoxy-	POCl$_3$	110	0	69
	P$_2$O$_5$	140	3	69
2,3,6,7-Tetramethoxy-9-methyl-	POCl$_3$	110	85	69
2,3,6,7-Tetramethoxy-9-ethyl-	POCl$_3$	110	85	69
2,3-Dimethyl-6,7-dimethoxy-9-phenyl-1,4,11,12-tetrahydro-	POCl$_3$	140	93	315
2,3,6,7-Tetramethoxy-9-phenyl-	POCl$_3$	110	90	69

B. *3,3a,5,6-Tetrahydro-4H-benz[d,e]isoquinolines*

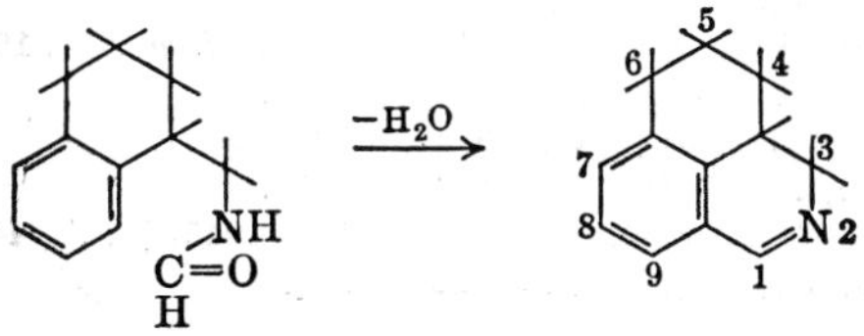

TABLE IX—Continued

BENZISOQUINOLINES

Substituents	Condensing Agent	Temperature °C.	Yield %	Reference
None	P_2O_5	110	26	316
1-Methyl-	P_2O_5	110	61	316
1-Ethyl-	P_2O_5	110	59	316
1-Phenyl-	P_2O_5	110	12	316
1-Benzyl-	P_2O_5	110	48	316
C. *ar-Benzisoquinolines*				
1-Methyl-5,6-benz- (from hydroxyamide)	P_2O_5	140	24	317
1-Methyl-3,4-dihydro-5,6-benz- (from oxime)	P_2O_5	110	22	318
1-Phenyl-3,4-dihydro-5,6-benz-	$POCl_3$	140	—	36, 319
1-Phenyl-3-methyl-5,6-benz- (from hydroxyamide)	P_2O_5 + $POCl_3$	140	12	86
1-Methyl-3,4-dihydro-6,7-benz-	$POCl_3$	140	55	36
1-*n*-Propyl-3,4-dihydro-6,7-benz-	$POCl_3$	140	50	36
1-Cyclohexyl-3,4-dihydro-6,7-benz-	$POCl_3$	140	—	36
1-Phenyl-3,4-dihydro-6,7-benz-	$POCl_3$	110	56	319
	$POCl_3$	140	56	36
1-(3,4-Diethoxyphenyl)-3,4-dihydro-6,7-benz-	$POCl_3$	140	83	36, 319
5,8-Diphenyl-3,4-dihydro-6,7-benz-	$POBr_3$	—	—	320
1-Benzyl-1′,2′,3,3′,4,4′-hexahydro-6,7-benz-	P_2O_5	110	78	38
1-Methyl-3,4-dihydro-7,8-benz- (from oxime)	P_2O_5	110	—	318

[300] Morgan and Walls, *J. Chem. Soc.*, **1932**, 2225.
[301] Morgan and Walls, Brit. pat. 372,859 [*C. A.*, **27**, 3483 (1933)].
[302] Morgan and Walls, Brit. pat. 511,353 [*C. A.*, **34**, 6020 (1940)].
[303] Koelsch, *J. Am. Chem. Soc.*, **58**, 1325 (1936).
[304] Petrow and Wragg, *J. Chem. Soc.*, **1947**, 1410.
[304a] Stepan and Hamilton, *J. Am. Chem. Soc.*, **71**, 2438 (1949).
[305] Ritchie, *J. Proc. Roy. Soc. N. S. Wales*, **78**, 169 (1945) [*C. A.*, **40**, 880 (1946)].
[306] Walls, *J. Chem. Soc.*, **1935**, 1405.
[307] Barber, Gregory, Major, Slack, and Woolman, *J. Chem. Soc.*, **1947**, 84.
[308] Walls, Brit. pat. 578,226 [*C. A.*, **41**, 2449 (1947)].
[309] Walls, U. S. pat. 2,397,391 [*C. A.*, **40**, 4086 (1946)].
[310] Walls, Brit. pat. 577,990 [*C. A.*, **41**, 2449 (1947)].
[311] Ritchie, *J. Proc. Roy. Soc. N. S. Wales*, **78**, 141 (1945) [*C. A.*, **40**, 876 (1946)].
[312] Walls, Brit. pat. 587,673 [*C. A.*, **42**, 622 (1948)].

Supplement to Table IX
Amides That Could Not Be Cyclized

Name	Condensing Agent	Temperature °C.	Reference
2-Formamido-4'-nitrobiphenyl	POCl₃	—	300
2-Formamido-5-nitrobiphenyl	POCl₃	—	300
2-Acetamido-4'-tosylamidobiphenyl	POCl₃	—	67
2-Dichloroacetamidobiphenyl	POCl₃	—	100
2-Trichloroacetamidobiphenyl	POCl₃	—	100
N-(2-Xenyl)-6-oxamic acid	POCl₃	—	100
2-Crotonamidobiphenyl	POCl₃	—	101
2-Acetoacetamidobiphenyl	POCl₃	—	101
2-(β-Carbomethoxy)propionamidobiphenyl	POCl₃	110	101
2-(β-Carboxy)propionamidobiphenyl	POCl₃	110	101
2-(β-Carboxy)acrylamidobiphenyl	POCl₃	—	101
2-(γ-Carboxy)butyramidobiphenyl	POCl₃	—	101
N,N'-Bis(o-xenyl)glutardiamide	POCl₃	110	101
1-Acetamidomethyl-2-methoxynaphthalene	POCl₃	110	53
	P₂O₅	140	53
1-Acetamidomethyl-2-acetoxynaphthalene	POCl₃	110–140	53
1-Acetamidomethyl-4-methoxynaphthalene	POCl₃	110–140	53
1-α-Acetamidoethyl-4-methoxynaphthalene	POCl₃	—	53
1-Benzamidomethyl-2-benzoyloxynaphthalene	POCl₃	—	53
1-(N-Acetylglycyl)naphthalene	POCl₃	110	53, 54
2-(N-Acetylglycyl)naphthalene	POCl₃	110	53, 54
1-(N-Acetylglycyl)-4-methoxynaphthalene	POCl₃	110	53, 54
1-Hippuryl-4-methoxynaphthalene	POCl₃	110	53, 54
N-Acetyl-β-hydroxy-β-(α-naphthyl)ethylamine	P₂O₅	140	53
N-Acetyl-β-hydroxy-β-(β-naphthyl)ethylamine	P₂O₅	140	53
	PCl₅	—	53
N-Formyl-β-(9-phenanthryl)ethylamine	—	—	76
N-Formyl-β-methoxy-β-(9-phenanthryl)ethylamine	—	—	76
N-Benzoyl-β-methoxy-β-(9-phenanthryl)ethylamine	—	—	76

[313] Ritchie, *J. Proc. Roy. Soc. N. S. Wales*, **78**, 177 (1945) [*C. A.*, **40**, 881 (1946)].
[314] Ritchie, *J. Proc. Roy. Soc. N. S. Wales*, **78**, 159 (1945) [*C. A.*, **40**, 879 (1946)].
[315] Sugasawa and Kodama, *Ber.*, **72**, 675 (1939).
[316] Späth and Kittel, *Ber.*, **73**, 478 (1940).
[317] Pictet and Manevitch, *Arch. sci. phys. nat.*, **35**, 40 [*C. A.*, **7**, 1713 (1913)].
[318] Gibson, Hariharan, Menon, and Simonsen, *J. Chem. Soc.*, **1926**, 2247.
[319] Kindler, Peschke, and Plüddemann, *Arch. Pharm.*, **277**, 25 (1939).
[320] Etienne and Robert, *Compt. rend.*, **223**, 331 (1946).

TABLE X

Naphthisoquinolines

Name	Condensing Agent	Temperature °C.	Yield %	Reference
6-Methylnaphth[1,2-c]isoquinoline	POCl₃	110	90	321
2,3,8,9-Tetramethoxy-4b,10b,11,12-tetrahydronaphth[1,2-c]isoquinoline	POCl₃	110	63	322
11-Methyl-5,6,8,9-tetrahydronaphth[2,1-g]isoquinoline	POCl₃	110	30	77
3-Methoxy-	POCl₃	110	28	77

Supplement to Table X

Amides That Could Not Be Cyclized

Name	Condensing Agent	Temperature °C.	Reference
N-Formyl-β-(3-phenanthryl)ethylamine	—	—	76
N-Formyl-β-methoxy-β-(3-phenanthryl)ethylamine	—	—	76

[321] Ritchie, *J. Proc. Roy. Soc. N. S. Wales*, **78**, 173 (1945) [*C. A.*, **40**, 880 (1946)].
[322] Richardson, Robinson, and Seijo, *J. Chem. Soc.*, **1937**, 835.

TABLE XI

BENZOQUINOLIZINES

Name	Condensing Agent	Temperature °C.	Yield %	Reference
8,9-Dimethoxy-6,7-dihydrobenzo[a]quinolizinium chloride	$POCl_3$	80	65	33
8,11-Dimethoxy-6,7-dihydrobenzo[a]quinolizinium chloride	$POCl_3$	80	60	33, 179
9,10-Methylenedioxy-6,7-dihydrobenzo[a]quinolizinium chloride	$POCl_3$	140	—	323
	$POCl_3$	140	69	324
8-Methyl-10,11-dimethoxy-6,7-dihydrobenzo[a]quinolizinium chloride	$POCl_3$	80	92	33
9,10-Dimethoxy-1,2,3,4,6,7-hexahydrobenzo[a]quinolizinium chloride	$POCl_3$	110	—	325
6-Methyl-9,10-methylenedioxy-1,2,3,4,6,7-hexahydrobenzo[a]quinolizinium chloride	$POCl_3$	110	—	325
1-Methyl-3-carbethoxy-9,10-dimethoxy-1,2,3,4,6,7-hexahydrobenzo[a]quinolizinium chloride	$POCl_3$	110	63	326, 179

[323] Sugasawa and Sugimoto, *Proc. Imp. Acad. Tokyo*, **15**, 49 (1939) [*C. A.*, **33**, 5401 (1939)].
[324] Sugasawa and Sugimoto, *Ber.*, **72**, 977 (1939).
[325] Sugasawa, Sakurai, and Sugimoto, *Proc. Imp. Acad. Tokyo*, **15**, 82 (1939) [*C. A.*, **33**, 6318 (1939)].
[326] Sugasawa, Sakurai, and Okayama, *Ber.*, **74**, 537 (1941).

TABLE XII

Dibenzoquinolizines

Name	Condensing Agent	Temperature °C.	Yield %	Reference
9,10-Methylenedioxy-6,7-dihydro-dibenzo[a,f]quinolizinium chloride	POCl$_3$	140	—	323
	POCl$_3$	140	36	324
2,3,9,10-Tetramethoxy-6,7,12,13-tetra-hydrodibenzo[a,f]quinolizinium chloride	POCl$_3$	140	—	327
	POCl$_3$	110	—	328
	POCl$_3$	140	40	329
2,3-Ethylenedioxy-9,10-dimethoxy-6,7-dihydrodibenzo[a,f]quinolizinium chloride	POCl$_3$	110	—	330
5,6-Dihydro-8H-dibenzo[a,g]quinolizine	POCl$_3$	110	>10	43
	P$_2$O$_5$	205	>38	136
	P$_2$O$_5$	205	>50	331
8-Hydroxy-5,6,13,13a-tetrahydro-8H-dibenzo[a,g]quinolizine	P$_2$O$_5$	205	—	332
2,3-Methylenedioxy-5,6-dihydro-8H-dibenzo[a,g]quinolizine	POCl$_3$	110	—	56
3,10-Dimethoxy-5,6-dihydro-8H-dibenzo[a,g]quinolizine	POCl$_3$	110	—	143
	POCl$_3$ or P$_2$O$_5$	—	20	144
3,11-Dimethoxy-5,6-dihydro-8H-dibenzo[a,g]quinolizine	POCl$_3$	110	66	122
3-Methoxy-8-oxo-5,6-dihydro-8H-dibenzo[a,g]quinolizine (after reduction)	POCl$_3$	100	23	333

TABLE XII—Continued

Dibenzoquinolizines

Name	Condensing Agent	Temperature °C.	Yield %	Reference
2,3-Methylenedioxy-8-oxo-5,6-dihydro-8H-dibenzo[a,g]quinolizine	POCl$_3$	110	—	56
3,10-Dimethoxy-8-oxo-5,6-dihydro-8H-dibenzo[a,g]quinolizine (after reduction)	POCl$_3$	100	67	334
2,3,10,11-Bis(methylenedioxy)-5,6-dihydro-8H-dibenzo[a,g]quinolizine (?)	PCl$_5$	25	25	335
2,3-Methylenedioxy-10,11-dimethoxy-5,6-dihydro-8H-dibenzo[a,g]quinolizine	POCl$_3$	110	—	35
2,3-Methylenedioxy-11,12-dimethoxy-5,6-dihydro-8H-dibenzo[a,g]quinolizine	POCl$_3$	110	>50	189
2,3,11,12-Tetramethoxy-5,6-dihydro-8H-dibenzo[a,g]quinolizine	POCl$_3$	110	—	189
2,3,9,10-Bis(methylenedioxy)-8-oxo-5,6-dihydro-8H-dibenzo[a,g]quinolizine	POCl$_3$	110	71	336
2,3-Methylenedioxy-9,10-dimethoxy-8-oxo-5,6-dihydro-8H-dibenzo[a,g]quinolizine	POCl$_3$	110	11	337
2,3-Dimethoxy-9,10-methylenedioxy-8-oxo-5,6-dihydro-8H-dibenzo[a,g]quinolizine	POCl$_3$	110	91	336
2,3,9,10-Tetramethoxy-8-oxo-5,6-dihydro-8H-dibenzo[a,g]quinolizine	POCl$_3$	110	—	337
2,3-Methylenedioxy-10,11-dimethoxy-8-oxo-5,6-dihydro-8H-dibenzo[a,g]quinolizine	POCl$_3$	110	—	56
5,6-Dihydro-8-oxo-8H-dibenzo[a,h]quinolizine	POCl$_3$	110	Poor	56
2,3-Methylenedioxy-5,6-dihydro-8-oxo-8H-dibenzo[a,h]quinolizine	POCl$_3$	110	—	56

TABLE XII—*Continued*

DIBENZOQUINOLIZINES

Name	Condensing Agent	Temperature °C.	Yield %	Reference
2,3-Methylenedioxy-11,12-dimethoxy-5,6,8,9-tetrahydrodibenzo[a,h]quinolizinium chloride	POCl$_3$ POCl$_3$ POCl$_3$	80 80 110	— 80 80	338 123 102
2,3,11,12-Tetramethoxy-5,6,8,9-tetrahydrodibenzo[a,h]quinolizinium chloride	POCl$_3$	110	—	123, 338
2,3,9,10-Tetramethoxy-7,12,12a,13-tetrahydrodibenzo[b,g]quinolizinium chloride	POCl$_3$	100	—	246
2,3,9,10-Tetramethoxy-5-veratryl-7,12,12a,13-tetrahydrodibenzo[b,g]-quinolizinium chloride	POCl$_3$	—	—	246

Supplement to Table XII

Amide That Could Not Be Cyclized

Name	Condensing Agent	Temperature °C.	Reference
α-[N-(β-Phenethyl)carbamyl]phthalide	—	—	333

[327] Sugasawa and Kakemi, *Ber.*, **71**, 1860 (1938).
[328] Sugasawa and Kakemi, *Proc. Imp. Acad. Tokyo*, **14**, 214 (1938) [*C. A.*, **32**, 8421 (1938)].
[329] Kakemi, *J. Pharm. Soc. Japan*, **60**, 2 (1940) [*C. A.*, **34**, 3747 (1940)].
[330] Sugasawa, *J. Pharm. Soc. Japan*, **57**, 1023 (1937) [*C. A.*, **32**, 3402 (1938)].
[331] Leithe, *Ber.*, **63**, 2343 (1930).
[332] Leithe, *Ber.*, **67**, 1261 (1934).
[333] Chakravarti and Nair, *J. Annamalai Univ.*, **1**, 186; *J. Indian Chem. Soc.*, **9**, 577 (1932).
[334] Chakravarti and Perkin, *J. Chem. Soc.*, **1929**, 196.
[335] Stevens, *J. Chem. Soc.*, **1935**, 663.
[336] Haworth and Perkin, *J. Chem. Soc.*, **1926**, 1769.
[337] Haworth, Koepfli, and Perkin, *J. Chem. Soc.*, **1927**, 548.
[338] Sugasawa and Kakemi, *Proc. Imp. Acad. Tokyo*, **15**, 52 (1939) [*C. A.*, **33**, 5401 (1939)].

TABLE XIII
2-CARBOLINES

Substituents	Condensing Agent	Temperature °C	Yield %	Reference

A. *3,4-Dihydro-2-carbolines*

$$\text{(indole-CH}_2\text{-C(CH}_3\text{)}_2\text{-NH-CHO)} \xrightarrow{-H_2O} \text{(3,4-dihydro-2-carboline)}$$

Substituents	Condensing Agent	Temperature °C	Yield %	Reference
None	P_2O_5	140	0	84
	P_2O_5	205	2	84
	P_2O_5	110	50	339
1-Methyl-	P_2O_5	140	56	83
1-Ethyl-	P_2O_5	140	56	84
1-*n*-Propyl-	P_2O_5	140	77	84
1-Isopropyl-	P_2O_5	140	40	84
1-*n*-Butyl-	$POCl_3$	110	92	340
1-*n*-Pentadecyl-	$POCl_3$	110	98	340
1-*n*-Heptadecyl-	$POCl_3$	110	80	340
1-Phenyl-	P_2O_5	140	36	341, 84
1-(3,4-Dimethoxyphenyl)-	P_2O_5	140	—	341
1-Benzyl-	P_2O_5	140	—	341
	$POCl_3$	80	90	81
1-(*o*-Methylbenzyl)-	P_2O_5	140	63	61
	$POCl_3$	80	89	342
1,1'-Tetramethylene-bis-	$POCl_3$	110	99	340
1,1'-Pentamethylene-bis-	$POCl_3$	110	94	340
1,1'-Hexamethylene-bis-	$POCl_3$	110	36	340
1,1'-Octamethylene-bis-	$POCl_3$	110	99	340
1-(N-Tosyl-3-piperidyl)-	$POCl_3$	60	80	343
1-(2-Chloro-3-lepidyl)-	$POCl_3$	100	85	343
1,9-Dimethyl-	P_2O_5	140	34	84
1-Methyl-3-carboxy-	$(CH_3CO)_2O$-$ZnCl_2$	—	—	111, 112, 344
(also methyl and ethyl esters)	Various	—	0	345
1-Methyl-6-methoxy-	P_2O_5	140	58	84
1-Methyl-7-methoxy-	P_2O_5	140	78	83
1-Methyl-8-methoxy-	P_2O_5	140	32	84
1-Ethyl-9-methyl-	P_2O_5	140	45	84
1-Phenyl-9-methyl-	$POCl_3$	60	—	346
1-Phenyl-3-ethyl-	P_2O_5	140	80	88

TABLE XIII—Continued
2-CARBOLINES

Substituents	Condensing Agent	Temperature °C.	Yield %	Reference
B. 3,4-Benzo-2-carbolines				
None	POCl$_3$	110	76	82
1-Methyl-	Various	—	0	82
	POCl$_3$	110	92	82
1-Ethyl-	POCl$_3$	110	61	82
1,9-Dimethyl-	POCl$_3$	110	69	82

C. 2-Carbolines

Substituents	Condensing Agent	Temperature °C.	Yield %	Reference
None	Polyphosphoric acid + POCl$_3$	125	36	65
1-Methyl-	Polyphosphoric acid + POCl$_3$	125	5–15	65

Supplement to Table XIII

Amides That Could Not Be Cyclized

Name	Condensing Agent	Temperature °C.	Reference
N-(Lepidyl-3-carboxy)tryptamine	Various	—	343
N-Formyltryptophan	POCl$_3$ or PCl$_5$	—	65

[339] Schöpf and Steuer, *Ann.*, **558**, 124 (1947).
[340] Hahn and Gudjons, *Ber.*, **71**, 2175 (1938).
[341] Asahina and Osada, *J. Pharm. Soc. Japan*, **534**, 63 (1926) [*Chem. Zentr.*, **I**, 1479 (1927)].
[342] Clemo and Swan, *J. Chem. Soc.*, **1946**, 617.
[343] Marion, Manske, and Kulka, *Can. J. Research*, **24B**, 224 (1946).
[344] Harvey, Miller, and Robson, *J. Chem. Soc.*, **1941**, 153.
[345] Snyder, Hansch, Katz, Parmerter, and Spaeth, *J. Am. Chem. Soc.*, **70**, 219 (1948).
[346] Manske, *Can. J. Research*, **5**, 592 (1931).

TABLE XIV
Miscellaneous Compounds

Name	Condensing Agent	Temperature °C.	Yield %	Reference
1-Phenylphthalazine	HCl	>100	—	24a
1-Phenyl-5-methoxyphthalazine	HCl	>100	—	24a
1-Phenyl-7-methoxyphthalazine	HCl	>100	—	24a
1-Benzyl-7-methoxyphthalazine	HCl	>100	—	24b
1-Phenyl-6,7-methylenedioxyphthalazine	POCl$_3$	65	—	24a
1-Phenyl-6,7-dimethoxyphthalazine	HCl	>100	50	24a
1-Benzyl-6,7-dimethoxyphthalazine	HCl	>100	53	24b
1-Veratryl-6,7-dimethoxyphthalazine	HCl	>100	—	24b
1-Methyl-3,4-dihydrothiopheno[2,3-c]pyridine	P$_2$O$_5$ + POCl$_3$	140	50	346a
1-Phenyl-3,4-dihydrothiopheno[2,3-c]pyridine	P$_2$O$_5$ + POCl$_3$	140	60	346a
8,9-Dimethoxy-2,3,5,6-tetrahydro-1H-benzo[g]pyrrocolinium chloride	POCl$_3$	110	65	325, 179
[Attempted to prepare 1-(γ-chloropropyl)-6,7-dimethoxy-3,4-dihydroisoquinoline]	POCl$_3$	110	95	57
5-Methyl-8,9-methylenedioxy-2,3,5,6-tetrahydro-1H-benzo[g]pyrrocolinium chloride	POCl$_3$	110	90	325, 179

TABLE XIV—*Continued*

Miscellaneous Compounds

Name	Condensing Agent	Temperature °C.	Yield %	Reference
1-Phenyl-3,4-dihydro-3,4-cyclopropanoisoquinoline	P_2O_5	110	21	347
1-Methyl-3,3a,4,5-tetrahydrocyclopent[d,e]isoquinoline	P_2O_5 $POCl_3$	140 110	18 Poor	348 348
1-Phenyl-3,3a,4,5-tetrahydrocyclopent[d,e]isoquinoline	P_2O_5	140	27	348
1-Methyl-3,4,7,8-tetrahydro-6H-cyclopent[g]isoquinoline	P_2O_5	110	—	38
1-Piperonyl-3,4,7,8-tetrahydro-6H-cyclopent[g]isoquinoline	P_2O_5	140	>15	38
1-(5-Indanylmethyl)-3,4,7,8-tetrahydro-6H-cyclopent[g]isoquinoline	P_2O_5	140	71	38
2-Oxo-3,4-dimethoxy-6-methyl-8,8a-dihydro-2H-furo[2,3,4-d,e]isoquinoline	P_2O_5	110	32	349

TABLE XIV—Continued
MISCELLANEOUS COMPOUNDS

Name	Condensing Agent	Temperature °C.	Yield %	Reference
2-Oxo-3,4-dimethoxy-6-phenyl-8,8a-dihydro-2H-furo[2,3,4-d,e]isoquinoline	POCl$_3$	—	21	349
10,11-Methylenedioxy-2,3,4,5,7,8-hexahydro-1H-azepo[a]isoquinolinium chloride	POCl$_3$	110	ca. 70	325, 179
7-Methyl-10,11-methylenedioxy-2,3,4,5,7,8-hexahydro-1H-azepo[a]isoquinolinium chloride	POCl$_3$	110	—	325, 179
3-Phenyl-8,9-dimethoxy-5,6-dihydroimidazo[5,1-a]isoquinoline	POCl$_3$	110	60	57
4,9-Dimethyl-1,2-benzo-3-carboline	POCl$_3$	110	—	85
1-Methyl-3,4-dihydrothianaphtheno[2,3-c]pyridine	P$_2$O$_5$ + POCl$_3$	140	55–60	346b

TABLE XIV—Continued
Miscellaneous Compounds

Name	Condensing Agent	Temperature °C.	Yield %	Reference
1-Phenyl-3,4-dihydrothianaphtheno[2,3-c]-pyridine	P_2O_5 + $POCl_3$	140	65–70	346b
5-Phenyldibenzo[b,h][1,5]naphthyridine	P_2O_5	270–280	60	350
4-Azapyrene	P_2O_5	140	33	78
5-Methyl-4-azapyrene	P_2O_5	—	47	78
5-Phenyl-4-azapyrene	P_2O_5	—	—	78
5,10-Di-(o-carboxyphenyl)-pyrido[2,3,4,5-l,m,n]phenanthridine	$AlCl_3$ + NaCl	200	25	132, 133, 134, 80

TABLE XIV—Continued

MISCELLANEOUS COMPOUNDS

Name	Condensing Agent	Temperature °C.	Yield %	Reference
9-Oxo-9H-indolo[3,2,1-*d,e*]-1-azapher.-anthridine	P_2O_5	200–230	<1	351
5-Oxo-5,7,8,13-tetrahydrobenz[*g*]-indolo[2,3-*a*]quinolizine	$POCl_3$ $POCl_3$	100 110	36 52	352 63
1,2,3,4,4*a*,14*a*-Hexahydro- 1-Methyl-	$POCl_3$ $POCl_3$ $POCl_3$ $POCl_3$	100 110 — 110	43–46 5–10 — 40–43	352 353 354 355, 63

Supplement to Table XIV

Amides That Could Not Be Cyclized

Name	Condensing Agent	Temperature °C.	Reference
3-(β-Homophthalimidoethyl)indole	POCl$_3$	110	342
3-[β-(o-Carboxyphenylacetamido)ethyl]indole	Vacuum	275	356
3-[β-(o-Carbomethoxyphenylacetamido)ethyl]indole	Various	—	342, 356
2-Formyl-1-benzyl-1,2,3,4-tetrahydro-2-carboline	—	—	342
2-(o-Benzamidophenyl)pyridine	Various	—	350
3-(o-Benzamidophenyl)pyridine	Various	—	350
2-Acetamido-3-phenylquinoline	Various	—	350
2-(o-Benzamidophenyl)quinoline	Various	—	350
1-Phenyl-6-(β-benzamidoethyl)-3,4-dihydroisoquinoline	POCl$_3$ or PCl$_5$	—	75
1-Phenyl-6-(β-benzamidoethyl)isoquinoline	POCl$_3$ or PCl$_5$	—	75
1-Phenyl-7-(β-benzamidoethyl)isoquinoline	POCl$_3$	135	75
2-Formamidomethylmeconin	POCl$_3$ P$_2$O$_5$ SOCl$_2$	—	349
6-(β-Formamidoethyl)-1,2,3,4-tetrahydrocarbazole	POCl$_3$	—	357
6-(β-Acetamidoethyl)-1,2,3,4-tetrahydrocarbazole	POCl$_3$	—	357
6-(β-Carbethoxyaminoethyl)-1,2,3,4-tetrahydrocarbazole	POCl$_3$	—	357
2-Phenyl-3-benzamidoindole	POCl$_3$	35	358

[346a] W. Herz, private communication.
[346b] Herz, *J. Am. Chem. Soc.*, **72**, 4999 (1950).
[347] Burger and Yost, *J. Am. Chem. Soc.*, **70**, 2198 (1948).
[348] Flack and Lions, *J. Proc. Roy. Soc. N. S. Wales*, **73**, 253 (1940) [*C. A.*, **34**, 5846 (1940)].
[349] Dey and Srinivasan, *Arch. Pharm.*, **275**, 397 (1937).
[350] Petrow, Stack, and Wragg, *J. Chem. Soc.*, **1943**, 316.
[351] Marion and Manske, *Can. J. Research*, **16B**, 432 (1938).
[352] Schlittler and Allemann, *Helv. Chim. Acta*, **31** 128 (1948).
[353] Jost, *Helv. Chim. Acta*, **32**, 1297 (1949).
[354] Julian, Karpel, Magnani, and Meyer, *J. Am. Chem. Soc.*, **70**, 2834 (1948).
[355] Schlittler and Speitel, *Helv. Chim. Acta*, **31**, 1199 (1948).
[356] Scholz, *Helv. Chim. Acta*, **18**, 923 (1935).
[357] Manske and Kulka, *Can. J. Research*, **25B**, 376 (1947).
[358] Robinson and Thornley, *J. Chem. Soc.*, **1926**, 3144.

CHAPTER 3

THE PICTET-SPENGLER SYNTHESIS OF TETRAHYDROISOQUINOLINES AND RELATED COMPOUNDS

WILSON M. WHALEY * and TUTICORIN R. GOVINDACHARI †

University of Illinois

CONTENTS

	PAGE
INTRODUCTION	152
THE COURSE OF THE REACTION	156
Mechanism of Cyclization	156
Direction of Ring Closure	157
Side Reactions	162
FACTORS AFFECTING THE EASE OF CYCLIZATION	164
Reactivity of the Aromatic Nucleus	164
Table I. Types of Carbonyl Components Used	166
Nature of the Carbonyl Component	167
EXPERIMENTAL CONDITIONS AND CONDENSING AGENTS	168
Laboratory Conditions	168
Physiological Conditions	169
EXPERIMENTAL PROCEDURES	172
6-Methoxy-1,2,3,4-tetrahydroisoquinoline	172
1-Methyl-6,7-dihydroxy-1,2,3,4-tetrahydroisoquinoline	172
1-Methyl-1-carboxy-6,7-dihydroxy-1,2,3,4-tetrahydroisoquinoline	173
2,3-Methylenedioxy-10,11-dimethoxy-5,6,13,13a-tetrahydro-8H-dibenzo[a,g]quinolizine	173
2,3,9,10-Tetramethoxy-7,12,12a,13-tetrahydro-5H-dibenzo[b,g]quinolizine	173
1-Methyl-1,2,3,4-tetrahydro-2-carboline	173
1-Benzyl-1,2,3,4-tetrahydro-2-carboline	174
1-Benzyl-1-carboxy-1,2,3,4-tetrahydro-2-carboline	174
TABULAR SURVEY OF THE PICTET-SPENGLER REACTION	174
Explanation of Tables	174
Table II. 1,2,3,4-Tetrahydroisoquinolines	175
Supplement to Table II	179
Table III. Benzisoquinolines and Naphthisoquinoline	180
Supplement to Table III	180

* Present address: University of Tennessee, Knoxville, Tennessee.
† Present address: 25, Thanikachalam Chetty Road, T. Nagar, Madras, India.

	PAGE
Table IV. Benzoquinolizine and Dibenzoquinolizines	181
Supplement to Table IV	184
Table V. 1,2,3,4-Tetrahydro-2-carbolines	185
Supplement to Table V	187
Table VI. Miscellaneous Compounds	188
Supplement to Table VI	190

INTRODUCTION

The Pictet-Spengler reaction, in its simplest form, consists in the condensation of a β-arylethylamine with a carbonyl compound to yield a tetrahydroisoquinoline, and is a special example of the Mannich reaction.[1] The condensation of phenethylamine with methylal in concentrated hydrochloric acid to form 1,2,3,4-tetrahydroisoquinoline was achieved in 1911 by Pictet and Spengler,[2] giving substance to an ingenious theory concerning the origin of isoquinoline alkaloids in plants. The reaction was immediately extended by Decker[3] to the condensation of substituted phenethylamines with various aldehydes including formaldehyde itself. Decker carried out the reaction in two steps as indicated by the following general equation.

$$\text{RO-C}_6\text{H}_3(\text{OR})\text{-CH}_2\text{-CH}_2\text{-NH}_2 \xrightarrow{\text{R'CHO}} \text{RO-C}_6\text{H}_3(\text{OR})\text{-CH}_2\text{-CH}_2\text{-N=CH-R'} \xrightarrow{\text{HCl}} \text{tetrahydroisoquinoline}$$

The intermediate azomethine is seldom isolated, though it is often formed before addition of the condensing agent.

The Pictet-Spengler reaction has been applied to the synthesis of other ring systems also, notably dibenzoquinolizines and 2-carbolines. Typical examples are the preparation of 2,3,10,11-tetramethoxy-8-methyl-5,6,13,13a-tetrahydro-8H-dibenzo[a,g]quinolizine (I)[4] and 1-methyl-1,2,3,4-tetrahydro-2-carboline (tetrahydroharman) (II).[5] Attempts to pre-

[1] Blicke, *Org. Reactions*, **1**, 303 (1942).
[2] Pictet and Spengler, *Ber.*, **44**, 2030 (1911).
[3] Decker and Becker, *Ann.*, **395**, 342 (1913).
[4] Hahn and Schuls, *Ber.*, **71**, 2135 (1938).
[5] Akabori and Saito, *Ber.*, **63**, 2245 (1930).

pare dihydrophenanthridines by condensing 2-aminobiphenyls with aldehydes have so far been inconclusive.[5a]

A minor variation of the reaction, which has been seldom employed, utilizes an N-hydroxymethyl or N-methoxymethyl [6] derivative of the amine as starting material.[7,8] These derivatives of homopiperonylethylamine are converted to N-ethylnorhydrohydrastinine (III) when heated with hydrochloric acid.[9]

The use of concentrated hydrochloric acid as a catalyst in preparing tetrahydroisoquinolines was not satisfactory to those who sought the key to nature's synthetical transformations, and it was very much desired to effect the condensation under physiologically possible (zellmöglich) conditions. In 1934 Schöpf and Bayerle [10] achieved a Pictet-Spengler type of reaction under conditions of temperature, concentration, and acidity comparable to those which exist in plants, and since

[5a] Whaley and White, unpublished results.
[6] Merck and Co., Ger. pat. 273,323 [Frdl., **12**, 761 (1914–1916)].
[7] Merck and Co., Ger. pat. 280,502 [Frdl., **12**, 760 (1914–1916)].
[8] Rosenmund, Ger. pat. 320,480 [Frdl., **13**, 883 (1916–1921)].
[9] Rosenmund, Ger. pat. 336,153 [Frdl., **13**, 884 (1916–1921)].
[10] Schöpf and Bayerle, Ann., **513**, 190 (1934).

then numerous applications have been recorded. For example, the previously mentioned reaction of tryptamine (β-indolylethylamine) with acetaldehyde to yield tetrahydroharman (II) may be carried out at pH 5–6 and 25° to give a 70% yield of product after three days.[11] Condensation of β-(3,4-dihydroxyphenyl)ethylamine with homopiperonal at pH 6 and 25° yielded 84% of 1-piperonyl-6,7-dihydroxy-1,2,3,4-tetrahydroisoquinoline (IV).[12]

Naturally occurring phenylacetaldehydes probably are derived from appropriate α-amino acids through the corresponding phenylpyruvic acids; Hahn [13, 14] thought it probable that the α-keto acids were the actual precursors during biogenesis of isoquinoline alkaloids. His hypothesis was supported by the preparation of 1-benzyl-1-carboxy-6,7-dihydroxy-1,2,3,4-tetrahydroisoquinoline (V) under conditions which he considered biologically plausible.[15] The reaction of pyruvic acids is

[11] Hahn and Ludewig, *Ber.*, **67**, 2031 (1934).
[12] Schöpf and Salzer, *Ann.*, **544**, 1 (1940).
[13] Hahn, Bärwarld, Schales, and Werner, *Ann.*, **520**, 107 (1935).
[14] Hahn and Werner, *Ann.*, **520**, 123 (1935).
[15] Hahn and Stiehl, *Ber.*, **69**, 2627 (1936).

much slower than the reaction of aldehydes, and it has not been possible to decarboxylate the 1-carboxy-1,2,3,4-tetrahydroisoquinolines under mild conditions, so that Hahn's hypothesis must be considered unlikely.

These reactions involved in the biogenesis of alkaloids are non-enzymatic and therefore depend entirely upon the use of extremely reactive intermediates. Frequently the reactive intermediates are difficult to prepare and store, the reaction is slow, and the yields are poor if the intermediates are not sufficiently reactive. Thus, the theoretical elegance of the method is offset considerably by the difficulty of its practical application, and at the present time it offers no threat to the popularity of the conventional Pictet-Spengler reaction for preparative syntheses with the possible exception of the 2-carbolines obtained from pyruvic acids.[16,17]

The Adamkiewicz, Hopkins and Cole, and Rosenheim tests for tryptophan may involve the Pictet-Spengler reaction, for they yield 3-carboxy-1,2,3,4-tetrahydro-2-carboline, which is then oxidized to a characteristic blue pigment of unknown structure.[18,19] Color tests performed on 2-methyltryptophan were positive in the presence of mercury or copper salts, casting some doubt upon this hypothesis.[20] The base obtained in 1903 by Hopkins and Cole from the oxidation of tryptophan with ferric chloride in the presence of alcohol has been shown to be harman (1-methyl-2-carboline) (VI).[21]

VI

The theory that proteins are the parent substances of alkaloids was tested by Pictet,[22,23] who heated casein with methylal and hydrochloric acid, obtaining a mixture of pyridine and isoquinoline bases. Very small yields were obtained, and most of the products were not definitely identified.

[16] Hahn, Ger. pat. 644,999 [*Frdl.*, **23**, 570 (1936)].
[17] Hahn and Hansel, *Ber.*, **71**, 2163 (1938).
[18] Homer, *Proc. Cambridge Phil. Soc.*, **16**, 405 (1912) [*C. A.*, **6**, 1611 (1912)].
[19] Harvey, Miller, and Robson, *J. Chem. Soc.*, **1941**, 153.
[20] Rydon, *J. Chem. Soc.*, **1948**, 705.
[21] Kermack, Perkin, and Robinson, *J. Chem. Soc.*, **119**, 1602 (1921).
[22] Pictet and Chou, *Compt. rend.*, **162**, 127 (1916).
[23] Pictet and Chou, *Ber.*, **49**, 376 (1916).

THE COURSE OF THE REACTION

Mechanism of Cyclization. There has been no direct work on the electronic mechanism of the Pictet-Spengler reaction,[23a] but it does not seem unlike other examples of aromatic substitution by electrophilic attack. The intermediate Schiff bases have been isolated in many reactions and then cyclized as a separate reaction catalyzed by acid. A probable over-all reaction mechanism is illustrated with the synthesis of norhydrohydrastinine (X) from homopiperonylamine. The reaction

may be carried out with secondary amines also, in which case the isolable intermediate is the hydroxymethyl derivative VII, and the Schiff base VIII must be by-passed because it cannot form; loss of water by the hydroxymethyl derivative under the influence of acid yields the ammonium compound IX directly.

The validity of such a scheme for reactions conducted at pH 7 may well be questioned, though of course aliphatic amines of the type used

---

[23a] The mechanism of the simpler Mannich reaction [*Organic Reactions*, **1**, 303 (1942)] has been studied by Alexander and Underhill, *J. Am. Chem. Soc.*, **71**, 4014 (1949).

have considerable tendency to form ammonium ions in the presence of water.

Since pyruvic acids having no β-hydrogen atom, for example phenylglyoxylic acid, will not enter into the Pictet-Spengler reaction, it has been postulated that those pyruvic acids which can react do so as a result of enolization followed by addition of the amine to the double bond in the enol (XI).[13]

Direction of Ring Closure. As in the Bischler-Napieralski reaction,[24] the *ortho* position involved in the ring closure is almost invariably the one of greater electron density as required by the mechanism of the reaction. Condensation of the phenethylamine XII with formaldehyde yielded only 6-methoxy-1,2,3,4-tetrahydroisoquinoline (XIII) and not the 8-methoxy compound which would have resulted from cyclization in the alternate *ortho* position. The structure of the product was proved by oxidation to 4-methoxyphthalic acid (XIV).[25]

A 3,4-dialkoxy-β-phenethylamine invariably yields the 6,7-dialkoxy product upon cyclization; the 7,8-dialkoxy compound is never formed. It has been reported that treatment of homopiperonylamine or N-methylhomopiperonylamine with formaldehyde and hydrochloric acid gave a product which was not identical with that obtained with the same

[24] Whaley and Govindachari, *Organic Reactions*, **6**, 74 (1951).
[25] Helfer, *Helv. Chim. Acta*, **7**, 945 (1924).

reactants and under the same conditions by earlier investigators; the suggestion [26] that such new products could be 5,6-methylenedioxy derivatives is untenable.

If both *ortho* positions are activated by *m*-alkoxyl groups, cyclization occurs in both directions to yield a mixture of the two possible tetrahydroisoquinoline derivatives. An example is found in the condensation of N-(3-methoxybenzyl)homomyristicylamine (XV) with formaldehyde.[27] The two products have different properties, but absolute assignment of structures by degradation was not attempted.

A historically significant example of the tendency for ring closure to occur *para* to an alkoxyl group is provided by the preparation of tetrahydro-ψ-berberine (XVII) from 1-veratrylnorhydrohydrastinine (XVI). Pictet and Gams [28, 29] claimed that the product was identical with tetrahydroberberine (XVIII) from natural sources, though they expressed surprise that the closure should have occurred at the position of lesser activation. Subsequently, Haworth, Perkin, and Rankin [30] disproved this claim and established conclusively that tetrahydro-ψ-berberine (XVII) is the only product of the reaction and that it is easily distinguished from the natural product. These findings have since been verified by Späth,[31] who discovered further that if the alkoxyl groups are replaced by hydroxyl groups the orientation rule becomes invalid and ring closure proceeds in both *ortho* positions with nearly equal facility.

[26] Buck, *J. Am. Chem. Soc.*, **56**, 1769 (1934).
[27] Redemann, Wisegarver, and Icke, *J. Org. Chem.*, **13**, 886 (1948).
[28] Pictet and Gams, *Ber.*, **44**, 2480 (1911).
[29] Pictet and Gams, *Compt. rend.*, **153**, 386 (1911).
[30] Haworth, Perkin, and Rankin, *J. Chem. Soc.*, **125**, 1686 (1924).
[31] Späth and Kruta, *Monatsh.*, **50**, 341 (1928).

Thus, treatment of tetrahydropapaveroline (XIX) with formaldehyde afforded equal parts of products XX and XXI (isolated after conversion to the tetramethoxy derivatives norcoralydine and tetrahydropalmatine, respectively). By carrying out the reaction under physiological conditions, Schöpf [32] obtained 80% of compound XXI. Apparently the presence of free hydroxyl groups in the benzyl residue activates the *ortho*

positions to such an extent that instantaneous reaction is possible at whichever position is made available by random oscillation of the benzyl

[32] Schöpf, *Angew. Chem.*, **50**, 797 (1937).

group. Identical results were obtained with 1-(α-methyl-3,4-dihydroxybenzyl)-6,7-dihydroxy-1,2,3,4-tetrahydroisoquinoline.[33]

Several *m*-hydroxyphenethylamines have been condensed with aldehydes and pyruvic acids, but in each instance only a single product has been isolated. The products have been tacitly assumed to be 6-hydroxy-1,2,3,4-tetrahydroisoquinolines without considering the possibility of ring closure in two directions as discussed in the previous paragraph. Such an assumption was made in the synthesis of anhalamine [34] but was withdrawn when anhalamine was shown by degradative studies to be 6,7-dimethoxy-8-hydroxy-1,2,3,4-tetrahydroisoquinoline.[35]

Condensation of β-(2-naphthyl)ethylamine with formaldehyde yielded only 1,2,3,4-tetrahydro-7,8-benzisoquinoline (XXII), whose structure was proved by oxidation to mellophanic acid (XXIII).[36] Under similar conditions β-(1-naphthyl)ethylamine (XXIV) could not be cyclized.[36] An attempted *peri*-cyclization of 1-aminomethyl-2-methoxynaphthalene (XXV) was also unsuccessful.[37] The examples cited indicate that cyclization to the *alpha* position in naphthalene is much more likely than cyclization to the *beta* or the *peri* positions. The reaction of β-(2-

[33] Späth and Kruta, *Ber.*, **62**, 1024 (1929).
[34] Späth and Röder, *Monatsh.*, **43**, 93 (1922).
[35] Späth, *Ber.*, **65**, 1778 (1932).
[36] Mayer and Schnecko, *Ber.*, **56**, 1408 (1923).
[37] Dey and Rajagopalan, *Arch. Pharm.*, **277**, 377 (1939).

phenanthryl)ethylamine with formaldehyde was assumed to yield 1,2,3,4-tetrahydronaphth[1,2-h]isoquinoline (XXVI),[38] and the assumption is probably correct because it is in line with the preference for reaction in the *alpha* position. On the other hand, β-(3-phenanthryl)-ethylamine would not condense with formaldehyde though there is an *alpha* position available.[38]

Non-occurrence of paraberine (5H-dibenzo[b,g]quinolizine) derivatives in nature has been attributed to the comparative difficulty of forming linear structures, and it was only with much labor that Perkin and co-workers [39] succeeded in obtaining a paraberine derivative (XXVII) in low yield from 3-veratryl-6,7-methylenedioxy-1,2,3,4-tetrahydroisoquinoline and formaldehyde. More recently, the tetramethoxy analog of XXVII was prepared in almost quantitative yield using similar conditions,[40,41,42] proving that formation of linearly condensed molecules is more facile than was at first supposed. Additional evidence of the ease of paraberine formation is supplied by the cyclization of 1,3-diveratryl-6,7-dimethoxy-1,2,3,4-tetrahydroisoquinoline (XXVIII), which is so constituted that condensation with formaldehyde could lead to either a tetrahydroparaberine or a tetrahydroprotoberberine. The only product isolated was 2,3,9,10-tetramethoxy-5-veratryl-7,12,12a,13-tetrahydro-5H-dibenzo[b,g]quinolizine (XXIX); the protoberberine XXX which was expected to form more easily was not isolated [40,42] but was probably present to some extent in a mixture of by-products.[43]

[38] Mosettig and May, *J. Am. Chem. Soc.*, **60**, 2962 (1938).
[39] Campbell, Haworth, and Perkin, *J. Chem. Soc.*, **1926**, 32.
[40] Kakemi, *J. Pharm. Soc. Japan*, **60**, 11 (1940) [*C. A.*, **34**, 3748 (1940)].
[41] Sugasawa, Kakemi, and Kazumi, *Proc. Imp. Acad. Tokyo*, **15**, 223 (1939) (in English) [*C. A.*, **33**, 8617 (1939)].
[42] Sugasawa, Kakemi, and Kazumi, *Ber.*, **73**, 782 (1940).
[43] S. Sugasawa, private communication.

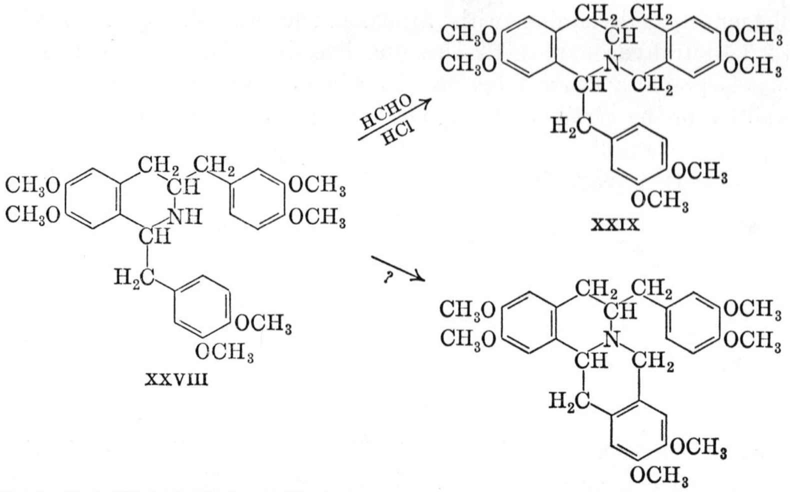

XXVIII XXIX XXX

Side Reactions. Although the Pictet-Spengler reaction employs the same reactants that are used to prepare phenolic resins and a host of less complex compounds, very few instances of definite side reactions have been recorded. Cyclization of β-phenethylamine with methylal and hydrochloric acid has been found to yield mostly bis(β-phenethylamino)-methane.[44] Decker [3] found that treatment of homopiperonylamine with methylal and hydrochloric acid gave as much as 70% of a polymeric base, which could also be obtained if the methylal was replaced by formaldehyde. At 130° an Eschweiler [45] reaction occurred and 88% of hydrohydrastinine (XXXI) was obtained from homopiperonylamine, formaldehyde, and hydrochloric acid. A normal reaction occurred only if the Schiff base was prepared before addition of acid.

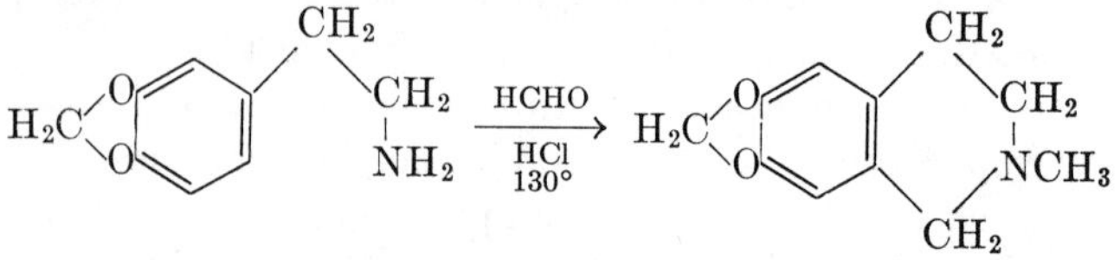

XXXI

In the preparation of 2-methyl-6-ethoxy-1,2,3,4-tetrahydroisoquinoline a small amount of methylene polymer was formed, being detected by the strong hypotensive activity that it conferred upon the major product.[46] Such polymers were produced from primary, secondary, and tertiary amines, indicating that separate benzene nuclei were being linked

[44] Kondo and Ochiai, *J. Pharm. Soc. Japan*, **495**, 313 (1923) [*C. A.*, **17**, 3032 (1923)].

[45] Moore, *Org. Reactions*, **5**, 301 (1949).

[46] Baltzly, Buck, deBeer, and Webb, *J. Am. Chem. Soc.*, **71**, 1301 (1949).

together by methylene groups from formaldehyde, as in the production of a phenolic resin. Fractions corresponding to a dimer, a trimer, and a tetramer were isolated (XXXII, $n = 0, 1$, and 2, respectively). Fractions having free $HOCH_2$— groups were also judged to be present.

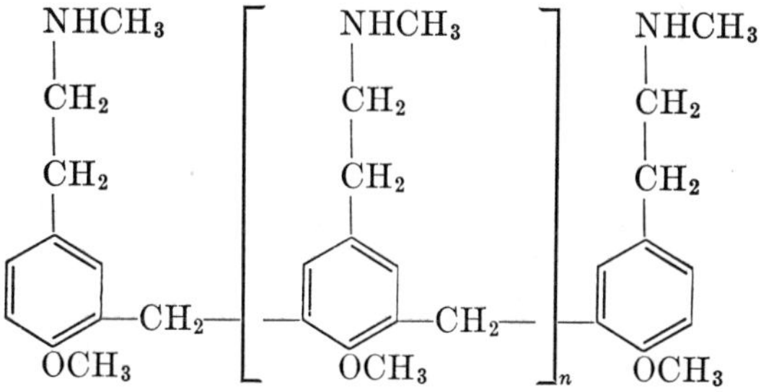

XXXII

Side reactions are most commonly encountered in the so-called biogenetic application of the Pictet-Spengler reaction because the intermediates used are extremely reactive. Phenylacetaldehydes are easily resinified, especially in the presence of acids; pyruvic acids are unstable in the presence of amines; hydroxy-β-phenethylamines and hydroxy-tetrahydroisoquinolines are susceptible to oxidation in the presence of air when in neutral or alkaline solution. As a result, the conditions of reaction are of primary importance in using these labile reactants, and the problems of their use are further mentioned under the heading, "Experimental Conditions and Condensing Agents."

A few secondary reactions have been encountered in which the heterocyclic system first formed was modified by further reaction of the 1-substituent, resulting in compounds such as the lactam XXXIII [15] and the quinolizine XXXIV.[17]

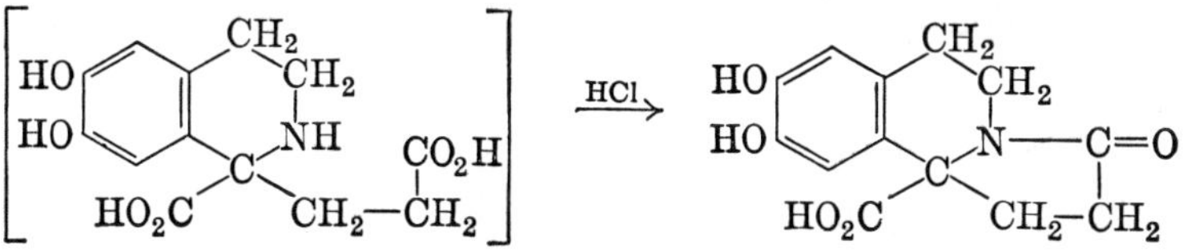

XXXIII

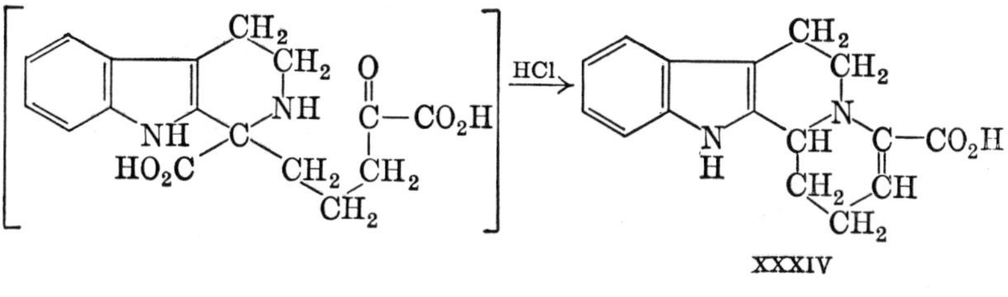

XXXIV

FACTORS AFFECTING THE EASE OF CYCLIZATION

The reactivity of the aromatic nucleus of the arylethylamine and the nature of the carbonyl component are important to the success of the Pictet-Spengler reaction. It might be supposed that substituents on the side chain of the arylethylamine would have an influence on the ease of cyclization comparable to that which they exert in the Bischler-Napieralski reaction, but the available data are insufficient even to predict the validity of the supposition.

Reactivity of the Aromatic Nucleus. It has been shown that the Pictet-Spengler reaction is one which is facilitated by increased electron density at the point of ring closure. Few phenethylamines lacking an alkoxyl or hydroxyl group *para* to the position of closure have been cyclized. β-Phenethylamine and phenylalanine were converted to the corresponding tetrahydroisoquinolines in approximately 35% yield by treatment with methylal and hydrochloric acid.[2] The first result has been disputed by Kondo and Ochiai,[44] who could obtain only a trace of the product. Cyclization of the hydroxy amine XXXV to the hydroxytetrahydroisoquinoline XXXVI took place quantitatively,[47] and

XXXV   XXXVI

tyramine [48] and tyrosine [49] have also been cyclized in good yield, indicating that the reaction does not require great activation. Contrariwise, β-(*o*-ethoxyphenyl)ethylamine (XXXVII) [50] and 1-(*p*-methoxy-

XXXVII   XXXVIII

[47] Kondo and Tanaka, *J. Pharm. Soc. Japan*, **50**, 119 (1930) (in English).
[48] Fränkel and Zeimer, *Biochem. Z.*, **110**, 234 (1920).
[49] Wellisch, *Biochem. Z.*, **49**, 173 (1913).
[50] Ide and Buck, *J. Am. Chem. Soc.*, **59**, 726 (1937).

benzyl)-6-methoxy-1,2,3,4-tetrahydroisoquinoline (XXXVIII) [51] could not be cyclized.

There have, indeed, been instances in which amines having methoxyl groups *para* to a position of possible ring closure have failed to cyclize under the ordinary conditions; ω-aminoacetoveratrone [39] (XXXIX) and β-(3-methoxy-4-hydroxyphenyl)isopropylamine [52] (XL) are two such amines.

Experiments conducted under physiological conditions require a very active nucleus (great electron density at the point of closure), a condition amply fulfilled in the β-(3-indolyl)ethylamines. In the benzenoid series even alkoxyl substituents do not furnish enough activation to promote the reaction satisfactorily, and Schöpf [10] stated that the reaction would not proceed in the absence of free hydroxyl groups. Hahn [53] successfully condensed homopiperonylamine and homopiperonal at pH 5 and 25°, but obtained the desired 1-piperonylnorhydrohydrastinine in only 5% yield. That he obtained the compound at all has been disputed.[54] In contrast, β-(3,4-dihydroxyphenyl)ethylamine and homopiperonal at pH 6 and 25° yielded 84% of 1-piperonyl-6,7-dihydroxy-1,2,3,4-tetrahydroisoquinoline (IV).[12] Thus, the thesis that either an indole nucleus or a hydroxylated benzene ring is necessary for cyclization under *quasi*-biological conditions is widely accepted. Nevertheless, there is one dissenting note in the reported preparation of norcoralydine (XLI) from tetrahydropapaverine and formaldehyde at pH 4 and 25° in more than 80% yield after eighteen hours.[4]

[51] Chakravarti, Vaidyanathan, and Venkatasubban, *J. Indian Chem. Soc.*, **9**, 573 (1932).
[52] Clemo and Turnbull, *J. Chem. Soc.*, **1945**, 533.
[53] Hahn and Schales, *Ber.*, **68**, 24 (1935).
[54] Späth, Kuffner, and Kesztler, *Ber.*, **69**, 378 (1936).

TABLE I

Types of Carbonyl Components Used in the Pictet-Spengler Reaction
(Mineral Acid Catalyst)

Carbonyl Component	Effectiveness	Reference
Formaldehyde	Excellent	Many
Acetaldehyde	Good	5, 4, 55
Chloral	—	56
Glycolaldehyde	—	19
Glyoxylic acid	Fair	19
Paraldol	—	57, 58
α-Ketoglutaric acid	Fair	15
Glutardialdehyde	Good	17
Benzaldehyde	Good	3, 55, 19
Salicylaldehyde	Good	59
o-Chlorobenzaldehyde	Good	59
o-Nitrobenzaldehyde	Poor	60
m-Nitrobenzaldehyde	Good	59
p-Nitrobenzaldehyde	Good	55
p-Methoxybenzaldehyde	Good	59
Piperonal	Excellent	61
p-Dimethylaminobenzaldehyde	Good	55
Phenylacetaldehyde	Good	55
Homopiperonal	Poor	62, 63
Homoveratraldehyde	Poor	62
m-Hydroxyphenylacetaldehyde	Fair	17
p-Methoxyphenylacetaldehyde	Excellent	55
Cinnamaldehyde	Good	59
Hydrocinnamaldehyde	Good	64
o-Hydroxyphenylpyruvic acid	Good	15
o-Nitrophenylpyruvic acid	Poor	60
o-Cyanophenylpyruvic acid	Poor	60
3,4-Dimethoxyphenylpyruvic acid	Fair	17

[55] Snyder, Hansch, Katz, Parmerter, and Spaeth, *J. Am. Chem. Soc.*, **70**, 219 (1948).
[56] Tatsui, *J. Pharm. Soc. Japan*, **49**, 116 (1929) (in English).
[57] Jacobs and Craig, *Science*, **82**, 421 (1935).
[58] Jacobs and Craig, *J. Biol. Chem.*, **113**, 759 (1936).
[59] Weinbach and Hartung, *J. Org. Chem.*, **15**, 676 (1950).
[60] Clemo and Swan, *J. Chem. Soc.*, **1949**, 487.
[61] Reichert and Hoffman, *Arch. Pharm.*, **274**, 153 (1936).
[62] Späth and Berger, *Ber.*, **63**, 2098 (1930).
[63] Späth, Kuffner, and Kesztler, *Ber.*, **70**, 1017 (1937).
[64] Külz and Schöpf, Ger. pat. 726,173 [*C. A.*, **37**, 6277 (1943)].

Nature of the Carbonyl Component. Formaldehyde and methylal have been the carbonyl compounds most frequently employed in the conventional Pictet-Spengler reaction. Formaldehyde has given excellent yields in a great number of instances and is definitely to be preferred to methylal.[3, 30] Tetrahydropapaverine was cyclized to norcoralydine (XLI) in 46% yield using methylal, whereas a 69% yield was obtained with formaldehyde under the same conditions.[31] In Table I are listed representative aldehydes and pyruvic acids that have been used in the Pictet-Spengler reaction with a mineral acid as catalyst.

In the second column of the table an attempt is made to indicate the general effectiveness of the carbonyl component in the cyclization, though the judgment in many cases is based on only one experiment.

Good yields are usually obtained with formaldehyde, which is apparently the most effective of the aldehydes. The very poor results with homopiperonal and homoveratraldehyde result from their instability in the presence of hydrochloric acid.[62, 54] The phenylacetaldehydes having fewer substituents give better results but are also easily resinified. Tryptophan failed to condense with crotonaldehyde,[57] chloral hydrate,[65] chloroacetal,[65] or formamide.[65]

The foregoing remarks have pertained to the conventional Pictet-Spengler reaction; the following are confined to the use of carbonyl compounds under simulated biological conditions.

The importance of formaldehyde in phytochemical processes is unquestionable, and it is surprising that there are only two recorded instances of its use in the Pictet-Spengler reaction under physiological conditions. Tryptophan and formaldehyde at pH 6.5 and 38° for 15 hours yielded 80% of product XLII.[19] The excellent yield of norcoraly-

dine (XLI) obtained by condensing tetrahydropapaverine with formaldehyde has been noted. Under the same conditions tetrahydropapaverine and acetaldehyde would not react.[4] No carbonyl compound other than formaldehyde has been condensed in acceptable yield with an arylethylamine activated only by alkoxyl groups. Less reactive aldehydes or ketones require the great activation of free hydroxyl groups or an indole nucleus.

[65] Snyder, Parmerter, and Katz, *J. Am. Chem. Soc.*, **70**, 222 (1948).

Hahn found that tryptamine condensed easily with acetaldehyde (70%) and phenylacetaldehyde (90%), but less easily with homopiperonal (15%), trimethoxyphenylacetaldehyde (16%), and benzaldehyde (48%).[13, 11] No condensation occurred with hydroxy and alkoxy benzaldehydes, glyoxal, D(+)glucose, and citral.[13] He considered that the condensation with aldehydes under physiological conditions was very much dependent upon the nature of the aldehyde.

Hahn and co-workers believed that pyruvic acids react much more easily with tryptamine and m-hydroxyphenethylamines than do aldehydes.[13] They found that nuclear alkoxyl substitution of arylpyruvic acids decreased their reactivity and that no reaction occurred if the pyruvic acid lacked a β-hydrogen atom (trimethylpyruvic acid and phenylglyoxylic acid) or contained a basic substituent (β-indolylpyruvic and 2-quinolylpyruvic acids).

The data in Fig. 5, p. 171, show that increased alkoxyl substitution of pyruvic acids does not always result in decreased yields.[66]

Convincing evidence has been presented by Schöpf to controvert the results of Hahn. Schöpf[12] pointed out that Hahn used higher concentrations than were likely to obtain in living cells, and that the reaction mixtures were not homogeneous. Self-condensation of the substituted phenylacetaldehydes to form resins was a natural result of their being outside the aqueous phase. Repetition of the experiments at proper dilutions showed that the aldehydes react hundreds of times faster than the pyruvic acids. Some of Schöpf's results have been plotted in Fig. 2, p. 170, to demonstrate that homopiperonal condenses much more rapidly with 3,4-dihydroxy-β-phenethylamine than does 3,4-methylenedioxyphenylpyruvic acid.

EXPERIMENTAL CONDITIONS AND CONDENSING AGENTS

Laboratory Conditions. The Pictet-Spengler reaction may be carried out by heating the amine with a slight excess of aldehyde and a considerable excess of 20–30% hydrochloric acid at 100° for one-half hour to six hours. Many amines and aldehydes have been heated together for an hour or so to form the azomethines, which were then heated at 100° with aqueous hydrochloric acid to effect cyclization. Some investigators have found the two-step method preferable.[3] In rarer instances the Schiff base has been isolated and purified before being cyclized with aqueous or ethanolic hydrochloric acid.[59]

As suggested in the previous paragraph, aqueous hydrochloric acid has been the favorite condensing agent for the preparation of tetra-

[66] Hahn and Rumpf, *Ber.*, **71**, 2141 (1938).

hydroisoquinolines. In a study of the condensation of Schiff bases derived from substituted benzaldehydes and homoveratrylamine, using three reaction media (hydrogen chloride in benzene, aqueous hydrochloric acid, and ethanolic hydrogen chloride), it has been shown that the optimum medium for condensation of a Schiff base can be determined only by trial.[59] Usually one reaction medium would give good results and the other two would cause hydrolysis of the azomethine or gum formation. For no obvious reason, aqueous sulfuric acid has enjoyed greater popularity in the synthesis of tetrahydro-2-carbolines. At times the hydrochloride of the amine has been used without further addition of acid, and in his condensation of tetrahydropapaveroline with formaldehyde Späth [31] did not use any condensing agent.

The occasional use of hydrobromic acid,[64] phosphorus oxychloride,[67] phosphorus pentoxide,[7] acetic anhydride,[49] or methyl iodide [68] has not conferred any special advantage.

Physiological Conditions. In carrying out the Pictet-Spengler reaction under physiological conditions the amine and aldehyde may be dissolved in an appropriate buffer solution, or, alternatively, the amine hydrohalide and aldehyde may be dissolved in water and the pH adjusted by addition of alkali. The solution is then set aside at a moderate temperature (25–40°) until the reaction has proceeded to maximum yield. The time allowed for reaction may vary from one day to several weeks, depending upon the reactivity of the system.

Although some workers have used concentrations of reactants as high as $0.3\ M$, it is believed by Schöpf that such concentrations are unnatural and that to ensure physiological conditions one must use 0.01–$0.04\ M$, solutions.[12] In fact, some of the reagents, especially substituted phenylacetaldehydes, are not sufficiently soluble to afford $0.3\ M$ solutions.[69] Workers using those concentrations have apparently had heterogeneous systems, and the data obtained therefrom are of questionable value.

The hydrogen-ion concentration of the mixture may lie between $pH\ 3$ and $pH\ 8$, and Figs. 2–5 show that there is no consistent relationship between pH and yield of product. In nearly all reactions, however, the optimum pH lies in the region of 5–7. The principal deterrent to the use of $pH\ 7$ and above is the danger of oxidation by atmospheric oxygen of the reactants and the products.[15,12] Though the curves suggest that an increased yield could often be obtained at higher values of pH, reactions conducted in the neighborhood of $pH\ 7$ must usually be of short duration if a product is to be isolated at all, and the increased speed of reaction cannot be put to practical use.[66,15]

[67] Decker, Ger. pat. 257,138 [*Frdl.*, **11**, 1001 (1912–1914)].
[68] Hoshina and Kotake, *Ann.*, **516**, 76 (1935).
[69] C. Schöpf, private communication.

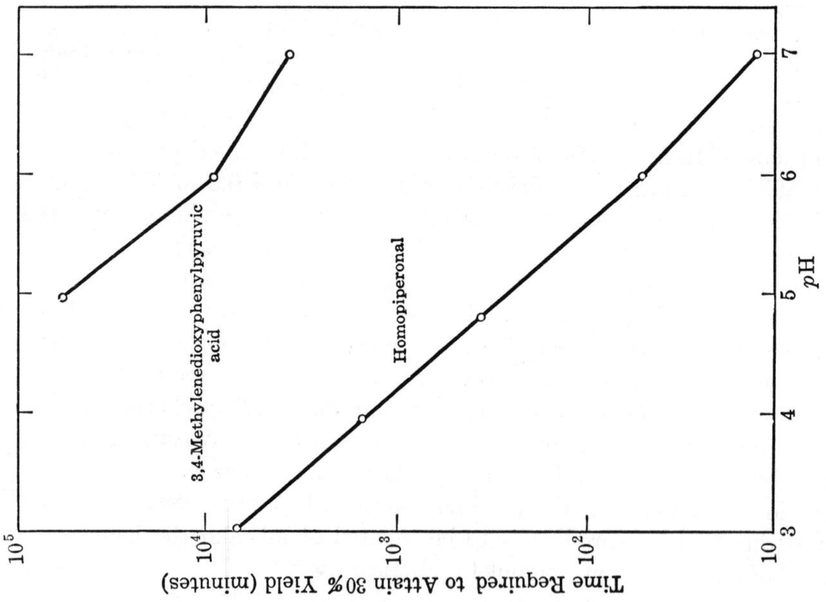

170 ORGANIC REACTIONS

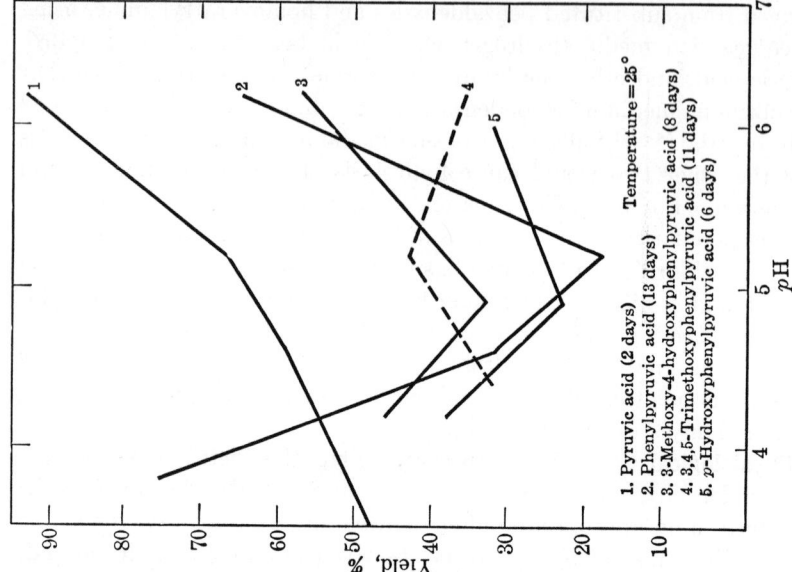

Fig. 3. Condensations with tryptamine.[13]

Fig. 2. Condensations of 3,4-dihydroxy-β-phenethylamine.[12]

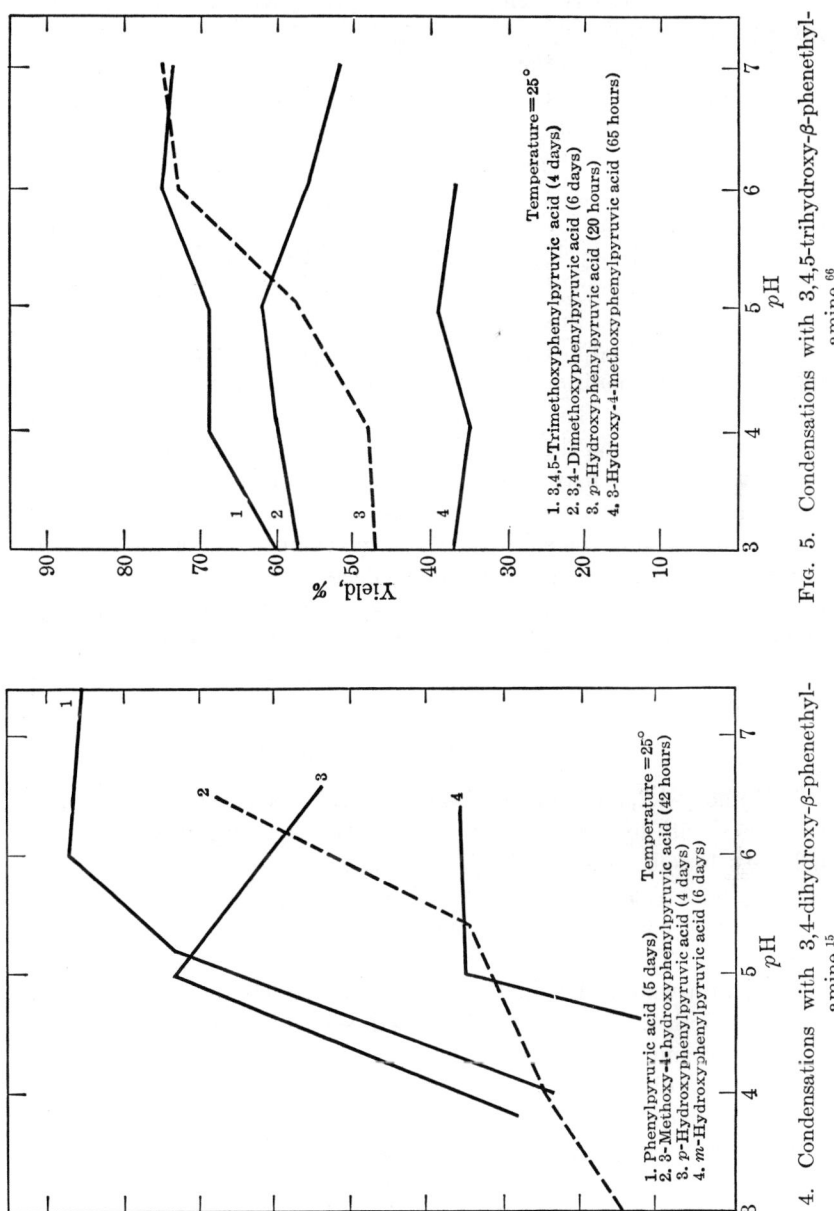

Fig. 4. Condensations with 3,4-dihydroxy-β-phenethylamine.[15]

Fig. 5. Condensations with 3,4,5-trihydroxy-β-phenethylamine.[66]

Ultraviolet light has been shown to have a definite catalytic effect upon the reaction.[17]

The extent of cyclization can be determined only by isolation and characterization of the expected product. Hahn [53] has demonstrated that whereas 90% of the homopiperonal disappeared in its reaction with homopiperonylamine at pH 5, only 5% of 1-piperonyl-6,7-methylenedioxy-1,2,3,4-tetrahydroisoquinoline could be isolated. Apparently the disappearance of the initial reactants is not accompanied to a corresponding degree by cyclization, because one of the ensuing steps of the process is slower than the initial formation of an aldehyde-ammonia. Schöpf [12] has verified this disclosure.

The curves plotted in Figs. 3–5 reveal a perplexing variation of yields under different conditions for several reactants. There is no simple correlation between the structure of similar carbonyl components and the effect of pH upon their reactivity with a single amine. A degree of explanation for this may be found in the consideration that the Pictet-Spengler reaction embodies several steps, all of which may not be equally affected by varying the substituents in the components of the reaction mixture. Superimposed upon these effects are the resinification of phenylacetaldehydes at high hydrogen-ion concentrations, the instability of pyruvic acids in the presence of amines, and the air oxidation of hydroxyphenethylamines and hydroxytetrahydroisoquinolines in neutral or alkaline solution.

EXPERIMENTAL PROCEDURES

6-Methoxy-1,2,3,4-tetrahydroisoquinoline.[25] (Schiff base isolated and condensed with hydrochloric acid.) Twenty-five grams of 20% formaldehyde solution was added dropwise to 24.5 g. of β-(*m*-methoxyphenyl)-ethylamine. The warm, clear solution soon deposited an oil and the reaction was completed by heating the mixture for one hour on the water bath. The oil was extracted with benzene, and the extract was washed with water. Distillation of the benzene left the azomethine, a viscous, colorless oil (100%), which was dissolved in 32 g. of 20% hydrochloric acid and evaporated to dryness on the water bath. The crystalline mass was dissolved in a little water, made alkaline with concentrated potassium hydroxide solution, and extracted with ether. Distillation of the extract yielded 21.3 g. (80%) of the pure tetrahydroisoquinoline, b.p. 143–144°/6 mm.

1-Methyl-6,7-dihydroxy-1,2,3,4-tetrahydroisoquinoline.[10] (Use of an aldehyde under physiological conditions.) A solution of 1.87 g. of β-(3,4-dihydroxyphenyl)ethylamine hydrobromide and 0.79 g. of acetaldehyde in 200 ml. of water was maintained at 25° for three days. The

pH of the unbuffered solution remained at 5 through the experiment. The solution was evaporated to dryness in vacuum at 25°, and the residual oil crystallized overnight after being seeded. It was recrystallized from constant-boiling hydrobromic acid to yield 1.73 g. (83%) of the tetrahydroisoquinoline hydrobromide, m.p. 182–184°.

1-Methyl-1-carboxy-6,7-dihydroxy-1,2,3,4-tetrahydroisoquinoline.[15] (Use of pyruvic acid under physiological conditions.) A solution of 0.9 g. of β-(3,4-dihydroxyphenyl)ethylamine and 0.4 g. of pyruvic acid in 5 ml. of water was adjusted to pH 4 by the addition of a few drops of concentrated aqueous ammonia. After four days at 25°, 0.84 g. of product crystallized in long needles. The mother liquor was concentrated to 2 ml., yielding a further 0.15 g. of product. In all, there was obtained 0.99 g. (93%) of the desired amino acid, decomposition temperature 230–235° (rapid heating).

2,3-Methylenedioxy-10,11-dimethoxy-5,6,13,13a-tetrahydro-8H-dibenzo[a,g]quinolizine.[30] (Cyclization with formaldehyde and concentrated hydrochloric acid.) A solution of 5 g. of 1-veratryl-6,7-methylenedioxy-1,2,3,4-tetrahydroisoquinoline in 10 ml. of methanol was warmed for a few minutes with 10 ml. of 40% formaldehyde solution. The azomethine was freed of excess formaldehyde by washing several times with water, then dissolved in concentrated hydrochloric acid and heated a few minutes on the steam bath. Filtration of the cooled mixture yielded the sparingly soluble hydrochloride of the product; it was decomposed by treatment with sodium hydroxide, and the resulting base was recrystallized from ethanol. The product obtained in 60% yield melted at 177°.

2,3,9,10-Tetramethoxy-7,12,12a,13-tetrahydro-5H-dibenzo[b,g]-quinolizine.[42] (Use of formalin and dilute hydrochloric acid.) A solution containing 1 g. of 3-veratryl-6,7-dimethoxy-1,2,3,4-tetrahydroisoquinoline hydrochloride in 6 ml. of 2 N hydrochloric acid was treated with 2 ml. of 40% formaldehyde solution and heated for thirty minutes on the water bath. The hydrochloride, which crystallized as the solution cooled, was recrystallized from dilute hydrochloric acid to furnish 1 g. (97%) of yellow needles, m.p. 272°.

1-Methyl-1,2,3,4-tetrahydro-2-carboline.[5] (Use of acetaldehyde in dilute sulfuric acid.) A solution of 5 g. of tryptamine in 100 ml. of water and 16 ml. of 2 N sulfuric acid was added to 100 ml. of a 10% acetaldehyde solution. The solution was gradually heated in an oil bath to 110° and maintained at that temperature for twenty minutes. The cooled solution was treated with excess sodium carbonate solution, which precipitated a crystalline solid. The solid was dissolved in dilute hydrochloric acid, filtered, and treated with sodium hydroxide; the precipitate was extracted with ether. Evaporation of the ether yielded the product as a crystalline residue weighing 5.0 g. (86%). After recrystallization from 50% ethanol, the carboline melted at 179–180°.

1-Benzyl-1,2,3,4-tetrahydro-2-carboline.[11] (Condensation with phenylacetaldehyde under physiological conditions.) A mixture of 4 ml. of 0.2 M tryptamine hydrochloride (150 mg.) and 4 ml. of phosphate buffer (pH 6.2) was treated with 150 mg. of phenylacetaldehyde, shaken vigorously, and then allowed to stand at 25° for twenty-four hours. The unreacted aldehyde was removed by extraction with ether, and the phosphate of the product was collected by filtration. It was dissolved in water, and the base was freed by addition of ammonia. The dried product was dissolved in methanol and converted to its hydrochloride by saturation with dry hydrogen chloride. The sparingly soluble salt weighed 180 mg. (90%) and melted at 278°.

1-Benzyl-1-carboxy-1,2,3,4-tetrahydro-2-carboline.[13] (Use of phenylpyruvic acid under physiological conditions.) A solution of 0.82 g. of phenylpyruvic acid and 1 g. of tryptamine hydrochloride in 25 ml. of water and 15 ml. of acetate buffer (pH 3.8) was placed in a thermostat at 37°. After a few hours a yellow precipitate began to separate; after seven days the precipitate weighed 0.9 g. (59%); after thirteen days the yield was 1.15 g. (75%). The amino acid was dissolved in aqueous ammonia and precipitated as fine needles by boiling off the ammonia; it decomposed at 253° with evolution of carbon dioxide.

TABULAR SURVEY OF THE PICTET-SPENGLER REACTION

Explanation of Tables. It has been intended to include in the following tables all examples of the Pictet-Spengler reaction published before July, 1949. The compounds in the tables are listed in order of increasing substitution upon the basic nucleus. Among compounds having the same number of substituents, precedence has been given those having a substituent at the point of ring closure (position 1 for isoquinolines and 2-carbolines). Compounds with a substituent at the point of cyclization have been arranged in order of increasing complexity of that substituent (alkyl, aryl, aralkyl, heterocyclic). Data for more than one preparation of a single compound are listed in order of increasing yield.

Duration of the reaction has been indicated, where possible, for syntheses under physiological conditions by an additional entry in the column "Condensing Agent"; the abbreviations used are hr. (hours), d. (days), wk. (weeks).

Nearly all patents were consulted in the original, although secondary references are given for the convenience of the reader.

TABLE II

1,2,3,4-Tetrahydroisoquinolines

$$\underset{\text{CH}_2}{\underset{|}{\overset{\text{CH}_2}{\underset{7}{\overset{6}{\bigcirc}}\underset{8}{\overset{5}{}}}\,}}\overset{\overset{\text{CH}_2}{|}}{\underset{1}{\overset{4}{}}\underset{2}{\overset{3}{\text{CH}_2}}}\text{NH}$$

Substituents	Condensing Agent	Temperature, °C.	Yield %	Reference
None	HCl	100	—	70
	HCl	140	Trace	44
	HCl	100	Poor	44
	HCl	100	36	2
3-Carboxy-	HCl	100	—	70, 71
	HCl	100	37	2
	HCl	100	61	72
6-Methoxy-	HCl	100	—	26
	HCl	—	80	25
6-Ethoxy-	HCl	100	—	50
7-Hydroxy-	HCl	100	—	48
2-Methyl-6-methoxy-	HCl	100	—	26
2-Methyl-6-ethoxy-	HCl	100	—	50, 46
3-Phenyl-6-methoxy-	HCl	—	95	61
3-Carboxy-6-methoxy-	HCl	100	—	73
3-Carboxy-7-hydroxy-	HCl	100	0	19
	HCl	100	70	2
	HCl	100	100	49
4-Hydroxy-5-methoxy-	HCl	60	100	47
5,6-Dimethoxy-	HCl	100	—	26
5-Ethoxy-6-methoxy-	HCl	100	—	50
6,7-Methylenedioxy-	HCl	—	—	74
	HCl	100	—	67, 70
	HCl	100	19	75
	HCl	100	60–70	30, 59
	HCl	100	85	3
6,7-Dimethoxy-	HCl	100	—	26
	HCl	100	61	59
6-Methoxy-7-ethoxy-	HCl	100	—	50
6-Ethoxy-7-methoxy-	HCl	100	—	50
6,7-Diethoxy-	HCl	100	—	50
x,x-Methylenedioxy-	HCl	100	—	26
1-Methyl-6,7-dihydroxy-	pH 5; 3 d.	25	83	10
1-Methyl-6,7-methylenedioxy-	HCl	80	—	67
	pH 5; 16 d.	25	0	59

TABLE II—Continued

1,2,3,4-Tetrahydroisoquinolines

Substituents	Condensing Agent	Temperature, °C.	Yield %	Reference
1-Phenyl-6,7-methylenedioxy-	POCl$_3$	80	—	67
	HCl	100	34	59
	HCl	—	Good	3
	BF$_3$ or SOCl$_2$	100	Trace	59
	pH 5; 16 d.	25	0	59
1-Phenyl-6,7-dimethoxy-	HCl	100	56	59
1-(o-Hydroxyphenyl)-6,7-dimethoxy-	HCl	100	78	59
1-(p-Hydroxyphenyl)-6,7-dimethoxy-	HCl	100	83	59
1-(p-Methoxyphenyl)-6,7-dimethoxy-	HCl	80	75	59
1-(o-Chlorophenyl)-6,7-dimethoxy-	HCl	100	74	59
1-(m-Nitrophenyl)-6,7-dimethoxy-	HCl	100	80	59
1-(2-Hydroxy-5-chlorophenyl)-6,7-dimethoxy-	HCl	100	68	59
1-(3,4-Methylenedioxyphenyl)-6,7-methylenedioxy-	POCl$_3$	110	—	67
1-(3,4-Diethoxyphenyl)-6,7-dimethoxy-	HCl	100	81	59
1-(3,4-Dihydroxybenzyl)-2-methyl-7-hydroxy-	pH 4.2	—	—	76
1-Piperonyl-6,7-dihydroxy-	pH 4; 14 d.	25	80	12
	pH 6; 1 d.	25	84	12
1-Piperonyl-6,7-methylenedioxy-	pH 5; 8 d.	25	5	53
	HCl	100	2	63, 54
1-Veratryl-6,7-dimethoxy-	HCl	100	7	62
1-(β-Phenethyl)-6,7-dihydroxy-	HBr	80	75–80	64
1-Styryl-6,7-methylenedioxy-	HCl	—	—	67
	HCl	80	70	59
2-(m-Methoxybenzyl)-3-carboxy-6-methoxy-	HCl	100	—	73
2-Methyl-5,6-dimethoxy-	HCl	100	—	26, 46
2-Methyl-5-ethoxy-6-methoxy-	HCl	100	—	50
2-Methyl-6,7-methylenedioxy-	H$_2$SO$_4$	100	—	9
	P$_2$O$_5$	110	—	7
	HCl	130	—	77, 78, 46
	HCl	130	88	3
2-Methyl-x,x-methylenedioxy-	HCl	100	—	26

SYNTHESIS OF ISOQUINOLINES 2

TABLE II—Continued

1,2,3,4-TETRAHYDROISOQUINOLINES

Substituents	Condensing Agent	Temperature, °C.	Yield %	Reference
2-Methyl-6,7-dimethoxy-	HCl	100	—	26
2-Methyl-6-methoxy-7-ethoxy-	HCl	100	—	50
2-Methyl-6-ethoxy-7-methoxy-	HCl	100	—	50
2-Methyl-6,7-diethoxy-	HCl	100	—	50
2-Ethyl-6,7-methylenedioxy-	H₂SO₄	90	—	9
	HCl	100	—	77, 7
3-Methyl-6,7-methylenedioxy-	HCl	100	—	8, 9, 7
3-Phenyl-6,7-methylenedioxy-	HCl	—	93	61
3-Phenyl-6,7-dimethoxy-	HCl	—	94	61
3-(3,4-Methylenedioxyphenyl)-6,7-methylenedioxy-	HCl	—	87	61
3-Veratroyl-6,7-methylenedioxy-	HCl	100	Good	39
6,7-Dimethoxy-8-hydroxy-	HCl	100	14–30	34
6,8-Dimethoxy-7-carbethoxy-	HCl	100	28	34
1-Methyl-1-carboxy-6,7-dihydroxy-	pH 4.2; 2 d.	25	95	79
	pH 4; 4 d.	25	92	15
1-Methyl-1-carboxy-6-hydroxy-7-methoxy-	pH 5; 20 hr.	25	85	66
1,2-Dimethyl-6,7-dihydroxy-	pH 4; 3 d.	25	—	10
1-Phenyl-2-ethyl-6,7-methylenedioxy-	HCl	150	—	77
	POCl₃	80	—	77
1-Benzyl-1-carboxy-6,7-dihydroxy-	pH 6; 5 d.	25	87	15
1-Benzyl-3-methyl-6,7-methylenedioxy-	HCl	—	—	80
1-(o-Hydroxybenzyl)-1-carboxy-6,7-dihydroxy-	HCl	100	71	15
1-(m-Hydroxybenzyl)-1-carboxy-6,7-dihydroxy-	pH 5; 12 d.	25	56	15
1-(m-Hydroxybenzyl)-1-carboxy-6-hydroxy-7-methoxy-	pH 7; 30 hr.	25	61	66
1-(p-Hydroxybenzyl)-1-carboxy-6,7-dihydroxy-	pH 3.8; 12 d.	25	41	79
	pH 5; 12 d.	25	84	79, 15
1-Vanillyl-1-carboxy-6,7-dihydroxy-	pH 4.2; 4 d.	25	67	79
	pH 6.5; 2 d.	25	68	15
1-Vanillyl-1-carboxy-6-hydroxy-7-methoxy-	pH 6.4; 9 d.	25	72	66
1-Piperonyl-3-methyl-6,7-methylenedioxy-	HCl	—	—	80

TABLE II—Continued

1,2,3,4-TETRAHYDROISOQUINOLINES

Substituents	Condensing Agent	Temperature, °C.	Yield %	Reference
1-(β-Phenethyl)-2-methyl-6,7-dihydroxy-	HBr	80	—	64
1-Styryl-2-methyl-6,7-dihydroxy-	HBr	80	—	64
1-Styryl-3-methyl-6,7-dihydroxy-	HBr	80	—	64
2,3-Dimethyl-6,7-methylenedioxy-	HCl	100	—	8, 9
2-(m-Methoxybenzyl)-6,7-methylenedioxy-8-methoxy- and 2-(m-Methoxybenzyl)-6-methoxy-7,8-methylenedioxy-	HCl	100	—	27
3-Phenyl-6,7,8-trimethoxy-	HCl	—	80	81
1-Methyl-1-carboxy-6,7,8-trihydroxy-	pH 4; 15 d.	25	70	66
	pH 7; 15 d.	25	88	66
1,3-Dimethyl-2-(γ-phenylpropyl)-6,7-dihydroxy-	HBr	90	—	64
1-Benzyl-1-carboxy-6,7,8-trihydroxy-	pH 8; 1 d.	25	78	66
1-(m-Hydroxybenzyl)-1-carboxy-6,7,8-trihydroxy-	pH 7; 1 d.	25	68	66
1-(p-Hydroxybenzyl)-1-carboxy-6,7,8-trihydroxy-	pH 3; 20 hr.	25	47	66
	pH 7; 20 hr.	25	75	66
1-Vanillyl-1-carboxy-6,7,8-trihydroxy-	pH 7.8; 2 d.	25	68	66
1-Isovanillyl-1-carboxy-6,7,8-trihydroxy-	pH 7; 2.5 d.	25	37	66
1-Veratryl-1-carboxy-6,7,8-trihydroxy-	pH 7; 6 d.	25	52	66
	pH 3; 6 d.	25	57	66
1-(3,4,5-Trimethoxybenzyl)-1-carboxy-6,7,8-trihydroxy-	pH 3; 4 d.	25	59	66
	pH 7; 4 d.	25	74	66

SUPPLEMENT TO TABLE II

UNSUCCESSFUL REACTIONS

Amine	Carbonyl Component	Conditions	Reference
β-(o-Ethoxyphenyl)ethylamine	Formaldehyde	HCl	50
β-(m-Aminoethylphenyl)ethylamine	Formaldehyde	HCl	82
	Benzaldehyde	Various	82
β-(p-Ethoxyphenyl)ethylamine	Formaldehyde	HCl	50
N-Benzylphenylalanine	Formaldehyde	HCl	73
	Methylal	HCl	73
N-Methyl-β-(o-ethoxyphenyl)-ethylamine	Formaldehyde	HCl	50
N-Methyl-β-(p-ethoxyphenyl)-ethylamine	Formaldehyde	HCl	50
β-(3-Methoxy-4-hydroxyphenyl)-isopropylamine	Methylal	HCl	52
ω-Aminoacetoveratrone	Formaldehyde	—	39
β-(p-Methoxyphenyl)-α-phenethylamine	Formaldehyde	HCl	61
β-(2,4-Dimethoxyphenyl)-α-phenethylamine	Formaldehyde	HCl	61
β-(p-Methoxyphenyl)ethylamine	Pyruvic acid	Physiol.	15
Homopiperonylamine	Pyruvic acid	Physiol.	15
Homoveratrylamine	3,4-Dimethoxyphenylpyruvic acid	Physiol.	15
	o-Ethoxybenzaldehyde	HCl	59
Adrenaline (and its ethers)	Pyruvic acid	Physiol.	15
	Pyruvic ester	Physiol.	15
	3,4-Dimethoxyphenylpyruvic ester	Physiol.	15
	Acetaldehyde	Physiol.	15
	Phenylacetaldehyde	Physiol.	15

[70] Clemo and Swan, *J. Chem. Soc.*, **1946**, 617.
[71] Pictet, Ger. pat. 241,425 [*Frdl.*, **10**, 1185 (1910–1912)].
[72] Julian, Karpel, Magnani, and Meyer, *J. Am. Chem. Soc.*, **70**, 180 (1948).
[73] Chakravarti and Rao, *J. Chem. Soc.*, **1938**, 172.
[74] Pictet and Gams, *Compt. rend.*, **152**, 1102 (1911).
[75] Pictet and Gams, *Ber.*, **44**, 2036 (1911).
[76] Schöpf, *Angew. Chem.*, **59**, 174 (1947).
[77] Decker, Ger. pat. 281,546 [*Frdl.*, **12**, 755 (1914–1916)].
[78] Decker, Ger. pat. 281,547 [*Frdl.*, **12**, 756 (1914–1916)].
[79] Hahn, Ger. pat. 646,706 [*Frdl.*, **24**, 414 (1937)].
[80] Wolfes, Ger. pat. 551,870 [*Frdl.*, **18**, 2766 (1931)].
[81] Reichert and Hoffmann, *Arch. Pharm.*, **274**, 217 (1936).
[82] Leupin and Dahn, *Helv. Chim. Acta*, **30**, 1945 (1947).

TABLE III

Benzisoquinolines and Naphthisoquinoline

Name	Condensing Agent	Temperature, °C.	Yield %	Reference
1,2,3,4-Tetrahydro-7,8-benzisoquinoline	HCl	100	11	36
1,2,3,4-Tetrahydro-5,6,7,8-dibenzisoquinoline	HCl	100	65	38
1,2,3,4-Tetrahydronaphth[1,2-*h*]-isoquinoline	HCl	100	70	38

Supplement to Table III
Amines That Would Not Condense with Formaldehyde

Name	Conditions	Reference
β-(1-Naphthyl)ethylamine	HCl	36
1-Aminomethyl-2-methoxynaphthalene	HCl	37
β-[2-(9,10-Dihydrophenanthryl)]ethylamine	HCl	83
β-[2-(7-Methoxy-9,10-dihydrophenanthryl)]ethylamine	HCl	83
β-(3-Phenanthryl)ethylamine	HCl	38

[83] Stuart and Mosettig, *J. Am. Chem. Soc.*, **62**, 1110 (1940).

TABLE IV
Benzoquinolizine and Dibenzoquinolizines

Name	Condensing Agent	Temperature, °C.	Yield %	Reference
1,2,3,4,6,11-Hexahydro-11aH-benzo-[b]quinolizine	HCl	—	Poor	84
5,6,13,13a-Tetrahydro-8H-dibenzo-[a,g]quinolizine	HCl	—	—	85
	HCl	—	0	86, 87
2,3-Methylenedioxy-	HCl	—	—	85
2,3,9,10-Tetrahydroxy- and 2,3,10,11-Tetrahydroxy-	—	100	15	31
	pH 5	—	90	32
2,3-Methylenedioxy-10-hydroxy-11-methoxy-	—	—	—	88
2,3,11-Trimethoxy-10-hydroxy-	HCl	—	—	85
2,3,10,11-Bis(methylenedioxy)-	HCl	100	55	54, 63
	HCl	—	80	89
2,3-Methylenedioxy-10,11-dimethoxy-	HCl	100	37	90, 30, 28, 29
	HCl	100	60	30
2,3-Dimethoxy-10,11-methylenedioxy-	—	—	—	91
2,3,10,11-Tetramethoxy-	HCl	100	46–69	31
	pH 4; 18 hr.	25	>80	4
	HCl	100	83–85	86, 92, 93
2,3,11,12-Tetramethoxy-	HCl	100	93	94
	H₂SO₄	100	—	95
	HCl	100	—	96

TABLE IV—Continued
Benzoquinolizines and Dibenzoquinolizines

Name	Condensing Agent	Temperature, °C.	Yield %	Reference
2,3,10,11-Tetramethoxy-8-methyl-	HCl	25	0	4
	HCl	100	82	4
	HCl	100	—	92, 97
2,3,10,11-Tetramethoxy-8-phenyl-	HCl	100	—	92
2,3,9,10-Tetrahydroxy-13-methyl- and 2,3,10,11-Tetrahydroxy-13-methyl-	—	100	—	33
2,3,-Methylenedioxy-10,11-dimethoxy-13-hydroxy-	HCl	—	—	90
2,3,9,10,11-Pentamethoxy-	HCl	100	53	98
2-Hydroxy-3,10,11-trimethoxy-13-methyl-5,6,13,13a-tetrahydro-8H-dibenzo[a,g]quinolizine	HCl	100	—	99
2,3,9,10-Tetramethoxy-5,6,7,12-tetrahydro-6aH-dibenzo[b,f]quinolizine	HCl	100	—	100
2,3-Dimethoxy-9,10-methylenedioxy-7,12,12a,13-tetrahydro-5H-dibenzo[b,g]quinolizine	HCl	100	20	39

TABLE IV—Continued
BENZOQUINOLIZINES AND DIBENZOQUINOLIZINES

Name	Condensing Agent	Temperature, °C.	Yield %	Reference
2,3,9,10-Tetramethoxy-7,12,12a,13-tetrahydro-5H-dibenzo[b,g]quinolizine	HCl HCl	— 100	— 97	40, 41 42
2,3,9,10-Tetramethoxy-5-veratryl-7,12,12a,13-tetrahydro-5H-dibenzo[b,g]quinolizine	HCl	100	—	42, 40

Supplement to Table IV

Unsuccessful Reactions

Name	Carbonyl Component	Conditions	Reference
1,3-Diphenylisopropylamine	Formaldehyde	—	101
1-(p-Methoxybenzyl)-6-methoxy-1,2,3,4-tetrahydroisoquinoline	Formaldehyde	—	51, 102
1-(2,3-Dimethoxybenzyl)-6,7-dimethoxy-1,2,3,4-tetrahydroisoquinoline	Acetal	—	95
1-Veratroyl-6,7-methylenedioxy-1,2,3,4-tetrahydroisoquinoline	Formaldehyde	—	39
1-(6-Nitroveratryl)-6,7-methylenedioxy-1,2,3,4-tetrahydroisoquinoline	Formaldehyde	--	103
1-(6-Bromoveratryl)-6,7-methylenedioxy-1,2,3,4-tetrahydroisoquinoline	Formaldehyde	—	103
1-Benzyl-6,7-methylenedioxy-1,2,3,4-tetrahydroisoquinoline	Formaldehyde	HCl	104

[84] v. Braun and Pinkernelle, *Ber.*, **64**, 1871 (1931).
[85] Kitasato, *Acta Phytochim.*, **3**, 215 (1927).
[86] Craig and Tarbell, *J. Am. Chem. Soc.*, **70**, 2783 (1948).
[87] Chakravarti, Haworth, and Perkin, *J. Chem. Soc.*, **1927**, 2275.
[88] Kitasato, *J. Pharm. Soc. Japan*, **523**, 791 (1925) [*C. A.*, **20**, 421 (1926)].
[89] Buck, Perkin, and Stevens, *J. Chem. Soc.*, **127**, 1462 (1925).
[90] Buck and Davis, *J. Am. Chem. Soc.*, **52**, 660 (1930).
[91] Buck and Perkin, *J. Chem. Soc.*, **125**, 1675 (1924).
[92] Pictet, Ger. pat. 281,047 (1913) [*Frdl.*, **12**, 749 (1914–1916)].
[93] Pictet and Chou, *Ber.*, **49**, 370 (1916).
[94] Hahn and Kley, *Ber.*, **70**, 685 (1937).
[95] Späth and Mosettig, *Ann.*, **433**, 138 (1923).
[96] Chakravarti and Swaminathan, *J. Indian Chem. Soc.*, **11**, 107 (1934).
[97] Pictet and Malinowski, *Ber.*, **46**, 2688 (1913).
[98] Späth and Meinhard, *Ber.*, **75**, 400 (1942).
[99] Haworth and Perkin, *J. Chem. Soc.*, **127**, 1453 (1925).
[100] Sugasawa, Kodama, and Inagaki, *Ber.*, **74**, 455 (1941).
[101] Chakravarti and Ganapati, *J. Annamalai Univ.*, **3**, 208 (1934) [*C. A.*, **29**, 1094 (1935)].
[102] Chakravarti, Vaidyanathan, and Venkatasubban, *J. Annamalai Univ.*, **1**, 190 (1932) [*C. A.*, **27**, 1351 (1933)].
[103] Haworth and Perkin, *J. Chem. Soc.*, **127**, 1448 (1925).
[104] Haworth, Perkin, and Pink, *J. Chem. Soc.*, **127**, 1709 (1925).

TABLE V

1,2,3,4-Tetrahydro-2-carbolines

Substituents	Condensing Agent	Temperature, °C.	Yield %	Reference
None	H_2SO_4	100	65	105
1-Methyl-	—	—	—	106
	pH 7; 3 d.	25	35	11
	pH 5–6; 3 d.	25	70	11
	H_2SO_4	110	86	5
1-Trichloromethyl-	—	—	—	56
1-Phenyl-	CH_3I	100	50	68
	pH 5.2; 3 wk.	25	48	13
1-Benzyl-	pH 6.2; 1 d.	25	90	11
1-(m-Hydroxybenzyl)-	HCl	100	36	17
1-Piperonyl-	pH 6.2; 8 d.	25	15	13
1-(3,4,5-Trimethoxybenzyl)-	pH 6.2; 10 d.	25	16	13
1,1′-Trimethylenebis-	HCl	45	60	17
1-Furyl-	Physiological conditions	25	10	13
3-Carboxy-	$(CH_3CO)_2O$	25	—	49
	H_2SO_4	25	—	57, 58, 21
	NaOH	37	—	107, 108
	pH 6.5	38	80	19
6-Methoxy-	H_2SO_4	70	23	105
8-Methoxy-	H_2SO_4	70	50	105
9-Methyl-	H_2SO_4	70	75	105
9-Ethyl-	H_2SO_4	70	82	109
1-Methyl-1-carboxy-	pH 5.2	37	66	13
	pH 6.2	25	100	16
1,2-Dimethyl-	H_2SO_4	100	58	110
	H_2SO_4	110	80	111
1-Methyl-3-carboxy-	H_2SO_4	100	—	57, 58, 21, 112
	—	60–80	62–67	113, 65
	H_2SO_4	100	66	55
	—	25	100	114
1-Methyl-7-methoxy-	H_2SO_4	110	85	5
1-Hydroxymethyl-3-carboxy-	H_2SO_4	100	—	19

TABLE V—Continued

1,2,3,4-Tetrahydro-2-carbolines

Substituents	Condensing Agent	Temperature, °C.	Yield %	Reference
1-Ethyl-3-carboxy-	H_2SO_4	—	—	109
1-(β-Hydroxypropyl)-3-carboxy-	H_2SO_4	100	—	57, 58
1-(β-Carboxyethyl)-1-carboxy-	pH 3.8; 2 d.	25	45	16, 17
1,3-Dicarboxy-	H_2SO_4	100	50	19
1-Phenyl-3-carboxy-	H_2SO_4	—	—	57, 58
	H_2SO_4	—	71	55
1-(p-Dimethylaminophenyl)-3-carboxy-	H_2SO_4	100	85	55
1-(p-Nitrophenyl)-3-carboxy-	H_2SO_4	100	88	55
1-Benzyl-1-carboxy-	pH 3.8; 13 d.	37	75	13
1-Benzyl-3-carboxy-	H_2SO_4	90	81	55
1-(m-Hydroxybenzyl)-1-carboxy-	HCl	100	10	17
	pH 4.2; 10 d.	25	85	16, 14
1-(p-Hydroxybenzyl)-1-carboxy-	pH 4.2; 10 d.	25	74	16, 14
1-(p-Methoxybenzyl)-3-carboxy-	H_2SO_4	—	92	55
1-Vanillyl-1-carboxy-	pH 6.2; 10 d.	25	57	16, 13
1-Piperonyl-1-carboxy-	pH 6.2; 7 d.	25	61	13
1-Veratryl-1-carboxy-	HCl	100	43	17
	pH 4.2; 28 d.	25	54	13
	pH 5.2; 10 d.	25	86	16
1-(3,4,5-Trimethoxybenzyl)-1-carboxy-	pH 5.3; 7 d.	25	41	13
1-(α-Phenethyl)-3-carboxy-	H_2SO_4	100	62	55
2-Methyl-3-carboxy-	—	38	76	19
1,2-Dimethyl-3-carboxy-	H_2SO_4	100	14	58
	HCl	25	90	19
1-Methyl-3-carboxy-6-bromo-	H_2SO_4	100	49	65
1-Methyl-3-carboxy-7-methoxy-	—	—	100	114
1-Phenyl-2-methyl-3-carboxy-	H_2SO_4	100	—	57, 58
	H_2SO_4	85	70	19
1-(m-Hydroxybenzyl)-1-carboxy-7-methoxy-	pH 4.2	25	80	16

Supplement to Table V

Unsuccessful Reactions

Amine	Carbonyl Component	Conditions	Reference
Tryptamine	o-Hydroxybenzaldehyde	Physiological	13
	p-Hydroxybenzaldehyde	Physiological	13
	Piperonal	Physiological	13
	Vanillin	Physiological	13
	o-Nitrobenzaldehyde	—	60
	Glyoxal	Physiological	13
	Methylglyoxal	Physiological	13
	D(+)Glucose	Physiological	13
	Citral	Physiological	13
	o-Nitrophenylpyruvic acid	—	60
	o-Cyanophenylpyruvic acid	HCl, 80°	60
	β-Indolylpyruvic acid	Physiological	115
Tryptophan	Crotonaldehyde	H_2SO_4	57
	Chloral hydrate	H_2SO_4	65
	Chloroacetal	H_2SO_4	65
	Formamide	H_2SO_4	65

[105] Späth and Lederer, *Ber.*, **63**, 2102 (1930).
[106] Tatsui, *J. Pharm. Soc. Japan*, (555) **48**, 92 (1928) (in English).
[107] Snyder, Walker, and Werber, *J. Am. Chem. Soc.*, **71**, 527 (1949).
[108] Speitel and Schlittler, *Helv. Chim. Acta*, **32**, 860 (1949).
[109] Leonard and Elderfield, *J. Org. Chem.*, **7**, 556 (1942).
[110] Barger, Jacob, and Madinaveitia, *Rec. trav. chim.*, **57**, 548 (1938).
[111] Yurashevskii, *J. Gen. Chem. U.S.S.R.*, **11**, 157 (1941) [*C. A.*, **35**, 5503 (1941)].
[112] Mookerjee, *J. Indian Chem. Soc.*, **20**, 11 (1943).
[113] Otani, *Z. physiol. Chem.*, **214**, 30 (1933).
[114] Harvey and Robson, *J. Chem. Soc.*, **1938**, 97.
[115] Gudjons, Dissertation, Frankfurt, 1938. Compare ref. 66.

TABLE VI
Miscellaneous Compounds

Name	Condensing Agent	Temperature, °C.	Yield %	Reference
4,5,6,7-Tetrahydro-1H-imidazo[c]-pyridine	HCl	100	—	48, 116
6-Carboxy-3-Oxo-8,9-dihydroxy-2,3,5,6-tetrahydro-10b-carboxy-1H-benzo[g]-pyrrocoline	HCl	100	75	49
	HCl	100	38	15
1,2,3,4-Tetrahydrobenzofuro[2,3-f]-isoquinoline	HCl	—	—	117
4-Carboxy-1,2,6,7,12,12b-hexahydro-indolo[2,3-a]quinolizine	HCl	45	58	17

TABLE VI—Continued

Miscellaneous Compounds

Name	Condensing Agent	Temperature, °C.	Yield %	Reference
2-Hydroxy-5,7,8,13,13b,14-hexahydrobenz[g]indolo[2,3-a]quinolizine	pH 4.4 pH 4.2; 1 d.	— —	88 67	14 118
2-Methoxy-3-hydroxy-5,7,8,13,13b,14-hexahydrobenz[g]indolo[2,3-a]quinolizine	HCl	—	72	118
2,3-Dimethoxy-5,7,8,13,13b,14-hexahydrobenz[g]indolo[2,3-a]quinolizine	HCl, 12 d.	—	57	118

SUPPLEMENT TO TABLE VI

UNSUCCESSFUL REACTIONS

Amine	Carbonyl Component	Condensing Agent	Reference
Histidine	Acetaldehyde	HCl	49
	Pyruvic acid	—	49

[116] Ges. Chem. Ind. Basel, Swiss pat. 92,297 [*C. A.*, **17**, 2119 (1923)].
[117] Kirkpatrick, *Iowa State Coll. J. Sci.*, **11**, 75 (1936) [*C. A.*, **31**, 1800 (1937)].
[118] Hahn and Hansel, *Ber.*, **71**, 2192 (1938).

CHAPTER 4

THE SYNTHESIS OF ISOQUINOLINES BY THE POMERANZ-FRITSCH REACTION

WALTER J. GENSLER

Boston University

CONTENTS

	PAGE
INTRODUCTION	192
MECHANISM OF CYCLIZATION	193
SCOPE AND LIMITATIONS	193
Formation of the Schiff Base	193
Cyclization	194
Factors Affecting Yield	194
Orientation	195
Extension and Variation of the Pomeranz-Fritsch Synthesis	196
Application	199
EXPERIMENTAL PROCEDURES	199
Aminoacetal from Chloroacetal	199
8-Bromoisoquinoline from o-Bromobenzaldehyde and Aminoacetal	200
7-Hydroxy-8-chloroisoquinoline from 2-Chloro-3-hydroxybenzaldehyde and Aminoacetal	200
1-Methyl-6,7-dimethoxy-8-hydroxyisoquinoline from 2-Benzyloxy-3,4-dimethoxyacetophenone and Aminoacetal	201
Isoquinoline from Benzylamine and Glyoxal Semiacetal	201
1-Methyl-7-methoxyisoquinoline from α-(3-Methoxyphenyl)ethylamine and Glyoxal Semiacetal	202
TABLES OF POMERANZ-FRITSCH SYNTHESES	202
I. Isoquinolines with No Substituent at the 1 Position	203
II. Isoquinolines Substituted at the 1 Position	204
III. Unsuccessful Variations	205

INTRODUCTION

Acid-catalyzed cyclization of benzalaminoacetal (I) results in formation of the isoquinoline nucleus. This reaction, first reported by

C₆H₅CHO + H₂NCH₂CH(OC₂H₅)₂ → C₆H₅CH=NCH₂CH(OC₂H₅)₂ (I) → isoquinoline

Pomeranz [1,2,3] and by Fritsch,[4,5] has been utilized in the synthesis of a variety of isoquinoline compounds.

The process is carried out in two stages: the first a condensation leading to the benzalaminoacetal, and the second a ring closure leading to the isoquinoline. In the first step, in which the Schiff base is formed by the reaction of an aromatic aldehyde and aminoacetal, the yields are generally high and the reaction smooth. An alternative route involves condensation of the corresponding benzylamine with glyoxal semiacetal.[6] Cyclization of the benzalaminoacetal prepared in either manner is effected with sulfuric acid, or with sulfuric acid mixed with other acidic reagents. The yield of the isoquinoline cyclization products varies widely.

Extension of the Pomeranz-Fritsch method to the use of a ketimine in place of an aldimine (and thus to the synthesis of 1-substituted isoquinolines) has been realized, but the results reported are either poor or negative. Various attempts to cyclize compounds more or less closely related in structure to benzalaminoacetal have failed to yield isoquinolines as products.

The Pomeranz-Fritsch synthesis offers the possibility of preparing isoquinolines with substituent groups in an orientation often difficult to attain in the Bischler-Napieralski or the Pictet-Spengler syntheses. The Pomeranz-Fritsch synthesis thus supplements these other two methods. Furthermore, it differs from them in that the product is a fully aromatic isoquinoline, whereas in most of the phenethylamine reactions the products are partially hydrogenated isoquinolines.

[1] Pomeranz, *Monatsh.*, **14**, 116 (1893).
[2] Pomeranz, *Monatsh.*, **15**, 299 (1894).
[3] Pomeranz, *Monatsh.*, **18**, 1 (1897).
[4] Fritsch, *Ber.*, **26**, 419 (1893).
[5] Fritsch, *Ann.*, **286**, 1 (1895).
[6] Schlittler and Müller, *Helv. Chim. Acta*, **31**, 914 (1948).

MECHANISM OF CYCLIZATION

Bradsher [7] has pointed out the relation between Pomeranz-Fritsch cyclizations and the general class of aromatic cyclodehydration reactions, and has proposed a mechanism involving intramolecular aromatic substitution. Certainly the use of strong acids in bond formation between the acetal carbon and the benzene nucleus suggests the operation of an electrophilic process. If so, ease of cyclization would depend on the susceptibility of the benzene ring to electrophilic attack. Thus we find that *meta* alkoxy and hydroxy derivatives (which possess active *para* positions accessible to the attacking group) react under relatively mild conditions; that benzaldehyde and halogen-substituted derivatives require higher temperatures and more concentrated acid; and that nitrobenzalaminoacetal with a nucleus of low activity fails to react at all.[8] One factor tending to deactivate the aromatic ring is operative in all cases, namely, the fact that in the aldimmonium grouping II, in which form the Schiff base would exist in strong acid solution, there is effective electron withdrawal from the ring and therefore deactivation to electrophilic attack.

II

Details of the cyclization process are not known; whether the Schiff base reacts as acetal, as a vinyl ether, or as the free aldehyde is a matter of speculation.

SCOPE AND LIMITATIONS

Formation of the Schiff Base. Condensation of aromatic aldehydes with aminoacetal occurs readily and in excellent yield. The product may be used in the cyclization step either directly or after purification by crystallization or distillation. The condensation can be carried out by allowing a mixture of aldehyde and aminoacetal to stand at room temperature or on the steam bath. An alternative method, first reported by Schlittler and Müller,[6] is available in the reaction of a benzylamine with glyoxal semiacetal.* Cyclization of the product so obtained

---

* Permanganate hydroxylation of acrolein acetal affords glyceraldehyde acetal which, on cleavage with lead tetraacetate, furnishes glyoxal semiacetal. The overall yield for the two steps is 18%. See Fischer and Baer, *Helv. Chim. Acta*, **18**, 514 (1935).

[7] Bradsher, *Chem. Revs.*, **38**, 447 (1946).

[8] Andersag, *Medicine in Its Chemical Aspects*, Vol. II, p. 359, I.G. Farbenindustrie A.G., Leverkusen, 1934. No experimental details are given.

(e.g., III) furnishes the same isoquinoline as that obtained from the Schiff base derived from the aromatic aldehyde and aminoacetal. The Schiff bases formed in either manner may be isomers, or mixtures of tautomeric forms,[9] or the same compound.[6]

PhCH$_2$NH$_2$ + CHOCH(OC$_2$H$_5$)$_2$ →

PhCH$_2$N=CHCH(OC$_2$H$_5$)$_2$ → isoquinoline

III

Cyclization. Although a variety of methods has been reported for the cyclization step, all involve the use of sulfuric acid. Sulfuric acid has been used alone, in concentrations ranging from fuming acid to approximately 70% sulfuric acid, and in admixture with such reagents as gaseous hydrogen chloride, acetic acid, phosphorus pentoxide, or phosphorus oxychloride. Pomeranz[2] reported that in the absence of sulfuric acid benzalaminoacetal is not cyclized by zinc chloride, phosphorus pentachloride, phosphorus oxychloride, phosphoric acid, acetic anhydride, or oxalic acid. Use of fluorosulfonic acid with m-chlorobenzalaminoacetal results only in polymeric materials.[10]

Temperatures at which the cyclization reactions have been carried out range from 0° or below (with reactive nuclei such as in alkoxy- or hydroxy-benzalaminoacetals) to 150–160° (with unreactive nuclei such as in halobenzalaminoacetals).

Factors Affecting Yield. The yields reported for Pomeranz-Fritsch syntheses vary from zero to more than 80%. However, for the most part, the yields are below 50%. Gratifying results are obtained with m-alkoxy-, m-hydroxy-, and m-halo-benzalaminoacetals. On the other hand o- or p-alkoxy or hydroxy derivatives form isoquinolines in low yield or fail altogether to furnish the product. 8-Chloro-, 5-(and 7-)-chloro-, and 6-chloro-isoquinoline are formed in yields of 9%, 50%, and 14%, respectively. The corresponding bromoisoquinolines are formed in yields of 29%, 65%, and 24%, respectively.

The yield of isoquinoline can vary markedly with the conditions employed in cyclization, and especially with the concentration of sulfuric acid. The sensitivity of yield to acid concentration is well illustrated

---

[9] Hsü, Ingold, and Wilson, *J. Chem. Soc.*, **1935**, 1778.
[10] Manske and Kulka, *Can. J. Research*, **27B**, 161 (1949).

by results obtained with m-ethoxybenzal-,[5] m-hydroxybenzal-,[11] and 3,4-methylenedioxybenzal-aminoacetal.[5] A small deviation from the optimum acid concentration results in appreciable decrease in the yield of isoquinoline as is shown in the accompanying table.

Yields of Isoquinoline Cyclization Products with Varying Sulfuric Acid Concentration

Sulfuric acid solutions of m-ethoxybenzalaminoacetal were held at 50° for five hours; m-hydroxybenzalaminoacetal was allowed to stand in acid, first at 3–5° (twelve hours) and then at room temperature; the methylenedioxy derivative in sulfuric acid saturated with hydrogen chloride was kept for ten days at 0° and then four days at room temperature.

$m\text{-}C_2H_5OC_6H_4CH\!=\!NCH_2CH(OC_2H_5)_2$ $m\text{-}HOC_6H_4CH\!=\!NCH_2CH(OC_2H_5)_2$

Acid Concentration %	Yield [5] %	Acid Concentration %	Yield [11] %
92.2	4.5	84	31
86.4	28.5	82	44
81.3	67.5	80	64
76.5	79.7	78	59
72.8	70.0	76	43
69.1	49.0	72	30
62.8	15.5		

$3,4\text{-}CH_2O_2C_6H_3CH\!=\!NCH_2CH(OC_2H_5)_2$

Acid Concentration %	Yield [5] %
73.6	19.1
72.6	23.6
69	18.3

Variation of yield with acid concentration may be attributed, at least in part, to the fact that hydrolytic cleavage of the Schiff base may occur under conditions of cyclization. It is possible that the effect is due to change in the relative rates of cyclization and hydrolysis, so that, when cyclization is slow compared to the competing hydrolysis, the yield of isoquinoline is low.

Other factors that must be taken into account include the possibility of disruption (aside from hydrolysis) of the starting material as well as the destruction of the product during the reaction.

Orientation. Cyclization of unsymmetrically substituted benzal-aminoacetals in which the two positions *ortho* to the aldimine group are

[11] Woodward and Doering, *J. Am. Chem. Soc.*, **67**, 860 (1945).

unoccupied may lead to one or both of two isomeric isoquinolines. In several such cases the composition of the product is known. For example, m-ethoxybenzalaminoacetal affords a single product in more than 80% yield.[5] That this material is 7-ethoxyisoquinoline and not 5-ethoxyisoquinoline is shown by oxidation of the isoquinoline to 4-ethoxyphthalic acid. m-Hydroxybenzalaminoacetal is transformed to a mixture consisting mainly of 7-hydroxyisoquinoline together with some 5-hydroxyisoquinoline.[5, 11] The structure of the former compound is demonstrated by its conversion to 7-ethoxyisoquinoline. The 5-hydroxyisoquinoline is identical with the product obtained by alkali fusion of isoquinoline-5-sulfonic acid.[11] A mixture of 5- and 7-chloroisoquinoline is obtained from m-chlorobenzalaminoacetal.[8, 10] In one experiment, the main product was found to be 5-chloroisoquinoline; in another experiment, the two isomers were obtained in equal amounts. m-Bromobenzalaminoacetal is transformed to 5- and 7-bromoisoquinoline in approximately equal amounts.[12]

3,4-Methylenedioxybenzalaminoacetal yields only 6,7-methylenedioxyisoquinoline,[5] the structure of which is shown by relating the compound to the 6,7-disubstituted reduced isoquinolines, hydrastinin and hydrohydrastinin. Similarly, 3,4-dimethoxybenzylaminoacetal yields 6,7-dimethoxyisoquinoline on oxidative cyclization.[13, 14] Orientation in the product is shown by comparing the reduced compound, 1,2,3,4-tetrahydro-6,7-dimethoxyisoquinoline, with a degradation product from papaverine.

Extension and Variation of the Pomeranz-Fritsch Synthesis. When a ketone is used in the Pomeranz-Fritsch synthesis in place of an aromatic aldehyde, the product is a 1-substituted isoquinoline. Acetophenone, for example, leads to 1-methylisoquinoline (IV). For the most part,

poor results are obtained in this extension of the synthesis. The difficulty may lie in the reluctance with which ketones combine with aminoacetal to yield Schiff bases. Attempts have been made to carry out

[12] Tyson, *J. Am. Chem. Soc.*, **61**, 183 (1939).
[13] Forsyth, Kelly, and Pyman, *J. Chem. Soc.*, **127**, 1659 (1925).
[14] Rügheimer and Schön, *Ber.*, **42**, 2374 (1909).

the synthesis with acetophenone and with benzophenone by adding a mixture of ketone and aminoacetal directly to hot sulfuric acid, thereby eliminating the separate condensation step.[2] The expected products were obtained, but in low yield.

It is in the synthesis of 1-substituted isoquinolines that the Schlittler-Müller preparation of the Schiff bases may offer real advantage. In place of the difficult ketone-aminoacetal condensation of the conventional method, a relatively facile amine-aldehyde condensation is employed. By this method, α-phenylethylamine (V) is first converted to the Schiff base with glyoxal semiacetal and then, on treatment with

$$\text{Ph-CH(CH}_3\text{)-NH}_2 + \text{CHOCH(OC}_2\text{H}_5)_2 \rightarrow$$

V

$$\text{Ph-CH(CH}_3\text{)-N=CH-CH(OC}_2\text{H}_5)_2 \rightarrow \text{1-methylisoquinoline}$$

concentrated sulfuric acid at 160°, to 1-methylisoquinoline.[6] The yield is given as 40%, a substantial improvement over the yield obtained starting with acetophenone and aminoacetal. Similarly, in the preparation of 1-methyl-7-methoxyisoquinoline the yield from α-(m-methoxyphenyl)-ethylamine is 37.5%, whereas the yield from m-methoxyacetophenone and aminoacetal is only 0.1%.[6]

Difficulty in formation of ketimines cannot be the only factor contributing to low yields in syntheses of 1-substituted isoquinolines. In at least one example in which a purified Schiff base is prepared from an acetophenone and aminoacetal,[15] and in several cases in which Schiff bases are prepared according to Schlittler and Müller,[6,10] the yields of cyclization products are either very low or nil.

Only one example has been found in which a substituted aminoacetal is successfully utilized in the Pomeranz-Fritsch synthesis. When 3-aminobutanone ketal is used with benzaldehyde, the expected product, 3,4-dimethylisoquinoline (VI), is obtained.[17] However, the yield is

[15] Späth and Becke, *Ber.*, **67**, 266 (1934).
[16] Schlittler and Müller, *Helv. Chim. Acta*, **31**, 1119 (1948).
[17] Witkop, *J. Am. Chem. Soc.*, **70**, 1424 (1948).

$$\text{C}_6\text{H}_5\text{CHO} + \text{H}_2\text{NCHC(OC}_2\text{H}_5)_2 \rightarrow$$
$$\underset{\text{H}_3\text{C}\ \text{CH}_3}{}$$

[Intermediate imine with C(OC$_2$H$_5$)$_2$, CHCH$_3$ groups on N=CH–C$_6$H$_5$] → 1,3-dimethylisoquinoline (VI)

evidently very small. It should be noted in this connection that condensations of benzylamines and substituted glyoxals have been attempted. This variation of the Schlittler-Müller procedure gives negative results. Thus, piperonylamine and phenylglyoxal do not condense to yield 4-phenyl-6,7-methylenedioxyisoquinoline.[18]

Fischer reported that cold fuming sulfuric acid, in an oxidative process, converts benzylaminoacetaldehyde (VII) to isoquinoline.[19,20] A similar

[Structure VII: C$_6$H$_5$–CH$_2$–NH–CH$_2$–CHO] → isoquinoline

reaction, with arsenic pentoxide in sulfuric acid as the oxidizing agent, has been used in the cyclization of 3,4-dimethoxybenzylaminoacetal to 6,7-dimethoxyisoquinoline.[13,14] It is noteworthy that none of this isoquinoline could be obtained from 3,4-dimethoxybenzalaminoacetal. Other attempts at oxidative cyclization have failed. Thus N-(3-methoxy-4,5-methylenedioxybenzyl)-,[21] N-[1,2-di-(3,4-dimethoxyphenyl)-ethyl]-, and N-[1,2-di-(3,4-methylenedioxyphenyl)ethyl]-aminoacetal[22] do not furnish the expected products. Judging from these results, this variation of the Pomeranz-Fritsch synthesis appears not particularly useful.

Many and indeed steadily recurring attempts have been made to form the isoquinoline system by methods that are related to the Pomeranz-Fritsch synthesis in so far as the pyridine ring is to be formed by juncture of the number-four carbon atom and the benzene nucleus.

[18] Dey and Govindachari, *Arch. Pharm.*, **275**, 383 (1937).
[19] Fischer, *Ber.*, **26**, 764 (1893).
[20] Fischer, *Ber.*, **27**, 165 (1894).
[21] Rügheimer and Ritter, *Ber.*, **45**, 1340 (1912).
[22] Allen and Buck, *J. Am. Chem. Soc.*, **52**, 310 (1930).

All such attempts, except the pyrolysis of benzalethylamine which yields isoquinoline,[22a] have proved futile (Table III). An incomplete list of compounds subjected to cyclizing conditions includes N-benzylethanolamine, N-benzyl-N-tosylglycyl chloride, hippuric acid, N-piperonyl-N-methyl-α-aminoacetophenone, ethyl 2,3-dimethoxybenzalaminoacetate, and N-benzyl-N-methyloxalamide.

Application. The usefulness of the Pomeranz-Fritsch isoquinoline synthesis as a general preparative method is severely limited by the yields obtained. Actually, only *m*-hydroxy-, *m*-alkoxy-, and *m*-halobenzaldehyde have been converted to isoquinolines in yields of 50% or better. For these isoquinolines the aminoacetal synthesis is more satisfactory than, for example, the synthesis of the corresponding 7-substituted tetrahydroisoquinoline by application of the phenethylamine-formaldehyde method. Where yield is not the primary consideration, the Pomeranz-Fritsch synthesis is applicable to the preparation of a variety of substituted isoquinolines.

A useful feature of the Pomeranz-Fritsch method is the possibility it affords of placing substituents on the isoquinoline nucleus in an orientation sometimes attainable only with difficulty by other syntheses. For example, in the Pomeranz-Fritsch synthesis, 8-substituted isoquinolines are the products from *ortho*-substituted benzaldehydes, whereas 8-substituted isoquinolines are not formed, as a rule, from *meta*-substituted arylethylamines. Further, the fact that 6-substituted isoquinolines are obtained unequivocally in the aminoacetal synthesis with *p*-substituted benzalaminoacetals assists in demonstrating the mode of ring closure with *m*-substituted phenethylamines.

Most of the syntheses involving the use of phenethylamine lead to partially hydrogenated isoquinoline systems. The aminoacetal method, by making the fully aromatic system available directly, may offer some advantage.

EXPERIMENTAL PROCEDURES

Aminoacetal from Chloroacetal.[11] Dry ammonia is passed into a solution of 38.2 g. of chloroacetal in 1 l. of absolute methanol at 0° until 283 g. is absorbed. The reaction mixture is then heated ten hours at 140° in the autoclave. The colored solution is concentrated on the steam bath to 500 ml.; 100 ml. of 5% aqueous potassium hydroxide is added, and concentration is continued until the vapors can no longer be ignited. The solution is saturated with sodium chloride, treated with 100 ml. of 50% aqueous potassium hydroxide, and extracted continuously with ether overnight. Concentration of the ether extract yields

[22a] Pictet and Popovici, *Ber.*, **25**, 733 (1892).

an oil from which, after fractionation in vacuum, 24.1 g. (72.5%) of aminoacetal, b.p. 99–103°/100 mm., is obtained.

If twice the quantity of chloroacetal and the same quantities of methanol and ammonia are used, 40.0 g. of aminoacetal (60%) is obtained.

Directions for the preparation of aminoacetal from bromoacetal in 32–39% yield are given in *Organic Syntheses*.[23] The use of chloroacetal in place of bromoacetal in the *Organic Syntheses* procedure increases the yield to 46%.

8-Bromoisoquinoline from *o*-Bromobenzaldehyde and Aminoacetal.[12] Aminoacetal in 15% excess is mixed with 50 g. of *o*-bromobenzaldehyde and heated on the steam bath for two hours. After the mixture cools, the water layer is removed and the crude product distilled under reduced pressure. The *o*-bromobenzalaminoacetal, b.p. 167–170°/6 mm., weighs 72 g. (89%).

To 180 g. of concentrated sulfuric acid maintained at 5° is added 20 g. of *o*-bromobenzalaminoacetal. The resulting mixture is added over a period of five minutes with mechanical stirring to 10 g. of concentrated sulfuric acid containing 20 g. of phosphoric anhydride. The temperature is held at 160°.

After the reaction mixture has been stirred and heated for an additional twenty-five minutes, it is allowed to cool, treated with ice, and filtered. The solid residue and the filtrate are extracted with ether in order to remove neutral and acidic material. Solid sodium carbonate in excess is added to the filtrate, and the alkaline mixture is steam-distilled. Toward the end of the distillation, the solid residue is added to the distillation flask and the distillation is continued.

The distillate, after acidification with hydrochloric acid, is evaporated to dryness on the steam bath. The residue is made alkaline with excess sodium hydroxide solution and is continuously extracted with ether. After removal of ether from the extract, the solid residue of 8-bromoisoquinoline is dried in a vacuum desiccator over calcium chloride. Crude 8-bromoisoquinoline prepared in this manner is a white, crystalline solid. The yield is 4 g. (29%).

The presence of phosphoric anhydride in the cyclization step results in a small but definite improvement in the yield.

7-Hydroxy-3-chloroisoquinoline from 2-Chloro-3-hydroxybenzaldehyde and Aminoacetal.[10] A mixture of 15 g. of 2-chloro-3-hydroxybenzaldehyde and an equal weight of aminoacetal is heated on the steam bath for one-half hour. Water formed in the reaction is then carefully removed by alternate addition and distillation of benzene. To the cold,

[23] Allen and Clark, *Org. Syntheses*, **24**, 3 (1944).

well-dried, dark brown residual liquid is added, with stirring, 100 ml. of 76% sulfuric acid previously cooled to 0°. The mixture is stirred at 2–5° for four hours and then is allowed to stand at 8° for forty hours and at room temperature for thirty hours. Water is added, and the resulting solution, after being made alkaline with aqueous ammonia, is buffered with sodium carbonate. The crude product which precipitates as a brown solid is collected by filtration. Sublimation at 175°/1 mm. furnishes 12 g. (64%) of white crystals of 7-hydroxy-8-chloroisoquinoline. Recrystallization of this material from methanol yields white needles, m.p. 230–231°.

1-Methyl-6,7-dimethoxy-8-hydroxyisoquinoline from 2-Benzyloxy-3,4-dimethoxyacetophenone and Aminoacetal.[15] A mixture of 15 g. of 2-benzyloxy-3,4-dimethoxyacetophenone and 10.5 g. of aminoacetal (50% excess) is heated at 165° for one and one-half hours. After the excess aminoacetal has been removed by distillation at 12 mm., the residue is distilled several times under 0.02 mm. pressure. The Schiff base is collected at 180–200°/0.02 mm. in amounts up to 22 g. (73%).

For conversion to the isoquinoline, the crude product is transferred to a flask provided with a well-fitting stopper and is treated (ice-salt cooling) with 90 g. of 73% sulfuric acid. The mixture is agitated for two days at 15–20°, then diluted with 95 ml. of water and warmed for one hour at 50°. Insoluble resinous material is removed at this point by filtering the cooled mixture. The filtrate is extracted with ether before and after being made alkaline with sodium carbonate. The ether extract from the alkaline solution is in turn extracted with 6 N hydrochloric acid, and the acidic aqueous phase is made alkaline with sodium carbonate and again extracted with ether. After removal of ether from the last extract, the residue is distilled at 0.02 mm. A fraction consisting of 2-hydroxy-3,4-dimethoxyacetophenone distils at 100–130°; the desired product is collected at 160–180°. 1-Methyl-6,7-dimethoxy-8-hydroxyisoquinoline, after high-vacuum sublimation at 155–165°, melts at 180–182°. The yield is 1.5 g. (13%).

Isoquinoline from Benzylamine and Glyoxal Semiacetal.[6] On mixing 1.06 g. of benzylamine and 1.4 g. of glyoxal semiacetal, the temperature of the mixture rises to 40–50°. The mixture is allowed to stand for one hour on the steam bath. The crude product is taken up in ether, and the ether solution is dried over anhydrous sodium sulfate. Removal of ether and distillation of the residue affords 1.85 g. (83%) of the Schiff base, b.p. 155–156°/16 mm.

The Schiff base is dissolved in 2 ml. of concentrated sulfuric acid at 0°, and the solution is slowly added to 3 ml. of concentrated sulfuric acid held at 160°. The black reaction mixture is made strongly alkaline and distilled with steam, and the product is extracted from the distillate with ether. Isoquinoline is isolated as the picrate, m.p. 225–227°, in 45% yield.

1-Methyl-7-methoxyisoquinoline from α-(3-Methoxyphenyl)ethylamine and Glyoxal Semiacetal.[6] The necessary starting material, α-(3-methoxyphenyl)ethylamine, is obtained in 47% overall yield from m-methoxyacetophenone by sodium-amalgam reduction of the oxime.

A mixture of 1.7 g. of α-(3-methoxyphenyl)ethylamine and 2.0 g. of glyoxal semiacetal in 5 ml. of anhydrous toluene containing 1 drop of piperidine is heated under reflux in a bath at 135–145° for one and one-half hours. More glyoxal semiacetal (0.4 g.) is added, and the heating is continued for another hour. During this hour, an air-cooled condenser is used so that toluene condenses but water slowly distils. The amount of water formed serves as a convenient measure of the extent of reaction. Finally, the last traces of water are removed by distilling the toluene under reduced pressure. The resulting pale-red mixture is distilled first up to 100°/15 mm. in order to remove low-boiling materials, and then under high vacuum. The Schiff base, b.p. 102–103°/0.04 mm., is obtained as a colorless oil; yield 2.24 g. (75%).

Dry hydrogen chloride gas is bubbled into 40 ml. of 72% sulfuric acid for about three minutes. The Schiff base is added to the acid at −10°, and the mixture is held at −10° for two days, at 0° for three days, and at 20° for twelve hours.

The resulting brown-red solution is diluted with 160 ml. of ice water and allowed to stand overnight. After removal of 0.8 g. of light-brown crystalline isoquinoline sulfate, the filtrate is neutralized with sodium carbonate and extracted with ether. The ethereal extract is washed twice with 2 N sodium hydroxide solution and twice with water, then dried over potassium carbonate, and distilled to remove solvent. The residual crude base is converted to its picrate. 1-Methyl-7-methoxyisoquinoline is obtained in the form of its sulfate and picrate in a 50% yield. The free base boils at 83–85°/0.04 mm. and melts at 32–34° after crystallization from petroleum ether.

TABLES OF POMERANZ-FRITSCH SYNTHESES

The literature through 1948 has been examined for examples of Pomeranz-Fritsch syntheses. The material has been arranged in three tables. Tables I and II cover examples of cyclizations leading to isoquinolines unsubstituted and substituted, respectively, at the 1 position. Unsuccessful isoquinoline syntheses related to the Pomeranz-Fritsch methods are listed in Table III.

TABLE I

Isoquinolines with No Substituent at the 1 Position

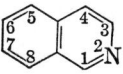

Isoquinoline	Schiff Base	Yield %	Reference
Isoquinoline	$C_6H_5CH=NCH_2CH(OC_2H_5)_2$	0, 2.5(?), 50	1, 2, 4, 24, 25
Isoquinoline	$C_6H_5CH_2N=CHCH(OC_2H_5)_2$	45	6
8-Methyl-	$2\text{-}CH_3C_6H_4CH=NCH_2CH(OC_2H_5)_2$	18–20	3
6-Methyl-	$4\text{-}CH_3C_6H_4CH=NCH_2CH(OC_2H_5)_2$	21	3, 8
8-Chloro-	$2\text{-}ClC_6H_4CH=NCH_2CH(OC_2H_5)_2$	9	3, 8, 26
5- and 7-Chloro-	$3\text{-}ClC_6H_4CH=NCH_2CH(OC_2H_5)_2$	0, 25–38, 50	8, 10
6-Chloro-	$4\text{-}ClC_6H_4CH=NCH_2CH(OC_2H_5)_2$	14	8
5,8-Dichloro-	$2,5\text{-}Cl_2C_6H_3CH=NCH_2CH(OC_2H_5)_2$	35	8
8-Bromo-	$2\text{-}BrC_6H_4CH=NCH_2CH(OC_2H_5)_2$	29	12
5- and 7-Bromo-	$3\text{-}BrC_6H_4CH=NCH_2CH(OC_2H_5)_2$	65	12
6-Bromo-	$4\text{-}BrC_6H_4CH=NCH_2CH(OC_2H_5)_2$	6, 24	8, 12
7(?)-Nitro-	$3\text{-}O_2NC_6H_4CH=NCH_2CH(OC_2H_5)_2$ (?)	0	8
8-Hydroxy-	$2\text{-}HOC_6H_4CH=NCH_2CH(OC_2H_5)_2$	0	27
5- and 7-Hydroxy-	$3\text{-}HOC_6H_4CH=NCH_2CH(OC_2H_5)_2$	69, 80	5, 10, 11, 27
6-Hydroxy-	$4\text{-}HOC_6H_4CH=NCH_2CH(OC_2H_5)_2$	0	27
7-Hydroxy-8-chloro-	$2\text{-}Cl\text{-}3\text{-}HO\text{-}C_6H_3CH=NCH_2CH(OC_2H_5)_2$	64	10
8-Methoxy-	$2\text{-}CH_3OC_6H_4CH=NCH_2CH(OC_2H_5)_2$	0	25, 27
7-Methoxy-	$3\text{-}CH_3OC_6H_4CH=NCH_2CH(OC_2H_5)_2$	80	5, 8, 25
7-Methoxy-	$3\text{-}CH_3OC_6H_4CH_2N=CHCH(OC_2H_5)_2$	70	6
6-Methoxy-	$4\text{-}CH_3OC_6H_4CH=NCH_2CH(OC_2H_5)_2$	0	25, 27
7-Ethoxy-	$3\text{-}C_2H_5OC_6H_4CH=NCH_2CH(OC_2H_5)_2$	80	5, 25
7-Diethylaminoethoxy-	$3\text{-}(C_2H_5)_2NCH_2CH_2OC_6H_4CH=NCH_2CH(OC_2H_5)_2$	70	8
7,8-Dimethoxy-	$2,3\text{-}(CH_3O)_2C_6H_3CH=NCH_2CH(OC_2H_5)_2$	5	8, 28
6,7-Dimethoxy-	$3,4\text{-}(CH_3O)_2C_6H_3CH=NCH_2CH(OC_2H_5)_2$	0	14
6,7-Methylenedioxy-	$3,4\text{-}(CH_2O_2)C_6H_3CH=NCH_2CH(OC_2H_5)_2$	23.6	5, 27
5,6-Methylenedioxy-7-methoxy-(?)	$3,4\text{-}(CH_2O_2)\text{-}5\text{-}CH_3O\text{-}C_6H_2CH=NCH_2CH(OC_2H_5)_2$	0	29

3,4-Dimethyl-	$C_6H_5CH=NCH\text{—}C(OC_2H_5)_2$
	                | |
	                CH_3 CH_3
		Low	17

[24] Farbwerke Meister Lucius and Brüning, Ger. pat. 80,044 [*Frdl.*, **4**, 1148 (1894–1897)].
[25] Fritsch, Ger. pat. 85,566 [*Frdl.*, **4**, 1149 (1894–1897)].
[26] Keilin and Cass, *J. Am. Chem. Soc.*, **64**, 2442 (1942).
[27] Fritsch, Ger. pat. 86,561 [*Frdl.*, **4**, 1150 (1894–1897)].
[28] Perkin and Robinson, *J. Chem. Soc.*, **105**, 2376 (1914).
[29] Salway, *J. Chem. Soc.*, **95**, 1204 (1909).

TABLE II

Isoquinolines Substituted at the 1 Position

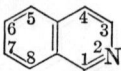

Isoquinoline	Reactant(s)	Yield %	Reference
1-Methyl-	$C_6H_5COCH_3 + H_2NCH_2CH(OC_2H_5)_2$	15	2, 24
1-Methyl-	$C_6H_5CH(CH_3)N{=}CHCH(OC_2H_5)_2$	40	6
1-Methyl-7-methoxy-	$3\text{-}CH_3OC_6H_4C(CH_3){=}NCH_2CH(OC_2H_5)_2$	0.1	6
1-Methyl-7-methoxy-	$3\text{-}CH_3OC_6H_4CH(CH_3)N{=}CHCH(OC_2H_5)_2$	50	6
1-Methyl-6,7-di-methoxy-8-hydroxy-	$2\text{-}C_6H_5CH_2O\text{-}3,4\text{-}(CH_3O)_2C_6H_2C(CH_3){=}NCH_2CH(OC_2H_5)_2$	14	15
1-Phenyl-	$C_6H_5COC_6H_5 + H_2NCH_2CH(OC_2H_5)_2$	Poor	2
1-Benzyl-	$C_6H_5C(CH_2C_6H_5){=}NCH_2CH(OC_2H_5)_2$	0	30
1-Dimethoxybenzyl-6,7-dimethoxy-	![structure with CH3O groups, C=NCH2CH(OC2H5)2, CH2 linker to dimethoxyphenyl]	0	30
1-Dimethoxybenzyl-6,7-dimethoxy-	![structure with CH3O groups, CHN=CHCH(OC2H5)2, CH2 linker to dimethoxyphenyl]	0	6
Isothebaine methyl ether	![structure with three CH3O groups on dihydrophenanthrene, N=CHCH(OC2H5)2]	0	16

[30] Fritsch, *Ann.*, **329**, 37 (1903).

TABLE III

Unsuccessful Variations

Reactant(s)	Reference
$C_6H_5CH_2NHCH_2CH_2OH$	31, 32
$C_6H_5CH_2N(CH_3)CH_2CH_2OH$	32
$3,4\text{-}(CH_2O_2)C_6H_3CH_2N(CH_3)CH_2CH_2OH$	33
$C_6H_5CH_2N(SO_2C_6H_5)CH_2CH_2OH$	34
$C_6H_5CH_2NHCH_2CHOHCH_3$	34
2-(2-hydroxyethyl)-1-phenyl-6,7-methylenedioxy-1,2,3,4-tetrahydroisoquinoline	35
2-(2-hydroxyethyl)-1-(3,4-methylenedioxyphenyl)-1,2,3,4-tetrahydroisoquinoline	35
N-methyl-N-[(3,4-methylenedioxyphenyl)methyl]-2-hydroxycyclohexylamine	36
N-[(3,4-methylenedioxyphenyl)methyl]-2-hydroxycyclohexylamine	36
N-acetyl-N-[(3,4-dimethoxyphenyl)methyl]-2-hydroxycyclohexylamine	36
$3,4\text{-}(CH_2O_2)C_6H_3CH_2NHCOCHOHCH_3$	18
$C_6H_5CH{=}NCH_2CHOHCH_3$	34
$C_6H_5CHO + H_2NCH_2CHOHCO_2H$	1
$3,4\text{-}(CH_2O_2)C_6H_3CH_2N(CH_3)CH_2CH_2Cl\cdot HCl$	33
$C_6H_5CH_2N(SO_2C_6H_5)CH_2CH_2Br$	34
$3,4\text{-}(CH_2O_2)C_6H_3CH_2NHCH_2CH(OC_2H_5)_2$	32
$3,4\text{-}(CH_2O_2)\text{-}5\text{-}CH_3O\text{-}C_6H_2NHCH_2CH(OC_2H_5)_2$	29
$3,4\text{-}(CH_2O_2)C_6H_3CH_2N(CH_3)CH_2CH(OC_2H_5)_2$	37
$C_6H_5CONHCH_2CH(OC_2H_5)_2$	4, 19, 20
$3,4\text{-}(CH_2O_2)C_6H_3CH_2NHCHOHCHO$ (?)	18
$3,4\text{-}(CH_2O_2)C_6H_3CH_2NHCHOHCOCH_3$ (?)	18
$3,4\text{-}(CH_2O_2)C_6H_3CH_2NHCHOHCOC_6H_5$ (?)	18
$C_6H_5CH_2NHCOCHCl_2$	32
$3,4\text{-}(CH_2O_2)C_6H_3CH_2N(CH_3)CH_2COC_6H_5$	37
$C_6H_5CH_2N(COCH_3)CH_2COC_6H_5$	34
$C_6H_5CH_2N(SO_2C_6H_5)CH_2COC_6H_5$	34
$C_6H_5CH_2NHCH_2COCl\cdot HCl$	38
$3,4\text{-}(CH_2O_2)C_6H_3CH_2NHCH_2COCl\cdot HCl$	38
$3,4\text{-}(CH_2O_2)C_6H_3CH_2NHCH_2CN$	18
$C_6H_5CH_2N(SO_2C_6H_4CH_3\text{-}4)CH_2CO_2H$	39
$C_6H_5CH_2N(SO_2C_6H_4CH_3\text{-}4)CH_2COCl$	39
$C_6H_5CH_2N(CH_3)CH_2COCl\cdot HCl$	38
$C_6H_5CH_2N(CH_3)COCONH_2$	32
$2,3\text{-}(CH_3O)_2C_6H_3CH{=}NCH_2CO_2C_2H_5$	28
$C_6H_5CONHCH_2CO_2H$	20, 40, 41

REFERENCES TO TABLE III

[31] Goldschmiedt and Jahoda, *Monatsh.*, **12,** 81 (1891).
[32] Mannich and Kuphal, *Arch. Pharm.*, **250,** 539 (1912).
[33] Kaufmann and Dürst, *Ber.*, **50,** 1630 (1917).
[34] Staub, *Helv. Chim. Acta*, **5,** 888 (1922).
[35] Reichert and Hoffmann, *Arch. Pharm.*, **274,** 153 (1936).
[36] Forrest, Haworth, Pinder, and Stevens, *J. Chem. Soc.*, **1949,** 1311.
[37] Young and Robinson, *J. Chem. Soc.*, **1933,** 275.
[38] Mannich and Kuphal, *Ber.*, **45,** 314 (1912).
[39] Clemo and Perkin, *J. Chem. Soc.*, **127,** 2297 (1925).
[40] Rügheimer, *Ber.*, **19,** 1169 (1886).
[41] Schwanert, *Ann.*, **112,** 59 (1859).

CHAPTER 5

THE OPPENAUER OXIDATION

CARL DJERASSI

Syntex, S. A., Mexico City, D. F.

CONTENTS

	PAGE
INTRODUCTION	208
MECHANISM OF THE REACTION	209
SCOPE OF THE REACTION	210
Saturated Alcohols	210
Unsaturated Alcohols	212
Polyhydroxyl Compounds	216
Nitrogen-Containing Alcohols	219
Oxidation of Primary Alcohols	222
Isolation of Aldehydes	222
Simultaneous Condensation of Resulting Aldehydes with Hydrogen Acceptors	223
Side Reactions	224
CHOICE OF EXPERIMENTAL CONDITIONS	225
Aluminum Alkoxides	225
Aluminum t-Butoxide	226
Aluminum Isopropoxide	226
Aluminum Phenoxide	227
Other Catalysts	227
Hydrogen Acceptors	228
Solvents	231
Time and Temperature	232
Ratio of Alkoxide to Alcohol	233
Isolation of Products	233
Miscellaneous Suggestions	234
EXPERIMENTAL PROCEDURES	234
cis-α-Decalone	235
$\Delta^{20,23}$-24,24-Diphenylcholadiene-3,11-dione	235
Desoxycorticosterone Acetate	235
Methyl $\Delta^{4,6}$-3-Ketoetiocholadienate	236
α-Cyclocitral	236
β-Ketobutyraldehyde 2-Methylpentane-2,4-diol Acetal	237
ψ-Ionone	237

		PAGE
Survey of Oppenauer Oxidations Reported in the Literature		238
Table I.	Oppenauer Oxidation of Saturated Secondary Alcohols	240
Table II.	Oppenauer Oxidation of Unsaturated Secondary Alcohols	244
Table III.	Oppenauer Oxidation of Polyhydroxyl Compounds	255
Table IV.	Selective Oppenauer Oxidation of Polyhydroxyl Compounds	257
Table V.	Oppenauer Oxidation of Alcohols Containing Halogen, Lactone, Acetal, or Ketal Groups	260
Table VI.	Oppenauer Oxidation of Nitrogen-Containing Alcohols	263
Table VII.	Oppenauer Oxidation of Primary Alcohols	266
Table VIII.	Oppenauer Oxidation of Primary Alcohols and Simultaneous Condensation of Resulting Aldehydes	268
Table IX.	Unsuccessful Oppenauer Oxidations	271

INTRODUCTION

The application of the reaction

$$\underset{R'}{\overset{R}{>}}C{=}O + \underset{R'''}{\overset{R''}{>}}CHO\frac{Al}{3} \rightleftarrows \underset{R'}{\overset{R}{>}}CHO\frac{Al}{3} + \underset{R'''}{\overset{R''}{>}}C{=}O$$

to the reduction of aldehydes and ketones has been reviewed in an earlier volume of this series [1] under the title "Reduction with Aluminum Alkoxides (The Meerwein-Pondorff-Verley Reduction)." The reversible nature of the above reaction was demonstrated by Verley [2] in 1925 and shortly thereafter by Pondorff,[3] but it was not until 1937 that Oppenauer [4] showed that unsaturated steroid alcohols could be oxidized to the corresponding ketones in excellent yields through the use of aluminum t-butoxide in the presence of a large amount of acetone, that compound functioning as the hydrogen acceptor and the large excess serving to shift the equilibrium in the desired direction. This reaction, which has been called the Oppenauer oxidation,[5] has been extremely useful in steroid chemistry, but so far it has been applied to only a limited extent elsewhere. As will be illustrated in the subsequent discussion, the Oppenauer oxidation employs very mild conditions which are applicable to a variety of sensitive compounds; and it will, undoubtedly, find extensive use in synthetic organic chemistry. The recent introduction of the experimental modifications outlined below has already increased the scope of the reaction appreciably.

[1] Wilds, *Org. Reactions*, **2**, 178 (1944).
[2] Verley, *Bull. soc. chim. France*, [4], **37**, 537 (1925).
[3] Pondorff, *Angew. Chem.*, **39**, 138 (1926).
[4] Oppenauer, *Rec. trav. chim.*, **56**, 137 (1937).
[5] Bersin, *Angew. Chem.*, **53**, 266 (1940). This review article has been translated and partly revised in *Newer Methods of Preparative Organic Chemistry*, pp. 143–158, Interscience Publishers, New York, 1948.

MECHANISM OF THE REACTION

In view of the reversible nature of the reaction, many statements as to the mechanism of the Meerwein-Pondorff-Verley reduction [1] are equally applicable to the Oppenauer oxidation. The earlier workers [2,3,6] postulated the formation of an acetal of type A, without giving an adequate explanation for the hydrogen transfer that must occur to account for the course of the reaction. Pondorff [3] postulated an unusual type of addition to the carbonyl group, and Verley's [2] mechanism required an unprecedented migration of an aluminum alkoxide radical. Meerwein's original hemiacetal structure A was revised [7] in favor of the noncommittal molecular addition compound B in order to rationalize the function of the aluminum alkoxide. Activation of the alcoholic hydrogen atom by the aluminum resulting in hydrogen bonding has also been proposed.[8]

$$\underset{R'}{\overset{R}{>}}\!\!\underset{|}{\overset{OAl(OCHR''R''')_2}{C}}\!\!-\!\!O\!\!-\!\!\underset{\underset{R'''}{\diagdown}}{\overset{R''}{\diagup}}\!\!CH \qquad \underset{R'}{\overset{R}{>}}\!\!C\!\!=\!\!O\cdots Al(OR'')_3$$

$$\quad\quad\quad\quad A \quad\quad\quad\quad\quad\quad\quad\quad\quad B$$

A mechanism employing a pseudo-cyclic intermediate has been suggested by Woodward [9] and Oppenauer.[10] Although the tendency to

[diagram of pseudo-cyclic mechanism showing equilibria between structures labeled C and D, with final products including $R\!>\!CHOAl_3$ and $R''\!>\!C=O$]

accept a pair of electrons, thus facilitating both step C and the hydrogen transfer D, is particularly pronounced in aluminum with its sextet of electrons, this mechanism is equally applicable to those oxidations in which alkali alkoxides can be employed in place of the aluminum compounds.[9] It will be noted that aluminum t-butoxide, or other alkoxide,

---

[6] Meerwein and Schmidt, *Ann.*, **444**, 221 (1925).

[7] Meerwein, v. Bock, Kirschnick, Lenz, and Migge, *J. prakt. Chem.*, [2], **147**, 211 (1936).

[8] Davies and Hodgson, *J. Soc. Chem. Ind.*, **62**, 109 (1943).

[9] Woodward, Wendler, and Brutschy, *J. Am. Chem. Soc.*, **67**, 1425 (1945); *cf.* also Jackman and Mills, *Nature*, **164**, 789 (1949); Lutz and Gillespie, *J. Am. Chem. Soc.*, **72**, 345 (1950); Doering and Young, *ibid.*, **72**, 631 (1950), and references cited therein.

[10] R. V. Oppenauer, private communication.

does not appear in the above reactions. It is assumed that their only role in the overall reaction is to provide a source of aluminum ion. Experiments [11] with deuterated 2-propanol indicate that no appreciable exchange of deuterium occurs during the Oppenauer oxidation.

SCOPE OF THE REACTION

Saturated Alcohols *

It has been implied that alcoholic groups not activated by unsaturation are resistant to oxidation by Oppenauer's method; this was believed to be true both for steroidal secondary alcohols [12] such as cholestanol and for aliphatic primary alcohols.[13] More recent work has proved this view to be incorrect although it is true that modified conditions may be required. A variety of steroid alcohols in which the double bond is three or more carbon atoms removed from the hydroxyl group can be oxidized under relatively mild conditions using benzene and acetone. γ-Cholestenol (I) [14] and a variety of similar ergosterol derivatives, e.g., α-ergostenol (II),[15,16] give 40–60% of the corresponding ketone. The steroid alcohol III which possesses the acid-labile dienone grouping in ring A is converted to the corresponding ketone in 55% yield.[17] Other

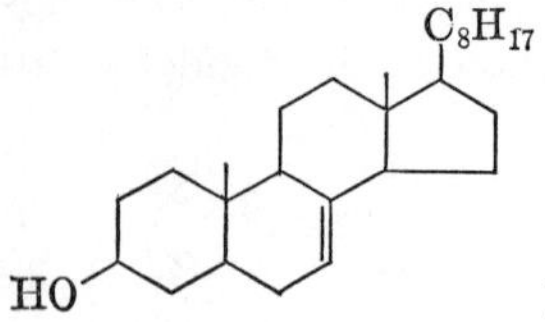

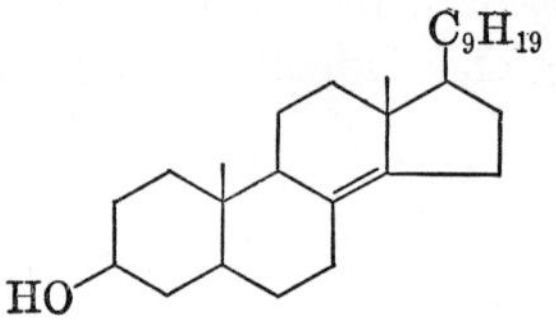

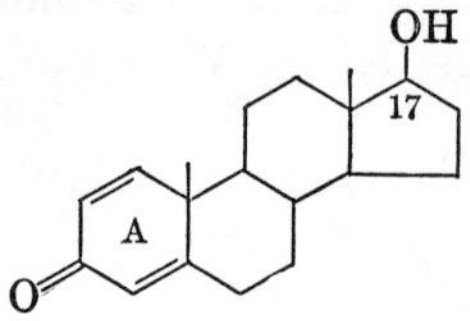

* Compounds not possessing a double bond or aromatic nucleus α,β or β,γ to a secondary hydroxyl group will be listed with the saturated alcohols. The Oppenauer oxidation of primary alcohols involves special conditions which are considered in a separate section.

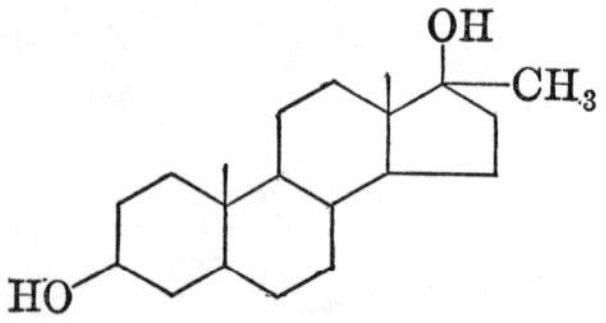

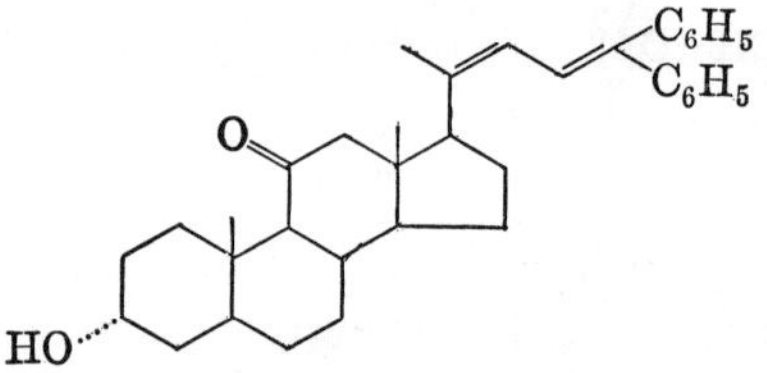

[11] Westheimer and Nicolaides, *J. Am. Chem. Soc.*, **71**, 26 (1949).
[12] Jones, Wilkinson, and Kerlogue, *J. Chem. Soc.*, **1942**, 391.
[13] Batty, Burawoy, Harper, Heilbron, and Jones, *J. Chem. Soc.*, **1938**, 175.
[14] Buser, *Helv. Chim. Acta*, **30**, 1390 (1947).
[15] Barton and Cox, *J. Chem. Soc.*, **1948**, 783.
[16] D. H. R. Barton, private communication.
[17] Inhoffen, Zühlsdorff, and Huang-Minlon, *Ber.*, **73**, 457 (1940).

examples in the steroid series where saturated ketones are produced in excellent yield (80–86%) are 17-methylandrostane-3β,17β-diol (IV) [18] and the diene V, which represents a key intermediate in a novel synthesis [19] of the cortical hormone 11-dehydrocorticosterone. The successful Oppenauer oxidation of "α" and "β" estradiol to estrone has formed the basis for a differential bioassay of the two C-17 epimeric estradiols.[20]

Among non-steroidal alcohols, both the *cis* and *trans* α-decalols (VI) give excellent yields of the corresponding decalones,[21] but chromic anhydride oxidation appears to be more economical on a larger scale. Since free phenolic groups are not attacked,[22] the direct oxidation of octahydrodiethylstilbestrol (VII) [23] with aluminum *t*-butoxide and acetone is more satisfactory than other methods where the phenolic group must be protected by benzoylation. Robinson and co-workers employed the Oppenauer reaction with a number of synthetic naphthalene and phenanthrene derivatives (Table I).[22, 24–28] An interesting example is the sensitive acetal VIII, which was smoothly oxidized [29] to the corresponding ketone by a modified Oppenauer oxidation (see Experimental Procedures); classical methods of oxidation failed in this instance.

VI

HO—⟨ ⟩—CH(C_2H_5)CH(C_2H_5)—⟨ ⟩—OH CH_3CHOHCH$_2$CH⟨O—CH(CH$_3$)/O—C(CH$_3$)$_2$⟩CH$_2$

VII   VIII

[18] St. André, unpublished observation.
[19] Wettstein and Meystre, *Helv. Chim. Acta*, **30**, 1267 (1947).
[20] Pearlman and Pincus, *J. Biol. Chem.*, **147**, 384 (1943); Pearlman and Pearlman, *Arch. Biochem.*, **4**, 97 (1944).
[21] J. English, private communication; see English and Cavaglieri, *J. Am. Chem. Soc.*, **65**, 1085 (1943).
[22] Cornforth and Robinson, *J. Chem. Soc.*, **1949**, 1855.
[23] II. E. Ungnade, private communication; see Ungnade and Ludutsky, *J. Am. Chem. Soc.*, **69**, 2630 (1947).
[24] Robinson and Walker, *J. Chem. Soc.*, **1938**, 185.
[25] Robinson and Slater, *J. Chem. Soc.*, **1941**, 381.
[26] McGinnis and Robinson, *J. Chem. Soc.*, **1941**, 404.
[27] King and Robinson, *J. Chem. Soc.*, **1941**, 469.
[28] Cornforth and Robinson, *J. Chem. Soc.*, **1942**, 688.
[29] E. Theimer, private communication; *cf.* Abstracts, North Jersey Section Meeting-in-Miniature, Jan. 10, 1949.

Unsaturated Alcohols

Oppenauer [4,30] first demonstrated the direct oxidation of Δ^5-3-hydroxy steroids (IX) * to the Δ^4-3-ketones (X) by means of aluminum t-butoxide and acetone in benzene solution. The steroid aluminate was formed by interchange *in situ* from the aluminum t-butoxide. As is apparent from Table II, this type of oxidation has been used extensively, and migration of the double bond from the β,γ to the α,β position was invariably observed; the shift also occurs when ring B is five-membered.[31]

This migration of the double bond, resulting in an α,β-unsaturated ketone with characteristic absorption in the ultraviolet region of the spectrum, has been used as a test for the homogeneity of phytosterols,[12,32,33] in the proof of structure [34] of dihydrovitamin D_3 (XI), where alternate positions were considered for the 5-10 double bond, as well as for the polarographic determination of Δ^5-3-hydroxy steroids (IX) [35] since the resulting Δ^4-3-keto portion exhibits a characteristic

wave. In the oxidation of steroid alcohols containing two conjugated double bonds (e.g., XII), only the β,γ-double bond migrates (XIII).[4,36]

The Oppenauer oxidation is superior, with respect to both yield (80–95%) and convenience, to methods previously used for transform-

---

* In the formulas of this and other 3-hydroxy steroids the 3β configuration is implied when the hydroxyl group is attached directly to the nucleus; the 3α configuration is indicated in the usual manner by a dotted line.

[30] Oppenauer, U. S. pat. 2,384,335 (1945) [*C. A.*, **40**, 178 (1946)].
[31] Sorm and Dykova, *Coll. Czechoslov. Chem. Commun.*, **13**, 418 (1948) [*C. A.*, **43**, 1789 (1949)].
[32] Heilbron, Jones, Roberts, and Wilkinson, *J. Chem. Soc.*, **1941**, 344.
[33] Barton and Jones, *J. Chem. Soc.*, **1943**, 599.
[34] Windaus and Roosen-Runge, *Z. physiol. Chem.*, **260**, 184 (1939).
[35] Hershberg, Wolfe, and Fieser, *J. Am. Chem. Soc.*, **62**, 3516 (1940).
[36] Windaus and Kaufmann, *Ann.*, **542**, 220 (1939).

ing β,γ-unsaturated steroidal alcohols such as IX into the related α,β-unsaturated steroidal ketones X, and it has found use in the manufacture [37] of a number of hormones such as testosterone (X, R = OH), progesterone (X, R = $COCH_3$), and desoxycorticosterone acetate (X, R = $COCH_2OCOCH_3$). The specificity of the reaction is illustrated by the successful oxidation of many compounds containing labile substituents such as allyl,[38] vinyl,[39,40] ethynyl,[37,40,41] benzal,[42,43] and various other unsaturated side chains.[44-52] An instructive example is the oxidation of the unsaturated alcohol XIV, which contains both nuclear and side-chain unsaturation, to the ketone XV in 95% yield in one-half hour through the use of cyclohexanone and aluminum isopropoxide in toluene.[45]

Halogen-containing alcohols, such as 22,23-dibromostigmasterol (XVI) [15,53] or 21-chloropregnenolone (XVII) [54,55] are oxidized in good yield; the chloro compound cannot be subjected to the alternative chromic anhydride oxidation, since removal of the bromine atoms added to protect the nuclear double bond also removes the chlorine atom in the side chain.

[37] *British Intelligence Objectives Subcommittee*, Final Report 996, H. M. Stationery Office, London, 1947.
[38] Butenandt and Peters, *Ber.*, **71**, 2688 (1938).
[39] Ruzicka, Hofmann, and Meldahl, *Helv. Chim. Acta*, **21**, 597 (1938).
[40] Inhoffen, Logemann, Hohlweg, and Serini, *Ber.*, **71**, 1024 (1938).
[41] Ruzicka, Hofmann, and Meldahl, *Helv. Chim. Acta*, **21**, 373 (1938).
[42] Marker, Wittle, Jones, and Crooks, *J. Am. Chem. Soc.*, **64**, 1283 (1942).
[43] Schmidlin and Miescher, *Helv. Chim. Acta*, **32**, 1797 (1949).
[44] Ruzicka, Goldberg, and Hardegger, *Helv. Chim. Acta*, **22**, 1297 (1939). The position of the Δ $^{17-20}$ double bond was established subsequently, *ibid.*, **25**, 1297 (1942).
[45] Meystre, Frey, Neher, Wettstein, and Miescher, *Helv. Chim. Acta*, **29**, 632 (1946).
[46] Meystre, Wettstein, and Miescher, *Helv. Chim. Acta*, **30**, 1025 (1947).
[47] Meystre and Wettstein, *Helv. Chim. Acta*, **30**, 1261 (1947).
[48] Wieland and Miescher, *Helv. Chim. Acta*, **32**, 1764 (1949).
[49] Levin, Spero, McIntosh, Heyl, and Thompson, *Abstracts*, p. 33L, A.C.S. San Francisco Meeting, April, 1949.
[50] Spero, McIntosh, and Levin, *J. Am. Chem. Soc.*, **71**, 834 (1949).
[51] Julian, Cole, Meyer, and Herness, *J. Am. Chem. Soc.*, **67**, 1375 (1945).
[52] Julian, Meyer, and Printy, *J. Am. Chem. Soc.*, **70**, 890 (1948).
[53] Fernholz and Stavely, *J. Am. Chem. Soc.*, **61**, 2956 (1939).
[54] Reich and Reichstein, *Helv. Chim. Acta*, **22**, 1124 (1939).
[55] Reichstein and v. Euw, *Helv. Chim. Acta*, **23**, 136 (1940).

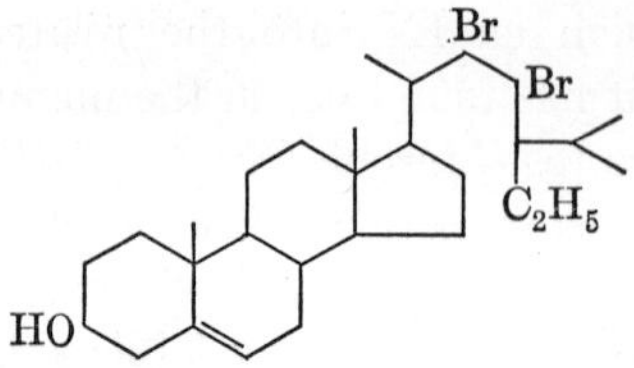

[Structures XVI and XVII]

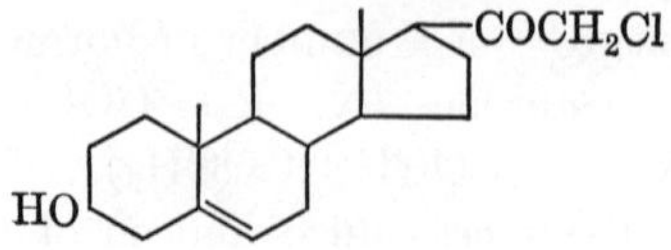

Acid-labile acetals (e.g., XVIII [56]), mercaptals, or ketals (e.g., XIX [57]) are amenable to oxidation via the Oppenauer procedure, and such examples have been collected separately in Table V.

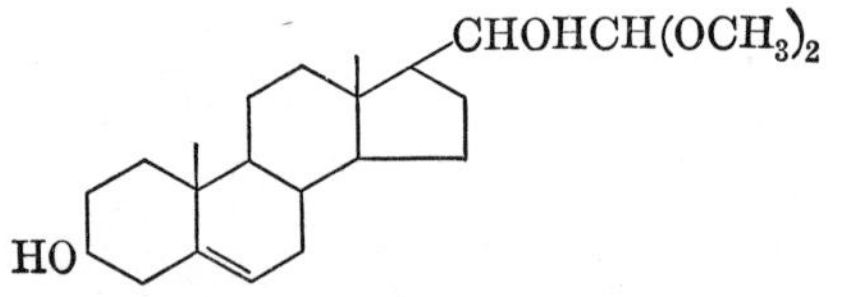

[Structures XVIII and XIX]

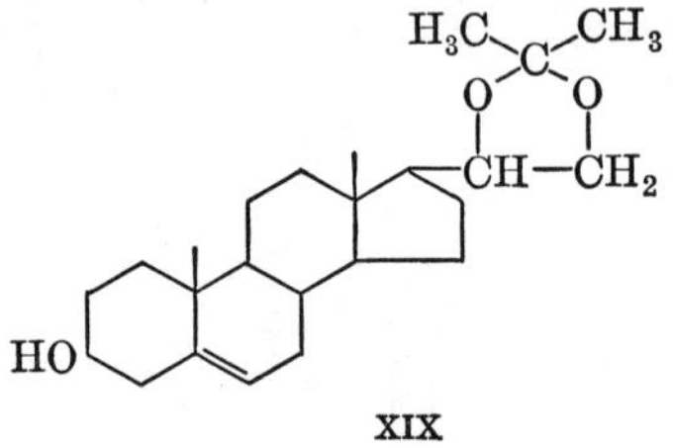

Additional examples illustrating the oxidation of unsaturated alcohols can be found in Tables II and V. Noteworthy are the unsaturated lactone XX,[58] the phenolic derivative XXI,[59] and the sensitive 16,17-oxido-20-keto derivative (XXII, R = H, OCOCH$_3$),[60,61] which are

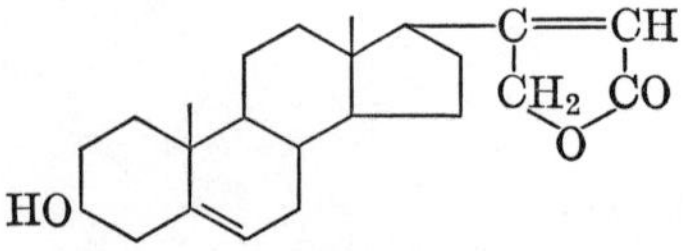

[Structures XX and XXI]

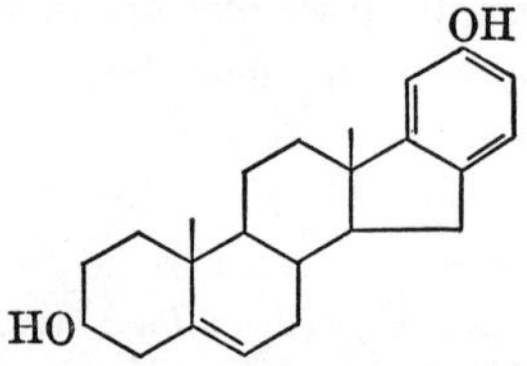

[Structure XXII]

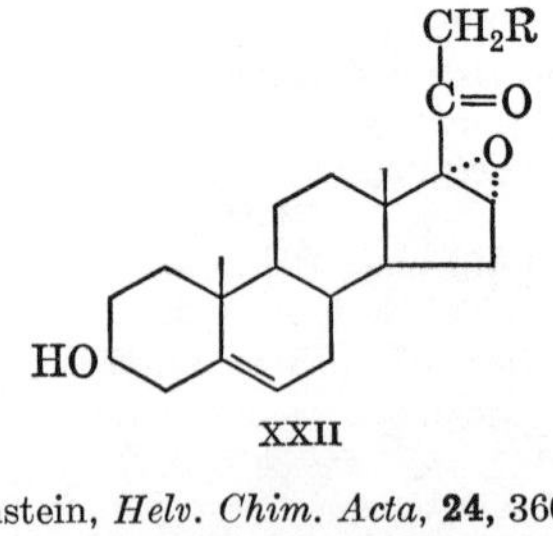

[56] Schindler, Frey, and Reichstein, *Helv. Chim. Acta*, **24**, 360 (1941).
[57] Steiger and Reichstein, *Helv. Chim. Acta*, **21**, 177 (1938).
[58] Ruzicka, Plattner, Fürst, and Heusser, *Helv. Chim. Acta*, **30**, 698 (1947).
[59] Ruzicka, Prelog, and Battegay, *Helv. Chim. Acta*, **31**, 1300 (1948).
[60] Julian, Meyer, Karpel, and Ryden, *J. Am. Chem. Soc.*, **71**, 3574 (1949); Julian, Meyer, Karpel, and Waller, *ibid.*, **72**, 5146 (1950).
[61] Julian, Meyer, and Ryden, *J. Am. Chem. Soc.*, **72**, 369 (1950).

oxidized in excellent yield to the corresponding α,β-unsaturated ketones.

A number of *non-steroid* unsaturated alcohols have been oxidized in one step by the Oppenauer procedure; alternative procedures would have been more cumbersome and would have resulted in lower yields. In common with phenanthrene derivatives (e.g., XXIII [62]) similar to the steroids discussed above, $\Delta^{8,9}$-1-octalone (XXV) is obtained [63] in 74% yield from $\Delta^{9,10}$-1-octalol (XXIV). This last example involves a rearrangement of a double bond from one α,β position to another, and this fact should be considered in structural studies since it is generally assumed that the shift of a double bond in an Oppenauer oxidation will involve migration from the β,γ to the α,β positions.[64] The thiopyran derivative XXVI was smoothly oxidized, presumably without migration of the double bond.[26]

The polyenes, α-ionol (XXVII, R = H) and a mixture of α-(XXVII, R = CH$_3$) and γ-irol [65] are oxidized to the corresponding ketones in good yield. A similar observation has been made in respect to the conjugated diene XXVIII,[66] although with the related β-ionol considerable resinification was encountered.[65] The Oppenauer oxidation of secondary alcohols similar to XXVII has proved to be of exceptional usefulness in the preparation of a number of vitamin A analogs.[67–70] The reaction is

---

[62] Köster and Logemann, *Ber.*, **73**, 298 (1940).
[63] Campbell and Harris, *J. Am. Chem. Soc.*, **63**, 2721 (1941).
[64] Ruzicka, Rey, Spillmann, and Baumgartner, *Helv. Chim. Acta*, **26**, 1653 (1943).
[65] Ruzicka, Seidel, Schinz, and Tavel, *Helv. Chim. Acta*, **31**, 277 (1948).
[66] Milas, Lee, Sakal, Wohlers, MacDonald, Grossi, and Wright, *J. Am. Chem. Soc.*, **70**, 1584 (1948).
[67] Chanley and Sobotka, *J. Am. Chem. Soc.*, **71**, 4141 (1949).
[68] Heilbron, Jones, and Richardson, *J. Chem. Soc.*, **1949**, 292.
[69] Heilbron, Jones, Lewis, Richardson, and Weedon, *J. Chem. Soc.*, **1949**, 742.
[70] Heilbron, Jones, Lewis, and Weedon, *J. Chem. Soc.*, **1949**, 2023.

equally applicable to open-chain, unsaturated alcohols such as the octatrienol XXIX,[71] which afforded 80% of the corresponding ketone.

CH=CHCH(CH$_3$)CHOHCH$_2$CH$_3$

XXVIII

CH$_3$CHOHCH=CHCH=C(CH$_3$)CH=CH$_2$
XXIX

Polyhydroxyl Compounds

The simultaneous oxidation of two hydroxyl groups can be accomplished in both saturated and unsaturated compounds unless steric factors intervene. Thus methyl hyodesoxycholate (XXX) is oxidized to methyl 3,6-diketo*allo*cholanate (XXXI, R = C$_6$H$_{11}$O$_2$),[72] inversion occurring at C-5 during the process. The unsaturated diol XXXII affords a good yield of the corresponding diketone.[73] Analogies from steroid chemistry cannot always be applied to simpler compounds, as is shown by the oxidation of Δ^4-cholestene-3β,6 (α and β)-diol (XXXIII),[74,75] which leads to the *saturated* diketone XXXI (R = C$_8$H$_{17}$), while Δ^{9,10}-octalin-1,5-diol (XXXIV) undergoes oxidation of both hydroxyl groups to the *unsaturated* diketone.[63] Of interest is the fact that the 3,5,19-trihydroxy steroid XXXV was recovered completely unchanged under a variety of conditions.[76] Since the hydroxyl groups at positions 3 and 5 are *cis* to each other, an aluminum complex involving both of them may be the interfering factor; supporting evidence for complex formation is afforded by the successful oxidation of XXXV when Raney nickel[77] was substituted for the aluminum alkoxide. When the hydroxyl groups are *trans* to each other, the C-5 substituent suffers dehydration.[78,79]

The Oppenauer reaction has been particularly useful in the preferential oxidation of polyhydroxyl compounds of the steroid series, and all such examples have been collected in Table IV. The order of oxidation appears to be almost the reverse of that found with chromic anhydride.

[71] Cheeseman, Heilbron, Jones, Sondheimer, and Weedon, *J. Chem. Soc.*, **1949**, 2031.
[72] Gallagher and Xenos, *J. Biol. Chem.*, **165**, 365 (1946).
[73] Levin, Spero, McIntosh, and Rayman, *J. Am. Chem. Soc.*, **70**, 2960 (1948).
[74] Butenandt and Hausmann, *Ber.*, **70**, 1159 (1937).
[75] Prelog and Tagmann, *Helv. Chim. Acta*, **27**, 1871 (1944).
[76] Ehrenstein, Johnson, Olmsted, Vivian, and Wagner, *J. Org. Chem.*, **15**, 264 (1950).
[77] Kleiderer and Kornfeld, *J. Org. Chem.*, **13**, 455 (1948).
[78] Ruzicka and Muhr, *Helv. Chim. Acta*, **27**, 509 (1944).
[79] Henbest and Jones, *J. Chem. Soc.*, **1948**, 1797.

In the cholic acid series, the following order prevails with chromic anhydride: C-7 > C-12 > C-3; in hyodesoxycholic acid (XXX), the C-6 hydroxyl group is oxidized in preference to the one at C-3, and similarly C-11 is oxidized before C-3. With the Oppenauer reagent, on the other hand, a C-3 hydroxyl group is always attacked first, while one at C-11 remains untouched. Referring to the type formula XXXVI, the following partial oxidations have been accomplished by Oppenauer's method: C-3 vs. C-12; [80-84] C-3 vs. C-17a; [85] C-3 vs. C-11; [86,87] C-3 vs. C-6; [72] C-3 vs. C-20; [72,86,88] C-3 vs. C-7 and C-12; [81,84,89] C-17 vs. C-11; [90] and C-20 vs. C-11 [91] in contrast to chromic anhydride (C-11 vs. C-20). The superiority of the Oppenauer procedure is exemplified by the one-step oxidation [83,84] of methyl desoxycholate (XXXVII) to the corresponding 3-ketone XXXVIII in 57–63% yield; the alternative method of partial saponification and chromic anhydride oxidation involves five steps.

When one hydroxyl group is activated by a double bond, preferential oxidation appears even easier. The unsaturated alcohol XXXIX is oxidized to the corresponding Δ^4-3-ketone in fifteen minutes; [92] in the mixed primary-secondary alcohol XL, the primary hydroxyl group

[80] Ehrenstein and Stevens, *J. Org. Chem.*, **5**, 671 (1940). The structure originally assigned to the triol was revised by Ehrenstein, *J. Org. Chem.*, **13**, 222 (1948).
[81] Gallagher, *J. Biol. Chem.*, **133**, XXXVI (1940).
[82] Fuchs and Reichstein, *Helv. Chim. Acta*, **26**, 523 (1943).
[83] Riegel and McIntosh, *J. Am. Chem. Soc.*, **66**, 1099 (1944).
[84] Jones, Webb, and Smith, *J. Chem. Soc.*, **1949**, 2164.
[85] Marker and Rohrmann, *J. Am. Chem. Soc.*, **61**, 2721 (1939). Klyne, *Nature*, **166**, 559 (1950), has shown that Marker's "urane-3,11-diol" is 17-methyl-D-homoandrostane-3β,17α-diol.
[86] Reich and Reichstein, *Arch. intern. pharmacodynamie*, **65**, 415 (1941) [*C. A.*, **35**, 5526 (1941)].
[87] v. Euw, Lardon, and Reichstein, *Helv. Chim. Acta*, **27**, 1293 (1944).
[88] Wieland and Miescher, *Helv. Chim. Acta*, **32**, 1922 (1949).
[89] Kuwada and Morimoto, *Bull. Chem. Soc. Japan*, **17**, 147 (1942) [*C. A.*, **41**, 4504 (1947)].
[90] Sarett, *J. Biol. Chem.*, **173**, 186 (1948).
[91] v. Euw, Lardon, and Reichstein, *Helv. Chim. Acta*, **27**, 821 (1944).
[92] Jeanloz and v. Euw, *Helv. Chim. Acta*, **30**, 803 (1947).

remains virtually untouched.[93] Nevertheless it should be possible to achieve the oxidation of another hydroxyl group in the presence of a Δ^5-3-hydroxy grouping without affecting the latter, by temporary protection through conversion to the i-steroid form, which appears to be resistant to aluminum isopropoxide.[94]

XXXIX   XL

Some of the examples of the specificity of the Oppenauer oxidation of polyhydroxyl compounds of the steroid series are probably due to the presence of unique steric factors. Steric hindrance undoubtedly is the reason why the C-11 hydroxyl group remains unattacked. More subtle configurational effects can also be noticed: in the C-17 epimeric Δ^5-androstene-3β,17-diols (IX, R = OH),[95] the 17α-isomer affords 65% of "*cis*"-testosterone (X, R = OH)[18] while the 17β-epimer yields only 40% of testosterone.[18, 96, 97] The resistance to oxidation thus parallels the saponification rates of the corresponding C-17 esters; the proximity of the C-12 methylene group appears to have a more pronounced effect on the C-17 substituent than a *cis* (β) or *trans* (α) relationship to the C-18 angular methyl group.

Frequently a choice of conditions will determine the extent of oxidation. With methyl hyodesoxycholate (XXX) complete oxidation to the diketone XXXI is achieved on refluxing, and selective oxidation of the C-3 hydroxyl group on carrying out the reaction at 40°.[72] Similarly with Δ^5-androstene-3β,17β-diol[18] the yield of partial oxidation product ("*cis*"-testosterone) is lowered by almost one-half by doubling the reaction time.

Nitrogen-Containing Alcohols

The Oppenauer oxidation has been used with both steroidal and nonsteroidal alkaloids. Retronecanol (XLI) can be oxidized to retroneca-

[93] Miescher and Wettstein, *Helv. Chim. Acta*, **22**, 1266 (1939).
[94] Riegel and Kaye, *J. Am. Chem. Soc.*, **66**, 724 (1944).
[95] Currently accepted conventions regarding the configuration of nuclear substituents in the steroid series are summarized by Fieser and Fieser, *Natural Products Related to Phenanthrene*, 3rd ed., Reinhold, New York, 1949, and by Petit, *Bull. soc. chim. France*, **1949**, 545.
[96] Kuwada and Joyama, *J. Pharm. Soc. Japan*, **57**, 914 (1937). [German summary p. 247; see *Chem. Zentr.*, II, **1938**, 1612.]
[97] Ushakov and Chinaeva, *J. Gen. Chem. U.S.S.R.*, **15**, 661 (1945) [*C. A.*, **40**, 5879 (1946)].

none [98] even though the latter compound is rather unstable. In the yohimbine (XLII) series,[99] the ketone yohimbone (XLIII) was obtained in nearly quantitative yield.[100] The previous synthesis of yohimbone involved alkali fusion under drastic conditions and gave only a 5% yield. With the stereoisomeric yohimbene, alkali fusion results in an inversion, giving yohimbone (XLIII); under the relatively mild Oppenauer conditions the isomeric yohimbenone is obtained. The corresponding free acids can be used with equal success.

Quinine (XLIV) has been recovered unchanged under the usual conditions of the Oppenauer oxidation,[101] and this has been ascribed [9] to complex formation with the nitrogen atom, $R_3N^+:\overline{AlR'_3}$. This explanation, if correct, would appear to apply to quinine only, since a considerable number of nitrogen-containing alcohols have been oxidized by the Oppenauer procedure (Table VI). Furthermore, the aluminum isopropoxide reduction of aminoketones in general and of quininone [102] (XLV) in particular can be realized, and complex formation should also interfere in these instances.[102a] By employing potassium t-butoxide and benzophenone in benzene solution, it is possible [9] to achieve a nearly quantitative conversion of quinine (XLIV) to quininone (XLV), and

---

[98] Adams and Hamlin, *J. Am. Chem. Soc.*, **64**, 2599 (1942). The position of the carbonyl group was proved by total synthesis: Adams and Leonard, *J. Am. Chem. Soc.*, **66**, 257 (1944).

[99] Witkop, *Ann.*, **554**, 83 (1943).

[100] Jost, *Helv. Chim. Acta*, **32**, 1301 (1949), and G. A. Swan (private communication) were unable to obtain more than 50% of the ketone XLIII.

[101] McKee and Henze, *J. Am. Chem. Soc.*, **66**, 2021 (1944).

[102] Doering, Cortes, and Knox, *J. Am. Chem. Soc.*, **69**, 1700 (1947).

[102a] The failure of a number of β-amino alcohols to undergo the conventional Oppenauer reaction [Lutz, Jordan, and Truett, *J. Am. Chem. Soc.*, **72**, 4085 (1950)] was rationalized in terms of either a stable, five-membered complex interfering with the hydrogen transfer step or "of simple electron displacements resulting from complex formation involving coordination between nitrogen and aluminum." Subsequent work by Lutz and Wayland (*J. Am. Chem. Soc.*, in press), in which it was found that neither *cis*- nor *trans*-1-amino-2-indanol could be oxidized, was considered evidence in favor of the latter explanation. In the morphine series, Rapoport, Naumann, Bissell, and Bonner [*J. Org. Chem.*, **15**, 1103 (1950)] observed stereospecificity in the Oppenauer oxidation: dihydrocodeine (OH *cis* to C—O bond at C-5) and dihydroallo-ψ-codeine (OH *cis* to 9-14 C—C bond) were oxidized successfully, while the corresponding *trans* epimers were recovered unchanged.

the process is equally applicable to other 9-rubanols,[9,102] or even the simple benzyl alcohol XLVI.[103] This modified Oppenauer oxidation should prove useful in the oxidation of other alcohols as well, provided the resulting carbonyl compound will not suffer condensation in the presence of the strongly basic potassium t-butoxide.

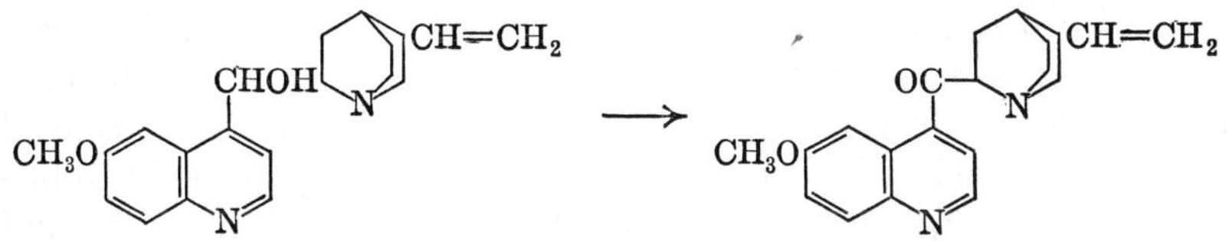

XLIV → XLV

XLVI

No difficulty has been encountered in the oxidation of both saturated [104,105] and Δ^5-unsaturated [104–109] 3-hydroxy steroidal alkaloids. Oxidation of the latter compounds is accompanied by migration of the double bond to the Δ^4 position as observed with other steroids.

Diazo ketones do not appear to be affected by aluminum isopropoxide,[110] and steroidal alcohols containing a diazo ketone group at C-17 have been oxidized by the Oppenauer procedure.[55,111,112] The mild conditions (twenty days at room temperature) employed for the conversion [55] of 21-diazopregnenolone (XLVII) to 21-diazoprogesterone (XLVIII) in 68% yield, though not necessary in this particular case because of the stability of the diazo ketone XLVII in boiling benzene, may prove useful for more sensitive compounds.

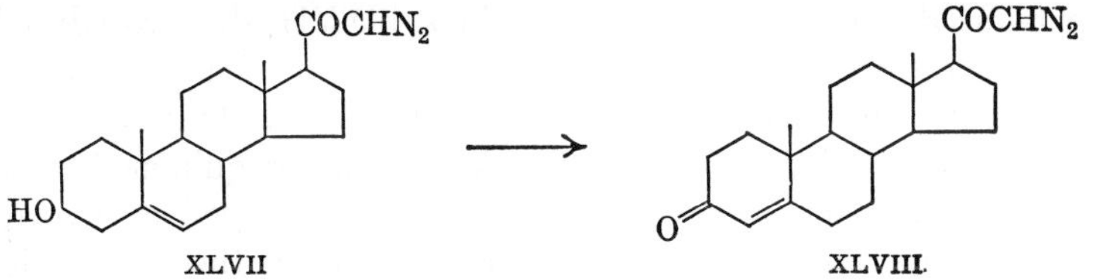

XLVII → XLVIII

[103] Woodward and Kornfeld, *J. Am. Chem. Soc.*, **70**, 2513 (1948).
[104] Prelog and Szpilfogel, *Helv. Chim. Acta*, **27**, 390 (1944).
[105] Rochelmeyer, *Arch Pharm.*, **277**, 340 (1939).
[106] Rochelmeyer, *Arch. Pharm.*, **277**, 339 (1939).
[107] Jacobs and Craig, *J. Biol. Chem.*, **159**, 617 (1945).
[108] Jacobs and Huebner, *J. Biol. Chem.*, **170**, 643 (1947).
[109] Jacobs and Sato, *J. Biol. Chem.*, **175**, 57 (1948).
[110] Lutz et al., *J. Am. Chem. Soc.*, **68**, 1818 (1946).
[111] Ehrenstein, *J. Org. Chem.*, **9**, 435 (1944).
[112] Reichstein, U. S. pat. 2,404,768 [*C. A.*, **40**, 6222 (1946)].

Oxidation of Primary Alcohols

Isolation of Aldehydes. Until very recently the Oppenauer reaction, except in isolated instances, has not been used as a preparative method for the oxidation of primary alcohols to aldehydes because the aldehydes condensed with the hydrogen acceptor (see below). In 1926, Pondorff [3] showed that 1-menthol could be oxidized to menthone with aluminum isopropoxide in the presence of cinnamaldehyde by continuous removal of the menthone. This procedure was subsequently extended [8] to primary alcohols, such as benzyl alcohol and 1-butanol, but has not found any general applicability because of the large excess of alcohol necessary.

By substituting quinone for acetone or cyclohexanone as the hydrogen acceptor, it has been found possible to oxidize unsaturated primary alcohols to the corresponding aldehydes.[113] Benzyl and anisyl alcohol gave 50–60% of the aromatic aldehyde, furfuryl alcohol 20% of furfural, and geraniol 38% of citral. Saturated alcohols, such as 1-heptanol or 3-phenyl-1-propanol, gave only very poor yields (5–8%) of aldehyde by this method. A special case is vitamin A aldehyde, which was obtained from vitamin A in the presence of acetaldehyde,[114] whereas with other hydrogen acceptors only side reactions were observed.[13,115]

Schinz and Lauchenauer [116,117] have developed a general preparative method for the Oppenauer oxidation of low-molecular-weight primary alcohols to aldehydes. The procedure is essentially a reversal of the Meerwein-Pondorff-Verley reduction [1] but does not require an excess of alcohol: [8] the alcohol to be oxidized is converted completely into its aluminate; an aldehyde (e.g., cinnamaldehyde or anisaldehyde) with a boiling point some 50° higher than that of the expected product is added to serve as the hydrogen acceptor, and the product is slowly distilled under reduced pressure. As illustrated in Table VII, this procedure has proved quite useful for the oxidation of a number of unsaturated primary alcohols and has succeeded even with alcohols (e.g., citronellol) where the conventional Oppenauer oxidation using quinone [113] failed. Ketones, such as benzophenone, can also be employed as hydrogen acceptors, and this experimental modification of the conventional Oppenauer oxidation promises to be of general use, even in the large-

---

[113] Yamashita and Matsumura, *J. Chem. Soc. Japan*, **64**, 506 (1943) [*C. A.*, **41**, 3753 (1947)]. This article and references 118 and 125 were kindly translated by Dr. Y. Sato of the Rockefeller Institute.

[114] Hawkins and Hunter, *J. Chem. Soc.*, **1944**, 411.

[115] Heilbron, Johnson, and Jones, *J. Chem. Soc.*, **1939**, 1560.

[116] Schinz, Lauchenauer, Jeger, and Rüegg, *Helv. Chim. Acta*, **31**, 2235 (1948); Rüegg and Jeger, *ibid.*, **31**, 1758 (1948).

[117] Lauchenauer and Schinz, *Helv. Chim. Acta*, **32**, 1265 (1949).

scale oxidation of low-molecular-weight secondary alcohols such as the acetal VIII derived from aldol.[29]

Simultaneous Condensation of Resulting Aldehydes with Hydrogen Acceptors. Initial attempts to apply the usual Oppenauer procedure to primary alcohols such as vitamin A [13,115] demonstrated that the aldehyde condensed with the acetone used as the hydrogen acceptor:

$$RCH_2OH \rightarrow RCHO \xrightarrow{(CH_3)_2CO} RCH=CHCOCH_3$$

As pointed out in the preceding section, this condensation can be prevented by the proper experimental modifications. However, in many instances this condensation is desirable. Geraniol (XLIX, R = H) in the presence of acetone and aluminum alkoxides affords ψ-ionone (L, R = H) in good yield,[13,118,119] and the reaction has been applied with conspicuous success to the methylated geraniols. [120-123] Thus, from 3-methylgeraniol (XLIX, R = CH$_3$), dl-ψ-irone (L, R = CH$_3$) was obtained. This latter compound on cyclization gave dl-α-irone (LI), providing a total synthesis of this important perfume.

<pre>
R   CHCH₂OH   R   CHCH=CHCOCH₃   H₃C   CH=CHCOCH₃

        XLIX   L   LI
</pre>

The one-step oxidation-condensation reaction has been studied with a number of primary alcohols such as phytol,[124] cinnamyl alcohol, and furfuryl alcohol,[13] using acetone,[13] diethyl ketone,[115] or methyl ethyl ketone [118] as the hydrogen acceptor. Activation of the hydroxyl group by adjacent unsaturation [13] does not seem necessary.[125] A variety of intermediates for polyene and isoprenoid syntheses has been prepared by such procedures,[71,126-132] and all such examples are collected in Table

[118] Yamashita and Honjo, *J. Chem. Soc. Japan*, **63**, 1335 (1942) [*C. A.*, **41**, 3041 (1947)].
[119] Tavel, Sc.D. Thesis, Eidgenöss. Techn. Hochschule, Zürich, 1946, pp. 54–59.
[120] Naves, Grampoloff, and Bachmann, *Helv. Chim. Acta*, **30**, 1607 (1947).
[121] Schinz, Ruzicka, Seidel, and Tavel, *Helv. Chim. Acta*, **30**, 1813 (1947); Seidel, Schinz, and Ruzicka, *ibid.*, **32**, 2113 (1949).
[122] Winter, Schinz, and Stoll, *Helv. Chim. Acta*, **30**, 2215 (1947).
[123] Rouvé and Stoll, *Helv. Chim. Acta*, **30**, 2220 (1947).
[124] Karrer and Epprecht, *Helv. Chim. Acta*, **24**, 1043 (1941).
[125] Yamashita and Shimano, *J. Chem. Soc. Japan*, **63**, 1338 (1942) [*C. A.*, **41**, 3042 (1947)].
[126] Milas and Harrington, *J. Am. Chem. Soc.*, **69**, 2248 (1947).
[127] Milas, Grossi, Penner, and Kahn, *J. Am. Chem. Soc.*, **70**, 1292 (1948).
[128] Karrer and Benz, *Helv. Chim. Acta*, **32**, 232 (1949).
[129] Karrer, Karanth, and Benz, *Helv. Chim. Acta*, **32**, 436 (1949).
[130] H. Schinz, private communication; *cf.* Simon, Thesis, Eidgenöss. Techn. Hochschule, Zürich, 1948.
[131] Zobrist and Schinz, *Helv. Chim. Acta*, **32**, 1195 (1949).
[132] H. Schinz, private communication; *cf.* Zobrist, Thesis, Eidgenöss. Techn. Hochschule, Zürich, 1948.

VIII. The aluminum alkoxide catalyzed condensation of carbonyl compounds involves very mild conditions,[133, 134] and side reactions, such as the loss of a formyl group,[135] rarely occur. In Oppenauer oxidations of primary alcohols where subsequent condensation of the aldehyde is desired (Table VIII), it may be necessary to use larger amounts of aluminum alkoxide since the water formed during the condensation will remove an equivalent amount of catalyst.

Side Reactions

Two common side reactions which have already been discussed are the migration of a double bond as observed in the oxidation of Δ^5-3-hydroxy steroids to the Δ^4-3-ketones, and the condensation of an aldehyde with the hydrogen acceptor. The loss of a 7-alkoxy group in cholesterol derivatives [136, 137, 138] is not encountered elsewhere and is probably associated with the unusual reactivity of the C-7 position of Δ^5-steroids as illustrated by the quinone oxidation of Δ^5-3-hydroxy steroids to the $\Delta^{4,6}$-dienones (LVI).[139] Occasionally the dehydration of secondary [140] and tertiary alcohols [78, 79, 141, 142] is noted, and partial hydrolysis of esters [84, 143] may occur although a choice of conditions may prevent it. Cholesterol acetate is hydrolyzed to a certain extent by aluminum isopropoxide,[4, 144] but not by the *t*-butoxide.[4] The apparent loss of the elements of acetic acid from a 3-acetoxy-4-hydroxy steroid has been reported.[145] Inversion of configuration of an asymmetric carbon atom adjacent to the hydroxyl group to be oxidized has been observed for both aluminum [72, 146] and potassium *t*-butoxide [9, 102] catalyzed Oppenauer

[133] Wayne and Adkins, *J. Am. Chem. Soc.*, **62**, 3401 (1940).
[134] Heilbron, Jones, and Lacey, *J. Chem. Soc.*, **1946**, 29; Heilbron, Johnson, Jones, and Spinks, *ibid.*, **1942**, 733.
[135] Wilds and Djerassi, *J. Am. Chem. Soc.*, **68**, 1718 (1946).
[136] Henbest and Jones, *Nature*, **158**, 950 (1946); *J. Chem. Soc.*, **1948**, 1798.
[137] Bergström and Wintersteiner, *J. Biol. Chem.*, **143**, 506 (1942); Bergström, *Arkiv Kemi, Mineral. Geol.*, **16A**, No. 10, p. 25 (1942). The alcohol was believed to be Δ^6-cholestene-3,5-diol, but the correct structure has since been shown (ref. 136) to be 7-ethoxycholesterol.
[138] Prelog, Ruzicka, and Stein, *Helv. Chim. Acta*, **26**, 2239 (1943); the structure originally assigned to the starting material has been corrected (ref. 136).
[139] Wettstein, *Helv. Chim. Acta*, **23**, 388 (1940).
[140] Marker and Turner, *J. Am. Chem. Soc.*, **62**, 2541 (1940).
[141] Marker and Turner, *J. Am. Chem. Soc.*, **64**, 482 (1942).
[142] Julian and Cole, U. S. pat. 2,394,551 [*C. A.*, **40**, 2593 (1946)].
[143] Reichstein, Meystre, and v. Euw., *Helv. Chim. Acta*, **22**, 1107 (1939).
[144] Windaus and Schenck, U. S. pat. 2,098,985 [*C. A.*, **32**, 196 (1938)].
[145] S. Lieberman and D. K. Fukushima, unpublished observation, and *J. Am. Chem. Soc.*, **72**, 5216 (1950). The acetoxyl group may have been hydrolyzed in working up the reaction mixture.
[146] Linstead, Whetstone, and Levine, *J. Am. Chem. Soc.*, **64**, 2021 (1942).

oxidations. Formyl groups may also be lost with either reagent.[103, 135] Ring enlargement appears to occur during the Oppenauer oxidation of steroids containing the 17-acetyl-17-hydroxy grouping (LII) with the formation of the D-homo compounds (LIII); [147–150] but, since alumina also promotes such rearrangements [147] and all but one of the reaction mixtures [150] were chromatographed over alumina, the results are not conclusive. The corresponding 16,17-oxido derivatives do not suffer ring enlargement.[60, 61] An unusual reaction is the Oppenauer oxidation of the triphenylmethyl ether of the triol LIV [38] to Δ^4-androstene-3,17-dione (LV) in 42% yield with loss of the entire side chain.

CHOICE OF EXPERIMENTAL CONDITIONS

Aluminum Alkoxides

The three most common catalysts in the Oppenauer oxidation are aluminum *t*-butoxide, isopropoxide, and phenoxide. The *t*-butoxide was used initially by Oppenauer,[4] and its use has persisted, but there are very few reactions in which it has proved superior to the others. Aluminum isopropoxide and, in particular, the phenoxide are much easier to prepare, although this may not have too much influence on the choice of reagent since the alkoxides are now commercially available. No

[147] Stavely, *J. Am. Chem. Soc.*, **63**, 3127 (1941).
[148] Hegner and Reichstein, *Helv. Chim. Acta*, **24**, 842 (1941).
[149] v. Euw and Reichstein, *Helv. Chim. Acta*, **24**, 889 (1941).
[150] Goldberg, Aeschbacher, and Hardegger, *Helv. Chim. Acta*, **26**, 684 (1943). The structure of the product was not definitely established.

thorough comparison has been carried out to determine whether one of the three alkoxides possesses special merit. Thus aluminum phenoxide is superior to the *t*-butoxide for the oxidation of certain saturated hydroxy steroids [86] in the presence of acetone and benzene, but no comparison was made [151] between aluminum phenoxide and isopropoxide in conjunction with cyclohexanone and toluene. Aluminum phenoxide has been reported [96] to be superior to all other alkoxides in the partial oxidation of Δ^5-androstene-3β,17β-diol, but in another laboratory [97] the *t*-butoxide appeared to be equally satisfactory.

Aluminum *t*-Butoxide. Detailed methods [4, 37] are described, particularly in *Organic Syntheses*,[152] for the preparation of the *t*-butoxide from aluminum, *t*-butyl alcohol, and mercuric chloride. Often the colloidal mercury is not separated,[16] and most preparations contain small amounts of mercury or mercuric chloride. High-vacuum sublimation affords a white powder [153, 154] free of metallic impurities; however, studies with such material [154] have indicated that zinc, aluminum, or mercuric chloride may exert a promoter effect in certain Oppenauer oxidations similar to that noted in the Tishchenko condensation.[155] If the promoter effect of certain impurities is established it might explain the sometimes conflicting reports from various laboratories on the advantages of certain alkoxides. The relatively short time employed in the oxidation of saturated steroid alcohols [15] with unpurified *t*-butoxide as compared to the longer time for comparable unsaturated alcohols [12, 33] with purified material may be due to a promoter effect of mercuric chloride.

The *t*-butoxide may be preserved in toluene solution, and aliquots added to the reaction mixture with prior centrifugation [51] to remove traces of aluminum hydroxide. Since aluminum *t*-butoxide decomposes slowly in solutions above 115°, xylene is about the highest-boiling solvent that can be recommended for use with this reagent.[133]

Aluminum Isopropoxide. Directions for the preparation of aluminum isopropoxide are given in an earlier volume of this series.[1] Material prepared in this manner appears to exist in various degrees of association, thus accounting for the numerous observed melting points.[156] A detailed study of several factors (aluminum particle size, moisture content, catalysts, etc.) entering in the preparation of aluminum alkoxides has been reported.[157] In a version [45, 46, 47] of the Oppenauer oxidation

[151] T. Reichstein, private communication.
[152] Wayne and Adkins, *Org. Syntheses*, **21**, 8 (1941).
[153] R. H. Baker, private communication.
[154] Baker and Abramovitch, unpublished observation.
[155] Child and Adkins, *J. Am. Chem. Soc.*, **45**, 3013 (1923); **47**, 798 (1925).
[156] Macbeth and Mills, *J. Chem. Soc.*, **1949**, 2648.
[157] Brown, *Abstracts*, p. 40M, A.C.S. Atlantic City Meeting, September, 1949.

described in detail in the experimental section, a solution of aluminum isopropoxide in toluene is used. It may be advantageous to store the reagent in that form, for 25–30 weight per cent solutions can be readily prepared [37] and material which crystallizes from such solutions on standing can be redissolved by warming. Since the isopropoxide is easy to prepare and has been used in a variety of reactions, it is probably the preferred catalyst. It is generally used in commercial operations.[37] Nothing seems to be known about possible promoter effects in aluminum isopropoxide-catalyzed Oppenauer reactions. Occasionally [158,159] aluminum isopropoxide has been added to a solution of the alcohol to be oxidized, all the isopropanol formed during the interchange with the alcohol being removed by distillation before introduction of the hydrogen acceptor. Such a procedure has proved to be especially advantageous in the oxidation of primary alcohols to the corresponding aldehydes.[116,117]

Aluminum Phenoxide. Aluminum phenoxide is particularly easy to prepare, although it is almost invariably contaminated by phenol. This is especially true of those procedures [99,119,160] in which aluminum foil or shavings are added to hot phenol in the absence of a solvent, and the cooled and crushed material is used directly. The reaction can be started by the addition of traces of iodine or mercuric chloride. A purer product is obtained [82] by conducting the reaction in benzene solution and isolating the product by concentration and precipitation with petroleum ether. No direct comparison among phenoxides of differing degrees of purity has been made, but material prepared without solvent was found to be satisfactory for the oxidation of geraniol to ψ-ionone,[119] and to compare favorably with aluminum t-butoxide. The phenoxide prepared in benzene solution is claimed [86] to be superior to other alkoxides in the oxidation of saturated alcohols.

Other Catalysts

Chloromagnesium alkoxides have been suggested in the patent literature [30,159] as catalysts for the Oppenauer oxidation, and they have been used occasionally for the reduction of carbonyl compounds.[6] Potassium t-butoxide has been proved to be superior to the aluminum derivative in the oxidation of quinine and related compounds [9,102,103] but it can be used only with carbonyl compounds and hydrogen acceptors that do

---

[158] Chinaeva, Ushakov, and Marchevskii, *J. Gen. Chem. U.S.S.R.*, **9**, 1865 (1939) [*C. A.*, **34**, 4073 (1940)].

[159] Serini, Köster, and Strassberger, U. S. pat. 2,379,832 [*C. A.*, **39**, 5053 (1945)]. The corresponding Fr. pat. 822,551 was granted in 1938.

[160] Cook, *J. Am. Chem. Soc.*, **28**, 608 (1906).

not undergo condensation in its presence. Although not within the scope of the Oppenauer reaction, it should be noted that a Raney nickel catalyst [77] was effective in the oxidation of a number of alcohols when substituted for aluminum alkoxides.

Hydrogen Acceptors

Acetone in conjunction with benzene as a solvent was used exclusively by Oppenauer [4] in his original studies, and this ketone has remained one of the most widely used hydrogen acceptors. However, with the introduction of cyclohexanone [159] and the concomitant use of toluene or xylene as solvents, higher reaction temperatures and shorter reaction times were achieved. In an extensive polarographic study, Adkins and co-workers [161-164] determined the apparent oxidation potentials and relative reactivities (based on diisopropyl ketone) of ninety ketones of various structures. On the basis of these results, the important features of a useful hydrogen acceptor in the Oppenauer oxidation were considered and five of the more readily available ketones were studied in detail.[165]

Although a high oxidation potential is desirable, a comparatively low one can be offset by using a large excess of the ketone. Acetone, with a relatively low potential (0.129 volt), is cheap and can thus be used economically in large excess. It is low boiling, and even its condensation product, mesityl oxide, always formed by the aluminum alkoxide-catalyzed self-condensation,[133] can be removed fairly readily.

Cyclohexanone not only has a higher oxidation potential (0.162 volt) than acetone, but the higher boiling point permits a shorter reaction time (about one-tenth of that necessary with acetone) and thus reduces side reactions due to condensation. Cyclohexanone is also readily available and is particularly useful with steroids, since it can be separated from the reaction product by steam distillation.

Methyl ethyl ketone and benzil have been studied by Adkins and Franklin;[165] the diketone would appear to be useful in the preparation of comparatively low-boiling carbonyl compounds, although to date it has been employed only for the oxidation of benzhydrol.[165]

Methyl ethyl ketone [118] and diethyl ketone [115] have been examined for use in the oxidation of primary alcohols but were found to undergo condensation with the resulting aldehyde, as is true with acetone. An unusual reaction observed when diethyl ketone was used as the hydrogen

---

[161] Adkins and Cox, *J. Am. Chem. Soc.*, **60**, 1151 (1938).
[162] Cox and Adkins, *J. Am. Chem. Soc.*, **61**, 3364 (1939).
[163] Baker and Adkins, *J. Am. Chem. Soc.*, **62**, 3305 (1940).
[164] Adkins, Elofson, Rossow, and Robinson, *J. Am. Chem. Soc.*, **71**, 3622 (1949).
[165] Adkins and Franklin, *J. Am. Chem. Soc.*, **63**, 2381 (1941).

acceptor in the oxidation of vitamin A was the apparent introduction of a double bond into the ionone ring.[166]

In the modified Oppenauer oxidation of quinine in which potassium t-butoxide was used,[9] benzophenone was found to be a satisfactory oxidizing agent since it cannot undergo condensation in the presence of the strongly basic catalyst. This ketone also was superior to all other hydrogen acceptors in the modified Oppenauer oxidation (continuous distillation) of the acetal of aldol (VIII).[29] It is interesting to note that fluorenone, with a lower oxidizing potential than benzophenone, was effective [102] in the oxidation of epi-quinidine, which could not be oxidized with benzophenone. The unusual reactivity of fluorenone was also observed in the polarographic studies.[163]

The use of catalytic amounts of anthraquinone in place of the usual large excess of hydrogen acceptor has been suggested,[167] since anthrahydroquinone is readily oxidized to the quinone by air. Test runs at room temperature in conjunction with polarographic determinations proved the feasibility of this suggestion in the oxidation of benzohydrol and fluorenol. Cholesterol was also attacked, although no definite product was isolated. Since the reactions at room temperature required from fifty to four hundred hours, an attempt was made to examine the usefulness of this catalytic method on a preparative scale by refluxing cholesterol for as long as sixteen hours.[168] Only a poor yield (7%) of Δ^4-cholesten-3-one was obtained.

p-Benzoquinone is of unusual interest as a hydrogen acceptor, and its very high oxidation potential (0.71 volt) has been ascribed [163] to isomerization of its reduced quinol form to the benzenoid hydroquinone. Although quinone and its reduction product, hydroquinone, introduce certain difficulties in the isolation of the reaction product, the rapid rate of reaction with quinone permits the use of relatively small quantities (1 to 3 moles) and low temperatures (25–60°).[165] Quinone is one of the few hydrogen acceptors that allows the isolation of aldehydes in the unmodified Oppenauer oxidation of primary alcohols.[113] Although the basis of its usage is largely empirical, quinone seems to be the best hydrogen acceptor for the oxidation of triterpenoid alcohols.[64, 169, 170, 171]

An unexpected extension of the Oppenauer oxidation was discovered by Wettstein,[139] who noted that replacement of acetone or cyclohexanone by quinone in the oxidation of Δ^5-3-hydroxysteroids (IX) resulted

[166] Haworth, Heilbron, Jones, Morrison, and Polya, *J. Chem. Soc.*, **1939**, 128.
[167] Baker and Stanonis, *J. Am. Chem. Soc.*, **70**, 2594 (1948).
[168] Djerassi, unpublished observation.
[169] Ruzicka and Rey, *Helv. Chim. Acta*, **24**, 529 (1941).
[170] Biedebach, *Arch. Pharm.*, **281**, 59 (1943).
[171] Heilbron, Jones, and Robins, *J. Chem. Soc.*, **1949**, 448.

in the formation of the corresponding $\Delta^{4,6}$-3-ketosteroids (LVI). The yields were not specified, but subsequent work [145, 172–175] has indicated that approximately 40% of the pure doubly unsaturated ketone LVI may be obtained. The nature of the C-17 substituent (R = CO_2CH_3, C_8H_{17}, $COCH_3$, $OCOC_6H_5$) does not seem to be critical although the ketol side chain, $COCH_2OCOCH_3$, is decomposed to a certain extent.

IX   LVI   LVII

Other satisfactory syntheses for such dienones from the same starting material (IX) involve at least three separate steps, one of which is usually an ordinary Oppenauer oxidation. The mechanism of this unusual reaction is not clear, but it appears that the steric peculiarities of the steroid molecule and the unusual reactivity of position 7 in Δ^5-unsaturated steroids are important factors. Under the same conditions saturated steroid alcohols give the saturated ketone [139] and Δ^4-3-ketosteroids remain unaltered.[139, 168] On the other hand, the Δ^5-3-*ketone* LVII (R = $OCOC_6H_5$) does afford [139] the dienone LVI in about 25% yield,[168] and it may well be the key intermediate in the reaction, since the usual Oppenauer oxidation of Δ^5-3-*hydroxy*steroids (IX) probably proceeds through such a compound [164] although the Δ^4-3-ketone (X) is invariably isolated. It is of interest to note that esters of Δ^5-3-hydroxy steroids when heated with quinone in a sealed tube give up to 30% of the $\Delta^{5,7}$-3-hydroxy derivative,[176] but the conditions employed are more drastic than those prevailing in the Wettstein-Oppenauer oxidation.

Until recently aldehydes have been used only infrequently as hydrogen acceptors. The use of benzaldehyde,[8, 165] cinnamaldehyde,[3, 8, 117] and anisaldehyde [117] has been cited, and acetaldehyde proved to be the only hydrogen acceptor effective in the Oppenauer oxidation of vitamin A.[114] The Tishchenko condensation of the aldehydes used as hydrogen acceptors and those arising from the oxidation presents a complication,[2]

---

[172] Marker and Turner, *J. Am. Chem. Soc.*, **63**, 771 (1941).

[173] Ushakov and Kosheleva, *J. Gen. Chem. U.S.S.R.*, **14**, 1138 (1944) [*C. A.*, **40**, 4071 (1946)].

[174] Wilds and Djerassi, *J. Am. Chem. Soc.*, **68**, 1713 (1946).

[175] Djerassi, *J. Am. Chem. Soc.*, **71**, 1009 (1949).

[176] Milas and Heggie, *J. Am. Chem. Soc.*, **60**, 984 (1938); Milas and Milone, *ibid.*, **68**, 738 (1946); Mazza and Migliardi, *Quad. Nutriz.*, **8**, 85 (1941) [*C. A.*, **37**, 3762 (1943)]; Sah, *Rec. trav. chim.*, **59**, 454 (1940).

but this difficulty can be circumvented by continuously distilling the oxidation product from the reaction mixture.[8,117] If such a procedure is employed, it is necessary to choose as a hydrogen acceptor an aldehyde with a boiling point higher than that of the product.[117]

With keto alcohols simultaneous oxidation and reduction may be achieved in the absence of additional hydrogen acceptor. Oppenauer[177] showed that when dehydroepiandrosterone (IX, R = O) is heated with aluminum t-butoxide in boiling benzene for fourteen hours approximately 10% of testosterone (X, R = OH) is obtained in addition to the completely oxidized Δ^4-androstene-3,17-dione and the reduced Δ^5-androstene-3,17-diol. Under similar conditions Δ^5-pregnen-3β-ol-20-one gave progesterone. With both compounds the keto group present in the steroid serves as the hydrogen acceptor; however, the yields are too low for preparative purposes. A polarographic investigation[154] of this dismutation indicates that the amount of the Δ^4-3-ketosteroid fraction (using Δ^4-cholestenone as the standard) could be raised significantly by the addition of small amounts of aluminum or zinc chloride. The promoter effect of such salts has also been observed in the aluminum alkoxide-catalyzed Tishchenko condensation of aldehydes.[155] Dismutation reactions similar to those considered above have been suggested as accounting for the abnormal products encountered in the aluminum isopropoxide reduction of 7-ketocholesteryl acetate[178] and of helenalin.[179]

Solvents

According to Oppenauer[4] a solvent such as benzene is necessary for the oxidation of secondary steroidal alcohols when acetone is used as the hydrogen acceptor. Although benzene is used most commonly in conjunction with acetone, toluene is employed occasionally.[42,180–183] Toluene is used almost invariably with cyclohexanone, and the reaction temperature can be raised even further by substituting xylene[99] for toluene. The choice of a solvent is at times critical; e.g., steroidal diazoketones are stable in boiling benzene solution but are decomposed slowly on refluxing in toluene.[55] Dioxane has been suggested[165] as a solvent, but it has been used only once and then with benzene.[78]

[177] Oppenauer, U. S. pat. 2,229,599 [*C. A.*, **35**, 3039 (1941)]; U. S. pat. 2,363,548 [*C. A.*, **39**, 3400 (1945)].
[178] Wintersteiner and Ruigh, *J. Am. Chem. Soc.*, **64**, 2455 (1942).
[179] Adams and Herz, *J. Am. Chem. Soc.*, **71**, 2550 (1949).
[180] Marker and Crooks, *J. Am. Chem. Soc.*, **64**, 1281 (1942).
[181] Marker, Crooks, Wagner, and Wittbecker, *J. Am. Chem. Soc.*, **64**, 2092 (1942).
[182] Marker, Wagner, and Wittbecker, *J. Am. Chem. Soc.*, **64**, 2096 (1942).
[183] Marker, Wagner, Ulshafer, Wittbecker, Goldsmith, and Ruof, *J. Am. Chem. Soc.*, **69**, 2185, 2209 (1947).

Although side reactions due to condensations of the mesityl oxide type are reduced by working in dilute solution,[165] several reports have indicated that a solvent can be dispensed with. Tavel [119] in a study of the oxidation of geraniol with acetone and aluminum phenoxide concluded that benzene had no beneficial influence on the yield of ψ-ionone. The conversion of primary alcohols to aldehydes, in which higher-boiling aldehydes were used as hydrogen acceptors [8,117] and the product removed by continuous distillation, has been carried out successfully without diluents. That solvents may not be necessary even in the case of steroidal alcohols is indicated in a patent [159] in which it is reported that nearly quantitative yields are obtained by heating the steroid in cyclohexanone in the presence of aluminum isopropoxide for a short time; independent confirmation is necessary to substantiate this claim.

Time and Temperature

Time and temperature can be varied over a wide range, depending on the alcohol to be oxidized, although the choice of solvent and hydrogen acceptor naturally controls the maximum temperature that can be reached. As a general but by no means universal rule, experiments in refluxing benzene and acetone are conducted for four to twenty hours, whereas with boiling toluene and cyclohexanone only fifteen minutes to two hours is required. There are obvious exceptions to the above generalization, but in most instances described in the literature the optimum length of time has not been determined. In a detailed study of the Oppenauer oxidization of geraniol [119] only a slight increase in yield was observed when the reaction time was increased from twenty-four to sixty-eight hours. A very useful variation, apparently applicable to both saturated [19,184] and unsaturated [45,46,47] alcohols, involves the dropwise addition of an aluminum isopropoxide solution over a period of thirty minutes to a slowly distilling solution of the alcohol in toluene and cyclohexanone. Sensitive compounds can be oxidized at room temperature for several days with acetone in benzene,[55,72] or with cyclohexanone or quinone in toluene.[165] Because of the rapid rate of reaction, oxidations with quinone often require lower temperatures than the corresponding oxidations with other hydrogen acceptors.[165] Reactions can also be carried out in a sealed tube.[87,91,114] For the simultaneous oxidation and condensation of primary alcohols with acetone, a reaction time of twenty-four to forty-eight hours appears necessary.

[184] Meystre and Wettstein, *Helv. Chim. Acta*, **30**, 1046 (1947); **31**, 1895 (1948).

Ratio of Alkoxide to Alcohol

Although only catalytic amounts of alkoxide are theoretically required in the Oppenauer oxidation, in practice at least 0.25 mole of alkoxide per mole of alcohol is used. Since an excess of alkoxide usually has no detrimental effect, 1 to 3 moles of alkoxide is recommended, particularly since water, either present in the reagents or formed during condensation reactions, will remove an equivalent amount of catalyst. The quantity of hydrogen acceptor to be used is in some measure dependent on its oxidation potential.[165] In the oxidation of steroids, acetone is employed in 50 to 200 molar excess, while 10 to 20 moles of cyclohexanone and 3 to 10 moles of quinone appear to be sufficient. These amounts can probably be reduced in the oxidation of simpler alcohols, although the optimum proportions have to be determined for each specific system. It should be noted that the scale of operation is limited only by the available equipment, and experiments in which the amount of alcohol ranged from 10 mg.[143] to 25.6 kg.[37] have been carried out successfully.

Isolation of Products

A number of procedures has been used for the isolation of the product of an Oppenauer oxidation. A preliminary steam distillation of the reaction mixture is desirable when the oxidation product is a non-volatile ketone and when toluene and cyclohexanone are used. It is also advantageous when acetone is employed since condensation products such as mesityl oxide are invariably formed and these products are removed to a large extent by steam distillation. With particularly sensitive compounds like the ketol acetates a small amount of acetic acid may be added before the steam distillation to neutralize the reaction mixture. At times the steam distillation may be preceded by the hydrolysis and removal of the aluminum compounds (see below).

The preliminary steam distillation may sometimes be replaced by a simple distillation at reduced pressure until a nearly dry residue is obtained. When the system acetone and benzene is used the initial distillations may be omitted; the reaction mixture may simply be transferred to a separatory funnel and extracted with a suitable organic solvent, and the solution washed several times with dilute acid. The residues remaining after the preliminary distillations are treated in a similar manner. Dilute alkali is often substituted for the acid washes in the isolation of amino ketones. With sensitive compounds, such as acetals or diazoketones, a solution of Rochelle salt (sodium potassium tartrate) is equally satisfactory.

When quinone is used as the hydrogen acceptor a thorough washing with alkali is necessary to remove the hydroquinone. Phenol (from aluminum phenoxide) may be removed in this manner or at a later stage by high-vacuum sublimation.[87, 91] Traces of cyclohexanone may be extracted by washing with 40% bisulfite solution.[18, 93] Low-boiling condensation products if not eliminated by a preliminary steam or vacuum distillation can be removed by storing the residue in a high vacuum or, better,[16, 136] by a codistillation with xylene. In the oxidation of amino alcohols, the resulting amino ketones can be separated from most of the by-products by extraction with dilute acid. It should be noted that neither carboxyl [99, 181] nor phenolic [22, 23, 59, 185] groups need be protected during oxidation, and that extraction with dilute alkali may be employed for the isolation of oxidation products containing these groups. Finally, the carbonyl compound may be isolated directly by crystallization or distillation, or if present in mixtures, via Girard complexes, chromatography, etc., or by a combination of several such methods. Unreacted alcohol is conveniently removed by formation of its mono ester with succinic acid.

Miscellaneous Suggestions

In order to free the reaction system of traces of water it is advisable to distil a small amount of solvent from the alcohol-hydrogen acceptor-solvent mixture before adding the alkoxide. The alkoxide may be added as a solid or in solution, solution being preferable. Experience has shown that better results are obtained when the reaction mixture is not cooled during the introduction of the alkoxide, which may be added in one portion or stepwise. The reaction mixture and alkoxide solution must be protected from atmospheric moisture by a calcium chloride drying tube or other suitable means.

EXPERIMENTAL PROCEDURES

The following examples have been selected because they illustrate a variety of typical procedures and because they have been repeated sufficiently to be considered reproducible. Detailed directions for the oxidation of cholesterol to cholestenone in 70–81% yield by the original Oppenauer procedure (aluminum *t*-butoxide, acetone, and benzene) are given in *Organic Syntheses*.[186]

[185] Ungnade and Tucker, *J. Am. Chem. Soc.*, **70**, 4134 (1948).
[186] Oppenauer, *Org. Syntheses*, **21**, 18 (1941).

cis-α-Decalone.[21,187] (Use of aluminum isopropoxide, acetone, and benzene for the oxidation of saturated alcohols.) To a solution of 1.5 g. of *cis*-α-decalol (m.p. 92°) in 150 ml. of dry, thiophene-free benzene and 100 ml. of dry acetone is added 3 g. of freshly distilled aluminum isopropoxide. The mixture, protected with a calcium chloride drying tube, is refluxed for twelve hours. After being cooled to room temperature, the reaction mixture is washed twice with 30% sulfuric acid, with water until neutral, then is dried over sodium sulfate and the solvent is removed under reduced pressure. Fractional distillation of the residue gives a fore-run of mesityl oxide, and 1.2 g. (80%) of *cis*-α-decalone b.p. 116°/18 mm., $n_D^{20°}$ 1.4939; the semicarbazone melts at 219–220° (dec.).

$\Delta^{20,23}$-24,24-Diphenylcholadiene-3,11-dione.[19] (Illustration of the addition of a solution of the alkoxide to a continuously distilling solution of the alcohol in cyclohexanone and toluene.) Three grams of $\Delta^{20,23}$-24,24-diphenylcholadien-3α-ol-11-one (V) is dissolved in 300 ml. of dry toluene and 30 ml. of freshly distilled cyclohexanone contained in a two-necked flask equipped with a dropping funnel and a condenser set downward for distillation. Both the dropping funnel and the receiver attached to the condenser are protected by calcium chloride tubes. A slow rate of distillation is maintained after 100 ml. of distillate has been collected, and a solution of 3 g. of aluminum isopropoxide in 100 ml. of dry toluene is added dropwise over a period of one-half hour. The flask is cooled slightly, 30 ml. of a concentrated solution of Rochelle salt is added, and steam is passed through the mixture for one hour. The cooled residue is extracted with chloroform, and the chloroform layer is washed well with water and dried, and the solvent is evaporated. Crystallization of the residue from ether or a mixture of methanol and acetone gives 2.6 g. (86%) of $\Delta^{20,23}$-24,24-diphenylcholadiene-3,11-dione, m.p. 227–230°. This procedure is equally applicable to the oxidation of unsaturated alcohols as illustrated by the oxidation [45] (using exactly the same conditions as specified above) of $\Delta^{5,20,23}$-24,24-diphenylcholatrien-3β-ol (XIV) in 95% yield.

Desoxycorticosterone Acetate.[37] (Use of cyclohexanone and toluene in a large-scale preparation.) Five liters of distillate is collected from a solution of 600 g. of Δ^5-pregnen-3β,21-diol-20-one 21-acetate (IX, R = $COCH_2OCOCH_3$) in 26 l. of toluene and 5.4 l. of cyclohexanone to ensure anhydrous conditions, and to the boiling reaction mixture contained in a 120-l. flask is added 2.4 l. of an aluminum isopropoxide solution prepared by dissolving 260 g. of aluminum isopropoxide in 2.6 l. of dry toluene and filtering. (At least 95% of the aluminum

[187] Cavaglieri, Ph.D. Thesis, Yale, 1943.

isopropoxide must dissolve in the toluene.) The reaction mixture is refluxed for thirty minutes. A solution of 72 ml. of glacial acetic acid in 720 ml. of toluene is added, the mixture is allowed to cool to 40°, and steam is passed through for four hours at such a rate that 10 l. of distillate is obtained every seven to ten minutes. After the addition of 1.7 kg. of sodium chloride and 440 g. of kieselgur, the reaction vessel is cooled and the solid collected and air dried. Material adhering to the walls of the flask is recovered by extraction with boiling acetone. The dry ketone-kieselgur mixture is extracted in a Soxhlet apparatus for fifteen hours with 15 l. of acetone; the extract is concentrated to 2 l. and cooled; 78% of desoxycorticosterone acetate, m.p. 152–156°, is obtained in the first crop, and an additional 5% from the mother liquors.

Methyl $\Delta^{4,6}$-3-Ketoetiocholadienate (LVI, R = CO_2CH_3).[175] (The use of quinone as hydrogen acceptor and the simultaneous introduction of a double bond.) A solution of 2 g. of methyl Δ^5-3-hydroxyetiocholenate (IX, R = CO_2CH_3) and 12 g. of benzoquinone in 120 ml. of dry toluene is concentrated under reduced pressure to a volume of about 100 ml., 2 g. of aluminum isopropoxide or t-butoxide is added, and the mixture is refluxed for forty-five minutes. Water (100 ml.) is added to the black solution, and steam is passed through it until about 1 l. of distillate is collected. The residual solution is acidified with dilute sulfuric acid and extracted exhaustively with ether. After washing three times with sulfuric acid and with water, 5% potassium hydroxide solution is added carefully without shaking and the black layer is drawn off. This treatment is repeated until the ether solution is reddish (otherwise a troublesome emulsion results), and it is then washed thoroughly by shaking with alkali until no more color is removed. The organic layer is then washed with water, dried, and evaporated. The brownish crystalline residue (1.95 g.) has a single maximum at 282.5 mμ, characteristic of $\Delta^{4,6}$-3-ketosteroids, and is purified by chromatographing on 40 g. of alumina. The colorless crystals obtained from the petroleum ether-benzene (25/27) and benzene eluates give colorless rosets (0.81 g., 41%) of methyl $\Delta^{4,6}$-3-ketoetiocholadienate, m.p. 165–165.5°, after recrystallization from methanol.

α-Cyclocitral.[117] (Typical procedure for the oxidation of an alicyclic, primary alcohol to the corresponding aldehyde by continuous distillation in the presence of a higher-boiling aldehyde as hydrogen acceptor.) To 3.75 g. of α-cyclogeraniol in a 20-ml. round-bottomed flask equipped with a 10-cm. Vigreux column is added 1.66 g. of aluminum isopropoxide. The isopropanol formed is removed over the course of forty-five minutes at a bath temperature of 70–100° and 12 mm. pressure. To the cyclogeraniol aluminate is then added 5.1 g. (155%) of anisaldehyde in one portion, and the solution is distilled at the rate of 5–12 drops per minute by raising the bath temperature (12 mm. pressure) from 122 to 170° during twenty-five minutes. Fractionation of the distillate yields 2.46 g. (66%) of pure α-cyclocitral with b.p. 75°/12 mm., $n_D^{20°}$ 1.4701; 2,4-dinitrophenylhydrazone, m.p. 157°. Cinnamaldehyde appears to be the hydrogen acceptor of choice for the oxidation of low-molecular-weight *aliphatic* primary alcohols.

β-Ketobutyraldehyde 2-Methylpentane-2,4-diol Acetal.[29] (Oxidation of a low-molecular-weight, acid-labile, secondary alcohol by continuous distillation, using benzophenone as the hydrogen acceptor.) A mixture of 2 kg. of acetaldol 2-methylpentane-2,4-diol acetal (VIII),* 4 kg. of benzophenone, and 100 g. of aluminum isopropoxide lumps in a 12-l. flask, equipped with condenser for distillation, is heated in an oil bath at 15 mm. At a bath temperature of 150°, the aluminate commences to form and isopropanol is collected in the distillate. After complete removal of isopropanol, the bath temperature is raised slowly to 200° over a period of one hour, during which time 2.2 kg. of distillate (b.p. up to 135°/15 mm.) is collected. Redistillation affords 1.73 kg. (86%) of a mixture of aldol acetal (45%) and keto acetal (55%). To separate the pure keto acetal, the mixture is heated with 173 g. of boric anhydride at 15 mm. for one hour, the bath temperature slowly being raised to 175–200°, at which point water starts to distil. The vacuum is then reduced to 3 mm., whereupon substantially pure keto acetal distils below 100°; yield, 740 g. (37%). Pure β-ketobutyraldehyde 2-methylpentane-2,4-diol acetal distils at 81°/3 mm., $n_D^{20°}$ 1.4421; its semicarbazone melts at 191–192°.

ψ-Ionone (L, R = H).[119] (Use of aluminum phenoxide in the absence of a solvent for the oxidation of a primary alcohol.) The aluminum phenoxide for this oxidation is prepared by adding 10 g. of aluminum shavings to 99.5 g. of hot phenol, heating until hydrogen evolution ceases, then cooling and crushing. A mixture of 13.5 g. of the phenoxide (aluminum *t*-butoxide can also be used), 6.05 g. of geraniol (XLIX, R = H) (b.p. 107–110°/10 mm.), and 200 ml. of acetone (distilled from calcium chloride) is refluxed with exclusion of moisture for twenty-six hours. After concentrating, hydrolysis is accomplished by refluxing with water for two hours, and the solution is then subjected to steam distillation. The ψ-ionone is isolated from the distillate by extraction with ether.† After the solvent has been dried and evaporated, the resi-

---

* The acetal is prepared from acetaldol and 2-methylpentane-2,4-diol in the presence of dry hydrogen chloride; b.p. 83–86°/3 mm.

† As an alternative hydrolysis procedure the reaction mixture is cooled in ice, ether is added, and the organic layer is washed with dilute sulfuric acid, then with sodium carbonate and water.

due is distilled through a Widmer column to give 1.7 g. of fore-run boiling at 55–90°/0.2 mm., and 4.3 g. (57%) of ψ-ionone (L, R = H), b.p. 90–102°/0.15 mm. A reduction of the reaction time or of the amounts of acetone and phenoxide results in lower yields; the use of benzene as solvent has no beneficial effect. In another laboratory,[13] a 70% yield of ψ-ionone was obtained by refluxing 14 g. of geraniol with 20 g. of aluminum t-butoxide, 200 ml. of acetone, and 500 ml. of benzene for thirty minutes.

SURVEY OF OPPENAUER OXIDATIONS REPORTED IN THE LITERATURE

In Tables I–IX are summarized all examples of the Oppenauer oxidation which have been noted in a survey of the literature up to and including the January, 1950, issue of *Chemical Abstracts*. Only those patents are cited that contain significant material adequately supported by experimental work and not described elsewhere in the literature.

In general, alcohols are listed in the tables in the order of increasing molecular weight. Yields in parentheses refer to crude material; unless specified otherwise, the time denotes the period of refluxing. In several instances, α and β prefixes for steroids were altered from the original to conform with nomenclature revisions in this field.[95]

TABLE I

OPPENAUER OXIDATION OF SATURATED SECONDARY ALCOHOLS *

Alcohol	Reaction Conditions	Product	Yield %	Reference
(a) *Non-Steroids*				
2-Propanol	Al(OC$_4$H$_9$-*t*)$_3$, fluorenone, toluene, 6 hr.	Acetone		11
2-Ethylcyclohexanol	Al(OC$_4$H$_9$-*t*)$_3$, quinone, toluene, 8 days room temperature	2-Ethylcyclohexanone	76	165
l-Menthol	Al(OC$_3$H$_7$-*i*)$_3$, cinnamaldehyde, 4 hr.	Menthone	75	3
cis-α-Decalol	Al(OC$_3$H$_7$-*i*)$_3$, acetone, benzene, 12 hr.	*cis*-α-Decalone	80	21
α-Decalol (mixture of *cis* and *trans*)	Al(OC$_3$H$_7$-*i*)$_3$, acetone, benzene, 12 hr.	α-Decalone (mixture of *cis* and *trans*)	*ca.* 90	21, 187
3-Hydroxydodecahydro-1,2-cyclopentenonaphthalene	Al(OC$_3$H$_7$-*i*)$_3$, acetone, benzene, 20 hr.	3-Ketododecahydro-1,2-cyclopentenonaphthalene	87	25
cis-syn-cis-Perhydro-9-phenanthrol	Al(OC$_4$H$_9$-*t*)$_3$, acetone, benzene, 8 hr.	Mixture of *c-s-c-* and *t-s-c-*9-ketoperhydrophenanthrene		146
1-Hydroxy-7-methoxyocta-hydrophenanthrene	Al(OC$_4$H$_9$-*t*)$_3$, acetone, benzene, 36 hr.	1-Keto-7-methoxyoctahydrophenanthrene; recovered starting material	43 38	24
1-Hydroxy-2-methyl-7-methoxy-1,2,3,4,9,10,11,12-octahydrophenanthrene	Al(OC$_4$H$_9$-*t*)$_3$, acetone, benzene, 48 hr.	1-Keto-2-methyl-7-methoxy-octahydrophenanthrene		27

1,7-Dihydroxy-13-methyl-octahydrophenanthrene	Al(OC$_4$H$_9$-i)$_3$, acetone, benzene, 24 hr.	1-Hydroxy-7-keto-13-methylocta-hydrophenanthrene		22
1-Hydroxy-7-acetoxy-13-methyl-perhydrophenanthrene	Al(OC$_4$H$_9$-i)$_3$, acetone, benzene, 45 hr.	1-Keto-7-acetoxy-13-methylper-hydrophenanthrene	25	22
3-Hydroxy-7-methoxy-1,2,3,4,9,10,11,12,5′,6′-decahy-drothiopyrano-(4′,3′:1,2)-phenanthrene	Al(OC$_3$H$_7$-i)$_3$, cyclohexanone, toluene, 10 hr.	3-Keto-7-methoxydecahydrothio-pyrano-(4′,3′:1,2)phenanthrene	50	26
1-(2′,6′,6′-Trimethylcyclo-hexan-1′-yl)-3-methylhexan-4-ol	Al(OC$_4$H$_9$-i)$_3$, acetone, benzene, 14 hr.	1-(2′,6′,6′-Trimethylcyclohexan-1′-yl)-3-methylhexan-4-one	70	66
3-(4-Hydroxyphenyl)-4-(4-hydroxycyclohexyl)-hexane (octahydrodiethyl-stilbestrol)	Al(OC$_4$H$_9$-i)$_3$, acetone, benzene, 8 hr.	3-(4-Hydroxyphenyl)-4-(4-keto-cyclohexyl)hexane	(70) 20	23
Hexahydro-*meso*-hexestrol	Al(OC$_4$H$_9$-i)$_3$, acetone, benzene, 8 hr.	3-(4-Hydroxyphenyl)-4-(4-keto-cyclohexyl)hexane	18	185
(b) Steroids				
Estradiol	Al(OC$_4$H$_9$-i)$_3$, acetone, benzene, 6–12 hr.	Estrone		20, 188
Δ^{1,4}-Androstadien-17-ol-3-one	Al(OC$_3$H$_7$-i)$_3$, cyclohexanone, toluene, 1 hr.	Δ^{1,4}-Androstadiene-3,17-dione	55	17
17-Methylandrostane-3β,17β-diol	Al(OC$_3$H$_7$-i)$_3$, cyclohexanone, toluene, 0.5 hr.	17-Methylandrostan-17β-ol-3-one	80	18
Methyl Δ^{11}-3α-hydroxy-etiocholenate	Al(OC$_3$H$_7$-i)$_3$, cyclohexanone, toluene	Methyl Δ^{11}-3-ketoetiocholenate		189

* "Saturated" refers to compounds not possessing a double bond or aromatic nucleus α,β or β,γ to the hydroxyl group. Polyhydroxyl compounds of this type are given in Tables III and IV.

TABLE I—Continued
OPPENAUER OXIDATION OF SATURATED SECONDARY ALCOHOLS

Alcohol	Reaction Conditions	Product	Yield %	Reference
(b) Steroids (Continued)				
Androstane-3β,17β-diol 17-hexahydrobenzoate	Al(OC$_4$H$_9$-t)$_3$, quinone, toluene, 1 hr.	Dihydrotestosterone hexahydrobenzoate		139
Δ^{17}-3β-Hydroxy*allo*pregnen-21-oic acid	Al(OC$_4$H$_9$-t)$_3$, acetone, toluene, 6 hr.	Δ^{17}-3-Keto*allo*pregnen-21-oic acid		181
Δ^{17}-3β-Hydroxypregnen-21-oic acid methyl ester	Al(OC$_3$H$_7$-i)$_3$, acetone, toluene, 6 hr.	Δ^{17}-3-Ketopregnen-21-oic acid methyl ester		182
Zymosterol (Δ^{8,24}-cholestadien-3β-ol)	Al(OC$_3$H$_7$-i)$_3$, cyclohexanone, toluene, 2 hr.	Zymostadien-3-one	60	190
	Al(OC$_3$H$_7$-i)$_3$, acetone, benzene, 10 hr.	Zymostadien-3-one	20	190
24,25-Dihydrozymosterol (Δ^8-cholesten-3β-ol)	Al(OC$_3$H$_7$-i)$_3$, cyclohexanone, toluene, 2 hr.	Zymosten-3-one	54	190
Δ^7-Cholesten-3β-ol	Al(OC$_6$H$_5$)$_3$, acetone, benzene, 20 hr.	Δ^7-Cholesten-3-one	60	14
Cholestan-4-ol	Al(OC$_6$H$_5$)$_3$, acetone, benzene, 28 hr.	Cholestan-4-one	20	191

Δ^{8,14}-Ergostadien-3β-ol	Al(OC$_4$H$_9$-t)$_3$, acetone, benzene, 4 hr.		Δ^{8,14}-Ergostadien-3-one	15, 16
Δ^{7,22}-Ergostadien-3β-ol	Al(OC$_4$H$_9$-t)$_3$, acetone, benzene, 4 hr.	30	Δ^{7,22}-Ergostadien-3-one	16, 192
Δ^{7}-Ergosten-3β-ol	Al(OC$_4$H$_9$-t)$_3$, acetone, benzene, 4 hr.		Δ^{7}-Ergosten-3-one	15, 16
Δ$^{8(14)}$-Ergosten-3β-ol	Al(OC$_4$H$_9$-t)$_3$, acetone, benzene, 4 hr.	40	Δ$^{8(14)}$-Ergosten-3-one	15, 16
Δ^{14}-Ergosten-3β-ol	Al(OC$_4$H$_9$-t)$_3$, acetone, benzene, 4 hr.	40	Δ^{14}-Ergosten-3-one	15, 16
Methylcholesterylcarbinol	Al(OC$_4$H$_9$-t)$_3$, acetone, benzene, 14 hr.	16	Methyl cholesteryl ketone semicarbazone	193
Δ^{20,23}-24,24-Diphenylcholadien-3α-ol-11-one	Al(OC$_3$H$_7$-i)$_3$, cyclohexanone, toluene, 0.5 hr.	86	Δ^{20,23}-24,24-Diphenylcholadiene-3,11-dione	19
Δ^{20,23}-3α-Hydroxy-12α-acetoxy-24,24-diphenylcholadiene	Al(OC$_3$H$_7$-i)$_3$, cyclohexanone, toluene, 0.5 hr.	58	Δ^{20,23}-3-Keto-12α-acetoxy-24,24-diphenylcholadiene	184

The position of the nuclear double bond was established by Barton and Cox, J. Chem. Soc., **1949**, 214.

[188] Velluz and Petit, *Bull. soc. chim. France*, **1948**, 1113.
[189] Reichstein, U. S. pat. 2,387,706 [*C. A.*, **40**, 994 (1946)].
[190] Wieland, Rath, and Benend, *Ann.*, **548**, 19 (1941).
[191] Butenandt and Ruhenstroth-Bauer, *Ber.*, **77**, 402 (1944).
[192] Barton and Cox, *J. Chem. Soc.*, **1948**, 1356.
[193] Baker and Squire, *J. Am. Chem. Soc.*, **70**, 1488 (1948).

TABLE II
Oppenauer Oxidation of Unsaturated Secondary Alcohols *

Alcohol	Reaction Conditions	Product	Yield %	Reference
(a) *Non-Steroids*				
3-(α-Hydroxyethyl)-3,4-dihydrothiopyran	Al(OC$_4$H$_9$-*t*)$_3$, acetone, benzene, 10 hr. 65°	3-Acetyl-3,4-dihydrothiopyran	75	26
6-Methylocta-3,5,7-trien-2-ol	Al(OC$_4$H$_9$-*t*)$_3$, acetone, benzene, 24 hr.;	6-Methylocta-3,5,7-trien-2-one;	80	71
	Al(OC$_6$H$_5$)$_3$, acetone, benzene, 24 hr.	6-methylocta-3,5,7-trien-2-one	60	71
Δ^{9,10}-1-Octalol	Al(OC$_4$H$_9$-*t*)$_3$, acetone, benzene, 8 hr.	Δ^{8,9}-1-Octalone	74	63
1-(Cyclohexen-1′-yl)-1-buten-3-ol	Al(OC$_4$H$_9$-*t*)$_3$, acetone, benzene, 48 hr.	1-(Cyclohexen-1′-yl)-1-buten-3-one	43	67
6-(Cyclohexen-1′-yl)hex-3-en-5-yn-2-ol	Al(OC$_4$H$_9$-*t*)$_3$, acetone, benzene, 60 hr.	6-(Cyclohexen-1′-yl)hex-3-en-5-yn-2-one	21	69
Benzohydrol	Al(OC$_4$H$_9$-*t*)$_3$, toluene, benzil, cyclohexanone, quinone, benzaldehyde, etc.	Benzophenone	80–90	165
	Al(OC$_4$H$_9$-*t*)$_3$, anthraquinone, toluene, 71 hr., 35°	Benzophenone	56	167
Fluorenol	Al(OC$_4$H$_9$-*t*)$_3$, anthraquinone, toluene, 114 hr., room temperature	Fluorenone	85	167
α-Ionol	Al(OC$_3$H$_7$-*i*)$_3$, acetone, benzene, 18 hr.	α-Ionone	(80)	65
β-Ionol	Al(OC$_3$H$_7$-*i*)$_3$, acetone, benzene, 18 hr.	β-Ionone	65	65

THE OPPENAUER OXIDATION

Irol (mixture of α and γ isomers)		Mixture of α- and γ-irone	(95)	65
4-Methyl-5-(1-methyl-2-hydroxypropyl)resorcinol dimethyl ether	Al(OC₃H₇-i)₃, acetone, benzene, 18 hr.	4-Methyl-5-(1-methyl-2-ketopropyl)resorcinol dimethyl ether		194
8-(Cyclohexen-1′-yl)octa-3,5-dien-7-yn-2-ol	Al(OC₃H₇-i)₃, acetone, benzene, 21 hr.	8-(Cyclohexen-1′-yl)octa-3,5-dien-7-yn-2-one	39	69
8-(Cyclohexen-1′-yl)-6-methylocta-3,5-dien-7-yn-2-ol	Al(OC₄H₉-t)₃, acetone, benzene, 48 hr.	8-(Cyclohexen-1′-yl)-6-methylocta-3,5-dien-7-yn-2-one	45	68
10-(Cyclohexen-1′-yl)deca-3,5,7-trien-9-yn-2-ol	Al(OC₄H₉-t)₃, acetone, benzene, 50 hr.	10-(Cyclohexen-1′-yl)deca-3,5,7-trien-9-yn-2-one	75	69
8-(2′-Methylcyclohexen-1′-yl)-6-methylocta-3,5-dien-7-yn-2-ol	Al(OC₄H₉-t)₃, acetone, benzene, 48 hr.	8-(2′-Methylcyclohexen-1′-yl)-6-methylocta-3,5-dien-7-yn-2-one	35	70
1-(2′,6′,6′-Trimethylcyclohexen-1′-yl)-3-methyl-1-hexen-4-ol	Al(OC₄H₉-t)₃, acetone, benzene, 14 hr.	1-(2′,6′,6′-Trimethylcyclohexen-1′-yl)-3-methyl-1-hexen-4-one	79	66
8-(6′,6′-Dimethylcyclohexen-1′-yl)-6-methylocta-3,5-dien-7-yn-2-ol	Al(OC₄H₉-t)₃, acetone, benzene, 48 hr.	8-(6′,6′-Dimethylcyclohexen-1′-yl)-6-methylocta-3,5-dien-7-yn-2-one	34	70
1-Keto-2,4b-dimethyl-7-hydroxy-Δ⁸ᵃ,⁹-dodecahydrophenanthrene	Al(OC₃H₇-i)₃, cyclohexanone, toluene, 1.5 hr.	1,7-Diketo-2,4b-dimethyl-Δ⁸,⁸ᵃ-dodecahydrophenanthrene		62
1,7-Dihydroxy-1,2,4b-trimethyl-Δ⁸ᵃ,⁹-dodecahydrophenanthrene	Al(OC₃H₇-i)₃, cyclohexanone, toluene, 1 hr.	1-Hydroxy-1,2,4b-trimethyl-7-keto-Δ⁸,⁸ᵃ-dodecahydrophenanthrene		62
1,7-Dihydroxy-1-ethynyl-2,4b-dimethyl-Δ⁸ᵃ,⁹-dodecahydrophenanthrene	Al(OC₃H₇-i)₃, cyclohexanone, toluene, 1 hr.	1-Hydroxy-1-ethynyl-2,4b-dimethyl-7-keto-Δ⁸,⁸ᵃ-dodecahydrophenanthrene		62

* Only compounds with a benzene nucleus or a double bond α,β or β,γ to the hydroxyl group are considered in this table.

TABLE II—Continued

OPPENAUER OXIDATION OF UNSATURATED SECONDARY ALCOHOLS

Alcohol	Reaction Conditions	Product	Yield %	Reference
(a) *Non-Steroids* (*Continued*)				
Methyl 1-ethyl-2,4b-dimethyl-7-hydroxy-$\Delta^{8a,9}$-dodecahydrophenanthrene-2-carboxylate	Al(OC$_3$H$_7$-i)$_3$, cyclohexanone, toluene, 0.5 hr.	Methyl 1-ethyl-2,4b-dimethyl-7-keto-$\Delta^{8,8a}$-dodecahydrophenanthrene-2-carboxylate	21	195
SeO$_2$ Oxidation product of isonorargathenol acetate	Al(OC$_6$H$_5$)$_3$, acetone, benzene, 12 hr.	Unsaturated ketone		196
9-Hydroxy-10-benzylidene-9,10-dihydroanthracene	Al(OC$_3$H$_7$-i)$_3$, cyclohexanone, toluene, 8 hr.	Benzalanthrone	80	197
Methyl elemadienolate	Al(OC$_4$H$_9$-t)$_3$, quinone, benzene, 24 hr.	Mixture of methyl elemadienonate and isoelemadienonate	71	64
Lupeol	Al(OC$_4$H$_9$-t)$_3$, quinone, toluene, 1 hr.	Lupeone		170
Quassin	Al(OC$_3$H$_7$-i)$_3$, cyclohexanone, toluene, 2 hr.	Isoquassin	51	198
Butyrospermol	Al(OC$_4$H$_9$-t)$_3$, quinone, benzene, 12 hr.	Butyrospermone	45	171
(b) *Steroids*				
Dehydroepiandrosterone	Al(OC$_4$H$_9$-t)$_3$, acetone, benzene, 14 hr.	Δ^4-Androstene-3,17-dione	85	4
Δ^5-Androstene-3β,17β-diol 17-acetate	Al(OC$_4$H$_9$-t)$_3$, acetone, benzene, 5–11 hr.	Testosterone acetate	74–90	4, 30
Δ^5-Androstene-3β,17β-diol 17-benzoate	Al(OC$_3$H$_7$-i)$_3$, cyclohexanone, toluene, 2.5 hr.	Testosterone benzoate	92	37
	Al(OC$_4$H$_9$-t)$_3$, quinone, benzene, 45 min.	Δ^6-Dehydrotestosterone benzoate		139, 199

Δ⁵-16-Methylandrostene-3β,17β-diol 17-acetate	Al(OC₃H₇-i)₃, cyclohexanone, toluene, 2 hr.	16-Methyltestosterone	87	200
Δ⁵-17-Methylandrostene-3β,17β-diol	Al(OC₃H₇-i)₃, acetone, benzene, 25 hr.	17-Methyltestosterone	40	158, 201
Δ⁵-17-Methylandrostene-3β,17β-diol	Al(OC₄H₉-t)₃, acetone, benzene, 14 hr.	17-Methyltestosterone	91	4
Δ⁵-17-Methylandrostene-3β,17α-diol	Al(OC₃H₇-i)₃, cyclohexanone, toluene, 2 hr.	17-Isomethyltestosterone	70–80	202
Δ⁵-17-Ethynylandrostene-3β,17β-diol	Al(OC₃H₇-i)₃, cyclohexanone, toluene, 0.5 hr.	17-Ethynyltestosterone	80	37
	Al(OC₃H₇-i)₃, or Al(OC₄H₉-t)₃, acetone, benzene, 15–20 hr.	17-Ethynyltestosterone	60	40, 41
Δ⁵-17-Vinylandrostene-3β,17β-diol	Al(OC₄H₉-t)₃, acetone, benzene, 20 hr.	17-Vinyltestosterone	(65)	39, 40
Δ⁵-17-Vinylandrostene-3β,17β-diol 17-acetate	Al(OC₄H₉-t)₃, acetone, benzene, 24 hr.	17-Vinyltestosterone acetate	28	203
Δ⁵-17-Ethylandrostene-3β,17β-diol	Al(OC₄H₉-t)₃, acetone, benzene, 20 hr.	17-Ethyltestosterone	70	39
Δ⁵-17-Allylandrostene-3β,17β-diol	Al(OC₃H₇-i)₃, cyclohexanone, toluene, 40 min.	17-Allyltestosterone	80	38
	Al(OC₄H₉-t)₃, acetone, benzene, 24 hr.	17-Allyltestosterone	70	204
Δ⁵,¹⁷-17α-Methyl-D-homoandrostadien-3β-ol	Al(OC₄H₉-t)₃, acetone, benzene, 15 hr.	Δ⁴,¹⁷-17α-Methyl-D-homoandrostadien-3-one	74	205
Δ⁵-17α-Methyl-D-homoandrostene-3β,17α-diol-17-one	Al(OC₄H₉-t)₃, acetone, benzene, 8 hr.	Δ⁴-17α-Methyl-D-homoandrostene-3,17-dione-17α-ol; recovered starting material	14	147, 206
Δ⁵-17α-Methyl-D-homoandrostene-3β,17α-diol-17-one 17α-acetate	Al(OC₄H₉-t)₃, acetone, benzene, 20 hr.	Δ⁴-17α-Methyl-17α-acetoxy-D-homoandrostene-3,17-dione	36	207

TABLE II—*Continued*

OPPENAUER OXIDATION OF UNSATURATED SECONDARY ALCOHOLS

Alcohol	Reaction Conditions	Product	Yield %	Reference
(b) *Steroids (Continued)*				
(Benzo-1',2':16,17-Δ^5-androsten)-3β,4'-diol	Al(OC$_3$H$_7$-i)$_3$, cyclohexanone, toluene, 2 hr.	(Benzo-1',2':16,17-Δ^4-androsten)-4'-ol-3-one	(100)	59
$\Delta^{5,17}$-Pregnadien-3β-ol	Al(OC$_4$H$_9$-t)$_3$, acetone, benzene, 14 hr.	$\Delta^{4,17}$-Pregnadien-3-one	87	44
$\Delta^{5,20}$-Pregnadien-3β-ol	Al(OC$_3$H$_7$-i)$_3$, cyclohexanone, toluene, 1 hr.	$\Delta^{4,20}$-Pregnadien-3-one	82	52
Δ^5-10-Norpregnen-3-ol-20-one	Al(OC$_4$H$_9$-t)$_3$, acetone, benzene, 10 hr.	10-Norprogesterone	34	11
$\Delta^{5,16}$-Pregnadien-3β-ol-20-one	Al(OC$_3$H$_7$-i)$_3$, cyclohexanone, toluene, 0.75 hr.	Δ^{16}-Dehydroprogesterone	47	208
Δ^5-Pregnen-3β-ol-20-one	Al(OC$_4$H$_9$-t)$_3$, acetone, benzene, 11 hr.	Progesterone	60–75	4, 37
	Al(OC$_3$H$_7$-i)$_3$, cyclohexanone, toluene, 0.5 hr.	Progesterone	83	37, 209
	Al(OC$_3$H$_7$-i)$_3$, quinone, toluene, 3 hr.	Δ^6-Dehydroprogesterone	40	139, 145
Δ^5-17-Isopregnen-3β-ol-20-one	Al(OC$_3$H$_7$-i)$_3$, cyclohexanone, toluene, 1 hr.	17-Isoprogesterone	40	210
Δ^5-14-*Allo*-17-isopregnen-3β-ol-20-one	Al(OC$_4$H$_9$-t)$_3$, acetone, benzene, 22 hr.	Δ^4-14-*Allo*-17-isopregnene-3,20-dione	42 (75)	211
Δ^5-16,17-Oxidopregnen-3β-ol-20-one	Al(OC$_3$H$_7$-i)$_3$, cyclohexanone, toluene, 0.7 hr.	16,17-Oxidoprogesterone	72	61

THE OPPENAUER OXIDATION

Substrate	Conditions	Product	Yield	Ref.
$\Delta^{5,17}$-Homo-(ω)-pregnadien-3β-ol-21-one	Al(OC$_4$H$_9$-t)$_3$, acetone, benzene, 12 hr.	$\Delta^{4,17}$-Homo-(ω)-pregnadiene-3,21-dione	76	212
Δ^5-Homo-(ω)-pregnen-3β-ol-21-one	Al(OC$_4$H$_9$-t)$_3$, acetone, benzene, 12 hr.	Δ^4-Homo-(ω)-pregnene-3,21-dione	82	212
$\Delta^{5,16}$-16-Methylpregnadien-3β-ol-20-one	Al(OC$_3$H$_7$-i)$_3$, cyclohexanone, toluene, 2 hr.	$\Delta^{4,16}$-16-Methylpregnadiene-3,20-dione		213
Δ^5-17-Methylpregnen-3β-ol-20-one	Al(OC$_4$H$_9$-t)$_3$, cyclohexanone, toluene, 15 hr.	17-Methylprogesterone	80	214
Δ^5-16-Methylpregnen-3β-ol-20-one (isomers)	Al(OC$_4$H$_9$-t)$_3$, acetone, toluene, 6 hr.	16-Methylprogesterone	61	180, 213
Δ^5-21-Methylpregnen-3β-ol-20-one	Al(OC$_3$H$_7$-i)$_3$, cyclohexanone, toluene, 1.5 hr.	21-Methylprogesterone		215
Δ^5-Pregnen-3β,17β-diol-20-one	Al(OC$_3$H$_7$-i)$_3$, cyclohexanone, toluene, 1 hr.	Δ^4-17α-Methyl-D-homoandrostene-3,17-dione-17α-ol †	34	147, 150
Δ^5-Pregnen-3β,17α-diol-20-one	Al(OC$_4$H$_9$-t)$_3$, acetone, benzene, 20 hr.	Δ^4-17α-Methyl-D-homoandrostene-3,17-dione-17α-ol †	10	148, 149
Δ^5-21-Methoxypregnen-5β-ol-20-one	Al(OC$_4$H$_9$-t)$_3$, cyclohexanone, benzene, 18 hr.	Desoxycorticosterone 21-methyl ether	(66)	216
$\Delta^{5,17}$-20-Cyanopregnadien-3β-ol	Not specified	$\Delta^{4,17}$-20-Cyanopregnadien-3-one		217
Δ^5-21-Ethylpregnen-3β-ol-20-one	Al(OC$_3$H$_7$-i)$_3$, cyclohexanone, toluene, 1.5 hr.	21-Ethylprogesterone		215
$\Delta^{5,17}$-3β-Hydroxypregnadien-21-oic acid methyl ester	Al(OC$_4$H$_9$-t)$_3$, acetone, benzene, 12 hr.	$\Delta^{4,17}$-3-Ketopregnadien-21-oic acid methyl ester		218
Δ^5-16-Isopropylpregnen-3β-ol-20-one	Al(OC$_4$H$_9$-t)$_3$, acetone, toluene, 6 hr.	16-Isopropylprogesterone	57	180
Δ^5-Pregnene-3β,21-diol-20-one 21-acetate	Al(OC$_3$H$_7$-i)$_3$, cyclohexanone, toluene, 0.5 hr.	Desoxycorticosterone acetate	83	37
	Al(OC$_4$H$_9$-t)$_3$, quinone, toluene, 1 hr.	Δ^6-Dehydrodesoxycorticosterone acetate	Poor	139

† The two products are the 17a epimers.

TABLE II—*Continued*

Oppenauer Oxidation of Unsaturated Secondary Alcohols

Alcohol	Reaction Conditions	Product	Yield %	Reference
(b) *Steroids* (*Continued*)				
Δ^5-16,17-Oxidopregnene-3β,21-diol-20-one 21-acetate	Al(OC$_4$H$_9$-t)$_3$, cyclohexanone, toluene, 8 hr.	Δ^4-16,17-Oxidopregnen-21-ol-3,20-dione 21-acetate	70	60
Δ^5-16-t-Butylpregnen-3β-ol-20-one	Al(OC$_4$H$_9$-t)$_3$, acetone, toluene, 6 hr.	16-t-Butylprogesterone	62	180
Δ^5-Pregnen-3β,21-diol-11,20-dione 21-acetate (crude)	Al(OC$_4$H$_9$-t)$_3$, acetone, benzene, 25 hr.	11-Dehydrocorticosterone acetate	15	219
$\Delta^{5,17}$-21-Benzalpregnadien-3β-ol	Al(OC$_3$H$_7$-i)$_3$, cyclohexanone, toluene, 0.5 hr.	$\Delta^{4,17}$-21-Benzalpregnadien-3-one	82	43
Δ^5-21-Benzalpregnen-3β-ol-20-one	Al(OC$_4$H$_9$-t)$_3$, acetone, toluene, 5 hr.	21-Benzalprogesterone		42
Δ^5-21-Benzylpregnen-3β-ol-20-one	Al(OC$_4$H$_9$-t)$_3$, acetone, toluene, 5 hr.	21-Benzylprogesterone		42
Methyl Δ^5-3β-hydroxyetiocholenate	Al(OC$_3$H$_7$-i)$_3$, cyclohexanone, toluene, 2.5 hr.	Methyl Δ^4-3-ketoetiocholenate	41	93
	Al(OC$_3$H$_7$-i)$_3$, quinone, toluene, 0.75 hr.	Methyl $\Delta^{4,6}$-3-ketoetiocholadienate	22	175
Methyl Δ^5-3β,17β-dihydroxyetiocholenate	Al(OC$_4$H$_9$-t)$_3$, acetone, benzene, 24 hr.	Methyl Δ^4-3-keto-17β-hydroxyetiocholenate	22	143
Methyl Δ^5-3β,17α-dihydroxyetiocholenate	Al(OC$_4$H$_9$-t)$_3$, acetone, benzene, 24 hr.	Methyl Δ^4-3-keto-17α-hydroxyetiocholenate	30	143
Methyl Δ^5-3β-hydroxybisnorcholenate	Al(OC$_3$H$_7$-i)$_3$, cyclohexanone, toluene, 0.5 hr.	Methyl Δ^4-3-ketobisnorcholenate	84	220

Substrate	Conditions	Product	Yield	Ref.
2-(Δ^5-3β-Hydroxyternorcholenyl)propene	Al(OC$_4$H$_9$-t)$_3$, cyclohexanone, toluene, 2 hr.	2-(Δ^4-3-Ketoternorcholenyl)propene (70% with 20-iso derivative)	66	51
Δ^5-3β-Hydroxyternorcholenyl methyl ketone	Al(OC$_4$H$_9$-t)$_3$, cyclohexanone, toluene, 0.3 hr.	Δ^4-3-Ketoternorcholenyl methyl ketone	(85)	221, 222
Δ^5-3β-Hydroxyternorcholenyl ethyl ketone	Al(OC$_4$H$_9$-t)$_3$, cyclohexanone, toluene, 1 hr.	Δ^4-3-Ketoternorcholenyl ethyl ketone	90	221
Δ^5-3β-Hydroxyternorcholenyl isoamyl ketone	Al(OC$_4$H$_9$-t)$_3$, cyclohexanone, toluene, 2 hr.	Δ^4-3-Ketoternorcholenyl isoamyl ketone	67	221
Δ^5-3β-Hydroxyternorcholenyl phenyl ketone	Al(OC$_4$H$_9$-t)$_3$, cyclohexanone, toluene, 0.7 hr.	Δ^4-3-Ketoternorcholenyl phenyl ketone	70	221
Δ^5-3β-Hydroxy-23-acetoxynorcholen-22-one	Al(OC$_3$H$_7$-i)$_3$, cyclohexanone, toluene, 0.75 hr.	Δ^4-23-Acetoxynorcholen-3,22-dione		222
1-(Δ^5-3β-Hydroxyetiochclenyl)-1-methyl-2,2-diphenylethylene	Al(OC$_4$H$_9$-t)$_3$, cyclohexanone, toluene, 0.3 hr.	1-(Δ^4-3-Ketoetiocholenyl)-1-methyl-2,2-diphenylethylene		142
Δ^5-3β-Hydroxyetiocholerzylethyl diphenyl carbinel	Al(OC$_4$H$_9$-t)$_3$, cyclohexanone, toluene, 0.3 hr., complete dehydration with acetic acid	1-(Δ^4-3-Ketoetiocholenyl)-1-methyl-2,2-diphenylethylene	Good	142
Δ^5-*Bis*norcholesten-3β-ol-24-one	Al(OC$_6$H$_5$)$_3$, acetone, benzene	Δ^4-*Bis*norcholestene-3,24-dione	Good	223
$\Delta^{5,20}$-22-Phenyl*bis*norcholadien-3β-ol	Al(OC$_3$H$_7$-i)$_3$, cyclohexanone, toluene, 1 hr.	$\Delta^{4,20}$-22-Phenyl*bis*norcholadien-3-one		48, 49
Δ^5-22-Phenyl*bis*norcholen-3β,22-diol 22-benzoate	Al(OC$_3$H$_7$-i)$_3$, cyclohexanone, toluene, 1 hr.	Δ^4-22-Phenyl*bis*norcholen-22-ol-3-one 22-benzoate	76	48
$\Delta^{5,23}$-24-Phenylcholadien-3β-ol	Al(OC$_3$H$_7$-i)$_3$, cyclohexanone, toluene, 2.5 hr.	$\Delta^{4,23}$-24-Phenylcholadien-3-one		50
Δ^5-24-Phenylcholen-3β-ol-24-one	Al(OC$_3$H$_7$-i)$_3$, cyclohexanone, toluene, 4 hr.	Δ^4-24-Phenylcholene-3,24-dione		73
$\Delta^{5,20,23}$-24,24-Diphenylcholatrien-3β-ol	Al(OC$_3$H$_7$-i)$_3$, cyclohexanone, toluene, 0.5 hr.	$\Delta^{4,20,23}$-24,24-Diphenylcholatrien-3-one	95	45
$\Delta^{5,23}$-24,24-Diphenylcholadien-3β-ol	Al(OC$_3$H$_7$-i)$_3$, cyclohexanone, toluene, 0.5 hr.	$\Delta^{4,23}$-24,24-Diphenylcholadien-3-one	46	46

TABLE II—Continued

OPPENAUER OXIDATION OF UNSATURATED SECONDARY ALCOHOLS

Alcohol	Reaction Conditions	Product	Yield %	Reference
(b) *Steroids (Continued)*				
$\Delta^{5,20,23}$-21-Methoxy-24,24-diphenylcholatrien-3-ol	Al(OC$_3$H$_7$-i)$_3$, cyclohexanone, toluene, 0.5 hr.	$\Delta^{4,20,23}$-21-Methoxy-24,24-diphenylcholatrien-3-one	(100)	47
B-Norcholesterol	Al(OC$_4$H$_9$-t)$_3$, acetone, benzene, 6 hr.	Δ^4-B-Norcholesten-3-one		31
$\Delta^{5,7}$-Cholestadien-3β-ol	Al(OC$_4$H$_9$-t)$_3$, acetone, benzene, 10 hr.	$\Delta^{4,7}$-Cholestadien-3-one		36
Cholesterol	Al(OC$_4$H$_9$-t)$_3$, acetone or methyl ethyl ketone, benzene, 8–18 hr.	Δ^4-Cholesten-3-one	70–89	4, 12, 165, 186, 224
	Al(OC$_3$H$_7$-i)$_3$, cyclohexanone, xylene, 4 hr.	Δ^4-Cholesten-3-one	90	159, 190
	Al(OC$_3$H$_7$-i)$_3$ or Al(OC$_4$H$_9$-t)$_3$, quinone, toluene, 0.75–3 hr.	$\Delta^{4,6}$-Cholestadien-3-one	36–44	145, 173, 174
Epicholesterol	Al(OC$_4$H$_9$-t)$_3$, acetone, benzene, 24 hr.	Δ^4-Cholesten-3-one		225
Dihydrovitamin D$_3$	Al(OC$_4$H$_9$-t)$_3$, acetone, benzene, 9 hr.	Corresponding $\Delta^{4,7}$-unsaturated ketone		34

Starting material	Conditions	Product	Yield (%)	Reference
$\Delta^{9(11)}$-Dehydroergosterol	Not specified	$\Delta^{4,7,9(11),22}$-Ergostatetraen-3-one		225a
Neoergosterol	Al(OC$_3$H$_7$-i)$_3$, cyclohexanone, toluene, 2.5 hr.	Neoergostenone		226
Ergosterol	Al(OC$_4$H$_9$-t)$_3$, acetone, benzene, 3.5 hr.	$\Delta^{4,7,22}$-Ergostatrien-3-one	57	4, 225a
Lumisterol	Not specified	$\Delta^{4,7,22}$-Lumistatrien-3-one		225a
Brassicasterol ($\Delta^{5,22}$-ergostadien-3β-ol)	Al(OC$_4$H$_9$-t)$_3$, acetone, benzene, 4 hr.	$\Delta^{4,22}$-Ergostadien-3-one		227
Fucosterol	Al(OC$_4$H$_9$-t)$_3$, acetone, benzene, 18 hr.	$\Delta^{4,23}$-Fucostadien-3-one	50	12
Stigmasterol	Al(OC$_4$H$_9$-t)$_3$, acetone, benzene, 18 hr.	$\Delta^{4,22}$-Stigmastadien-3-one	58	12
	Al(OC$_3$H$_7$-i)$_3$, cyclohexanone, toluene, 2–10 hr.	Stigmastadien-3-one	50	53, 159
Sitosterol (tall-öl)	Al(OC$_4$H$_9$-t)$_3$, acetone, benzene, 18 hr.	Δ^4-Sitosten-3-one; sitostan-3-one	66 3	33
Clionasterol	Al(OC$_3$H$_7$-i)$_3$, cyclohexanone, toluene, 2 hr.	Δ^4-Clionasten-3-one		228
Poriferasterol	Al(OC$_3$H$_7$-i)$_3$, cyclohexanone, toluene, 4 hr.	Δ^4-Poriferasten-3-one		229
7-Methoxycholesterol	Al(OC$_6$H$_5$)$_3$, acetone, benzene, 40 hr.	$\Delta^{4,6}$-Cholestadien-3-one; $\Delta^{4,6}$-7-methoxycholesten-3-one	35 55	136, 138
7-Ethoxycholesterol	Al(OC$_6$H$_5$)$_3$, acetone, benzene, 12 hr.	$\Delta^{4,6}$-Cholestadien-3-one	68	137

REFERENCES TO TABLE II

[194] Frye, Wallis, and Dougherty, *J. Org. Chem.*, **14**, 403 (1949).
[195] Heer and Miescher, *Helv. Chim. Acta*, **30**, 792 (1947).
[196] Ruzicka and Bernold, *Helv. Chim. Acta*, **24**, 1177 (1941). The structure of the oxidation product is doubtful.
[197] Julian, Cole, Diemer, and Schafer, *J. Am. Chem. Soc.*, **71**, 2061 (1949).
[198] Adams and Whaley, *J. Am. Chem. Soc.*, **72**, 378 (1950).
[199] Inhoffen and Zühlsdorff, *Ber.*, **76**, 245 (1943).
[200] Julian, Meyer, and Printy, *J. Am. Chem. Soc.*, **70**, 3875 (1948).
[201] Kiprianov and Frenkel, *J. Gen. Chem. U.S.S.R.*, **9**, 1682 (1939) [*C. A.*, **34**, 3756 (1940)].
[202] Miescher and Klarer, *Helv. Chim. Acta*, **22**, 967 (1939).
[203] Prins and Reichstein, *Helv. Chim. Acta*, **25**, 317 (1942).
[204] v. Euw and Reichstein, *Helv. Chim. Acta*, **23**, 1118 (1940).
[205] Ruzicka and Meldahl, *Helv. Chim. Acta*, **23**, 517 (1940).
[206] Shoppee and Prins, *Helv. Chim. Acta*, **26**, 216 (1943).
[207] Ruzicka and Meldahl, *Helv. Chim. Acta*, **21**, 1768 (**1938**). The D-homo structure of these compounds was demonstrated by Ruzicka and Meldahl, *Helv. Chim. Acta*, **23**, 364 (1940).
[208] Butenandt and Schmidt-Thomé, *Ber.*, **72**, 186 (1939).
[209] MacPhillamy and Scholz, *J. Biol. Chem.*, **178**, 37 (1949).
[210] Butenandt, U. S. pat. 2,341,594 [*C. A.*, **38**, 4386 (1944)].
[211] Plattner, Heusser, and Segre, *Helv. Chim. Acta*, **31**, 256 (1948).
[212] Plattner and Schreck, *Helv. Chim. Acta*, **24**, 472 (1941); for nomenclature, see v. Euw and Reichstein, *Helv. Chim. Acta*, **24**, 403 (1941).
[213] Wettstein, *Helv. Chim. Acta*, **27**, 1803 (1944).
[214] Plattner, Heusser, and Herzig, *Helv. Chim. Acta*, **32**, 270 (1949).
[215] Wettstein, *Helv. Chim. Acta*, **23**, 1371 (1940).
[216] Heusser, Engel, and Plattner, *Helv. Chim. Acta*, **32**, 2478 (1949).
[217] Sarett, *J. Am. Chem. Soc.*, **70**, 1455 (1948).
[218] Plattner and Schreck, *Helv. Chim. Acta*, **22**, 1182 (1939).
[219] v. Euw and Reichstein, *Helv. Chim. Acta*, **29**, 1919 (1946).
[220] Meystre and Miescher, *Helv. Chim. Acta*, **32**, 1761 (1949).
[221] Cole and Julian, *J. Am. Chem. Soc.*, **67**, 1369 (1945).
[222] Wettstein, *Helv. Chim. Acta*, **24**, 311 (1941).
[223] Kuwada and Yoshiki, *J. Pharm. Soc. Japan*, **58**, 669 (1938) [*C. A.*, **32**, 8432 (1938)].
[224] Barton and Jones, *J. Chem. Soc.*, **1943**, 602.
[225] Barnett, Heilbron, Jones, and Verrill, *J. Chem. Soc.*, **1940**, 1392.
[225a] Heilbron, Kennedy, Spring, and Swain, *J. Chem. Soc.*, **1938**, 869.
[226] Marker, Turner, Oakwood, Rohrmann, and Ulshafer, *J. Am. Chem. Soc.*, **64**, 721 (1942).
[227] Barton, Cox, and Holness, *J. Chem. Soc.*, **1949**, 1771.
[228] Kind and Bergmann, *J. Org. Chem.*, **7**, 341 (1942); Bergmann and Kind, *J. Am. Chem. Soc.*, **64**, 473 (1942).
[229] Lyon and Bergmann, *J. Org. Chem.*, **7**, 429 (1942).

TABLE III

Oppenauer Oxidation of Polyhydroxyl Compounds *

Alcohol	Reaction Conditions	Product	Yield %	Reference
(a) *Non-Steroids*				
$\Delta^{9,10}$-Octaline-1,5-diol	Al(OC$_4$H$_9$-t)$_3$, acetone, benzene, 8 hr.	$\Delta^{9,10}$-Octaline-1,5-dione	30	63
Betulin	Al(OC$_4$H$_9$-t)$_3$, quinone, benzene, 15 hr.	Betulonaldehyde; lupenol-2-one	33 20	169
1-(3-Hydroxybutyl)-2-hydroxy-1,2,3,4-tetrahydronaphthalene	Al(OC$_4$H$_9$-t)$_3$, methyl ethyl ketone, benzene, 36 hr.	1-(3-Ketobutyl)-2-keto-1,2,3,4-tetrahydronaphthalene		28
(b) *Steroids*				
Androstane-3β,5α-diol-17-one	Al(OC$_4$H$_9$-t)$_3$, acetone, dioxane, benzene, 21 hr.	Δ^4-Androstene-3,17-dione		78
Δ^5-Androstene-3β,17α-diol	Al(OC$_3$H$_7$-i)$_3$, cyclohexanone, toluene, 2 hr.	Δ^4-Androstene-3,17-dione; "*cis*"-testosterone	23 37	18
Δ^4-Androstene-3β,6β-diol-17-one	Al(OC$_4$H$_9$-t)$_3$, acetone, benzene, 17 hr.	Androstane-3,6,17-trione	20	230
Δ^4-Androstene-3β,6α-diol-17-one	Al(OC$_4$H$_9$-t)$_3$, acetone, benzene, 35 hr.	Androstane-3,6,17-trione		230
Δ^5-Pregnene-3β,20α-diol	Al(OC$_3$H$_7$-i)$_3$, cyclohexanone, toluene, 1.5 hr.	Progesterone, Δ^4-pregnene-20α-ol-3-one	37	88
Pregnane-3,16,20-triol	Al(OC$_3$H$_7$-i)$_3$, cyclohexanone, toluene, 18 hr.	Δ^{17}-Pregnene-3,16-dione(?)		140

* Only those examples where all oxidizable hydroxyl groups reacted are collected in this table. For partial oxidations, see Table IV.

TABLE III—Continued
Oppenauer Oxidation of Polyhydroxyl Compounds

Alcohol	Reaction Conditions	Product	Yield %	Reference
(b) *Steroids (Continued)*				
*Allo*pregnane-3,16,20-triol	Al(OC$_3$H$_7$-*i*)$_3$, cyclohexanone, toluene, 18 hr.	Δ^{16}-*Allo*pregnene-3,20-dione	46	140
20-Methylpregnane-3,16,20-triol	Al(OC$_4$H$_9$-*t*)$_3$, acetone, benzene, 30 hr.	Δ^{17}-20-Methylpregnene-3,16-dione(?)		141
Methyl hyodesoxycholate	Al(OC$_4$H$_9$-*t*)$_3$, cyclohexanone, benzene, 15 hr.	Methyl 3,6-diketo*allo*cholanate		72
Δ^4-Cholestene-3β,6β-diol	Al(OC$_3$H$_7$-*i*)$_3$, acetone, benzene, 10 hr.	Cholestane-3,6-dione	70	74
Δ^4-Cholestene-3β,6α-diol	Al(OC$_4$H$_9$-*t*)$_3$, acetone, benzene, 15 hr.	Cholestane-3,6-dione		75
Δ^6-Cholestene-3β,5α-diol	Al(OC$_4$H$_9$-*t*)$_3$, acetone, benzene, 24 hr.	Δ^{4,6}-Cholestadien-3-one	(65)	79
Clionastane-3,5,6-triol	Al(OC$_3$H$_7$-*i*)$_3$, acetone, benzene, 4 hr.	Clionastane-3,6-dione-5-ol		228
Δ^5-Cholestene-3,4,7-triol 3-acetate	Al(OC$_4$H$_9$-*t*)$_3$, cyclohexanone, toluene, 5 hr.	Δ^{4,6}-Cholestadien-3-one		145
Δ^5-24-Phenylcholene-3β,24-diol (isomers A and B)	Al(OC$_3$H$_7$-*i*)$_3$, cyclohexanone, toluene, 4 hr.	Δ^4-24-Phenylcholene-3,24-dione From isomer A From isomer B	40 80	73

[230] Davis and Petrow, *J. Chem. Soc.*, **1949**, 2539.

TABLE IV

Selective Oppenauer Oxidation of Polyhydroxyl Compounds *

Alcohol	Reaction Conditions	Product	Yield %	Reference
(a) Saturated				
Betulin	Al(OC$_4$H$_9$-t)$_3$, quinone, benzene, 15 hr.	Lupenol-2-one; betulonaldehyde	20 33	169
17-Methyl-D-homoandrostane-3β,17α-diol	Al(OC$_3$H$_7$-i)$_3$, cyclohexanone, toluene, 10 hr.	17-Methyl-D-homoandrostan-17α-ol-3-one	(25)	85
Δ^8-5-Methyl-10-norandrostene-3,6-diol-17-one	Al(OC$_3$H$_7$-i)$_3$, acetone, benzene, 52 hr.	Δ^8-5-Methyl-10-norandrosten-6-ol-3,17-dione	51	231
Androstane-3,11-diol-17-one	Al(OC$_6$H$_5$)$_3$, acetone, benzene, 22 hr.	Androstane-3,17-dione-11-ol; starting material	64 20	86
Etiocholane-3α,11,17-triol 3-acetate	Al(OC$_3$H$_7$-i)$_3$, acetone, benzene, 12 hr.	Etiocholane-3,11-diol-17-one 3-acetate	15 (29)	90
Pregnane-3α,20-diol	Not specified	Pregnan-20-ol-3-one	47	72
Pregnane-3α,7α,12α-triol-20-one 7-acetate	Al(OC$_3$H$_7$-i)$_3$, cyclohexanone, toluene, 2 hr.	Pregnane-7α,12α-diol-3,20-dione 7-acetate	47	80
Pregnane-3β,11β,20-triol 3-acetate	Al(OC$_6$H$_5$)$_3$, acetone, benzene, 40 hr.	Pregnane-3β,11β-diol-20-one 3-acetate	46	91
Pregnane-3α,11β,20-triol 3-acetate	Al(OC$_6$H$_5$)$_3$, acetone, benzene, 40 hr.	Pregnane-3α,11β-diol-20-one 3-acetate; starting material	32 57	91

* This table includes compounds in which at least two oxidizable hydroxyl groups are present; a substance containing one secondary and one tertiary alcoholic function thus does not fall within the scope of this definition.

TABLE IV—*Continued*

SELECTIVE OPPENAUER OXIDATION OF POLYHYDROXYL COMPOUNDS

Alcohol	Reaction Conditions	Product	Yield %	Reference
(a) Saturated (Continued)				
Pregnane-3β,11β,21-triol-20-one 21-acetate	Al(OC₆H₅)₃, acetone, benzene, 26 hr.	Pregnane-11β,21-diol-3,20-dione 21-acetate	42	87
Pregnane-3α,12α,21-triol-20-one 21-acetate	Al(OC₆H₅)₃, acetone, benzene, 20 hr.	Pregnane-12α,21-diol-3,20-dione 21-acetate	41	82
*Allo*pregnane-3β,17α,20β-triol	Al(OC₆H₅)₃, acetone, benzene, 16 hr.	*Allo*pregnane-17α,20β-diol-3-one 20-acetate	75	86
	Al(OC₄H₉-*t*)₃, acetone, benzene, 26 hr.	Diol; starting material	30 40	86
	Al(OC₄H₉-*t*)₃, cyclohexanone, toluene, 2 hr.	Diol; starting material	38 8	86
Methyl etiodesoxycholate	Al(OC₄H₉-*t*)₃, cyclohexanone, toluene, 2 hr.	Methyl 3-keto-12α-hydroxyetiocholanate	37	83
Methyl *bis*nordesoxycholate	Al(OC₄H₉-*t*)₃, cyclohexanone, toluene, 2.5 hr.	Methyl 3-keto-12α-hydroxy*bis*norcholanate	78	83
Methyl nordesoxycholate	Al(OC₄H₉-*t*)₃, cyclohexanone, toluene, 2.5 hr.	Methyl 3-keto-12α-hydroxynorcholanate	65	83
Methyl desoxycholate	Al(OC₄H₉-*t*)₃, cyclohexanone, toluene, 4.5 hr.	Methyl 3-keto-12α-hydroxycholanate	63	81, 83, 84

Methyl hyodesoxycholate	Al(OC$_4$H$_9$-t)$_3$, acetone, benzene, 119 hr., 40°	3-Keto-6-hydroxycholanic acid; methyl 3,6-diketo*allo*cholanate; starting material	17 (32) 7 23	72
Methyl cholate	Al(OC$_4$H$_9$-t)$_3$, acetone, benzene, 18 hr.	Methyl 3-keto-7α,12α-dihydroxycholanate	63	81, 84, 89
(b) Unsaturated				
Δ^5-Androstene-3β,17α-diol	Al(OC$_3$H$_7$-i)$_3$, cyclohexanone, toluene, 1 hr.	"*cis*"-Testosterone	65	18
	Al(OC$_3$H$_7$-i)$_3$, cyclohexanone, toluene, 2 hr.	"*cis*"-Testosterone; Δ^4-androstene-3,17-dione	37 23	18
Δ^5-Androstene-3β,17β-diol	Al(OC$_6$H$_5$)$_3$ or Al(OC$_4$H$_9$-t)$_3$, acetone, benzene, 18 hr.	Testosterone; Δ^4-androstenedione	40	96, 97
Δ^5-17-Hydroxymethylandrosten-3β-ol	Al(OC$_3$H$_7$-i)$_3$, cyclohexanone, toluene, 1.5 hr.	Δ^4-17-Hydroxymethylandrosten-3-one	(71)	93
Δ^5-Pregnen-3β,20α-diol	Al(OC$_3$H$_7$-i)$_3$, cyclohexanone, toluene, 1.5 hr.	Δ^4-Pregnen-20α-ol-3-one; progesterone	37	88
Δ^5-Pregnene-3β,12α,21-triol-20-one 21-acetate	Al(OC$_3$H$_7$-i)$_3$, cyclohexanone, toluene, 0.25 hr.	Δ^4-Pregnene-12α,21-diol-3,20-dione 21-acetate	50	92
Δ^5-Pregnene-3β,20-diol-21-al dimethyl acetal	Al(OC$_4$H$_9$-t)$_3$, acetone, benzene, 24 hr.	Δ^4-Pregnen-20-ol-3-one-21-al dimethyl acetal	60	56
	Al(OC$_3$H$_7$-i)$_3$, cyclohexanone, toluene, 1.25 hr.	Δ^4-Pregnen-20-ol-3-one-21-al dimethyl acetal	51	56
Δ^5-22-Phenyl*bis*norcholen-3β,22-diol	Al(OC$_3$H$_7$-i)$_3$, cyclohexanone, toluene	Δ^4-22-Phenyl*bis*norcholen-22-ol-3-one	49	

[231] Davis and Petrow, *J. Chem. Soc.*, **1949**, 2975.

TABLE V
Oppenauer Oxidation of Alcohols Containing Halogen, Lactone, Acetal, or Ketal Groups

Alcohol	Reaction Conditions	Product	Yield %	Reference
Acetaldol 2-methylpentane-2,4-diol acetal	Al(OC$_3$H$_7$-i)$_3$, benzophenone, 1 hr. 190–200°	β-Ketobutyraldehyde 2-methylpentane-2,4-diol acetal	37	29
Δ^5-21-Chloropregnen-3β-ol-20-one	Al(OC$_4$H$_9$-t)$_3$, acetone, benzene, 24 hr.	21-Chloroprogesterone	46	54
	Al(OC$_4$H$_9$-t)$_3$, acetone, benzene, 20 days, room temperature	21-Chloroprogesterone	29	55
Δ^5-21-Bromopregnen-3β-ol-20-one	Al(OC$_4$H$_9$-t)$_3$, acetone, benzene, 24 hr.	21-Bromoprogesterone		54
Δ^5-17-Methyl-21-chloropregnen-3β-ol-20-one	Al(OC$_4$H$_9$-t)$_3$, cyclohexanone, benzene, 16 hr.	17-Methyl-21-chloroprogesterone	67	214
Δ^{5,22}-Stigmastadien-3β-ol-22,23-dibromide	Al(OC$_4$H$_9$-t)$_3$, acetone, benzene, 12 hr.	Δ^{4,22}-Stigmastadien-3-one-22,23-dibromide	72	15, 53
Dehydroisoandrololactone	Al(OC$_3$H$_7$-i)$_3$, cyclohexanone, toluene, 48 hr.	Testololactone	(91)	232
Isoandrololactone	Al(OC$_3$H$_7$-i)$_3$, cyclohexanone, toluene, 6 hr.	Dihydrotestololactone	(49)	232
2,13-Dimethyl-2-hydroxymethyl-7-hydroxy-1,2,3,4,5,6,7,8,10,11,12,13-dodecahydrophenanthryl-1-acetic acid lactone	Not specified	2,13-Dimethyl-2-hydroxymethyl-7-keto-1,2,3,4,5,6,7,9,10,11,12,13-dodecahydrophenanthryl-1-acetic acid lactone		233

THE OPPENAUER OXIDATION

Starting material	Conditions	Product	Yield	Reference
$\Delta^{5:20,22}$-3β,21-Dihydroxynorcholadienic acid lactone (23→21)	Al(OC$_4$H$_9$-t)$_3$, cyclohexanone, toluene, 4 hr.	$\Delta^{4:20,22}$-3-Keto-21-hydroxynorcholadienic acid lactone (23→21)	70 (86)	58
Δ^5-Pregnene-3β,17α-diol-20-one ethylene ketal	Al(OC$_4$H$_7$-t)$_3$, cyclohexanone, toluene, 0.5 hr.	17α-Hydroxyprogesterone ethylene ketal	70	61
Δ^5-Pregnen-3β-ol-20-one-21-al dimethyl acetal	Al(OC$_4$H$_9$-t)$_3$, acetone, benzene, 24 hr.	Δ^4-Pregnene-3,20-dione-21-al dimethyl acetal	36	54
Δ^5-Pregnene-3β,20-diol-21-al dimethyl acetal	Al(OC$_4$H$_9$-t)$_3$, acetone, benzene, 24 hr.	Δ^4-Pregnen-20-ol-3-one-21-al dimethyl acetal	60	56
	Al(OC$_3$H$_7$-t)$_3$, cyclohexanone, toluene, 1.25 hr.	Δ^4-Pregnen-20-ol-3-one-21-al dimethyl acetal	51	56
Δ^5-Pregnen-3β,20-diol-21-al dimethyl acetal 20-acetate	Al(OC$_4$H$_9$-t)$_3$, acetone, benzene, 24 hr.	Δ^4-Pregnen-20-ol-3-one-21-al dimethyl acetal 20-acetate	67	56
Δ^5-Pregnen-3β-ol-20-one-21-al diethyl mercaptal	Al(OC$_4$H$_9$-t)$_3$, acetone, benzene, 30 hr.	Δ^4-Pregnene-3,20-dione-21-al diethyl mercaptal	65	56
Diosgenin	Al(OC$_3$H$_7$-t)$_3$, cyclohexanone, toluene, 10 hr.	Δ^4-Diosgen-3-one	74	234
	Al(OC$_3$H$_7$-t)$_3$, quinone, toluene, 1 hr.	Δ^6-Dehydrodiosgen-3-one	30	172
Yamogenin	Al(OC$_4$H$_9$-t)$_3$, acetone, toluene, 16 hr.	Δ^4-Neodiosgen-3-one (Δ^4-yamogenone)	60	183
Pennogenin	Al(OC$_4$H$_9$-t)$_3$, acetone, toluene, 10 hr.	Δ^4-Pennogenone	70	183
Δ^5-Androstene-3β,16,17-triol 16,17-acetonide	Al(OC$_3$H$_7$-t)$_3$, cyclohexanone, toluene, 10 hr.	16-Hydroxytestosterone acetonide	63	235

TABLE V—Continued

OPPENAUER OXIDATION OF ALCOHOLS CONTAINING HALOGEN, LACTONE, ACETAL, OR KETAL GROUPS

Alcohol	Reaction Conditions	Product	Yield %	Reference
Δ⁵-Pregnene-3β,20,21-triol 20,21-acetonide	Al(OC₄H₉-t)₃, acetone, benzene, 14 hr.	Δ⁴-Pregnene-20,21-diol-3-one 20,21-acetonide	74	57
Δ⁵-Pregnene-3β,17α,20α, 21-tetrol 20,21-acetonide	Al(OC₄H₉-t)₃, acetone, benzene, 24 hr.	Δ⁴-Pregnene-17,20,21-triol-3-one 20,21-acetonide; starting material	55 10	236
Δ⁵-Pregnene-3β,17β,20β,21-tetrol 20,21-acetonide	Al(OC₄H₉-t)₃, acetone, benzene, 24 hr.	Δ⁴-Pregnene-17,20,21-triol-3-one 20,21-acetonide	45	237
Pregnane-3α,17α,20β,21-tetrol-11-one 20,21-acetonide	Al(OC₃H₇-i)₃, acetone, benzene, 12 hr.	Pregnane-17α,20β,21-triol-3,11-dione 20,21-acetonide; recovered alcohol	20 60	238

[232] Levy and Jacobsen, *J. Biol. Chem.*, **171**, 71 (1947).
[233] Huffman, Lott, and Ashmore, *J. Am. Chem. Soc.*, **70**, 4268 (1948).
[234] Marker, Tsukamoto, and Turner, *J. Am. Chem. Soc.*, **62**, 2529 (1940).
[235] Butenandt, Schmidt-Thomé, and Weiss, *Ber.*, **72**, 423 (1939).
[236] Reich, Montigel, and Reichstein, *Helv. Chim. Acta*, **24**, 983 (1941).
[237] Koechlin and Reichstein, *Helv. Chim. Acta*, **26**, 1332 (1943).
[238] Sarett, *J. Am. Chem. Soc.*, **71**, 1174 (1949).

TABLE VI
Oppenauer Oxidation of Nitrogen-Containing Alcohols

Alcohol	Reaction Conditions	Product	Yield %	Reference
(a) *Saturated*				
1-Methyl-7-hydroxy-pyrrolizidine (retronecanol)	Al(OC$_4$H$_9$-*t*)$_3$, cyclohexanone, toluene, 6 hr.	1-Methyl-7-ketopyrrolizidine	30	98
Yohimbine	Al(OC$_6$H$_5$)$_3$, cyclohexanone, xylene, 40 hr.	Yohimbone	90	99, 100
Yohimbic acid	Al(OC$_6$H$_5$)$_3$, cyclohexanone, xylene, 40 hr.	Yohimbone	90	99
Corynanthine	Al(OC$_6$H$_5$)$_3$, cyclohexanone, xylene, 40 hr.	Yohimbone	39	238a
*Allo*yohimbine	Al(OC$_6$H$_5$)$_3$, cyclohexanone, xylene, 10 hr.	*Allo*yohimbone		99
*Allo*yohimbic acid	Al(OC$_6$H$_5$)$_3$, cyclohexanone, xylene, 10 hr.	*Allo*yohimbone		99
Yohimbene	Al(OC$_6$H$_5$)$_3$, cyclohexanone, xylene, 8 hr.	Yohimbenone		99
Yohimbenic acid	Al(OC$_6$H$_5$)$_3$, cyclohexanone, xylene, 8 hr.	Yohimbenone		99
Δ^9-21-Diazopregnen-3α-ol-20-one	Al(OC$_6$H$_5$)$_3$, acetone, benzene, 20 days, room temperature,	Δ^9-21-Diazopregnene-3,20-dione		112

238a Janot and Goutarel, *Bull. soc. chim. France*, **1949**, 509.

TABLE VI—Continued
OPPENAUER OXIDATION OF NITROGEN-CONTAINING ALCOHOLS

Alcohol	Reaction Conditions	Product	Yield %	Reference
Solanidan-3β-ol	Al(OC$_6$H$_5$)$_3$, acetone, benzene, 17 hr.	Solanidan-3-one		104
Solatuban-3-ol	Al(OC$_4$H$_9$-t)$_3$, acetone, benzene, 10 hr.	Solatuban-3-one	93	105
(b) *Unsaturated*				
Quinine	KOC$_4$H$_9$-t, benzophenone, benzene, 18 hr.	Quininone	95–98	9
Quinidine	KOC$_4$H$_9$-t, benzophenone, benzene, 18 hr.	Quininone	95	9
Epiquinidine	KOC$_4$H$_9$-t, fluorenone, benzene, 48 hr.	Quininone	79	102
Dihydroquinine	KOC$_4$H$_9$-t, benzophenone, benzene, 18 hr.	Dihydroquininone	95	9
Dihydrocinchonine	KOC$_4$H$_9$-t, benzophenone, benzene, 18 hr.	Dihydrocinchoninone	95	9
Δ^5-21-Diazo-10-*nor*pregnen-3-ol-20-one	Al(OC$_4$H$_9$-t)$_3$, acetone, benzene, 10 hr.	21-Diazo-10-*nor*progesterone		111
Δ^5-21-Diazopregnen-3β-ol-20-one	Al(OC$_4$H$_9$-t)$_3$, acetone, benzene, 20 days room temperature or 14 hr.	21-Diazoprogesterone	68	55

Δ⁵-Pregnene-3β,17-diol-20-one 20-anil	Al(OC₄H₉-t)₃, acetone, benzene, 15 hr.	Δ⁴-Pregnene-3,20-dione-17-ol 20-anil	70	239
Solanidine	Al(OC₄H₉-t)₃ or Al(OC₆H₅)₃, acetone, benzene, 18.5 hr.	Δ⁴-Solaniden-3-one	91	104
Solasodine	Al(OC₄H₉-t)₃, acetone, benzene, 10 hr.	Δ⁴-Solasoden-3-one		106
Solatubine	Al(OC₃H₇-i)₃, acetone, benzene, 8 hr.	Δ⁴-Solatuben-3-one; starting material	70 25	105
	Al(OC₄H₉-t)₃, acetone, benzene, 10 hr.	Δ⁴-Solatuben-3-one	90	105
Rubijervine	Al(OC₄H₉-t)₃, acetone, benzene, 6.5 hr.	Rubijervone	55	107
Isorubijervine	Al(OC₄H₉-t)₃, acetone, benzene, 4.5 hr.	Isorubijervone	55	107
Jervine	Al(OC₄H₉-t)₃, acetone, benzene, 23 hr.	Δ⁴-Jervone	60	108
Dihydrojervine	Al(OC₄H₉-t)₃, acetone, benzene, 5.5 hr.	Δ⁴-Dihydrojervone	39	109
β-Dihydrojervinol	Al(OC₄H₉-t)₃, acetone, benzene, 4 hr.	Δ⁴-β-Dihydrojervonol	50	109

[239] Goldberg and Aeschbacher, *Helv. Chim. Acta*, **22**, 1190 (1939). The structure of the anil was not definitely established.

TABLE VII

OPPENAUER OXIDATION OF PRIMARY ALCOHOLS

Alcohol	Reaction Conditions	Product	Yield %	Reference
(a) *Saturated*				
1-Butanol	Al(OC$_4$H$_9$-t)$_3$, cinnamaldehyde, continuous distillation	Butyraldehyde	72	8
1-Heptanol	Al(OC$_6$H$_5$)$_3$, quinone, benzene, 1 day room temperature, 2 hr.	Heptaldehyde	5	113
β-Phenoxyethanol	Al(OC$_6$H$_5$)$_3$, quinone, benzene, 3 hr.	Phenoxyacetaldehyde		113
3-Phenyl-1-propanol	Al(OC$_6$H$_5$)$_3$, quinone, benzene, 1 day room temperature, 2 hr.	β-Phenylpropionaldehyde		113
Dihydrocyclogeraniol	Al(OC$_3$H$_7$-i)$_3$, anisaldehyde, continuous distillation	Dihydrocyclocitral	64	117
Betulin	Al(OC$_4$H$_9$-t)$_3$, quinone, benzene, 15 hr.	Betulonaldehyde; lupenol-2-one	33 20	169
(b) *Unsaturated*				
Furfuryl alcohol	Al(OC$_6$H$_5$)$_3$, quinone, benzene, 1 week room temperature	Furfuraldehyde	20	113
Benzyl alcohol	Al(OCH$_2$C$_6$H$_5$)$_3$, cinnamaldehyde, continuous distillation	Benzaldehyde	94	8
	Al(OC$_6$H$_5$)$_3$, quinone, benzene, 1 day room temperature, 0.5 hr.	Benzaldehyde	50	113
β-Phenylethanol	Al(OC$_6$H$_5$)$_3$, quinone, benzene, 1 day room temperature, 0.5 hr.	Phenylacetaldehyde	3	113
Anisyl alcohol	Al(OC$_6$H$_5$)$_3$, quinone, benzene, 1 day room temperature, 0.5 hr.	Anisaldehyde	60	113

Substrate	Conditions	Product	Yield	Ref.
Cinnamyl alcohol	Al(OC$_6$H$_5$)$_3$, quinone, benzene, 1 day room temperature, 0.6 hr. 60°, 0.3 hr.	Cinnamaldehyde	13	113
5-Methyl-4-hexen-1-ol	Al(OC$_3$H$_7$-i)$_3$, cinnamaldehyde, continuous distillation	5-Methyl-4-hexen-1-al	37	117
α-Cyclogeraniol	Al(OC$_3$H$_7$-i)$_3$, anisaldehyde, continuous distillation	α-Cyclocitral	66	117
β-Cyclogeraniol	Al(OC$_3$H$_7$-i)$_3$, anisaldehyde, continuous distillation	β-Cyclocitral	78	117
Geraniol	Al(OC$_3$H$_7$-i)$_3$, piperonal, continuous distillation	Citral	30	117
	Al(OC$_6$H$_5$)$_3$, quinone, benzene, 1 day room temperature, 4 hr. 60°, 0.3 hr.	Citral	38	113
Δ^{3,4}-2,2,4-Trimethyltetrahydrobenzyl alcohol	Al(OC$_3$H$_7$-i)$_3$, anisaldehyde, continuous distillation	Δ^{3,4}-2,2,4-Trimethyltetrahydrobenzaldehyde	58	117
Δ^{2,3}-1-Methyl-3-isopropyl-1-hydroxymethylcyclopentene	Al(OC$_3$H$_7$-i)$_3$, anisaldehyde, continuous distillation	Δ^{2,3}-1-Methyl-3-isopropylcyclopenten-1-aldehyde	(26)	116
Lavandulol	Al(OC$_3$H$_7$-i)$_3$, cinnamaldehyde, continuous distillation	Isolavandulal		117
Citronellol	Al(OC$_2$H$_7$-i)$_3$, cinnamaldehyde, continuous distillation	Citronellal	25–42	117
Dihydrocitronellol	Al(OC$_3$H$_7$-i)$_3$, cinnamaldehyde, continuous distillation	Dihydrocitronellal	27–42	117
2-Dimethylamino-5-methylbenzyl alcohol	KOC$_4$H$_9$-t, benzophenone, benzene, 23 hr.	2-Dimethylamino-5-methylbenzaldehyde; dimethyl-p-toluidine	40	103
Vitamin A	Al(OC$_3$H$_7$-i)$_3$, acetaldehyde, benzene, 48 hr., 70°	Vitamin A aldehyde	35	114
	Al(OC$_4$H$_9$-t)$_3$, diethyl ketone, benzene, 48 hr.	Dehydrovitamin A aldehyde	(43)	166

TABLE VIII
OPPENAUER OXIDATION OF PRIMARY ALCOHOLS AND SIMULTANEOUS CONDENSATION OF RESULTING ALDEHYDES

Alcohol	Reaction Conditions	Product	Yield %	Reference
Furfuryl alcohol	Al(OC$_4$H$_9$-t)$_3$, acetone, benzene, 24 hr.	Furfurylideneacetone	15	13
	Al(OC$_4$H$_9$-t)$_3$, diethyl ketone, benzene, 48 hr.	α-Furfurylidenediethyl ketone	40	115
3-Methylpenta-2,4-dien-1-ol	Al(OC$_4$H$_9$-t)$_3$, acetone, benzene, 36 hr.	6-Methylocta-3,5,7-trien-2-one	30	71
Benzyl alcohol	Al(OC$_4$H$_9$-t)$_3$, acetone, benzene, 24 hr.	Benzylideneacetone	28	13
	Al(OC$_4$H$_9$-t)$_3$, diethyl ketone, benzene, 48 hr.	α-Benzylidenediethyl ketone	36	115
Heptyl alcohol	Al(OC$_4$H$_9$-t)$_3$, acetone, benzene, 30 hr.	Heptylideneacetone	17	125
Octyl alcohol	Al(OC$_4$H$_9$-t)$_3$, acetone, benzene, 30 hr.	Octylideneacetone	30	125
Cinnamyl alcohol	Al(OC$_4$H$_9$-t)$_3$, acetone, benzene, 24 hr.	Cinnamylideneacetone	(48)	13
	Al(OC$_4$H$_9$-t)$_3$, diethyl ketone, benzene, 48 hr.	α-Cinnamylidenediethyl ketone	35	115
3-Methyl-2,6-heptadien-1-ol	Al(OC$_3$H$_7$-i)$_3$, acetone, benzene, 72 hr.	6-Methyl-3,5,9-decatrien-2-one	74	132
Geraniol	Al(OC$_4$H$_9$-t)$_3$, acetone, benzene, 30 hr.	ψ-Ionone	70	13

	Al(OC$_6$H$_5$)$_3$, acetone, benzene, 26 hr.	ψ-Ionone	57	118, 119
2,6-Nonadien-1-ol	Al(OC$_6$H$_5$)$_3$ or Al(OC$_4$H$_9$-t)$_3$, methyl ethyl ketone, benzene, 30 hr.	Methyl-ψ-ionone (2 isomers)	25	118
3,6-Dimethyl-2,6-heptadien-1-ol	Al(OC$_3$H$_7$-i)$_3$, acetone, benzene, 40 hr.	3,5,9-Dodecatrien-2-one; recovered alcohol	28 51	132
	Al(OC$_3$H$_7$-i)$_3$, acetone, benzene, 48 hr.	6,9-Dimethyl-3,5,9-decatrien-2-one	59	132
Cyclogeraniol (mixture of α and β isomers)	Al(OC$_4$H$_9$-t)$_3$, acetone, benzene, 30 hr.	Mixture of α- and β-ionone	28	118
	Al(OC$_4$H$_9$-t)$_3$, methyl ethyl ketone, benzene, 30 hr.	Methylionone (4 isomers)	24	118
Isolavandulol	Al(OC$_4$H$_9$-t)$_3$, acetone, benzene	Isolavandulideneacetone		130
2,3,6-Trimethyl-2,6-octadien-8-ol (3-methylgeraniol)	Al(OC$_3$H$_7$-i)$_3$, acetone, benzene, 60 hr.	2,3,6-Trimethyl-2,6,8-undecatrien-10-one (dl-ψ-irone or 3-methyl-ψ-ionone)	35	121
	Al(OC$_4$H$_9$-t)$_3$, acetone, benzene	2,3,6-Trimethyl-2,6,8-undecatrien-10-one (dl-ψ-irone or 3-methyl-ψ-ionone)	75 *	120
5-Phenylpent-2-en-4-yn-1-ol	Al(OC$_4$H$_9$-t)$_3$, acetone, benzene, 42 hr.	8-Phenylocta-3,5-dien-7-yn-2-one	15	240
2,4,6-Trimethyl-2,6-octadien-8-ol	Al(OC$_3$H$_7$-i)$_3$, acetone, benzene, 60 hr.	2,4,6-Trimethyl-2,6,8-undecatrien-10-one	67	123
2,5,6-Trimethyl-2,6-octadien-8-ol	Al(OC$_3$H$_7$-i)$_3$, acetone, benzene, 60 hr.	2,5,6-Trimethyl-2,6,8-undecatrien-10-one	68	122
1-(Cyclohexen-1′-yl)-3-methyl-1,3-pentadien-5-ol	Al(OC$_4$H$_9$-t)$_3$, acetone, benzene, 44 hr.	1-(Cyclohexen-1′-yl)-3-methyl-1,3,5-octatrien-7-one	(90)	127

* This yield is based on the starting material consumed in the reaction.

TABLE VIII—*Continued*

OPPENAUER OXIDATION OF PRIMARY ALCOHOLS AND SIMULTANEOUS CONDENSATION OF RESULTING ALDEHYDES

Alcohol	Reaction Conditions	Product	Yield %	Reference
Lauryl alcohol	Al(OC$_4$H$_9$-t)$_3$, acetone, benzene, 33 hr.	Laurylideneacetone	8	125
α-Ionylidene ethanol	Al(OC$_4$H$_9$-t)$_3$, acetone, benzene, 50 hr.	1-(2′,6′,6′-Trimethylcyclohexen-2′-yl)-3-methyl-1,3,5-octatrien-7-one	26	129
β-Ionylidene ethanol	Al(OC$_4$H$_9$-t)$_3$, acetone, benzene, 44 hr.	1-(2′,6′,6′-Trimethylcyclohexen-1′-yl)-3-methyl-1,3,5-octatrien-7-one	50	126
	Al(OC$_4$H$_9$-t)$_3$, methyl ethyl ketone, benzene	1-(2′,6′,6′-Trimethylcyclohexen-1′-yl)-3,6-dimethyl-1,3,5-octatrien-7-one		128
Farnesol	Al(OC$_3$H$_7$-i)$_3$, acetone, benzene, 8 days	Farnesylideneacetone	73	131
Cetyl alcohol	Al(OC$_4$H$_9$-t)$_3$, acetone, benzene, 33 hr.	Cetylideneacetone	12	125
Δ^5-17-Hydroxymethylandrosten-3-ol	Al(OC$_3$H$_7$-i)$_3$, cyclohexanone, toluene, 1.5 hr.	Δ^4-Androsten-3-one-17-methylene-cyclohexanone	Small amount	93
Phytol	Al(OC$_4$H$_9$-t)$_3$, acetone, benzene, 10 hr.	6,10,14,18-Tetramethylnonadeca-3,5-dien-2-one	46	124
Vitamin A (axerophthol)	Al(OC$_3$H$_7$-i)$_3$, acetone, benzene, 48 hr.	Axerophthylideneacetone	70	13, 115

[240] Heilbron, Jones, and Sondheimer, *J. Chem. Soc.*, **1949**, 606.

TABLE IX
Unsuccessful Oppenauer Oxidations

Alcohol	Reaction Conditions	Reference
2-Butyne-1,4-diol	Al(OC$_3$H$_7$-i)$_3$, acetone, benzene	241
Glycerol α-monomethyl ether	Al(OC$_4$H$_9$-t)$_3$, acetone, benzene, 8 hr.	242
Pent-2-en-4-yn-1-ol	Al(OC$_4$H$_9$-t)$_3$, acetone, benzene	240
Ethyl β-hydroxybutyrate	Al(OC$_4$H$_9$-t)$_3$, fluorenone, benzene, 48 hr.	153
β-Phenylethanol	Al(OC$_4$H$_9$-t)$_3$, acetone, benzene, 24 hr.*	13
1-(β-Hydroxyethyl)cyclohexen	Al(OC$_3$H$_7$-i)$_3$, anisaldehyde	117
γ-Phenylpropanol	Al(OC$_4$H$_9$-t)$_3$, acetone, benzene, 24 hr.*	13
α-Cyclogeraniol	Al(OC$_3$H$_7$-i)$_3$, acetone, benzene, 30 hr.*†	243
Geraniol	Al(OC$_4$H$_9$-t)$_3$, diethyl ketone, benzene *†	115
Lavandulol	Al(OC$_4$H$_9$-t)$_3$, acetone, benzene *	130
Tetrahydrogeraniol	Al(OC$_4$H$_9$-t)$_3$, acetone, benzene, 24 hr.	13
Citronellol	Al(OC$_3$H$_7$-i)$_3$, acetone, benzene, 150° *	244
2,7-Dimethylocta-2,4,6-triene-1,8-diol	Not specified	245
3-Hydroxymethylheptan-2-one	Al(OC$_3$H$_7$-i)$_3$, cinnamaldehyde	117
1-Methyl-1,2,-dihydroxy-1,2,3,4-tetrahydro-naphthalene	Al(OC$_3$H$_7$-i)$_3$	187
trans-γ-(p-Hydroxycyclohexyl)butyric acid methyl ester	Al(OC$_4$H$_9$-t)$_3$, acetone or cyclohexanone	246
1,4-Diphenylbutanetetrol	Al(OC$_6$H$_5$)$_3$, acetone or quinone	247
1-(2′,6′,6′-Trimethylcyclohexen-1′-yl)-3-methyl-1-hexen-5-yl-4-ol(?)	Al(OC$_4$H$_9$-t)$_3$, acetone, benzene, 18 hr.	66
1-(2′,6′,6′-Trimethylcyclohexen-1′-yl)-3-methyl-1-hexen-3,5-diol	Al(OC$_3$H$_7$-i)$_3$ or Al(OC$_4$H$_9$-t)$_3$	248
Quinine	Al(OC$_4$H$_9$-t)$_3$, acetone or quinone, benzene, 12–24 hr. ‡	101
Ethyl 3β,5,19-trihydroxy-cholanate	Al(OC$_4$H$_9$-t)$_3$, acetone, benzene	76
9-ω-Hydroxybenzylanthracene	Al(OC$_3$H$_7$-i)$_3$, cyclohexanone, toluene	197
Alkaloid A (*B. sempervirens* L.)	KOC$_4$H$_9$-t, benzophenone, benzene	249
Lanosterol	Al(OC$_4$H$_9$-t)$_3$, acetone, 10 hr.	250
Tetrahydroanhydro-aucubigenin	Al(OC$_4$H$_9$-t)$_3$, acetone, benzene	251
2,3,6-Trimethyl-5-(3′,7′,11′,15′-tetramethyl-3′-hydroxyhexadecan-1′-yl)-1,4-benzoquinone	Al(OC$_4$H$_9$-t)$_3$	252
Δ^5-22-Isospirosten-2α,3β-diol (yuccagenin)	Al(OC$_4$H$_9$-t)$_3$, cyclohexanone, toluene, 8 hr.	253

* Successfully oxidized under different conditions; see Table VII.
† Successfully oxidized under different conditions; see Table VIII.
‡ Successfully oxidized under different conditions; see Table VI.

REFERENCES TO TABLE IX

[241] Johnson, *J. Chem. Soc.*, **1946**, 1011.
[242] Abouzeid and Linnell, *J. Pharmacy and Pharmacol.*, **1**, 235 (1949).
[243] H. Schinz, private communication.
[244] H. Schinz, private communication; *cf.* Vodoz, Thesis, Eidgenöss. Techn. Hochschule, Zürich, 1948.
[245] Deemer, Lutwak, and Strong, *J. Am. Chem. Soc.*, **70**, 155 (1948).
[246] Dauben and Adams, *J. Am. Chem. Soc.*, **70**, 1760 (1948).
[247] Ruggli, Dahn, and Fries, *Helv. Chim. Acta*, **29**, 312 (1946).
[248] Shantz, *J. Am. Chem. Soc.*, **68**, 2554 (1946).
[249] Heusler and Schlittler, *Helv. Chim. Acta*, **32**, 2227 (1949).
[250] Bellamy and Doree, *J. Chem. Soc.*, **1941**, 176.
[251] Karrer and Schmid, *Helv. Chim. Acta*, **29**, 550 (1946).
[252] Karrer, *Helv. Chim. Acta*, **22**, 343 (1939).
[253] Djerassi and Yashin, unpublished observation.

CHAPTER 6

THE SYNTHESIS OF PHOSPHONIC AND PHOSPHINIC ACIDS

Gennady M. Kosolapoff

Alabama Polytechnic Institute

CONTENTS

	PAGE
Introduction	275
Alkylation of the Phosphorus Atom in Phosphorous Esters	276
Mechanism	278
Scope and Limitations	279
Synthesis of Phosphonic Acids and Esters	279
Synthesis of Phosphinic Acids and Esters	283
Selection of Experimental Conditions and Procedures	284
Experimental Procedures	286
The Trialkyl Phosphite Procedure	286
Diethyl Ethanephosphonate	286
1-Naphthylmethanephosphonic Acid	286
Diethyl α-Oxo-α-toluenephosphonate	286
3-Bromopropane-1-phosphonic Acid	287
Di-β-chloroethyl β-Chloroethanephosphonate (Intramolecular Isomerization)	287
Tetraphenyl Ethane-1,2-diphosphonate	288
Isopropyl Isopropylphenylphosphinate	288
The Sodium Salt Procedure	289
Triethyl β-Phosphonopropionate	289
Dibutyl Alkanephosphonates	289
Triarylmethane Derivatives	290
Triphenylmethanephosphonic Acid	290
Special Methods	290
Diethyl 2-Hydroxyethanephosphonate	290
Stearamidomethanephosphonic Acid	290
Addition of Phosphorus Pentachloride to Unsaturated Compounds	291
Scope and Limitations	291
Experimental Procedures	292
β-Styrenephosphonic Acid	292
Phenylethynephosphonic Acid	293
The Grignard Reaction	293
Experimental Procedures	296
Diphenylphosphinic Acid (Michaelis-Wegner Procedure)	296

	PAGE
Diphenylphosphinic Acid (Kosolapoff Procedure)	296
Ethanephosphonic Acid (Malatesta-Pizzotti Procedure)	297

THE FRIEDEL-CRAFTS REACTION ... 297
  Scope and Limitations ... 300
  Experimental Procedures ... 301
    p-Toluenephosphonic Acid (Michaelis Procedure) ... 301
    Benzenephosphonic Acid (Kosolapoff Procedure) ... 302

THE ADDITION OF PHOSPHORUS COMPOUNDS TO THE CARBONYL GROUP ... 303
  Addition of Compounds Containing the Phosphorus-Hydrogen Linkage ... 304
  Experimental Procedures ... 306
    α-Toluenephosphonic and Dibenzylphosphinic Acids ... 306
    Di(α-hydroxyisopropyl)phosphinic Acid ... 306
    α-Hydroxyethanephosphonic Acid ... 307
  Addition of Phosphorus Chlorides ... 308
  Scope and Limitations ... 311
  Experimental Procedures ... 313
    α-Hydroxy-α-toluenephosphonic Acid (Fossek Procedure) ... 313
    Conant Procedures (Saturated Carbonyl Compounds) ... 313
      α-Chloro-α-phenylethanephosphonic Acid ... 313
      α-Hydroxy-α-phenylethanephosphonic Acid ... 314
      α,α-Diphenyl-α-hydroxymethanephosphonic Acid ... 314
      α-Hydroxy-α-toluenephosphonic Acid ... 314
      1-(4'-Methoxyphenyl)-2-benzoylethane-1-phosphonic Acid ... 315
      α-Phosphono-α-hydroxypropionic Acid ... 315

THERMAL DECOMPOSITION REACTIONS ... 315
  Pyrolytic Reactions ... 316
  Displacements with Organomercury Compounds ... 317
  Experimental Procedures ... 318
    p-Tolylphenylchlorophosphine ... 318
    Phenyldichlorophosphine ... 318
    n-Butyldichlorophosphine ... 319
    Phenyl-p-bromophenylchlorophosphine ... 319
  Thermal Decomposition of Phosphonium Compounds ... 319
  Experimental Procedures ... 321
    Di-n-propylchlorophosphine ... 321
    Diphenylchlorophosphine ... 321
  Disproportionation of Phosphonous Acids ... 321

MISCELLANEOUS SYNTHESES ... 322
  Oxidation of Phosphines and Phosphonous Acids ... 322
  Syntheses from Dialkylanilines ... 323
  Experimental Procedures ... 324
    Bis(4-dimethylaminophenyl)-phosphonic and -phosphinic Acids (Bourneuf Procedures) ... 324
    Dibenzophosphazinic Acid ... 325
  Oxidative Phosphonation ... 325
  Wurtz Reaction ... 326
  Direct Phosphonation of a Nitrogen Heterocycle ... 326

	PAGE
Tabulation of Compounds Reported Before May, 1948	327
Table I. Derivatives of Phosphonic and Phosphinic Acids Prepared by Alkylation of Phosphites or Other Trivalent Esters	328
Table II. Phosphonic Acids Prepared by Addition of Phosphorus Pentachloride to Unsaturated Compounds	333
Table III. Phosphonic and Phosphinic Acids Prepared by the Friedel-Crafts Reaction	334
Table IV. Phosphonic and Phosphinic Acids Prepared by the Addition to Carbonyl Compounds	
a. By Addition of P-H Linked Compounds	335
b. By Addition of Phosphorus Chlorides	335
Table V. Chlorophosphines Prepared from Organomercury Intermediates	337
Table VI. Chlorophosphines Prepared by Thermal Decomposition of Phosphonium Compounds	338

INTRODUCTION

The phosphonic acids, $RP(O)(OH)_2$, and the phosphinic acids, $RR'P(O)OH$, may be regarded as derivatives of phosphoric acid in which one or two hydroxyls are replaced by organic radicals.* Derivatives of the type $RP(O)HOH$, which are termed phosphonous acids by the current *Chemical Abstracts* system, contain phosphorus in a lower state of oxidation than is present in the phosphonic and the phosphinic acids and have chemical properties decidedly different from those of the latter classes. Phosphonous acids are not considered in this chapter, except as they are involved in the synthesis of phosphonic and phosphinic acids.

Although individual phosphonic and phosphinic acids have been known for several decades, the syntheses of these two classes of compounds have not been so well developed as have the methods for the corresponding arsenic compounds. Much of the work has been devoted to the parent substances of the various possible series, and there is but little information concerning the syntheses of compounds with a high degree of substitution.

This chapter is concerned only with the introduction of the phosphorus-containing functions, that is to say, with the synthesis of acids or their functional derivatives which can be isolated and hydrolyzed to

* *Note on nomenclature.* Phosphonic acids are named with reference to the *hydrocarbons* from which they are derived, whereas phosphinic acids are named with reference to the *alkyl and/or aryl groups* which they contain. Thus $C_6H_5PO(OH)_2$ is benzenephosphonic acid, but $(C_6H_5)_2P(O)OH$ is diphenylphosphinic acid. Esters of both series have names ending in *-ate*, e.g., diethyl (or ethyl) benzenephosphonate, ethyl diphenylphosphinate. If it is desirable to indicate the phosphonic group by means of a prefix, *phosphono-* is used. Thus, $(HO)_2P(O)CH_2CO_2H$ may be called phosphonoacetic acid and $(C_2H_5O)_2P(O)CH_2CO_2C_2H_5$ triethyl phosphonoacetate. Esters of phosphonous acids are given names ending in *-ite*. Thus, $C_6H_5P(OC_2H_5)_2$ is diethyl benzenephosphonite.

the acids. It does not cover the further possible modifications of the organic portions of the molecule, because most such modifications are quite similar to those of comparable carbon compounds with strongly electronegative substituents.

ALKYLATION OF THE PHOSPHORUS ATOM IN PHOSPHOROUS ESTERS

One of the most versatile methods for the synthesis of esters of phosphonic acids is based on the reaction of a trialkyl phosphite with an alkyl halide.[1] If the alkyl groups of the two reagents are identical, the process amounts to an isomerization of the phosphite, as illustrated in the accompanying equation. The general procedure often is referred

$$C_2H_5I + P(OC_2H_5)_3 \rightarrow$$
$$[C_2H_5P(OC_2H_5)_3]^+I^- \rightarrow C_2H_5PO(OC_2H_5)_2 + C_2H_5I$$

to as the "isomerization method," whether or not the several alkyl groups are identical; it is also called the Arbuzov transformation.

When the alkyl groups of the phosphite and of the halide are identical, as in the above example, only one phosphonate can be formed. When the alkyl halide employed is not identical with that eliminated in the second stage of the reaction, a mixture obviously may be formed. Even so, the reaction may be controlled to give a high yield of the desired phosphonate. For example, 1-chloromethylnaphthalene reacts with triethyl phosphite (in small excess) at 150–160° to give diethyl 1-naphthylmethanephosphonate in 87% yield (p. 286).

$$C_{10}H_7CH_2Cl + (C_2H_5O)_3P \rightarrow C_{10}H_7CH_2PO(OC_2H_5)_2 + C_2H_5Cl$$

Presumably, the success of the reaction is related both to the greater reactivity of the arylmethyl chloride, as compared to ethyl chloride, and to the volatility of the ethyl chloride, most of which escapes from the hot mixture through the condenser. When the alkyl halide employed and that formed are of approximately the same reactivity, the control of the reaction may be aided by the use of a large excess of the reagent. Thus, when a mixture of 5 moles of trimethylene bromide and 1 mole of triethyl phosphite is refluxed under a fractionating column (for removal of ethyl bromide), the ester of 3-bromopropanephosphonic acid is obtained in 90% yield (p. 287).

[1] Michaelis and Kaehne, *Ber.*, **31**, 1048 (1898).

Derivatives of phosphinic acids are obtained when a phosphonite is substituted for the phosphite. The preparation of ethyl ethylphenylphosphinate is an example.[2] Only a few phosphinates have been prepared in this way (see Table I) owing to the relatively difficult preparation of the necessary phosphonites.

$$C_6H_5P(OC_2H_5)_2 + C_2H_5I \rightarrow C_6H_5PO(OC_2H_5) + C_2H_5I$$
$$|$$
$$C_2H_5$$

Since only one alkoxy group of a phosphite participates in the reaction leading to the phosphonates, it might be anticipated that partial esters of phosphorous acid and esters of amidophosphorous acids would react in the same way. Such variations of the process were developed by Michaelis,[3,4] and the use of the salts of dialkyl acid phosphites has proved particularly satisfactory. This method is illustrated by the preparation of dibutyl 1-decanephosphonate from sodium dibutyl phosphite and decyl bromide.[5]

$$(C_4H_9O)_2PONa + C_{10}H_{21}Br \rightarrow C_{10}H_{21}PO(OC_4H_9)_2 + NaBr$$

An example of the use of amidophosphites is provided by the synthesis of the bisdiethylamide of methanephosphonic acid.[4]

$$C_2H_5OP[N(C_2H_5)_2]_2 + CH_3I \rightarrow CH_3P(O)[N(C_2H_5)_2]_2 + C_2H_5I$$

Several other syntheses of phosphonic acids and their derivatives probably can be included in the general category of alkylation of phosphite derivatives. They represent rather isolated examples and warrant further study and confirmation. These syntheses include the isomerization of triaryl phosphites by alcohols at high temperatures,[6] the formation of phosphonic acid derivatives from methylol derivatives of acyl amides and phosphorus trichloride,[7] and the formation of triarylmethanephosphonic acid derivatives from triarylcarbinols and phosphorus trichloride.[8,9] If these reactions can be formulated, as shown

[2] Arbuzov, *J. Russ. Phys. Chem. Soc.*, **42**, 395 (1910) [*C. A.*, **5**, 1397 (1911)].
[3] Michaelis and Becker, *Ber.*, **30**, 1003 (1897).
[4] Michaelis, *Ann.*, **326**, 129 (1903).
[5] Kosolapoff, *J. Am. Chem. Soc.*, **67**, 1180 (1945).
[6] Milobendzki and Szulgin, *Chem. Polsk.*, **15**, 66 (1917) [*C. A.*, **13**, 2867 (1919)].
[7] Pikl, U. S. pat. 2,304,156 [*C. A.*, **37**, 3262 (1943)].
[8] Arbuzov and Arbuzov, *J. Russ. Phys. Chem. Soc.*, **61**, 217 (1929) [*C. A.*, **23**, 3921 (1929)].
[9] Boyd and Smith, *J. Chem. Soc.*, **1926**, 2323; **125**, 1477 (1924); Boyd and Chignell, *ibid.*, **123**, 813 (1923).

below, as proceeding through a phosphonium-type addition complex, the analogy to the previously discussed "normal" isomerization of phosphite esters is obvious.

$$(ArO)_3P + 4ROH \rightarrow [RP(OR)_3]OH \rightarrow RPO(OR)_2 + ROH$$
$$+$$
$$3ArOH$$

$$RCONHCH_2OH + PCl_3 \rightarrow RCONHCH_2OPCl_2 \rightarrow RCONHCH_2POCl_2$$
$$+$$
$$HCl$$

$$Ar_3COH + PCl_3 \rightarrow Ar_3COP(H)Cl_2 \rightarrow Ar_3CP(O)Cl_2 + HCl$$
$$\quad\quad\quad\quad\quad\quad\quad\quad\quad | $$
$$\quad\quad\quad\quad\quad\quad\quad\quad\quad Cl$$

Mechanism

The mechanisms that have been proposed for the reactions are illustrated in the first equation given on p. 276 above. The principal feature is the formation of an intermediate salt, shown in brackets above, which undergoes the loss of a molecule of a simple halide. Michaelis and Kaehne [1] isolated the methiodides of triphenyl phosphite and tri-m-cresyl phosphite, and they reported the formation of a solid product, which could not be crystallized, from triphenyl phosphite and benzyl chloride. There is no direct evidence for the formation of an intermediate salt from an aliphatic phosphite and an alkyl halide; however, the induction period observed in such a reaction has been considered a measure of the stability of the intermediate salt. Likewise, there is no direct evidence for an intermediate in the reaction of the sodium salt of a dialkyl acid phosphite and an alkyl halide. The existence of such an intermediate has been inferred from induction periods during which no sodium halide forms. The fact that prolonged heating of such a reaction mixture may result in the formation of the sodium salt of an acid phosphonate and a molecule of alkyl halide [10] has been cited as an argument for the re-formation of the intermediate salt.

$$(RO)_2PONa + R'Br \rightarrow [R'P(ONa)(OR)_2]^+Br^- \rightarrow R'PO(ONa)OR + RBr$$

Neither of the arguments concerning the reactions of the salts seems very persuasive. It is possible that the alkylation of the acid phosphite is merely a displacement, most readily portrayed as involving the anion of the tautomeric form of the acid phosphite, as follows.

$$(RO)_2PO^- \rightleftarrows {}^-P(O)(OR)_2 \xrightarrow{R'Br} R'P(O)(OR)_2 + Br^-$$

[10] P. Nylen, dissertation, Uppsala, 1930.

Analogously, the reaction of the phosphonate with sodium halide may be a simple alkylation of the halide ion, operating through a displacement rather than through an addition.

$$R'P(O)(OR)_2 + X^- \rightarrow R'P(O)(OR)O^- + RX$$

Scope and Limitations

Synthesis of Phosphonic Acids and Esters

Only one aryl halide has been used successfully in the preparation of a phosphonic acid derivative by these methods; 9-phosphonoacridine was obtained in 60% yield by hydrolysis of the ester from 9-chloroacridine and triethyl phosphite. Simple aryl halides evidently are too unreactive, but from the example just cited it would appear that activated aryl halides, and especially those containing heterocyclic nuclei, might be employed.

The aliphatic halides used have been almost invariably primary halides. Only two secondary halides reacted satisfactorily; isopropyl isopropylphenylphosphinate has been obtained from isopropyl iodide and diisopropyl benzenephosphonite,[11] and α-phenylphosphonopropionic acid has been obtained from ethyl α-bromopropionate and diisobutyl benzenephosphonite.[12] Other secondary halides and simple

$$C_6H_5P[OCH(CH_3)_2]_2 + (CH_3)_2CHI \rightarrow$$

$$(CH_3)_2CHP(O)OCH(CH_3)_2 + (CH_3)_2CHI$$
$$|$$
$$C_6H_5$$

$$C_6H_5P[OCH_2CH(CH_3)_2]_2 + C_2H_5O_2CCHBrCH_3 \rightarrow$$

$$C_6H_5P(O)OCH_2CH(CH_3)_2 + (CH_3)_2CHCH_2Br$$
$$|$$
$$C_2H_5O_2CCHCH_3$$

tertiary halides either fail to react or give olefins. However, triarylmethyl halides react normally with triethyl phosphite to give esters of triarylmethanephosphonic acids.[8] These compounds cannot be prepared by the use of the sodium salt of the dialkyl acid phosphite; with this reagent an abnormal reaction occurs and the hexaarylethane (or triarylmethyl) is produced.

[11] Arbuzov, Kamai, and Belorossova, *J. Gen. Chem. U.S.S.R.*, **15**, 766 (1945) [*C. A.*, **41**, 105 (1947)].

[12] Arbuzov and Arbuzov, *J. Russ. Phys. Chem. Soc.*, **61**, 1599 (1929) [*C. A.*, **24**, 5289 (1930)].

The order of reactivity of the simple primary alkyl halides is the usual one, iodides being most and chlorides least reactive. Bromides have been used most often.

A considerable number of alkyl halides has been used in both the ester and the sodium salt procedures.[1, 3, 5, 10, 13, 14] Various functional substituents can be present in the alkyl halide. Neutral phosphite esters have been alkylated with the chloromethyl derivatives of various aromatic hydrocarbons and with 2-chloromethylthiophene, with triarylmethyl chlorides, with chloromethyl ethers and with one β-bromo ether, with ethyl chloroacetate and with esters of various ω-halo acids, with N,N-diphenylchloroacetamide, with 3-cyanopropyl chloride, with a bromomethyl ketone, with N-(bromoalkyl)phthalimides, and, as mentioned above, with 9-chloroacridine. α-Bromo nitro compounds, however, do not give the expected nitroalkane phosphonates; oxidation-reduction reactions intervene, the triethyl phosphite being oxidized to the phosphate, apparently with reduction of the nitro group. The exact nature of the reactions that occur is not understood.[15]

Esters of 2-chloroethanol may be converted to phosphonates by heat alone; thus, tri-β-chloroethyl phosphite yields an ester of 2-chloroethanephosphonic acid.

$$P(OCH_2CH_2Cl)_3 \rightarrow ClCH_2CH_2PO(OCH_2CH_2Cl)_2$$

Tri-β-bromoethyl phosphite isomerizes similarly. Evidently only one haloalkyl group is necessary, for the mixed ester, diethyl 2-chloroethyl phosphite, was converted to diethyl 2-chloroethanephosphonate.

$$ClCH_2CH_2OP(OC_2H_5)_2 \rightarrow ClCH_2CH_2PO(OC_2H_5)_2$$

However, when the phosphite contains two aryloxy residues the reaction takes a different course and produces esters of ethane-1,2-diphosphonic acids.[16] Though the nature of the process is not clear, the overall result may be shown in the formulation given by Kabachnik.

$$2(ArO)_2POCH_2CH_2Cl \rightarrow (ArO)_2P(O)CH_2CH_2PO(OAr)_2 + ClCH_2CH_2Cl$$

From experiments with ethylene bromide and trimethylene bromide it appears that the reaction of primary dihalides can be controlled to give either haloalkanephosphonates or alkanediphosphonates. The course of the reaction is determined by the ratio of phosphite to dihalide,

---

[13] Arbuzov, *J. Russ. Phys. Chem. Soc.*, **38**, 687 (1906).

[14] Ford-Moore and Williams, *J. Chem. Soc.*, **1947**, 1465.

[15] Arbuzov, Arbuzov, and Lugovkin, *Bull. acad. sci. U.R.S.S., classe sci. chim.*, **1947**, 535 [*C. A.*, **42**, 1886 (1948)].

[16] Kabachnik and Rossiiskaya, *Bull. acad. sci. U.R.S.S., classe sci. chim.*, **1947**, 631 [*C. A.*, **42**, 5845 (1948)].

the dihalide being used in considerable excess when the haloalkanephosphonate is desired. Methylene iodide reacts with triethyl phosphite, and both the iodomethanephosphonate [14,17] and the methanediphosphonate [14] have been isolated. Carbon tetrachloride reacts readily with trialkyl phosphites, yielding the esters of trichloromethanephosphonic acid; chloroform does not react at the reflux temperature.[18,19]

1,4-Dichloro-2-butene is the simplest allylic halide that has been treated with a phosphite. The product obtained with an excess of the halide was dehydrohalogenated and hydrolyzed by treatment with potassium hydroxide. As 1,3-butadiene-1-phosphonic acid was obtained, evidently the alkylation did not occur with allylic rearrangement.

$ClCH_2CH=CHCH_2Cl + P(OC_2H_5)_3 \rightarrow$

$ClCH_2CH=CHCH_2PO(OC_2H_5)_2 + C_2H_5Cl$

Examples of allylic rearrangement have been reported, however; 1-methoxy-3-chloro-4-pentene reacts with phosphites to give esters of the straight-chain phosphonic acid in good yield; [20] details of this work have not been published.

$CH_3OCH_2CH_2CHClCH=CH_2 + P(OR)_3 \rightarrow$

$CH_3OCH_2CH_2CH=CHCH_2PO(OR)_2 + RCl$

An unsaturated phosphonic acid derivative is formed when propylene bromide is heated with triethyl phosphite, evidently as a result of dehydrohalogenation of the primary reaction product.[14] The yield is poor.

$CH_3CHBrCH_2Br + P(OC_2H_5)_3 \rightarrow CH_3CH=CHPO(OC_2H_5)_2 + HBr$

Acid chlorides react readily with triethyl phosphite to yield α-ketophosphonic esters.[21] These compounds cannot be hydrolyzed to the free acids, the phosphono group being eliminated from the molecule under all hydrolytic conditions that have been tested.

If the reaction of triarylcarbinols with phosphorus trichloride is to be considered a variant of the phosphite isomerization reaction, as mentioned earlier, the following successful examples of its application may be mentioned here: triphenylcarbinol, *p*-chlorophenyldiphenylcarbinol,

[17] Arbuzov and Kushkova, *J. Gen. Chem. (U.S.S.R.)*, **6**, 283 (1936) [*C. A.*, **30**, 4813 (1936)].

[18] Kosolapoff, *J. Am. Chem. Soc.*, **69**, 1002 (1947).

[19] Kamai and Egorova, *J. Gen. Chem. U.S.S.R.*, **16**, 1521 (1946) [*C. A.*, **41**, 5439 (1947)].

[20] A. N. Pudovik, Report at the October, 1947, meeting of the Chemical Section of the Academy of Sciences, U.S.S.R., in Kazan.

[21] Kabachnik and Rossiiskaya, *Bull. acad. sci. U.R.S.S., classe sci. chim.*, **1945**, 364 [*C. A.*, **40**, 4688 (1946)].

p-bromophenyldiphenylcarbinol, p-anisyldiphenylcarbinol, m-anisyldiphenylcarbinol, 1-naphthyldiphenylcarbinol, 2-naphthyldiphenylcarbinol, p-nitrophenyldiphenylcarbinol, and p-tolyldiphenylcarbinol,[8, 9] all of which give the corresponding triarylmethanephosphonic acids after hydrolysis.

The reaction of methylol acylamides with phosphorus trichloride has been described only in the patent literature;[7] several compounds so reported were not well enough characterized for inclusion in this chapter. Sufficient information is given about the preparation and the properties of stearamidomethanephosphonic acid (see p. 290).

The reaction of alkyl halides with salts of dialkyl acid phosphites has been employed somewhat less frequently than the reaction with neutral phosphites. A number of simple primary alkyl halides have been converted to phosphonates. Primary halides having other functional groups which have been employed successfully include arylmethyl chlorides, α-halo ethers, α-halo ketones, ethyl chloroacetate and ethyl β-bromopropionate, N-(bromoalkyl)phthalimides,[22] and the hydrobromide of 2-aminoethyl bromide.[22] Methylene iodide reacted with sodium diethyl phosphite, but only methanediphosphonic acid was isolated.[10] Evidently the intermediate ester reacted with the sodium iodide, as discussed above (p. 278). Ethylene bromide is dehydrohalogenated by sodium dialkyl phosphites. The tetraethyl ester of propane-1,3-diphosphonic acid has been obtained from trimethylene dibromide and sodium diethyl phosphite, but in unrecorded yield. Only dehydrohalogenation occurs when the same sodium salt is treated with 1,2-dibromopropane, 2,3-dibromobutane, or 1,2-dibromo-2-methylpropane.[10]

When 1-methoxy-3-chloro-4-pentene is treated with a sodium dialkyl phosphite in slight excess, reaction occurs with allylic rearrangement.

$CH_3OCH_2CH_2CHClCH=CH_2 + NaOP(OR)_2 \rightarrow$

$CH_3OCH_2CH_2CH=CHCH_2PO(OR)_2 + NaCl$

If the phosphite derivative is not in excess, a complex mixture is produced.[20]

When the ethyl ester of a haloacetic acid is treated with sodium diethyl phosphite the expected phosphonate is produced in yields of 45–50% along with ethyl succinate in about 5% yield.[10, 23] Esters of higher α-bromo acids yield the coupling products in unspecified yields, but none of the phosphonates. The coupling has been explained on the basis of an exchange of bromine and sodium atoms between the reactants.[23]

[22] Chavane, *Compt. rend.*, **224**, 406 (1947).
[23] Chavane and Rumpf, *Compt. rend.*, **225**, 1322 (1947).

As mentioned above, triarylmethyl halides do not give phosphonates when they react with sodium dialkyl phosphites. Acid chlorides, which react normally with trialkyl phosphites (p. 281), give complex mixtures with sodium dialkyl phosphites.

Ethylene oxide [24] evidently is the only halogen-free alkylating agent that has been used successfully on a salt of a dialkyl phosphite. Moderately good yields (40%) of diethyl β-hydroxyethanephosphonate can be obtained from this reagent.

$$CH_2\underset{O}{\overset{}{\diagdown\diagup}}CH_2 + (C_2H_5O)_2PONa \rightarrow NaOCH_2CH_2PO(OC_2H_5)_2 \xrightarrow{CH_3CO_2H}$$

$$HOCH_2CH_2PO(OC_2H_5)_2 + CH_3CO_2Na$$

Synthesis of Phosphinic Acids and Esters

A number of mixed aliphatic-aromatic phosphinic acids and their esters have been prepared in excellent yields from dialkyl arylphosphonites and alkyl halides. The products that have been reported are methyl phenylmethylphosphinate from methyl iodide and dimethyl benzenephosphonite,[25, 26] ethyl phenylethylphosphinate from ethyl iodide and diethyl benzenephosphonite,[2] isobutyl isobutylphenylphosphinate from diisobutyl benzenephosphonite and isobutyl iodide,[27] and isobutyl phenyltritylphosphinate from triphenylbromomethane and diisobutyl benzenephosphonite.[27] Similarly successful were the preparations of the corresponding phosphinates from dialkyl benzenephosphonites with propyl iodide,[26] chloromethyl ethyl ether,[26] chloromethyl methyl ether,[26] isopropyl iodide,[11] ethyl chloroacetate,[12] and ethyl α-bromopropionate.[12]

Although di-n-alkyl aryl phosphonites react readily with alkyl halides, the *iso* esters exhibit a tendency to yield the free acids, rather than the expected alkyl phosphinates.[2, 11, 25] This reaction occurs especially when the reactants are heated and the resulting phosphinate esters break down to the free acid and the corresponding olefin. This difficulty is avoided if the reactants are mixed at room temperature. The addition of a trace of dimethylaniline serves to catalyze the normal reaction to a remarkable degree.[11]

No instance of the preparation of a phosphinate by the alkylation of the sodium salt of a phosphonite has been reported.

[24] Chelintsev and Kuskov, *J. Gen. Chem. U.S.S.R.*, **16**, 1481 (1946) [*C. A.*, **41**, 5441 (1947)].

[25] Arbuzov, *J. Gen. Chem. U.S.S.R.*, **4**, 898 (1934) [*C. A.*, **29**, 2146 (1935)].

[26] Arbuzov and Razumov, *Bull. acad. sci. U.R.S.S., classe sci. chim.*, **1945**, 167 [*C. A.*, **40**, 3411 (1946)].

[27] Arbuzov and Arbuzova, *J. Russ. Phys. Chem. Soc.*, **61**, 1905 (1929) [*C. A.*, **24**, 5289 (1930)].

Selection of Experimental Conditions and Procedures

Although it is impossible to give the specific conditions in which the use of either the neutral ester or the salt of a dialkyl acid phosphite is to be preferred, it may be said that the ester variant yields the normally expected products when the reaction can be made to take place at all. The salt variant tends to give abnormal results in the instances discussed above; in addition, if the sodium halide formed in the reaction is not removed before the distillation of the product, the phosphonate may react with the sodium halide (p. 278) to give a mixed ester-salt, with elimination of a molecule of alkyl halide.[28]

The removal of sodium halide can be effected in a number of ways. Filtration is practicable at times, although one frequently encounters a non-filterable dispersion of the salt which either runs through the filter or clogs its pores. In such cases the use of a centrifuge is indicated. If the precipitate does not separate cleanly in the process, the addition of a small amount of water to the mixture generally aids the coagulation. If the phosphonate ester is not appreciably soluble in water and is not rapidly hydrolyzed by it, it is possible to wash the sodium halide out of the mixture with cold water. It may be pointed out that the use of alkyl phosphites with three or four carbon atoms in each alkyl group suffices to bring about this condition of water insolubility. The esters made from the higher phosphites are essentially insoluble in water and have but little tendency to hydrolyze at room temperature.

The choice of the size of alkyl groups in the phosphite derivative is conditioned chiefly by the choice of either the neutral ester or the sodium dialkyl phosphite as the reagent. In the trialkyl phosphite variant, increasing size of the alkyl groups decreases the reactivity of the phosphite and requires progressively higher operating temperatures.[13] In the salt variant, the reagents do not show any particularly noticeable difference in reactivity when higher alkyl phosphites are substituted for diethyl phosphite. The use of dibutyl phosphite may be recommended because its sodium salt is readily soluble in hydrocarbon solvents such as petroleum ether and benzene, whereas sodium diethyl phosphite has only a limited solubility in such solvents, particularly at lower temperatures.[5] The resulting tendency to crystallize on cooling may be troublesome in reactions in which cooling is necessary, because of the vigor of the interaction of the phosphite reagent with the halide, as in the reaction with ethyl chloroacetate.

The ester variant of the reaction is generally carried out by heating

[28] Abramov, Sergeeva, and Chelpanova, *J. Gen. Chem. U.S.S.R.*, **14**, 1030 (1944) [*C. A.*, **41**, 700 (1947)].

the reactants to the necessary temperature until the reaction is complete. When low-boiling materials are used, sealed tubes or autoclaves are advisable, although there is insufficient evidence at hand that sealed vessels are necessary for many of the preparations so described in the older literature. The mixtures resulting from the reactions are usually subjected to fractional distillation to isolate the products, which, in turn, are readily converted to the free acids by hydrolysis with acids or bases. It is generally advantageous to distil the generated alkyl halide as it is formed in hydrolysis with hydrochloric or hydrobromic acids. It is decidedly advantageous to distil the alkyl halide generated during the isomerization reaction itself; this serves to suppress the side reaction that may result from its interaction with the as yet unreacted phosphite ester (see p. 276). For this reason, the use of apparatus suitable for slow distillation is recommended for many of the preparations.[14, 29]

The sodium salt reaction is carried out generally by heating a solution of the halogen derivative with an equimolar amount of the sodium dialkyl phosphite in an inert solvent until the precipitation of sodium halide is complete. The latter is then removed by filtering, centrifuging, or washing with water, and the product is isolated by fractional distillation. The ester can be converted to the free acid by acidic or alkaline hydrolysis.

Non-distillable esters, principally those of high molecular weight, may be hydrolyzed directly without purification since the resulting phosphonic or phosphinic acids are readily separable from the crude hydrolyzates by virtue of their alkali solubility. This procedure is frequently satisfactory because the isomerization reaction gives very good yields, often approaching the theoretical.

Most of the work on this reaction has been done with alkyl phosphites, which lead to esters of phosphonic acids. The examples of the use of alkyl phosphonites have been relatively few, principally because of the lack of simple syntheses for these esters.

The hydrolysis of the phosphonic esters to the free acids is readily performed by boiling hydrochloric or hydrobromic acids. Although the older publications favor the use of sealed tubes for such hydrolyses, in which dilute hydrochloric acid was generally used at 130–150°, the present author has found that the hydrolyses can be readily done in excellent yields by refluxing with the concentrated acids at atmospheric pressure. If the ester is resistant to hydrochloric acid, the use of 48% hydrobromic acid serves to accomplish the desired result in a few hours. The notable exceptions to the normal hydrolyses are phosphonates in which the phosphono group is adjacent to a carbonyl or a carboxyl

[29] Kosolapoff, *J. Am. Chem. Soc.*, **66**, 109 (1944).

group; hydrolyses of esters with such structures lead to complete dephosphonation under any conditions. Similarly, acidic hydrolysis of diethyl benzyloxymethanephosphonate leads not only to de-esterification but also to the cleavage of the ether bridge to yield hydroxymethanephosphonic acid.[28] Although the use of an alkaline hydrolytic agent is not reported in this instance, the use of 10% sodium hydroxide solution at 150–160° in a sealed tube led to a smooth de-esterification of an analogous diethyl 2-phenoxyethanephosphonate.[30]

Experimental Procedures

THE TRIALKYL PHOSPHITE PROCEDURE *

Diethyl Ethanephosphonate.[14] A mixture of 50 g. of triethyl phosphite and 46.8 g. of ethyl iodide is refluxed for four hours. Distillation of the mixture gives 48 g. (95%) of diethyl ethanephosphonate, b.p. 62°/2 mm.

1-Naphthylmethanephosphonic Acid.[31] A mixture of 43 g. of 1-chloromethylnaphthalene and 41 g. of triethyl phosphite is heated for four hours at 150–160°. Distillation of the mixture gives 58 g. (87%) of diethyl 1-naphthylmethanephosphonate, b.p. 205–206°/5 mm. The ester is refluxed for eight hours with 200 ml. of concentrated hydrochloric acid, and the precipitated 1-naphthylmethanephosphonic acid is filtered from the cooled mixture. After recrystallization from hot water, the pure acid, m.p. 212–212.5°, is obtained in the form of small lustrous plates (90% yield).

Diethyl α-Oxo-α-toluenephosphonate.[21] To 13.7 g. of benzoyl chloride contained in a flask equipped with a dropping funnel and a reflux con-

---

* There is but one practical method of preparation of trialkyl phosphites: the addition of 1 mole of phosphorus trichloride to a solution of 3 moles of the appropriate alcohol in an inert solvent in the presence of 3 moles of a tertiary amine.

$$PCl_3 + 3ROH + 3B = P(OR)_3 + 3B \cdot HCl$$

The principle of the reaction, laid down by Milobendzki and Sachnowski, *Chem. Polsk.*, **15**, 34 (1917) [*C. A.*, **13**, 2865 (1919)], has not been changed by later investigators.

The reaction is best conducted with cooling, at 10° to 15°, in the presence of dry pyridine or diethylaniline. Diethylaniline is somewhat better than dimethylaniline, since its hydrochloride is less hygroscopic. The hygroscopicity of the hydrochloride and the difficulty of obtaining the completely anhydrous base make pyridine less desirable than the dialkylanilines. The solvent may be ether, benzene, or the lower kerosene fractions. The last are best from the standpoint of clean-cut removal of the base hydrochloride; ether and benzene retain appreciable amounts of the latter.

After filtration of the hydrochloride, the solution is distilled under reduced pressure to yield the phosphite in conversions which usually range from 80% to 95%.

[30] Mikhailova, *Uchenye Zapiski Kazan. Gosudarst. Univ.*, **2**, 58 (1941) [*C. A.*, **40**, 555 (1946)].

[31] Kosolapoff, *J. Am. Chem. Soc.*, **67**, 2259 (1945).

denser protected by a calcium chloride tube, there is added in the course of thirty minutes 16.2 g. of triethyl phosphite at room temperature. The solution turns yellow-green and begins to evolve ethyl chloride. The mixture is heated on a steam bath for forty-five minutes and is distilled in vacuum to yield 15.7 g. (66.5%) of diethyl α-oxo-α-toluenephosphonate, as a yellowish liquid, b.p. 141°/2.5 mm.

3-Bromopropane-1-phosphonic Acid.[32] A mixture of 16.6 g. of triethyl phosphite and 101 g. of trimethylene bromide is placed in a flask equipped with a 12-in. Vigreux column. The mixture is heated by means of an oil bath kept at 150°, and ethyl bromide is collected in a graduated receiver. When 8.0 ml. of ethyl bromide is collected (approximately eighty minutes is required), the oil bath is removed and 100 ml. of 48% hydrobromic acid is added to the cooled reaction mixture. Heating is resumed, after the addition of boiling chips to reduce bumping, and the excess trimethylene bromide is distilled, along with ethyl bromide and hydrobromic acid, in the course of four hours. The distillation is continued until the solution in the reaction flask is concentrated to approximately 30 ml. The residual solution is poured into a beaker and evaporation is continued by means of an infra-red lamp until constant weight is attained. The dark gum is chilled in an ice-water mixture and rubbed vigorously until crystallization occurs. The product is sucked dry on a fritted-glass filter, dissolved in a small amount of warm water, decolorized with 0.5 g. of activated charcoal, filtered, and concentrated on a steam bath until crystallization begins. After cooling, filtering, and drying in a vacuum desiccator, 3-bromopropane-1-phosphonic acid, m.p. 107–108°, is obtained in 80–90% yield.

Di-β-chloroethyl β-Chloroethanephosphonate (Intramolecular Isomerization).[33] In a three-necked flask, equipped with a gas inlet tube, a stirrer, and a calcium chloride tube, there is placed 137.5 g. of phosphorus trichloride. Ethylene oxide is passed into the flask with vigorous stirring and effective cooling by means of an ice bath. The temperature of the solution is kept below 10–15°. The reaction is highly exothermic, but it may be kept under precise control by regulation of the rate of addition of ethylene oxide. When the temperature of the mixture no longer tends to rise (after somewhat more than 132 g. of ethylene oxide has been absorbed) the ethylene oxide supply is disconnected, the gas inlet tube is replaced with a stopper, and the mixture is allowed to stand overnight at room temperature without stirring.

The solution is then warmed with stirring to expel any residual eth-

---

[32] Kosolapoff, *J. Am. Chem. Soc.*, **66**, 1511 (1944).
[33] Kabachnik and Rossiiskaya, *Bull. acad. sci. U.R.S.S., classe sci. chim.*, **1946**, 403 [*C. A.*, **42**, 7242 (1948)].

ylene oxide. The steam bath is replaced by an oil bath, and the mixture is slowly heated with stirring to 150–160°. The ensuing isomerization reaction is rather exothermic and careful control of temperature is necessary. The temperature of the solution should not be allowed to rise above 165–170°, for secondary reactions begin to take place at higher temperatures. Heating is continued for five hours, after which the drying tube is replaced by a distillation head and the mixture is distilled in vacuum. The distillate is redistilled, and the fraction boiling at 170–172°/5 mm. is collected as bis-β-chloroethyl β-chloroethanephosphonate. The yield is generally over 40% (110 g. or more). If the temperature prescribed is closely followed, yields in excess of 70% are common. The product may be induced to crystallize by cooling and scratching. It forms colorless crystals, m.p. 37°.

It is possible to isolate the intermediate tris-β-chloroethyl phosphite, after the reaction mixture has been allowed to stand overnight, by distilling it at a pressure of not more than 2–3 mm. Under conditions of rapid distillation it is possible to recover the phosphite as a mobile liquid, b.p. 112–112.5°/2.5 mm. However, the ester tends to isomerize during the distillation, and accurate fractionation is impossible. The yields of the phosphite are variable, because of the isomerization, but it is possible to obtain 30–40% yields of rather pure product. In this connection it is interesting to note the patent disclosure of the addition of 3 moles of ethylene oxide to phosphorus trichloride under conditions similar to those given above. The product, described as tris-β-chloroethyl phosphite, is stated to boil at 50°/12 mm. and no mention is made of the occurrence of isomerization.[34]

Tetraphenyl Ethane-1,2-diphosphonate.[16] A flask equipped with a calcium chloride tube is charged with 3.6 g. of diphenyl 2-chloroethyl phosphite. After being heated to 250° for three and a half hours the mass is allowed to cool. Recrystallization from toluene gives 2.1 g. (60%) of tetraphenyl ethane-1,2-diphosphonate as colorless needles, m.p. 155–155.5°.

Hydrolysis may be effected by heating 0.5 g. of the ester and 10 ml. of 1:1 hydrochloric acid in a sealed tube for eight hours at 130°, then for thirty minutes at 140°. On cooling, the mixture is freed of phenol by extraction with ether and the aqueous layer is evaporated to dryness. Recrystallization of the residual solid from acetic acid yields 0.15 g. (90%) of ethane-1,2-diphosphonic acid, m.p. 220–221°.

Isopropyl Isopropylphenylphosphinate.[11] A mixture of 12 g. of diisopropyl benzenephosphonite and 9 g. of isopropyl iodide is allowed to stand for ten days in a closed vessel. Distillation of the mixture yields

[34] I.G. Farbenindustrie A.G., U. S. pat. 1,936,985 [*C. A.*, **28**, 1151 (1934)].

5.3 g. (44%) of isopropyl isopropylphenylphosphinate, b.p. 145–146°/10 mm. However, the addition of a drop of dimethylaniline to the original mixture catalyzes the isomerization to such an extent that after only two days' standing the yield is 95%.

The Sodium Salt Procedure

Triethyl β-Phosphonopropionate.[35] To 68 g. of dry sodium ethoxide in 500 ml. of dry xylene is added with stirring 138 g. of diethyl phosphite, the mixture being protected from moisture by a calcium chloride tube. To the resulting salt 181 g. of ethyl β-bromopropionate is added dropwise with stirring and cooling by an ice-salt bath. After standing overnight the mixture is heated for two hours on a steam bath, after which the precipitated sodium chloride is filtered. Distillation of the filtrate gives 193 g. (78%) of triethyl β-phosphonopropionate, b.p. 141–143°/9 mm.

Dibutyl Alkanephosphonates.[5] One-tenth mole of dibutyl phosphite is added dropwise to a suspension of 0.1 atom of sodium in 300–500 ml. of a dry hydrocarbon solvent (petroleum ether, benzene, toluene, or xylene), with stirring and heating at gentle reflux until the sodium dissolves. The alkyl halide (bromides are most satisfactory) is then added dropwise during thirty to sixty minutes. The amount of the halide need not exceed the theoretical 0.1 mole. After fifteen or twenty minutes the precipitation of sodium halide begins. It is completed by refluxing the mixture with stirring for two to six hours. The end of the reaction is indicated by a clean separation of the salt from the organic solution. On cooling, the mixture is shaken with two or three portions of cold water and the organic layer is run through a dry filter paper to remove the bulk of moisture. The filtrate is then freed of solvent at water-pump vacuum at approximately room temperature. This also serves to remove the residual moisture without an additional drying step. Distillation of the residue under reduced pressure (oil-pump vacuum for the higher members of the series) results in the isolation of 80–95% yields of dibutyl alkanephosphonates as colorless liquids. These may be hydrolyzed by refluxing with 2–3 volumes of concentrated hydrochloric acid. This is most satisfactorily done in a flask provided with a Vigreux distillation column which permits the continuous removal of butyl chloride. When the latter is completely removed, as indicated by the temperature of the condensing vapor in the still head, the bulk of the hydrochloric acid is distilled and the phosphonic acid is allowed to crystallize on cooling the residual mixture. Purification by crystal-

[35] Finkelstein, *J. Am. Chem. Soc.*, **68**, 2397 (1946).

lization from petroleum ether gives substantially quantitative yields of the alkanephosphonic acids.

TRIARYLMETHANE DERIVATIVES

Triphenylmethanephosphonic Acid.[8] A solution of 42.5 g. of triphenylcarbinol in boiling benzene is added in two or three portions to 50 g. of phosphorus trichloride contained in a flask provided with a reflux condenser and a calcium chloride tube for protection from moisture. The reaction is conducted at reflux temperature, and the additions are timed so that uncontrollable reflux is avoided. After the addition, the mixture is refluxed for one hour, the solvent is removed in vacuum, and the solid residue of triphenylmethylphosphonyl chloride is washed with dry ether and dried in a vacuum desiccator. The product is obtained in 95% yield in the form of colorless crystals, m.p. 189.5–190°.

Five grams of the above chloride is heated on a steam bath with 3.8 g. of potassium hydroxide in 38 ml. of ethanol until the precipitation of potassium chloride is complete. An equal volume of water is added to the mixture, and the ethanol is almost completely removed by evaporation. The cooled solution is filtered, and the clear filtrate is acidified with hydrochloric acid to precipitate the monoethyl ester of the phosphonic acid. This is separated, dried, and refluxed for one hour with 25 ml. of acetic acid and 12 ml. of hydriodic acid. On cooling, the product is filtered, washed with dilute hydrochloric acid, ethanol, and ether, in succession, and recrystallized from benzene to give 4–4.1 g. (91%) of triphenylmethanephosphonic acid, m.p. 275°.

SPECIAL METHODS

Diethyl 2-Hydroxyethanephosphonate.[24] To 2.3 g. of powdered sodium suspended in 120 ml. of dry ether is added with stirring 13.9 g. of diethyl phosphite. The mixture is stirred with gentle warming until the sodium has reacted, and the mixture is treated with 4.5 g. of ethylene oxide with stirring. The clear solution is stirred for one hour, and then 6.1 g. of glacial acetic acid is added dropwise. The precipitated sodium acetate is collected by filtration, and the filtrate is evaporated under reduced pressure. Traces of sodium acetate are removed by filtration, and the residual oil is dried in a desiccator over sulfuric acid. There is obtained 7.6 g. (42%) of diethyl 2-hydroxyethanephosphonate, which can be distilled with some decomposition at 120–130°/9 mm.

Stearamidomethanephosphonic Acid.[7] One hundred grams of N-methylolstearamide is added to a solution of 91.0 g. of phosphorus trichloride in 45 g. of carbon tetrachloride contained in a flask protected with a calcium chloride tube. After standing for one hour, the mixture is treated with 40 g. of glacial acetic acid, and the flask is allowed to stand at room temperature for four days. The resulting viscous mass is warmed to 50° with 8% hydrochloric acid until it changes to a crystalline solid, which is separated by filtration. Crystallization from ethanol yields 67 g. (40%) of stearamidomethanephosphonic acid, a colorless crystalline solid, which has an indefinite melting point, softening at 108°.

ADDITION OF PHOSPHORUS PENTACHLORIDE TO UNSATURATED COMPOUNDS

Olefins having reactive double bonds undergo the addition of phosphorus pentachloride to give substances that can be regarded as the chlorides of phosphonic acids.[36] Hydrolysis converts the addition products to phosphonic acids, usually with simultaneous dehydrochlorination as illustrated in the reaction with styrene.

$$C_6H_5CH{=}CH_2 + PCl_5 \rightarrow C_6H_5CHClCH_2PCl_4 \xrightarrow{3H_2O}$$

$$C_6H_5CH{=}CHPO(OH)_2 + 5HCl$$

Branched-chain olefins sometimes lead to chloroalkanephosphonic acids. Acetylenes yield phosphonic acids containing the chlorovinyl group. Such compounds, which do not undergo spontaneous loss of hydrogen chloride during hydrolysis, can be dehydrohalogenated by treatment with an alkaline reagent like potassium hydroxide.[37] The initial addition reaction takes place under mild conditions in an inert solvent, and the yields of α,β-unsaturated phosphonic acids usually range between 40 and 50%.

Scope and Limitations

The most obvious limitation to the reaction is the fact that groups capable of reacting with phosphorus pentachloride must either be absent or be protected. Such reactive groups are the hydroxyl, amino, sulfhydryl, and carboxyl.

The reaction has been successfully applied to the following unsaturated compounds: styrene,[36,38,39] α-methylstyrene,[38] α-chlorostyrene,[37] in-

---

[36] Thiele, *Chem. Ztg.*, **36**, 657 (1912); K. Harnist, dissertation, Strassburg, 1910; F. Bulle, dissertation, Strassburg, 1912.
[37] Bergmann and Bondi, *Ber.*, **66**, 278 (1933).
[38] Bergmann and Bondi, *Ber.*, **63**, 1158 (1930).
[39] Kosolapoff and Huber, *J. Am. Chem. Soc.*, **68**, 2540 (1946).

dene,[36,38] 1,1-diphenylethylene,[38] 1-phenyl-1-o-tolylethylene,[40] 1-phenyl-1-p-methoxyphenylethylene,[41] 1-phenyl-1-p-chlorophenylethylene,[41] 1,1-di-p-chlorophenylethylene,[41] 1-p-chlorophenyl-1-p-methoxyphenylethylene,[41] 1-phenyl-1-o-fluorophenylethylene,[41] 1-phenyl-1-p-biphenylylethylene,[41] 1-p-tolyl-1-p-biphenylylethylene,[41] 1-phenyl-1-m-chlorophenylethylene,[41] 1-phenyl-1-(α- and β-)-naphthylethylenes,[40] 1,4-phenylenebis(α-phenylethylene),[41] 4-phenyl-1,3-butadiene,[39,41] isobutylene,[39] butadiene,[42] 10,10-diphenyl-9-methyleneanthracene,[40] 2,4-dimethylstyrene,[39] 2,4,6-trimethylstyrene,[39] p-ethylstyrene,[39] o-tert-butylstyrene,[39] p-tert-butylstyrene,[39] o-, m-, and p-phenylstyrene,[39] 2-vinylnaphthalene,[39] 2-vinylfluorene,[39] phenylacetylene,[37] o-chlorophenylacetylene,[37] o-methoxyphenylacetylene,[37] p-methoxyphenylacetylene,[37] phenylmethylacetylene,[37] and 1-heptyne.[37]

Although the earlier work [36,38] indicated that lack of symmetry of the starting compound is a necessary condition for this reaction, it has been shown that symmetrical compounds may be capable of normal addition.[41,42] It appears, however, that the reaction is limited to unsaturated compounds that contain a terminal unsaturated carbon-carbon bond. Indene, which has a cyclic double bond, is an exception and appears to be reactive mainly because of the exposed, unhindered position of this double bond.

The steric factors that may further limit the reaction have not been satisfactorily clarified to permit any generalizations. The following compounds failed to yield phosphonic acids, although many of them have a double bond which is not apparently blocked by steric factors: benzylstilbene,[38] 1,1-diphenyl-1-propene,[41] 1,1-diphenyl-1-butene,[41] 1,1,3-triphenyl-1-propene,[41] 1,1-diphenyl-2,2-dimethylethylene,[41] α-benzylstyrene,[41] 1-phenyl-1-ethyl-2-methylethylene,[41] allylbenzene,[41] 1-phenyl-1,3-pentadiene,[41] 1,4-diphenylbutadiene,[41] isoeugenol methyl ether,[41] isosafrole,[41] triphenylethylene,[41] stilbene,[41] isostilbene,[41] 1,1-bis-o-methoxyphenylethylene,[41] 1-phenyl-1-(o-chloro- and o-bromo-phenyl)ethylenes,[41] 1-phenyl-1-o-biphenylylethylene,[41] 1,1-phenyl-(o- and m-methoxyphenyl)ethylenes,[41] tolan,[37] diphenylacetylene,[37] p-nitrophenylacetylene,[37] and phenylethylacetylene.[37]

Experimental Procedures

β-Styrenephosphonic Acid. (a) [38] To an ice-cooled stirred suspension of 104 g. of phosphorus pentachloride in dry benzene, 26.2 g. of styrene

---

[40] Bergmann and Bondi, *Ber.*, **66**, 286 (1933).
[41] Bergmann and Bondi, *Ber.*, **64**, 1455 (1931).
[42] Kosolapoff, U. S. pat. 2,389,576 [*C. A.*, **40**, 1536 (1946)].

is added dropwise in one hour. The mixture is protected from moisture by means of a calcium chloride tube. After stirring for two or three hours, the creamy suspension of the adduct is allowed to stand for twenty-four hours, after which it is poured into ice water. The mixture is allowed to stand for two or three days, with spontaneous evaporation of benzene. Shiny colorless crystals of the product gradually appear at the interface of the two layers. The yield of the crude product is 27 g. It is a mixture of *cis-trans* isomers and is composed of 3 g. of needles, m.p. 146°, and 24 g. of a granular solid, m.p. 150°. These can be readily separated mechanically. Recrystallization from ethylene bromide gives the same product from either isomer. The final product is obtained in the form of needles, m.p. 146°, and represents the stable isomer.

(b) [39] Dry chlorine gas is introduced slowly into an ice-cold stirred solution of 52.1 g. of styrene in 68.7 g. of phosphorus trichloride and 500 ml. of dry benzene until the solution becomes yellow from an excess of chlorine. Hydrolysis of the mixture as described in (a) results in 32.9 g. of crude 2′-styrenephosphonic acid. This is purified by dissolving it in dilute sodium hydroxide and pouring the solution slowly into warm, stirred, dilute hydrochloric acid. Crystallization of the precipitate from water gives 29–31 g. of the pure acid, m.p. 154.5–155.5°.

Phenylethynephosphonic Acid.[37] To an ice-cold, stirred suspension of 83 g. of phosphorus pentachloride in 150 ml. of dry benzene is added slowly 20.4 g. of phenylacetylene. After standing for two days, the mixture is poured into an ice-water mixture, the organic layer is diluted with ether, and the aqueous layer is discarded. Evaporation of the organic layer gives 1.5 g. (3%) of α-chloro-β-styrenephosphonic acid, m.p. 162° (from 1:1 hydrochloric acid). Five grams of this acid is refluxed for six hours with 80 ml. of 5% potassium hydroxide. On cooling, the mixture is treated with an excess of hydrochloric acid and the product is taken up in ether. Evaporation of the solvent gives a substantially quantitative conversion to the phenylethynephosphonic acid, m.p. 142°.

THE GRIGNARD REACTION

The application of the usually versatile Grignard reaction to the synthesis of phosphonic and phosphinic acids has not received the attention it probably deserves. References to its use in this connection are few, and the precise conditions for optimum yields have not been explored adequately.

The obvious advantage of the Grignard reaction resides in the mild conditions necessary for its use. The method favors the preservation of sensitive substituents, which might be destroyed in a more drastic reaction such as a Friedel-Crafts synthesis.

The earliest reference to the formation of phosphonic or phosphinic acids by the Grignard method is that of Auger and Billy,[43] who added methylmagnesium bromide to an excess of phosphorus trichloride at $-30°$ and hydrolyzed the resulting mixture of trimethylphosphine and methylchlorophosphines; oxidation of the crude mixture with nitric acid resulted in isolation of traces of methanephosphonic acid and dimethylphosphinic acid. Sauvage[44] treated phosphorus oxychloride with one-third of the molar quantity of Grignard reagents from bromobenzene, benzyl chloride, and 1-bromonaphthalene. The reaction products consisted of mixtures of, predominantly, triarylphosphine oxides and small amounts of the corresponding phosphinic acids, i.e., dibenzylphosphinic acid, diphenylphosphinic acid, and di-1-naphthylphosphinic acid.

Michaelis and Wegner[45] made the first step toward control of the reaction by using substituted phosphorus oxychlorides, thus leaving only two available chlorine atoms for the reaction. They found, however, that blocking by a phenoxyl group was ineffective; the use of phenoxyphosphoryl dichloride gave mostly the trisubstituted phosphine oxides in reactions with Grignard reagents. In other words, the Grignard reagents displace the phenoxyl group as readily as they displace the chlorine atoms in the phosphoryl chloride derivative. Unfortunately, the ease of such displacement has not been investigated for radicals other than phenyl. In the light of modern knowledge of the behavior of phosphate esters, such displacement of the phenoxyl group may be connected with the ready cleavage of phenyl phosphates by hydrogenation. The reductive action of the Grignard reagents used may well have been responsible for this failure of the Michaelis-Wegner attempt. If this explanation is correct, attempts to effect blocking by means of the benzyloxy groups should be fruitless for the same reason.

However, Michaelis and Wegner found a more suitable reagent in N-piperidylphosphonyl dichloride. The piperidine residue resisted attack by the Grignard reagents, which therefore could react with only the two available chlorine atoms. The resulting N-piperidides of diarylphosphinic acids were readily hydrolyzed by hydrochloric acid to the corresponding free acids. The reaction sequence is shown in the accompanying equation.

$$C_5H_{10}NP(O)Cl_2 + 2ArMgBr \rightarrow C_5H_{10}NP(O)Ar_2 \xrightarrow{HCl} Ar_2P(O)OH$$

[43] Auger and Billy, *Compt. rend.*, **139**, 597 (1904).
[44] Sauvage, *Compt. rend.*, **139**, 674 (1904).
[45] Michaelis and Wegner, *Ber.*, **48**, 316 (1915).

These authors applied this method to the Grignard reagents from bromobenzene, o- and p-bromotoluene, 1-bromonaphthalene, and benzyl chloride. Although the yields are stated to be "good," no numerical data are given. However, yields in excess of 50% may be expected.

There is no recorded instance of an attempt to extend such a blocking modification of phosphoryl chloride in order to synthesize phosphonic acids. This would require a doubly blocked reagent of a type B_2POCl, where B is a blocking group.

The main difficulty with the use of the Grignard reaction results from the tendency for complete substitution of phosphorus oxychloride. Therefore, an attempt was made by the present author to counteract this tendency by reversing the mode of mixing the reactants, that is, by adding the Grignard reagent to a moderate excess of phosphorus oxychloride solution. This method of addition, combined with the additional favorable factor of very dilute solutions, gave 50–55% yields of phosphinic acids from phenyl- and p-chlorophenyl-magnesium bromides.[46]

Mingoia[47] used the magnesium derivatives of α-methylindole, indole, and pyrrole in a reaction analogous to that of Sauvage to obtain low yields of di-3-(2-methylindolyl)phosphinic acid, di-3-indolylphosphinic acid, and di-2-pyrrylphosphinic acid.

A modification of the blocking procedure of Michaelis and Wegner has been reported in the work of Bode and Bach,[48] who treated phosphonitrilic chloride, $(PNCl_2)_3$, with a large excess of phenylmagnesium bromide and hydrolyzed the resulting product, $(C_6H_5)_7P_3N_3H \cdot HBr$, with hydrochloric acid to the diphenylphosphinic acid. The yield of the intermediate product was less than 10% and, although the hydrolysis step is essentially quantitative, the overall yields do not compare with those from the Michaelis-Wegner method. The tedious preparation of the phosphonitrilic chloride is an additional drawback to this procedure.

An entirely different approach was made by Malatesta and Pizzotti,[49] who treated phosphorus pentasulfide * with Grignard reagents from ethyl bromide, isopropyl bromide, and bromobenzene, and obtained mixtures of the corresponding tertiary phosphine sulfides, thiophosphinic acids, and thiophosphonic acids. The thio acids were readily oxidized to the oxygen analogs by treatment with nitric acid or bromine. This procedure appears to be the first reasonably practical method of preparation of phosphonic acids by the Grignard reaction. As mentioned above, the reaction gave the products of all three possible types. The course

* See the footnote on p. 412 concerning the formulas of the phosphorus sulfides.
[46] Kosolapoff, *J. Am. Chem. Soc.*, **64**, 2982 (1942).
[47] Mingoia, *Gazz. chim. ital.*, **62**, 333 (1932).
[48] Bode and Bach, *Ber.*, **75**, 215 (1942).
[49] Malatesta and Pizzotti, *Gazz. chim. ital.*, **76**, 167, 182 (1946).

of the reaction may be represented by the three equations given by Malatesta and Pizzotti.

$$P_2S_5 + 4RMgBr \rightarrow 2R_2P(S)SMgBr + MgS + MgBr_2 \quad (1)$$

$$P_2S_5 + 6RMgBr \rightarrow 2R_3PS + 3MgBr_2 + 3MgS \quad (2)$$

$$P_2S_5 + 2RMgBr \rightarrow BrMgS(S)(R)PSP(R)(S)SMgBr \quad (3)$$

Although reaction 1 takes place best at moderately low temperatures, all three reactions always take place and the yields of the acidic derivatives do not exceed 20% for any class under the best conditions. The mixtures of the thiophosphonic and thiophosphinic acids were separated by virtue of the different solubility of the nickel salts of the sulfur and oxygen acids. The reaction is best carried out with a suspension of phosphorus pentasulfide in an inert solvent, usually ether. The heterogeneous character of the reaction under such conditions may be responsible to a large extent for the difficulty of the control of the reaction.

The information given above includes all the pertinent data on the use of the Grignard reaction. It is readily seen that the scope of the reaction cannot be limited to the few examples that have been tried to date. Probably the reaction can be used with any substance capable of forming a Grignard reagent.

Experimental Procedures

Diphenylphosphinic Acid (Michaelis-Wegner Procedure).[45] The Grignard reagent from 31.4 g. of bromobenzene and 5 g. of magnesium, in ether solution, is treated slowly with 20.2 g. of N-piperidylphosphoryl dichloride. The mixture is refluxed until reaction is complete. After addition to water, the organic layer is separated. Evaporation of the solvent on a steam bath leaves a viscous residue of the amide, which is boiled with concentrated hydrochloric acid until solution is complete. Dilution with cold water causes the separation of the crude diphenylphosphinic acid. Purification by solution in sodium carbonate solution, followed by precipitation with hydrochloric acid and crystallization from ethanol, gives the pure compound, m.p. 190–191°. The yield is reported as "good."

Diphenylphosphinic Acid (Kosolapoff Procedure).[46] The Grignard reagent from 31.4 g. of bromobenzene and 4.86 g. of magnesium in 500 ml. of dry ether is filtered with exclusion of atmospheric moisture and is then added during three and a half hours to a gently refluxing, stirred solution of 30.6 g. of phosphorus oxychloride in 500 ml. of dry ether.

After standing overnight, the clear solution is decanted from the yellow precipitate. The precipitate is digested with ice water, and the insoluble residue is washed thoroughly with water. Extraction of the solid with 1 l. of warm dilute sodium hydroxide solution and acidification of the extract with hydrochloric acid, followed by crystallization from dilute ethanol, gives 12 g. (55%) of diphenylphosphinic acid, m.p. 190–192°.

Ethanephosphonic Acid (Malatesta-Pizzotti Procedure).[49] Two hundred milliliters of 1 M ethylmagnesium bromide in dry ether solution is added dropwise to a stirred suspension of 22 g. of phosphorus pentasulfide in dry ether. The mixture is refluxed for a brief time after the addition is complete. Cold water is added to the mixture, and the aqueous layer is separated. After treatment with charcoal, followed by filtration, an excess of nickel sulfate solution is added and the mixture is acidified to Congo red with dilute hydrochloric acid. Extraction with benzene removes any nickel diethyldithiophosphinate. The aqueous solution is extracted with ether, the extract is evaporated to dryness, and the residue is taken up in water. Bromine water is added to oxidize the sulfur compound to the corresponding oxygen analog. The addition is continued until a permanent color is attained. After filtration and evaporation, the residue is dissolved in dilute aqueous ammonia. Evaporation to dryness to remove the excess ammonia, followed by treatment of the residue with hydrogen sulfide in aqueous solution, serves to remove any residual nickel. Acidification of the filtrate with nitric acid, after the removal of nickel sulfide, and evaporation to dryness give crude ethanephosphonic acid. This is distilled under reduced pressure to give the pure product, b.p. 330–340°/8 mm.; m.p. 30–35°. The yield is approximately 15%.

THE FRIEDEL-CRAFTS REACTION

The preparation of aromatic dichlorophosphines by the interaction of aromatic hydrocarbons with phosphorus trichloride in the presence of aluminum chloride was accomplished for the first time by Michaelis.[50] This reaction, which takes place according to the accompanying equation, was subsequently used for the conversion of aromatic hydrocarbons into a variety of phosphonic acids. The conversion of the dichloro-

$$ArH + PCl_3 \rightarrow ArPCl_2 + HCl$$

phosphines into the phosphonic acids was effected by chlorination, which yields the corresponding tetrachlorides, followed by hydrolysis.

$$ArPCl_2 \xrightarrow{Cl_2} ArPCl_4 \xrightarrow{H_2O} ArP(O)(OH)_2$$

[50] Michaelis, *Ber.*, **12**, 1009 (1879).

Instead of the direct hydrolysis of the tetrachlorides, the latter can be converted to the corresponding oxychlorides, which on hydrolysis also give phosphonic acids. Usually there is little choice between the two alternatives.

$$RPCl_4 \xrightarrow{SO_2} RPOCl_2 \xrightarrow{H_2O} RP(O)(OH)_2$$

A number of the aromatic dichlorophosphines have been prepared by later workers without significant changes of the original procedure of Michaelis. The formation of small amounts of diaryl monochlorophosphines, R_2PCl, in the original reaction mixtures has been also observed for a few compounds.[51-54] These could be isolated in small amounts only, and the Friedel-Crafts reaction was not regarded as a suitable source of the disubstituted products by the Michaelis school. The diaryl monochlorophosphines can be converted to the corresponding diarylphosphinic acids by a reaction sequence analogous to that above. Later work by the present writer [55] indicated that the diaryl chlorophosphines are formed as a result of a general reaction, which is apparently catalyzed by aluminum chloride, and which proceeds through disproportionation of the monoaryl derivatives.

$$2ArPCl_2(AlCl_3) \rightarrow Ar_2PCl + PCl_3$$

The difficulties encountered in the isolation procedure used by the Michaelis school for the chlorophosphines prevented the discovery of the generality of this reaction. The isolation procedure of Michaelis is extremely inefficient. It is performed by extraction of the reaction mixture with an inert hydrocarbon solvent (petroleum ether has been generally favored). The extract is concentrated, and the residual chlorophosphines are distilled under reduced pressure. The bulk of the reaction products, however, remains in the rather intractable evil-smelling aluminum chloride complex layer which is insoluble in petroleum ether. The actual yields of the isolated dichlorophosphines rarely exceed 15–20% of the theoretical. The dichlorophosphines, after isolation, are treated with an equimolar quantity of dry chlorine gas, which may be added in solution in a suitable solvent (carbon tetrachloride has been usually employed) or may be introduced in the gaseous state into the dichlorophosphine, which is preferably dissolved in an inert solvent. The use of a solvent with external cooling moderates the very vigorous reaction. The resulting tetrachlorophosphine may be added directly to water to yield the corresponding phosphonic acid or may be treated

[51] Michaelis, *Ann.*, **315**, 43 (1901).
[52] Michaelis, *Ann.*, **293**, 193 (1896); **294**, 1 (1897).
[53] Sachs, *Ber.*, **25**, 1514 (1893).
[54] Lindner and Strecker, *Monatsh.*, **53/54**, 263 (1929).
[55] Kosolapoff and Huber, *J. Am. Chem. Soc.*, **69**, 2020 (1947).

with gaseous sulfur dioxide which converts it to the oxychloride, $RPOCl_2$, which may be purified by distillation. Treatment with warm water converts the oxychloride to the desired phosphonic acid.

A very useful modification of the Michaelis procedure has been developed by Dye.[56] In this procedure the dichlorophosphines are isolated by the removal of the aluminum chloride in the form of very stable complexes, either with water or with phosphorus oxychloride. In the first instance, the cooled mixture, after the Friedel-Crafts reaction proper has been completed, is treated with cold water added dropwise. The amount of water used is three times the molar amount of aluminum chloride, and it is advisable to remove the excess phosphorus trichloride before the hydrolysis. The resulting solid complex, which contains all the aluminum chloride, is removed, and the filtrate is used for the recovery of the aromatic dichlorophosphine by distillation. In the second variant, the reaction mixture is treated with phosphorus oxychloride, the molar quantity of which is slightly greater than that of the aluminum chloride used in the reaction. The mixture is warmed to approximately 50° with stirring to aid the formation of the $AlCl_3 \cdot POCl_3$ complex, which separates as a solid. The separation is assisted by the addition of petroleum ether to the mixture. After filtration of the mixture, the filtrate is used for isolation of the dichlorophosphines in the usual way. The yields by either procedure have been studied with benzene; consistent values of 60–70% of the theoretical can be attained. There are indications that the procedure can be used for other aromatic hydrocarbons.

A different variation of the Michaelis procedure has been developed by the present writer.[55] In this procedure the chlorophosphines are not isolated, but the entire reaction mixture is treated with chlorine in an inert solvent and the resulting mixture is esterified. Aluminum chloride is then removed by washing with water, and the resulting esters of phosphonic and phosphinic acids are readily recovered and isolated by vacuum distillation. Hydrolysis of the esters yields the corresponding free acids. This procedure not only eliminates the handling and the isolation of malodorous and sensitive chlorophosphines but also serves to produce the phosphonic and the phosphinic acid derivatives in much higher yields than those obtained by the Michaelis method. The yields are frequently nearly theoretical, based on the amount of the aromatic hydrocarbon used. The overall scheme of this method may be illustrated by the accompanying representation.

$$ArH + PCl_3 \xrightarrow{AlCl_3} (ArPCl_2 + Ar_2PCl) \xrightarrow{Cl_2}$$
$$(ArPCl_4 + Ar_2PCl_3) \xrightarrow{ROH} ArP(O)(OR)_2 + Ar_2P(O)OR$$

[56] Dye, *J. Am. Chem. Soc.*, **70**, 2595 (1948).

The use of this procedure, with its excellent recoveries, established that the formation of the disubstituted products is a general reaction, but that it can be essentially suppressed if the reaction period is relatively short (three to eight hours).

A variation of the above-described procedures has been reported by Bode and Bach.[48] They reacted trimeric phosphonitrilic chloride, $(PNCl_2)_3$, with benzene in the presence of aluminum chloride. The resulting diphenyl derivative, $(C_6H_5)_2P_3N_3Cl_4$, was hydrolyzed to diphenylphosphinic acid by heating with water to 150–160° in sealed tubes for twenty-four hours. Since the yield of the intermediate is poor, the significance of this procedure as a synthetic tool appears to be slight.

Scope and Limitations

The Friedel-Crafts reaction has been successfully applied to the preparation of phosphonic and phosphinic acid derivatives of the following aromatic compounds: benzene,[50, 55, 57] chlorobenzene,[52, 55] o-chlorotoluene,[58] bromobenzene,[52] toluene,[50, 51, 52, 54, 55, 59] ethylbenzene,[52] isopropylbenzene,[52] cymene,[52, 59] anisole,[52, 60] phenetole,[52] m- and p-xylene,[52, 59, 61] the trimethylbenzenes,[52, 62] naphthalene,[54] diphenylmethane,[51, 52] sym-diphenylethane,[51, 52] o- and p-dichlorobenzene,[55] biphenyl,[51, 52, 63] diphenyl ether,[64] thiophene,[53] and dimethylaniline.[65] In addition, monoaryl dichlorophosphines were prepared from N,N-diethylaniline, N,N-methylethylaniline, N,N-methylbenzylaniline, and N,N-ethylbenzylaniline,[65] but the products were not converted to the phosphonic acids.

The reaction failed to take place to a detectable extent with trichlorobenzene,[55] benzonitrile,[52] iodobenzene,[52] benzophenone,[52] ethyl benzoate,[52] and x-bromotoluene.[52]

The rather limited number of the compounds listed above cannot be considered as the true scope of the reaction. The reaction can probably be applied to all aromatic compounds that can undergo the acylation-type Friedel-Crafts reaction. However, the reaction has some inherent limitations which restrict its usefulness, particularly in the attempts to obtain compounds with a specific structure. Thus, the work done to

---

[57] Kamai, *J. Russ. Phys. Chem. Soc.*, **64**, 524 (1932) [*C. A.*, **27**, 966 (1933)].
[58] Melchiker, *Ber.*, **31**, 2915 (1898).
[59] Michaelis and Panek, *Ann.*, **212**, 203 (1882).
[60] Kamai, *J. Gen. Chem. U.S.S.R.*, **4**, 192 (1934) [*C. A.*, **29**, 464 (1935)].
[61] Weller, *Ber.*, **20**, 1718 (1887); **21**, 1492 (1888).
[62] Davies, *J. Chem. Soc.*, **1935**, 462.
[63] Lindner, Wirth, and Zannbauer, *Monctsh.*, **70**, 1 (1937).
[64] Davies and Morris, *J. Chem. Soc.*, **1932**, 2880.
[65] Michaelis and Schenk, *Ber.*, **21**, 1497 (1888); *Ann.*, **260**, 1 (1890).

date does not include the study of the possible isomerizations or migrations of the alkyl substituents on the aromatic nucleus. Such changes may be expected to take place in reactions involving the use of aluminum chloride at elevated temperatures. It will be noted that the compounds studied had comparatively short side chains, whose isomerization is rather improbable. The reaction with bromobenzene gives a poor yield of the desired product because of extensive debromination by the aluminum chloride. The identity of the dichlorophosphine obtained by Michaelis [52] from anisole has been seriously questioned by Kamai,[60] who showed that anisole suffers an extensive cleavage of the ether linkage and that the yield of the pure p-methoxy derivative is but 26%. It was also shown that the successful application of the Friedel-Crafts reaction to anisole and to phenetole requires the use of partially hydrated aluminum chloride,[52] because pure aluminum chloride, which is necessary for all the other reactions, yields phenyl dichlorophosphite, $C_6H_5OPCl_2$, instead of the dichlorophosphine.

The phosphorus residue enters the aromatic nucleus in orientations that are normally expected for the compounds that have been tried. Thus, the *para* isomer of the toluene derivative has been isolated and the presence of the *ortho* isomer has been deduced from the low melting point of the dichlorophosphine which remains after the removal of the *para* isomer by freezing.[52] The formation of two isomeric products has been established in the reaction of *meta*-xylene,[52] but the products from chlorobenzene and from the phenyl ethers have been assigned the *para* structure exclusively.[52] It is possible that a closer study of the products, which will be available in good yields as a result of the modifications of the isolation procedure, will reveal the presence of other isomers. Apparently an isomer of unknown structure has been isolated from bromobenzene,[52] besides the authentic *para* isomer. No definite assignment of structure has been given to the derivatives of naphthalene, biphenyl, diphenylmethane, or diphenylethane, although the last three products may be expected to be largely the *para* isomers.

Experimental Procedures

p-Toluenephosphonic Acid (Michaelis Procedure).[50, 52] A mixture of 150 g. of toluene, 200 g. of phosphorus trichloride, and 30 g. of aluminum chloride is refluxed for thirty-six hours with protection from atmospheric moisture. The cooled reaction mixture is mixed with 2 volumes of a hydrocarbon solvent (preferably petroleum ether), and the mixture is allowed to stand in a loosely stoppered separatory funnel until the layers separate cleanly. This may require a day. The extract is separated and transferred carefully to a distillation apparatus, and the mixture of isomeric tolyldichlorophosphines is recovered by distillation under reduced pressure, preferably in an inert atmosphere to prevent oxidation. Approximately 50 g. of the mixture is recovered in the form of a fraction which boils at 236–260° at atmospheric pressure. The *para* isomer may be largely recovered by freezing the mixture, and up to 25 g. of the pure product may be obtained. The liquid fraction is not the pure *ortho* isomer, and attempts to purify it have not been successful. The *para* isomer melts at 25°. The dichlorophosphine is treated with chlorine, either in carbon tetrachloride solution or without dilution,[59] until the absorption of an equimolar amount of chlorine takes place. The resulting tetrachlorophosphine is treated with dry sulfur dioxide until the conversion to the oxychloride is complete as shown by the liquefaction of the solid tetrachloride. The resulting product is treated with ice water and boiled briefly to complete the hydrolysis, and *p*-toluenephosphonic acid is isolated by cooling the solution. After recrystallization from aqueous ethanol the acid melts at 189°. The conversion from the dichlorophosphine to the acid is substantially quantitative.

Small amounts of the crude ditolyl derivatives are left behind after the isolation of the dichlorophosphines. They may be isolated by treating the viscous aluminum chloride complex residue, after the hydrocarbon extraction, with water, separating the semisolid insoluble mass, washing it with water, and extracting it with dilute aqueous ammonia. Acidification of the alkaline extract gives variable amounts of the ditolyl derivatives as a non-crystalline viscous mass.

Benzenephosphonic Acid (Kosolapoff Procedure).[55] A mixture of 78 g. (1 mole) of benzene, 411 g. (3 moles) of phosphorus trichloride, and 133 g. (1 mole) of aluminum chloride is refluxed for three hours with protection from atmospheric moisture. The excess phosphorus trichloride is removed under reduced pressure (water pump) with stirring, with the bath temperature below 60°. The residue is dissolved in 250 ml. of dry tetrachloroethane, and, with efficient stirring and ice-water cooling, dry chlorine is led into the solution until its absorption ceases, as indicated by escaping chlorine. This requires one to two hours. The gas inlet tube is replaced with a dropping funnel, and the flask is evacuated (water pump) by means of a connection to the top of the reflux condenser. With stirring and ice-water cooling, 230 g. (5 moles) of dry ethanol is added to the mixture in one to two hours, the mixture being kept at 10–15°. The connection to the water pump is maintained for one or two hours after the addition to facilitate the removal of the bulk of hydrogen chloride. The nearly colorless solution is then poured into ice water, and the organic layer is separated. After washing with two or three portions of water until the wash water is essentially free of aluminum ions, the solvent is stripped off at the water pump, the residual moisture being removed as an azeotrope at the lowest possible temperature. It is advisable to add 50–100 ml. of dry carbon tetrachloride to the solution before the distillation in order to facilitate the removal of moisture at room temperature. Distillation of the residue gives 174 g. (80.4%) of diethyl benzenephosphonate, as a colorless oil, b.p. 117–118°/1.5 mm. Refluxing the ester for eight hours with 2 volumes of concentrated hydrochloric acid, followed by evaporation of the clear solution, gives a quantitative conversion to benzenephosphonic acid, m.p. 158–159°.

If the refluxing period is extended to forty hours, the formation of the diphenylphosphinic acid derivative takes place to a substantial extent and the procedure as given above yields only 59% of the diethyl benzenephosphonate together with 30% of ethyl diphenylphosphinate, which boils at 173–175°/1.5 mm. The phosphinate is readily hydrolyzed by concentrated hydrochloric acid, as described above, to yield diphenylphosphinic acid, m.p. 190.5–192°, after crystallization from dilute ethanol. The distillation residue after the removal of the two esters consists of 11 g. of crude bis-substituted acid. The total amount of recovered products accounts for 99% of the benzene used.

The high molecular ratio of phosphorus trichloride used above favors the monosubstitution reaction in the short refluxing period and increases the rate of the reaction. In addition, it facilitates the stirring of the mixture. The amounts of aluminum chloride are not very critical, and the most suitable range is 0.25–1.0 mole per mole of hydrocarbon. Smaller amounts of aluminum chloride give lower yields; larger amounts are not economical. Although other alcohols may be used for esterification, ethanol is preferred because of the moderate temperatures which suffice for the distillation of the resulting esters.

THE ADDITION OF PHOSPHORUS COMPOUNDS TO THE CARBONYL GROUP

The addition of certain phosphorus-containing reagents to carbonyl compounds affords valuable methods for the synthesis of a variety of phosphonic and phosphinic acids. The reagents customarily are divided into two groups: those containing the phosphorus-hydrogen linkage, and the phosphorus halides. The details of the mechanisms of the reactions are unknown, and it is possible that the two processes are more closely related than this division suggests.

Addition of Compounds Containing the Phosphorus-Hydrogen Linkage

The synthesis of phosphonic acids by the addition of substances containing the phosphorus-hydrogen linkage to a carbonyl compound may be represented by an aldol-like condensation, with the formation of a phosphorus-containing acid having a hydroxyl group in the α position to the phosphorus atom. An example is the formation of α-hydroxy-α-toluenephosphonic acid from benzaldehyde and phosphorous acid.

$$C_6H_5CHO + HP(O)(OH)_2 \rightarrow C_5H_5CH(OH)P(O)(OH)_2$$

The reaction in its most primitive form was used by Litthauer,[66] who heated a mixture of phosphonium iodide, PH_4I, and benzaldehyde to 100° in a sealed tube and obtained a mixture of α-toluenephosphonic acid, $C_6H_5CH_2PO(OH)_2$, dibenzylphosphinic acid, $(C_6H_5CH_2)_2PO(OH)$, and tribenzylphosphine oxide, $(C_6H_5CH_2)_3PO$. It is evident that the hydrogen atoms of phosphonium iodide participated in the reaction and that the resulting α-hydroxy derivatives were reduced by hydriodic acid. Such reduction of α-hydroxyphosphonic acids has been observed by Fossek.[67]

A rather extensive series of experiments by Ville [68–71] and by Marie [72–83] between 1889 and 1904 established the general nature of this reaction of ketones and aldehydes. In reactions with hypophosphorous acid, phosphorous acid, and various phosphonous acids a large number of phosphonic and phosphinic acids were prepared. These compounds are listed in Table IV.

The reaction is conducted by heating a mixture of the carbonyl compound with the desired phosphorous acid for a prolonged period of

---

[66] Litthauer, *Ber.*, **22**, 2144 (1889).
[67] Fossek, *Monatsh.*, **5**, 121 (1884); **7**, 20 (1886).
[68] Ville, *Compt. rend.*, **109**, 71 (1889).
[69] Ville, *Compt. rend.*, **107**, 659 (1888).
[70] Ville, *Compt. rend.*, **110**, 348 (1890).
[71] Ville, *Ann. chim. phys.*, (6), **23**, 289 (1891).
[72] Marie, *Compt. rend.*, **133**, 219 (1901).
[73] Marie, *Compt. rend.*, **135**, 106 (1902).
[74] Marie, *Compt. rend.*, **135**, 1118 (1902).
[75] Marie, *Compt. rend.*, **133**, 818 (1901).
[76] Marie, *Compt. rend.*, **134**, 286 (1902).
[77] Marie, *Compt. rend.*, **134**, 847 (1902).
[78] Marie, *Compt. rend.*, **136**, 508 (1903).
[79] Marie, *Compt. rend.*, **136**, 48 (1903).
[80] Marie, *Compt. rend.*, **136**, 234 (1903).
[81] Marie, *Compt. rend.*, **138**, 1707 (1904).
[82] Marie, *Ann. phys. chim.*, (8), **3**, 335 (1904).
[83] Marie, *Compt. rend.*, **137**, 124 (1903).

time. Evaporation of the mixture yields the crude reaction product, which may be isolated by crystallization from suitable solvents.

When hypophosphorous acid is used, both its hydrogen atoms bound to phosphorus, i.e., hydrogens which are not titratable, can participate in the reaction. Such disubstitution is favored, as might be expected, by an excess of the carbonyl compound and by prolonged reaction time. The final product is a phosphinic acid, as illustrated by the following representation of the reaction of acetone.

$$2CH_3COCH_3 + H_2PO(OH) \rightarrow (CH_3)_2C(OH)P(O)(OH)C(OH)(CH_3)_2$$

If the reaction is interrupted before the completion of disubstitution, it is possible to isolate both the disubstituted (phosphinic) acid, shown above, and the monosubstituted (phosphonous) acid, which is formed in the primary reaction in which only one hydrogen atom of hypophosphorous acid is involved. Usually the reaction mixture contains appre-

$$CH_3COCH_3 + H_2P(O)OH \rightarrow (CH_3)_2C(OH)P(O)(H)OH$$

ciable amounts of a phosphonic acid, which is produced by oxidation of the phosphonous acid, probably by the action of atmospheric oxygen. In the reaction described above, this acid is 2-hydroxy-2-propanephosphonic acid, $(CH_3)_2C(OH)PO(OH)_2$.

The α-hydroxy phosphonous acids, obtained at the intermediate stage of the reaction, are obviously capable of further condensation with carbonyl compounds, because they still have one phosphorus-hydrogen linkage. It is possible to isolate these phosphonous acids and to use them in condensations with carbonyl compounds which are different from those used in the first stage. Such a procedure results in the formation of unsymmetrical phosphinic acids.

When phosphorous acid or a phosphonous acid is used in the carbonyl condensation, only one hydrogen atom is available for the reaction and, hence, the formation of a single product is assured. The products made with the aid of phosphorous acid are phosphonic acids; those made with the aid of a phosphonous acid are phosphinic acids.

Although the formation of the phosphinic acids by the condensations with hypophosphorous acid can be made to proceed almost quantitatively, there is no information about the yields of the intermediate phosphonous acids under such conditions. Similarly, there has not appeared any information about the variations of experimental conditions, such as temperature.

The reaction has been applied to a variety of aldehydes and ketones, including acetaldehyde (in the form of paraldehyde), isovaleraldehyde, heptanal, benzaldehyde, acetone, methyl ethyl ketone, diethyl ketone, methyl propyl ketone, acetophenone, and benzophenone.

The main difficulty in this reaction is the necessity of working with phosphorus acids which generate some phosphine on heating. This tendency is particularly pronounced with hypophosphorous acid. *Proper ventilation is required for this work in order to reduce the health hazard.*

The reaction is conducted with crystalline, essentially anhydrous, acids merely by heating them with the carbonyl compounds on a steam bath with suitable protection from atmospheric moisture. The duration of each reaction must be determined empirically, because no precise information can be found in the literature. When the phosphinic acids are being prepared, it suffices to purify the final product by removing any excess carbonyl compound and crystallizing the residual matter from a suitable solvent; water and ethanol have been favored. The α-hydroxyphosphonic acids, prepared from phosphorous acid, are purified similarly, although the purification through a salt of a heavy metal (usually lead) may be necessary to remove inorganic impurities. The α-hydroxy phosphonous acids are usually isolated from the filtrates after the removal of phosphinic acids, which are less soluble; such recovery may involve either a simple evaporation or, more commonly, a purification through a lead salt. The lead phosphonites are water soluble, in contrast to the lead salts of the corresponding phosphonic acids. The lead salts are readily converted to the free acids by treatment with hydrogen sulfide. The α-hydroxy phosphonous acids can be oxidized to the corresponding α-hydroxy phosphonic acids by mercuric chloride or, preferably, by a small excess of bromine water.

Experimental Procedures

α-Toluenephosphonic and Dibenzylphosphinic Acids.[66] A mixture of 10 g. of phosphonium iodide and 5 g. of benzaldehyde is heated in a sealed tube to 100° for four to five hours. On cooling, the tube is opened and appreciable amounts of phosphine and hydrogen iodide are allowed to escape. The reaction mixture is warmed with a small amount of water, and the warm solution is filtered. Evaporation of the solution gives α-toluenephosphonic acid, m.p. 166°. The yield is variable, averaging 10–15%. The water-insoluble mass is triturated with dilute potassium hydroxide, and the filtrate is acidified with hydrochloric acid to give 15–20% of dibenzylphosphinic acid, which, after crystallization from ethanol, melts at 191°.

Di(α-hydroxyisopropyl)phosphinic Acid.[72] A mixture of 400 g. of dry acetone and 250 g. of crystalline hypophosphorous acid is refluxed with protection from atmospheric moisture. After seventy hours, the mixture attains a boiling point of 69°, at which time the reaction is stopped by cooling the mixture. After standing in an ice bath for several hours, the mixture is filtered and the crude di(α-hydroxyisopropyl)phosphinic acid is washed with a little cold acetone. The filtrate is again heated as described above and the process is repeated until complete conversion of hypophosphorous acid is accomplished. The product is recrystallized from hot ethanol and melts at 185–186° with decomposition.

If the filtrate from the initial isolation of the phosphinic acid is worked up, the phosphonous and phosphonic acids can be recovered as follows. The filtrate is freed of excess acetone by vacuum distillation, and the residual syrup is dissolved in water. The solution is neutralized with lead carbonate, and after filtration of the insoluble lead salts the filtrate is evaporated carefully to dryness. The dry residue is extracted with hot 95% ethanol. On cooling, the lead salt of α-hydroxyisopropylphosphonous acid separates. It is taken up in water, and hydrogen sulfide is passed into the solution until the precipitation of lead sulfide is complete. The filtrate is evaporated cautiously on a water bath, and the residue is made to crystallize by chilling. 2-Hydroxy-2-propanephosphonous acid is obtained in the form of extremely hygroscopic colorless crystals, which melt at 40–41°. The water-soluble fraction of the lead salts being removed, the insoluble residue of the lead salts of the phosphonic acid is suspended in water and the mixture is treated with hydrogen sulfide as described above. Evaporation of the filtrate and cooling yield 2-hydroxy-2-propanephosphonic acid, which after crystallization from acetic acid melts at 169–170°.

The phosphonous acid, obtained above, may be readily oxidized to the corresponding phosphonic acid by treating its water solution with 2 molecular equivalents of mercuric chloride,[76] ferric chloride,[76] or, preferably, bromine water.[74] If the metal salts are used for oxidation, the mixture is treated with hydrogen sulfide, and the filtrate is evaporated to recover the product. The use of bromine water simplifies the recovery, because it is merely added to the aqueous solution of the phosphonous acid until a permanent color is obtained and the resulting solution is evaporated to dryness. Usually an additional evaporation with water is advisable in order to remove the residual hydrobromic acid.

α-Hydroxyethanephosphonic Acid.[79] Crystalline phosphorous acid is heated on a steam bath with a large excess of paraldehyde under a reflux condenser which is protected by a calcium chloride tube. After one hundred hours, the dark mixture is poured into cold water and the tarry matter is removed by filtration. The residual phosphorous acid is destroyed by the a dition of bromine water until permanent color is established. The excess bromine is removed by bubbling air through the solution, and, after the solution is made alkaline with aqueous ammonia and the phosphate ion is precipitated by magnesia mixture, the precipitate is discarded and the filtrate is evaporated to dryness. The residue is taken up in water and is neutralized with acetic acid. Lead acetate solution is added to precipitate the lead salt of the desired product. The lead salt is collected, washed with cold water, and suspended in water into which hydrogen sulfide is passed until the precipitation of lead sulfide is complete. The sulfide is removed by filtration, and the filtrate is evaporated to dryness to give, after standing in a vacuum desiccator, colorless crystals of α-hydroxyethanephosphonic acid, m.p. 74–78°. The yield varies, averaging 25–35%.

Addition of Phosphorus Chlorides

The synthesis of phosphonic and phosphinic acids by the addition of certain phosphorus chlorides to carbonyl compounds provides an alternative method for the preparation of α-hydroxy derivatives. In addition, this reaction serves as a source of certain β-keto phosphonic and phosphinic acids.

The reaction was discovered by Fossek,[67] who found that a number of aldehydes, including acetaldehyde, propionaldehyde, isobutyraldehyde, heptanal, and benzaldehyde, react with phosphorus trichloride, forming substances containing 3 units of the aldehyde to 1 unit of phosphorus trichloride. When one of these products was treated with water, 2 molecular equivalents of the aldehyde and 3 equivalents of hydrogen chloride were liberated. Evaporation of the aqueous solution gave a crystalline acid which was identified as the corresponding α-hydroxy phosphonic acid. Fossek visualized the reaction in the following manner.

$$\text{PCl}_3 + 3\text{RCHO} \rightarrow \underset{\underset{\text{RCHClO}}{\diagup}}{\overset{\overset{\text{Cl}}{\diagdown}}{\text{P}}}\underset{\underset{\text{OCHClR}}{\diagdown}}{\overset{\overset{\text{O}}{\diagdown}}{\text{—CHR}}} \xrightarrow{\text{H}_2\text{O}} \text{RCH(OH)PO(OH)}_2$$

Several years later, Michaelis [52] showed that the same reaction can be used with phenyldichlorophosphine. He reported the reactions of this substance with acetaldehyde and with benzaldehyde. The hydroxy phosphinic acids produced had the structures shown below.

$$\text{C}_6\text{H}_5\text{P(O)(OH)CHOHCH}_3 \quad \text{and} \quad \text{C}_6\text{H}_5\text{P(O)(OH)CHOHC}_6\text{H}_5$$

Except for minor variations,[84] the subject was dormant for twenty years, when Conant resumed a study of this reaction under somewhat different experimental conditions. He showed that mixtures of phosphorus trichloride with an essentially equimolar amount of saturated aldehyde or ketone, on treatment with an excess of acetic acid or acetic anhydride and then with water, give α-hydroxyphosphonic acids in yields comparable to those obtained by Fossek. The main difference between the procedures used by these investigators was that in the later work of Conant the excess of the carbonyl compound, which was advocated by Fossek, was replaced by the acetic acid or anhydride. In the course of this work it was also found that α,β-unsaturated ketones undergo an analogous reaction, yielding on hydrolysis the corresponding β-keto phosphonic acids. The overall reaction is shown for benzalacetophenone.

$$C_6H_5CH{=}CHCOC_6H_5 + PCl_3 \xrightarrow{CH_3CO_2H} C_6H_5CH(PO_3H_2)CH_2COC_6H_5$$

Similar reactions were successfully conducted when phosphorus trichloride was replaced by substituted trivalent phosphorus chlorides. These included phenyldichlorophosphine $(C_6H_5PCl_2)$,[85, 86] diphenylchlorophosphine $(C_6H_5)_2PCl$,[87] phenyl dichlorophosphite $(C_6H_5OPCl_2)$,[88] methyl dichlorophosphite (CH_3OPCl_2),[88] and ethyl dichlorophosphite $(C_2H_5OPCl_2)$.[88] In a subsequent paper by Drake and Marvel [89] it was shown that butyl dichlorophosphine, $C_4H_9PCl_2$, also reacts in the expected manner. It may be said, qualitatively at least, that this reaction is general for trivalent phosphorus chlorides. The quantitative aspect of the problem has not been explored adequately, but there are indications of some inexplicably low yields with several substituted phosphorus chlorides.[88] It was also found that benzophenone and camphor fail to react under the conditions cited above. Attempts to raise the reaction temperature above approximately 30–35° led to a vigorous reaction between phosphorus trichloride and acetic acid (or anhydride) which took precedence over the other reaction. It was found, however, that when benzoic acid was used instead of acetic acid the normal reaction could be carried out at higher temperatures.

When acetic anhydride is used in the reaction of an α,β-unsaturated ketone, evaporation of the reaction mixture leaves a residue of a very reactive substance, which on heating with phenol or an alcohol forms

---

[84] Page, *J. Chem. Soc.*, **101**, 423 (1912).
[85] Conant and Pollack, *J. Am. Chem. Soc.*, **43**, 1665 (1921).
[86] Conant, Bump, and Holt, *J. Am. Chem. Soc.*, **43**, 1677 (1921).
[87] Conant, Braverman, and Hussey, *J. Am. Chem. Soc.*, **45**, 165 (1923).
[88] Conant, Wallingford, and Gandheker, *J. Am. Chem. Soc.*, **45**, 762 (1923).
[89] Drake and Marvel, *J. Org. Chem.*, **2**, 387 (1937).

an ester of the β-keto phosphonic acid which would be normally obtained by hydrolysis of the reaction mixture. This behavior of the intermediate suggested to Conant that its structure is that of a cyclic mixed chloride anhydride, containing phosphorus, oxygen, and carbon atoms in the ring, which is formed by a 1,4 addition across the carbonyl group and the double bond. (See below.)

The subsequent work of Drake and Marvel [89] showed that the above-mentioned intermediate reacts with long-chain alcohols to yield monoalkyl esters of the type mentioned above. The alcohols used in this work included 1-decanol, 1-dodecanol, 1-tetradecanol, 1-hexadecanol, 1-octadecanol, and octadec-9-en-1-ol. Although the products were insoluble in alkali, they were assigned the structures of mono esters shown in the accompanying formula (R'' is the alcohol residue), because the analyses and the determination of active hydrogen by the Grignard reagents indicated the existence of a reactive hydrogen atom.

$$\text{RCHCH}_2\text{COR}'$$
$$|$$
$$\text{R}''\text{OP(O)OH}$$

The behavior of mixtures of phosphorus trichloride and saturated carbonyl compounds was explained by Conant [90] by the formation of a 1,2 addition product across the carbonyl group with the consequent formation of a three-membered ring structure. Reaction of such a compound with water would be expected to give the α-hydroxy phosphonic acids. Such a reaction scheme for benzaldehyde is shown in the accompanying formulation originated by Conant.

$$\text{RCHO} + \text{PCl}_3 \rightleftarrows \text{RCH}\!\!-\!\!\text{O} \xrightarrow{\text{H}_2\text{O}} \text{RCHOH}$$
$$\qquad\qquad\qquad\;\;\backslash\;/\qquad\qquad\quad\; |$$
$$\qquad\qquad\qquad\;\;\text{PCl}_3\qquad\qquad\;\; \text{PO(OH)}_2$$

A similar mechanism, involving the 1,4 addition, was proposed for the reaction of α,β-unsaturated ketones.

$$\text{RCH:CHCOR}' + \text{PCl}_3 \rightarrow \quad \text{RCHCH}\!=\!\text{CR}'$$
$$\qquad\qquad\qquad\qquad\qquad\qquad\;\; |\qquad\qquad\; |$$
$$\qquad\qquad\qquad\qquad\qquad\;\; \text{Cl}_3\text{P}\!\!-\!\!\!-\!\!\!-\!\!\text{O}$$

It was believed that acetic acid, or acetic anhydride, could react with the primary adduct more readily than with more phosphorus trichloride. This was taken to be the reason for the fact that the reaction goes to completion instead of coming to a definite equilibrium, such as is attained by mixtures of phosphorus trichloride and the carbonyl compounds without added reagents.

[90] Conant and Cook, *J. Am. Chem. Soc.*, **42**, 830 (1920).

A precise study of the reaction rates, however, forced Conant to abandon the mechanisms shown above as untenable.[91] Not only did the kinetic studies show the improbability of the above reaction mechanism, but also the existence of the cyclic intermediates was shown to be the result of a secondary reaction. It was further shown that the very slow addition of 1 mole of water to a mixture of benzaldehyde and phosphorus trichloride, followed by hydrolysis of the mixture with cold water, leads to good yields of α-hydroxy-α-toluenephosphonic acid. If the reaction mixture after the addition of a mole of water was heated, a mole of hydrogen chloride was evolved and the resulting syrup behaved like a lactone, i.e., like the products obtained by the older technique when acetic anhydride was used in the reaction mixture. As a result of this work, Conant was unable to supply a satisfactory alternative mechanism. He suggested that the overall reaction may be best represented by a trimolecular interaction.

$$RCHO + PCl_3 + CH_3CO_2H \rightarrow R(POCl_2)CHOH + CH_3COCl$$

or

$$RCHO + PCl_3 + (CH_3CO)_2O \rightarrow R(POCl_2)CHOCOCH_3 + CH_3COCl$$

The phosphonyl chlorides shown above may be expected to give the free acids on treatment with water, or esters upon treatment with alcohols.

Scope and Limitations

The reaction performed according to Conant's procedures has been used with success with the following carbonyl compounds: acetone,[92] methyl ethyl ketone,[92] ethyl propyl ketone,[92] methyl *tert*-butyl ketone,[92] acetophenone,[92] dibenzyl ketone,[92] benzylacetophenone,[92] dibenzylacetone,[92] and benzophenone,[92] as well as acetaldehyde,[52] heptanal,[92] and benzaldehyde.[52, 93] The Fossek procedure was successfully used with formaldehyde,[84] acetaldehyde,[52, 67] propionaldehyde,[67] isobutyraldehyde,[67] isovaleraldehyde,[67, 84] hexanal,[67] and benzaldehyde.[52, 67] A procedure similar to that of Conant was successfully used with pyruvic acid [94] to produce α-hydroxy-α-phosphonopropionic acid.

Successful additions to the following α,β-unsaturated ketones were reported: benzalacetophenone,[89, 90, 95] *p*-methoxybenzalacetophenone,[95] dibenzalacetone,[86, 95] cinnamylideneacetophenone,[86] *p*-chlorobenzalaceto-

---

[91] Conant and Wallingford, *J. Am. Chem. Soc.*, **46**, 192 (1924).
[92] Conant, MacDonald, and Kinney, *J. Am. Chem. Soc.*, **43**, 1928 (1921).
[93] Conant and MacDonald, *J. Am. Chem. Soc.*, **42**, 2337 (1920).
[94] Bernton, *Ber.*, **58**, 661 (1925).
[95] Conant, *J. Am. Chem. Soc.*, **39**, 2679 (1917).

phenone,[96] mesityl oxide,[89] *sym*-dibenzoylethylene,[89] and 5-ethyl-3-nonen-2-one.[89]

Whereas aldehydes react satisfactorily in this reaction, ketones tend to yield mixtures from which appreciable amounts of the corresponding unsaturated phosphonic acids can be isolated. These result from dehydration or dehydrohalogenation of the primary reaction products. Thus, acetophenone readily yields the corresponding styrenephosphonic acid derivative, $C_6H_5C(PO_3H_2)=CH_2$,[92, 97] and aliphatic ketones yield α-hydroxy phosphonic acids contaminated with varying amounts of similar by-products. This leads to considerable difficulty in crystallization of the reaction mixtures, and the products often have to be purified through metallic salts. Conant[92] recommends the use of lead salts. The reaction of acetophenone was studied in some detail, and it was shown that, besides the normally expected α-hydroxy acid and the styrene derivative, it is possible to secure good yields of the corresponding α-chloro phosphonic acid if the primary reaction mixture is saturated with hydrogen chloride.

As was mentioned earlier, benzophenone is too sluggish for the usual reaction in the presence of acetic acid, and the reaction must be run at approximately 150° in benzoic acid. A similar procedure was necessary for camphor, although the final product was not obtained in a pure state. Benzil and anthraquinone failed to react even under these conditions.[92]

Although Drake and Marvel[89] showed that phosphorus trichloride can be made to add to 9-ethyltridec-7-en-6-one, 5-ethylhept-3-en-2-one, 3,9-diethylhendec-4,7-dien-6-one, 3-ethyldodec-4-en-6-one, and 3-ethylhendec-4-en-6-one, pure products could not be isolated.

The tendency of the ketones to yield unsaturated products of the type discussed above was successfully utilized by Hamilton,[98] who found that the crude products can readily be converted to the pure unsaturated derivatives by passage through a tube heated to 190–220°, or by heating the mixtures with acetic anhydride to 150°. Heating with phosphorus pentachloride serves not only to yield the unsaturated acids but also to convert them to the corresponding unsaturated phosphonyl dichlorides, which can be readily purified by distillation under reduced pressure. The procedure is most clearly described for the product of the acetone-phosphorus trichloride reaction; the dehydration treatment described above gives 70–80% yields of 1-propene-2-phosphonyl dichloride, $CH_2=C(CH_3)POCl_2$, which can be readily purified by vacuum distillation.

[96] Conant and Jackson, *J. Am. Chem. Soc.*, **46**, 1003 (1924).
[97] Conant and Coyne, *J. Am. Chem. Soc.*, **44**, 2530 (1922).
[98] Hamilton, U. S. pat. 2,365,466 [*C. A.*, **39**, 4619 (1945)].

The final modification of the reaction, introduced by Conant, i.e., the slow addition of water to the reaction mixture, was used by him only for benzaldehyde. The scope and the limitations of this very simple procedure cannot be estimated because it succeeds probably by the virtue of differential reactivities of the reaction intermediates with water.

The nature of the phosphorus chloride derivative seems to be unimportant in this reaction provided that it is a chloride of trivalent phosphorus.

Experimental Procedures

α-Hydroxy-α-toluenephosphonic Acid (Fossek Procedure).[84] Thirty-seven grams of phosphorus trichloride is slowly added to 114 g. of benzaldehyde. The mixture is allowed to stand overnight with protection from moisture. The resulting oil is poured into 3 l. of cold water, and the aqueous layer is separated, filtered, and evaporated on a steam bath. After the addition of 500 ml. of water to the residue, the solution is re-evaporated to dryness to expel the residual hydrochloric acid. The resulting syrup is rubbed with dry ether to induce crystallization, and the product is recrystallized from a 2:1 mixture of benzene and acetic acid to give 42 g. (84%) of α-hydroxy-α-toluenephosphonic acid, m.p. 170°.

A similar reaction with formaldehyde is too vigorous to control. The use of paraformaldehyde, however, in a procedure similar to the above readily gives a 93% yield of hydroxymethanephosphonic acid, m.p. 85°.

Conant Procedures (Saturated Carbonyl Compounds).[92] The carbonyl compound is mixed with a 10% molar excess of phosphorus trichloride at 30–35°, and, after standing for two or three hours, the solution is treated with 3 moles of acetic acid, which is added with cooling at 20–30°. The mixture is allowed to stand for six to twelve hours at room temperature with protection from atmospheric moisture. It is then poured into cold water, and the solution is evaporated to dryness. If the product fails to crystallize, it is converted to the lead salt with lead acetate, after the removal of inorganic phosphorus with magnesium nitrate and aqueous ammonia. This procedure gives, when 10 g. of acetone and 30 g. of phosphorus trichloride are used, a 91% yield of 2-hydroxy-2-propanephosphonic acid, m.p. 167–169° after crystallization from acetic acid.

α-Chloro-α-phenylethanephosphonic Acid.[97] Ten grams of acetophenone and 14.2 g. of phosphorus trichloride are mixed at room temperature, and after standing for two hours with protection from atmospheric moisture the solution is treated with 25 g. of glacial acetic acid at 25°. The solution is allowed to stand overnight, after which a stream of dry hydrogen chloride is passed through it for two hours. The resulting solid is sucked dry on a sintered-glass filter. Recrystallization from ether gives 16 g. (87%) of α-chloro-α-phenylethane-α-phosphonic acid, m.p. 174–175°.

α-Hydroxy-α-phenylethanephosphonic Acid. The normally expected hydroxy acid is readily obtained only by careful hydrolysis of the above chloro acid. It cannot be obtained by the normal procedure, because it is too readily attacked by hydrochloric acid on heating. The hydrolysis procedure is as follows: Ten grams of the chloro acid, obtained above, is dissolved in 200 ml. of cold water, and the solution is allowed to stand at room temperature for two days. The solution is evaporated without warming by means of an air jet, and the residual syrup is placed in a vacuum desiccator, where it crystallizes after several days. There is obtained 7.5 g. (81%) of α-hydroxy-α-phenylethane-α-phosphonic acid, which melts at 154–155° after crystallization from a chloroform-ether mixture.

The chloro acid is also the best source of the styrene derivative, $C_6H_5C(PO_3H_2)=CH_2$. The chloro acid evolves hydrogen chloride on being heated to 180°; the cooled product on crystallization from a chloroform-ether mixture gives 80–90% yields of the unsaturated acid, m.p. 112–113°.

α,α-Diphenyl-α-hydroxymethanephosphonic Acid. Sluggish compounds like benzophenone can be phosphonated at elevated temperatures. A mixture of 10 g. of benzophenone and 20 g. of benzoic acid is melted on a steam bath, and 10 g. of phosphorus trichloride (55% excess) is added to the hot mixture during five to ten minutes. The mixture is heated to 155° in the course of ten minutes and is then allowed to cool to 130°, at which temperature it is kept for two or three hours. After cooling to 90°, the mixture is poured into 500 ml. of water. Sodium hydroxide solution is added to faint alkalinity, and the mixture is heated on a steam bath for four to five hours. The mixture is diluted to 750 ml., cooled, and extracted with ether to remove the unreacted ketone. The aqueous solution is acidified with hydrochloric acid and cooled in ice water, and the precipitated benzoic acid is filtered. The filtrate is evaporated to 250 ml., and the residual solution is extracted with ether. Evaporation of the extract gives a rapidly solidifying oil, which is fractionally crystallized from water slightly acidified with hydrochloric acid. There is obtained 7 g. (50%) of α,α-diphenyl-α-hydroxymethanephosphonic acid,[92] m.p. 171–172°.

α-Hydroxy-α-toluenephosphonic Acid.[91] A mixture of 10 g. of benzaldehyde and 13 g. of phosphorus trichloride is cooled by means of an ice bath, and, with vigorous stirring, 1.7 g. (1 mol. eq.) of water is added in small droplets in the course of fifteen minutes. The solution is kept at 10–15° until the evolution of hydrogen chloride subsides. The resulting yellow oil is poured into cold water, and the solution is evaporated to dryness at room temperature by means of an air jet. The residual oily product is converted to the aniline salt by treatment with aniline in ether solution. There is obtained 13.5 g. (57%) of the aniline salt of α-hydroxy-α-toluenephosphonic acid, which melts at 201–202° after crystallization from ethanol.

1-(4'-Methoxyphenyl)-2-benzoylethane-1-phosphonic Acid.[95] A suspension of 19 g. of p-anisalacetophenone in 40 ml. of glacial acetic acid is treated with 14 g. of phosphorus trichloride. The solution becomes cool and turns red. On standing overnight the color fades to yellow. The solution is poured into 500 ml. of water, and the rapidly solidifying oil is collected. It is redissolved in dilute sodium carbonate solution, the solution is extracted with ether to remove the unreacted ketone, and the aqueous solution is acidified with hydrochloric acid to give 23 g. (89%) of the keto phosphonic acid, m.p. 189°, after crystallization from dilute ethanol.

α-Phosphono-α-hydroxypropionic Acid.[94] Ten grams of pyruvic acid is treated with 15.5 g. of phosphorus trichloride with efficient stirring and cooling. Considerable amounts of hydrogen chloride are evolved. The mixture is stirred until a homogeneous solution is formed. This is allowed to stand overnight with protection from moisture. Then, 20.4 g. of acetic acid is added with stirring and cooling, and the mixture is allowed to stand for twelve hours. The resulting oil is poured into water, and the solution is evaporated under reduced pressure at 30–40°. The oily residue crystallizes on standing in a vacuum desiccator. Recrystallization from acetic acid gives 5–6 g. (about 40%) of somewhat hygroscopic colorless crystals, m.p. 165–170°.

THERMAL DECOMPOSITION REACTIONS

The preparation of certain intermediates for the synthesis of phosphonic and phosphinic acids by reactions which involve thermally induced dissociation or displacement may be divided for convenience into four categories. It must be understood that this division is arbitrary and that it does not necessarily imply that a different reaction mechanism operates in each category. As a matter of fact, the exact mechanisms involved in these reactions are essentially unknown, and only fragmentary uncorrelated observations have been made on most of them. The preparations have been conducted in a purely empirical manner; undoubtedly, considerable improvements in the yields and on the procedures may be expected in the future. The arbitrary classification adopted here is as follows.

    a. Pyrolytic reactions.
    b. Displacements with organomercury compounds.
    c. Thermal decomposition of phosphonium compounds.
    d. Disproportionation of phosphonous acids.

Pyrolytic Reactions

Pyrolysis is used to prepare a very limited number of monosubstituted dichlorophosphines of the aromatic series. The dichlorophosphines can be converted to the corresponding phosphonic acids by reactions which were discussed under the Friedel-Crafts reaction.

Michaelis [99] observed that when vapors of a mixture of phosphorus trichloride and benzene are allowed to contact a hot surface (at red heat) phenyldichlorophosphine is formed in accordance with the following equation.

$$C_6H_6 + PCl_3 \rightarrow C_6H_5PCl_2 + HCl$$

This observation led to the construction of various pieces of equipment in which the reaction could be carried out in a convenient manner.[100–103] The essential feature of all of them is a provision for leading the vapor mixture over the heated surface. The phenyldichlorophosphine, prepared as indicated above, is purified by distillation of the reaction mixture in a carbon dioxide atmosphere. The distillate, b.p. 225°, usually contains some free phosphorus and phosphine. It is best purified by heating for several hours at nearly reflux temperature in a stream of carbon dioxide.[101]

The use of the pyrolysis for preparative purposes appears to be limited to benzene. Only two other substances, thiophene [53] and toluene,[52, 104] have been converted to the corresponding dichlorophosphines by this method. With both compounds the yields were discouragingly poor. In a run of eight days' duration only 14 g. of the dichlorophosphine was obtained from 100 g. of thiophene. The use of toluene gave principally pyrolytic products of toluene, free phosphorus, and a trace of a tolyldichlorophosphine, which was obtained in such a

[99] Michaelis, *Ber.*, **6**, 601, 816 (1873).
[100] Michaelis, *Ann.*, **181**, 265 (1876).
[101] A. E. Arbuzov, dissertation, Kazan, 1914.
[102] Lecoq, *Bull. soc. chim. Belg.*, **42**, 199 (1933).
[103] Bowles and James, *J. Am. Chem. Soc.*, **51**, 1406 (1929).
[104] Michaelis and Lange, *Ber.*, **8**, 1313 (1875).

small amount that it could not be characterized as such, but was converted to the acid, which appeared to be the *meta* isomer.

The above summary indicates the present scope of this reaction when phosphorus trichloride is used. However, a related reaction has been extended to phenyldichlorophosphine.[101,105] This substance on being heated to 300° undergoes disproportionation to phosphorus trichloride and diphenylchlorophosphine, $(C_6H_5)_2PCl$; this reaction is best carried out by heating the dichlorophosphine in a sealed tube for seventy-two hours to 300°.[101] The cooled solution is filtered, and the filtrate is fractionated to give 40–50% yields of diphenylchlorophosphine, b.p. 178°/14 mm. The reaction does not seem to be applicable to other dichlorophosphines.

Displacements with Organomercury Compounds

Michaelis[100] found that phenyldichlorophosphine can be obtained from phosphorus trichloride and diphenylmercury.

$$PCl_3 + (C_6H_5)_2Hg \rightarrow C_6H_5PCl_2 + C_6H_5HgCl$$

The reaction has been used by many later workers for the preparation of mono- and di-substituted chlorophosphines. It appears to be a two-step reaction, with both reactions usually occurring, as shown below.

$$R_2Hg + PCl_3 \rightarrow RPCl_2 + RHgCl \qquad (a)$$

$$RHgCl + PCl_3 \rightarrow RPCl_2 + HgCl_2 \qquad (b)$$

The second step is favored by higher temperature and by an excess of phosphorus trichloride.[106] The reaction can be used for the synthesis of disubstituted chlorophosphines by employing monosubstituted dichlorophosphines instead of phosphorus trichloride. It is obvious that a mixture from such a reaction contains mono- and di-substitution products and may even contain traces of tertiary phosphines. The presence of the higher substitution products has been recognized for a long time,[100,107] but no detailed study of the extent of the side reactions has been reported. The desired product can be readily separated from the products of higher degree of substitution by distillation. However, the chlorophosphines obtained in this manner are usually contaminated with considerable amounts of organomercury compounds, which are extremely difficult to remove by distillation.[89,100,107] It is doubtful that a product completely free of mercury can be isolated from reaction

[105] Dörken, *Ber.*, **21**, 1505 (1888).
[106] Michaelis, *Ber.*, **13**, 2174 (1880).
[107] Guichard, *Ber.*, **32**, 1572 (1899).

mixtures of this category. This difficulty is of little importance, however, if the product is to be converted to a phosphonic or a phosphinic acid by methods given in the earlier sections. The mercury contaminants are best removed after such conversion.

The most favorable factor in this reaction lies in its capacity to produce definite products having the same carbon structures as those of the organomercury intermediates. Phosphorus enters the molecule at the site of the mercury attachment. For this reason, the reaction has been used for identification, by providing compounds of definite structures as reference substances suitable for comparison with compounds obtainable from the Friedel-Crafts reaction.

The present scope of the reaction includes both aliphatic and aromatic compounds, the list of which is given in Table V. The yield has been reported for only one alkyldichlorophosphine, butyldichlorophosphine, 61%.[89] The yield of aromatic dichlorophosphines usually exceeds 50%, except for the 2,4-dimethylphenyl derivative (20%)[61] and the 2,4,5-trimethylphenyl derivative (20%).[52] The yields of diarylmonochlorophosphines vary between 30% and 64%.

The reactions are conducted at approximately 200°, using either sealed tubes or ordinary reflux apparatus with a provision for an inert atmosphere. Although many compounds appear to react at fairly low temperatures, it is generally advisable to heat the mixture near the end of the reaction to temperatures in excess of 150°, principally to convert the mercury compounds to mercuric chloride, the bulk of which may be removed by filtration. The necessity for high temperatures limits the usefulness of this reaction to compounds that can withstand such heat.

Experimental Procedures

p-Tolylphenylchlorophosphine.[108] A mixture of 78 g. of phenyldichlorophosphine and 60 g. of di-*p*-tolylmercury is heated in a reflux apparatus in a carbon dioxide atmosphere for two or three hours to 270°. On cooling, the mixture is extracted with benzene and filtered, and the filtrate is distilled to give 30 g. (63.5%) of *p*-tolylphenylchlorophosphine, b.p. 230–240°/100 mm.

Phenyldichlorophosphine.[100] Ten grams of diphenylmercury and 34 g. of phosphorus trichloride are heated in a sealed tube for five hours at 180°. On cooling, the mixture is filtered and the filtrate, on distillation, gives crude phenyldichlorophosphine, which is allowed to stand until the metallic mercury droplets settle. Filtration and distillation

---

[108] Pope and Gibson, *J. Chem. Soc.*, **101**, 735 (1912).

give 5 g. of essentially mercury-free product, b.p. 216–220°; the yield is nearly quantitative, if calculated by equation (a), p. 317.

n-Butyldichlorophosphine.[89] Fifty grams of di-n-butylmercury was placed in a Pyrex tube, which was flushed with dry nitrogen, and 100 g. of phosphorus trichloride was slowly run in. A white precipitate began to form almost immediately. The sealed tube was heated at 200° for nine hours. The cooled mixture was washed out with phosphorus trichloride and, upon distillation, gave 17 g. of n-butyldichlorophosphine, b.p. 157–160°, which still contained mercury, probably as butylmercury chloride.

Phenyl-*p*-bromophenylchlorophosphine.[109] A mixture of 98 g. of *p*-bromophenyldichlorophosphine and 85 g. of diphenylmercury was heated to 210° for seventy-five minutes in a nitrogen atmosphere. The cooled mixture was shaken with 200 ml. of dry petroleum ether and filtered. Distillation of the filtrate gave 35–40 g. (47–53%) of the product, b.p. 203–204°/11 mm.

Thermal Decomposition of Phosphonium Compounds

Thermal decomposition of true quaternary phosphonium halides gives derivatives of tertiary phosphines. However, when the related tertiary phosphine dichlorides are heated, the products contain disubstituted chlorophosphines which can be converted to phosphinic acids by methods indicated in previous sections. The general reaction scheme may be illustrated by the following representation.

$$RR'R''PCl_2 \rightarrow R''Cl + RR'PCl$$

The reaction was observed by Michaelis and Soden [110] for the triphenyl derivative and by Collie and Reynolds [111] for the triethyl compound, although neither group was able to use the reaction for practical syntheses. Plets [112] developed a workable procedure based on this reaction. The results of his work are outlined below in some detail, because of the inaccessibility of the original publication.

The tertiary phosphine dichlorides are prepared either by addition of chlorine to the corresponding tertiary phosphines in an inert solvent (preferred for non-alkylated aromatic compounds), or by the reaction of phosphorus pentachloride with tertiary phosphine oxides (preferred

---

[109] Davies and Mann, *J. Chem. Soc.*, **1944**, 276.
[110] Michaelis and Soden, *Ann.*, **229**, 295 (1885).
[111] Collie and Reynolds, *J. Chem. Soc.*, **107**, 367 (1915).
[112] V. M. Plets, dissertation, Kazan, 1938.

for alkyl or alkaryl compounds). It is possible to use thionyl chloride, sulfuryl chloride, sulfur monochloride, chlorosulfonic acid, or titanium

$$(C_6H_5)_3P + Cl_2 \rightarrow (C_6H_5)_3PCl_2$$

$$(C_2H_5)_3PO + PCl_5 \rightarrow (C_2H_5)_3PCl_2 + POCl_3$$

chloride in the second reaction instead of phosphorus pentachloride, but the yields are poor.

The resulting dichlorides are decomposed by heating at 150–220° in a distilling apparatus in an inert atmosphere. The products are distilled, and the yields of disubstituted chlorophosphines obtained in this manner range from 40% to 70%. The compounds prepared by Plets are listed in Table VI.

The present scope of this reaction is indicated by the extent of the table. It is probable, however, that the reaction can be run with many other tertiary phosphine derivatives. The variety of the compounds listed in the table indicates good versatility.

Although Plets indicates that the reaction possibly can be extended to the preparation of monosubstituted dichlorophosphines by a repetition of the reaction sequence on the monochloro compounds obtained as indicated above, no experimental proof has been presented. From the material on hand, it is impossible to set down the specific order of the ease of cleavage of the substituent groups from tertiary phosphine dichlorides in this reaction, as has been done for the true quaternary phosphonium compounds by Ingold and co-workers.[113]

A related reaction was used by Michaelis and others [52, 108, 114] for the preparation of methylphenyl- and methyl-*p*-tolyl-phosphinic acids by thermal decomposition of the corresponding dipiperidylphosphonium hydroxides, according to the following scheme.

$$RR'(C_5H_{10}N)_2POH \rightarrow RR'P(O)OH$$

The exact mechanism of this reaction is obscure; it has been used only for the two compounds listed above. The procedure may be illustrated by the following example. Phenyldipiperidylphosphine (prepared from phenyldichlorophosphine and piperidine) is treated with an equimolar amount of methyl iodide; the resulting phosphonium iodide is treated with an excess of moist silver oxide and filtered; and the filtrate is evaporated to dryness. The residue is dissolved in water and is re-evaporated to dryness, and the residue is heated to 150° for three or four hours. It is dissolved in water, treated with aqueous ammonia, and evaporated to dryness, and the residue is taken up in a little water. The solution is

[113] Hey and Ingold, *J. Chem. Soc.*, **1933**, 531; Fenton, Hey, and Ingold, *ibid.*, **1933**, 989.
[114] Michaelis, *Ber.*, **31**, 1037 (1898).

treated with silver nitrate; the silver salt is separated, suspended in water, and decomposed with the calculated amount of hydrochloric acid. After filtration from silver chloride, the solution is evaporated to give a 75% yield of methylphenylphosphinic acid, m.p. 133–134°.

With only the two examples in existence, it is impossible to define the scope or the limitations of this reaction.

Experimental Procedures

Di-n-propylchlorophosphine.[112] A distillation apparatus, with a provision for the introduction of dry carbon dioxide, is charged with 17.6 g. of tri-n-propylphosphine oxide; 25 g. of phosphorus pentachloride is added, and the mixture is heated to 190–220°, at which point a considerable degree of foaming occurs. Heating must be carefully regulated, because any overheating produces appreciable amounts of yellow phosphorus and phosphines. When the reaction subsides, distillation under reduced pressure gives 9.1 g. (60%) of di-n-propylchlorophosphine, b.p. 99–101°/15 mm.

Diphenylchlorophosphine.[112] A solution of 26.2 g. of triphenylphosphine in 100 ml. of freshly distilled chloroform or carbon tetrachloride is treated with dry chlorine until the absorption is complete. A distillation condenser is attached, and while the apparatus is being swept by carbon dioxide the solvent is distilled; the residue is carefully heated to 190–210°, when a vigorous reaction commences. When the reaction subsides, the product is distilled and the distillate is redistilled in vacuum, yielding 9.9 g. (40%) of diphenylchlorophosphine, b.p. 178°/14 mm. (300–320°/760 mm.).

Disproportionation of Phosphonous Acids

Thermal disproportionation of phosphonous acids is a reaction common to oxygen acids of phosphorus which have a P-H linkage. The reaction has been observed with the aliphatic [107] and the aromatic [115] compounds. It may be presented as a mutual oxidation-reduction reaction which proceeds according to the accompanying equation.

$$3RPO_2H_2 \rightarrow 2RPO_3H_2 + RPH_2$$

The reaction occurs on heating and generally requires temperatures above 100°. It is of no significance for preparative purposes, since the phosphonous acids employed in it usually are made from dichlorophosphines which can be converted directly and in almost quantitative yields

---

[115] Michaelis and Ananoff, *Ber.*, **7**, 1688 (1874).

to phosphonic acids by reactions indicated in earlier sections. The simultaneous formation of the phosphines is another serious disadvantage to this reaction; *phosphines are generally very toxic*, and they possess disagreeable odors. The procedure is essentially that of dry distillation until the elimination of the phosphine is complete. Recrystallization of the residue from water yields the phosphonic acid. The yields are variable because the phosphonic acid may undergo dephosphonation at the temperatures employed. In most cases the reaction has been observed only qualitatively. Apparently it cannot be applied to α-hydroxy derivatives, for they suffer decomposition, with the loss of a carbonyl compound, before the oxidation-reduction can set in.

MISCELLANEOUS SYNTHESES

This section deals with synthetic methods that cannot be classified with the previously described procedures. Most of the reactions cited here have been explored but little, and their usefulness cannot be estimated accurately.

Oxidation of Phosphines and Phosphonous Acids

Primary and secondary phosphines, RPH_2 and $RR'PH$, can be oxidized by a variety of means to the corresponding phosphonic and phosphinic acids. However, the possible usefulness of this method is limited by a number of important factors. At the present time there is no satisfactory and safe way to prepare the phosphines in high purity. The venerable synthesis by heating mixtures of alkyl iodides, zinc oxide, and phosphonium iodide in sealed tubes gives poor yields of complex mixtures. It has been used only for very small-scale preparations. A possible solution is the synthesis of phosphines from sodium hydrogen phosphides (i.e., sodium derivatives of phosphine) and organic halides.[116] The information about this synthesis is too meager to be evaluated here. Under any circumstances, the oxidation of phosphines presents serious difficulties because of *the toxicity of phosphines* and the inflammable nature of the lower members of the series. A number of acids have been prepared on a minute scale by evaporation of solutions of the phosphines in nitric acid.[107] There is information neither about the best conditions nor about the yields.

Oxidation of phosphonous acids is not a particularly good source of phosphonic acids, as mentioned in the preceding section on the disproportionation reactions. The only important exception to this gen-

---

[116] Walling, U. S. pats. 2,437,795–8, [*C. A.*, **42**, 4198–4199 (1948)].

eralization is the oxidation of the α-hydroxy derivatives, which can be readily obtained from condensation reactions of carbonyl compounds. The methods of oxidation of these compounds have been discussed (p. 306). Another possible exception is the oxidation of N-dialkylaminobenzenephosphonous acids; these substances cannot be oxidized without decomposition by acidic reagents such as nitric acid, which is a common oxidant for other phosphonous acids.[107] Although the dimethylamino compound could be oxidized to the corresponding phosphonic acid [65] by 2 moles of mercuric chloride in aqueous solution, the higher members of the series suffered decomposition under these circumstances. Warming the acid with oxygenated water was found to be a good method for oxidizing both the dimethylamino and the diethylamino compounds, but there is no detailed information about the experimental conditions.[117]

Syntheses from Dialkylanilines

It will be recalled that in the section dealing with the Friedel-Crafts reaction mention was made of an early synthesis of p-dimethylaminobenzenephosphonic acid by that method.[65] It was found later by Bourneuf [117] and Raudnitz [118] that aluminum chloride is not necessary in the primary reaction, and that both the mono- and the di-substituted phosphine chlorides can be made by a direct reaction with phosphorus trichloride. Raudnitz worked only with dimethylaniline; Bourneuf used both dimethylaniline and diethylaniline, and he devised means for the direct synthesis of phosphonic and phosphinic acids in this series by using phosphorus oxychloride instead of phosphorus trichloride. The reactions used are shown in the accompanying equations. It will

$$2R_2NC_6H_5 + POCl_3 \rightarrow p\text{-}R_2NC_6H_4POCl_2 + R_2NC_6H_5 \cdot HCl$$

$$4R_2NC_6H_5 + POCl_3 \rightarrow (p\text{-}R_2NC_6H_4)_2POCl + 2R_2NC_6H_5 \cdot HCl$$

be noted that an excess of the amine is used to take up the hydrogen chloride generated in the reaction. The chlorides obtained in the reaction are then hydrolyzed to the corresponding acids.

A related reaction which depends on the reactivity of the *ortho* hydrogen atoms in diphenylamine has been used to prepare a derivative of the heterocyclic compound, dibenzophosphazine, in the form of the corresponding phosphinic acid.[119]

[117] Bourneuf, *Bull. soc. chim. France*, **33**, 1808 (1923).
[118] Raudnitz, *Ber.*, **60**, 743 (1927).
[119] Sergeev and Kudryashov, *J. Gen. Chem. U.S.S.R.*, **8**, 266 (1938) [*C. A.*, **32**, 5403 (1938)].

$(C_6H_5)_2NH + PCl_3 \rightarrow$ [phenazaphosphine with PCl] $\xrightarrow{H_2O}$

[phenazaphosphine with POH] $\xrightarrow{[O]}$ [phenazaphosphine with P(O)OH]

The above interactions of phosphorus trichloride and phosphorus oxychloride with the dialkylanilines have been described only for dimethylaniline and diethylaniline. It is probable that other dialkylanilines can be used. The formation of the heterocyclic phosphinic acid described above is the only instance reported. It may be expected that substituted diphenylamines will produce the corresponding cyclic compounds.

EXPERIMENTAL PROCEDURES

Bis(4-dimethylaminophenyl)-phosphonic and -phosphinic Acids (Bourneuf Procedures).[117] A mixture of 242 g. of dimethylaniline and 137 g. of phosphorus trichloride is heated on a steam bath for three hours under a reflux condenser with protection from moisture. The cooled mixture is added to a solution of 320 g. of sodium carbonate in 1 l. of water, and the excess dimethylaniline is removed by steam distillation. The residue is cooled for twenty-four hours, and 40 g. of insoluble solid is removed. The solution is treated with barium chloride until the precipitation of barium phosphate is complete, and the filtered solution is treated with excess saturated copper sulfate solution. The copper salt of the phosphonic acid is collected, suspended in water, and treated with hydrogen sulfide. Evaporation of the filtrate gives 110 g. (60%) of 4-dimethylaminobenzenephosphonic acid, m.p. 163°. The alkali-insoluble solid is extracted with boiling benzene, in which almost all of it dissolves. Cooling of the extract gives bis(4-dimethylaminophenyl)-phosphine oxide (or phosphinous acid), m.p. 169°, in 10% yield. This oxide is readily converted to the corresponding phosphinic acid by allowing a mixture of 4 g. of the oxide and 50 ml. of oxygenated water, to which just enough dilute sulfuric acid is added to effect solution, to stand for two days. Sodium carbonate is added until complete solution is attained, and the solution is acidified with acetic acid to give 4.2 g. (100%) of pure bis(4-dimethylaminophenyl)phosphinic acid, m.p. 249°.

The oxidation reactions may be avoided if phosphorus oxychloride is used. Thus, 75 g. of phosphorus oxychloride and 142 g. of dimethylaniline are heated to 130° for eight to nine hours in a reflux apparatus protected from moisture. Heating is stopped when the open-arm manometer attached to the reflux condenser begins to indicate a partial vacuum in the flask. The cooled mixture is carefully added to 1.5 l. of 10% sodium hydroxide solution, and the excess amine is removed by steam distillation. The residual solution is filtered after standing for twenty-four hours. The alkaline filtrate is acidified with acetic acid to give 85 g. (40%) of bis(4-dimethylaminophenyl)phosphinic acid, which is recrystallized from a mixture of benzene and methanol; m.p. 249° (on a copper block).

Dibenzophosphazinic Acid.[119] A mixture of 21 g. of diphenylamine and 17 g. of phosphorus trichloride is heated in a reflux apparatus which is protected from atmospheric moisture. The heating is effected by an oil bath, the temperature of which is raised to 200–220° in the course of six hours. The hot solution is poured into 1 l. of cold water. (It is advisable to conduct this operation in an open-top box in which several lumps of Dry Ice are placed, so as to provide an inert atmosphere. The reaction mixture contains some free phosphorus which will burst into flame on exposure to air.) The solidified reddish mass is broken up under water and is repeatedly extracted with a total of 4 l. of hot water. On cooling, the hydroxyphosphine is collected, dried in a vacuum desiccator, and suspended in 200 ml. of tetralin which is contained in a reflux apparatus. The tetralin is brought to the boiling point, and air is bubbled slowly through the solution for one to two hours. The cooled mixture is filtered, and the dibenzophosphazinic acid is purified by precipitation from dilute sodium hydroxide solution by hydrochloric acid. The substance does not melt at 250°. The yield is approximately 17%.

Oxidative Phosphonation

This reaction is less well understood than any of the other procedures discussed in this chapter. The structures of the compounds formed have not been proved, but the potentialities of the reaction are sufficiently interesting to justify its mention. The process involves the reaction of unsaturated compounds with elemental phosphorus and oxygen simultaneously. The primary reaction product appears to be an adduct of phosphorus tetroxide to the double bond. Hydrolysis of this adduct yields a substance with two acidic phosphorus-containing groups at the previous site of the double bond. Drastic hydrolysis removes one acidic group, indicating that it is connected to the hydrocarbon by an ester linkage. The remaining group is stable to hydrolysis and is a phosphonous acid group which can be oxidized to a phosphonic acid group.

The reaction was discovered by Willstätter and Sonnenfeld,[120] who applied it to cyclohexene, menthene, pinene, trimethylethylene, allyl alcohol, ethyl cinnamate, oleic acid, and olive oil. The products of hydrolysis were not characterized in detail, nor were the positions taken by the phosphono group and the ester phosphate group established. The phosphonous acid from cyclohexene was oxidized to the phosphonic acid by nitric acid. The reaction was also used by Montignie,[121] who obtained an alkali-soluble product from a similar reaction of cholesterol. This product, which retains the hydroxyl group of cholesterol, was characterized as the acetate, which melted at 250°.

It is impossible to define the scope and the limitations of this reaction from the limited information available. However, it is one of the mildest methods for introduction of a phosphono group into an organic molecule.

Wurtz Reaction

Although the Wurtz reaction has been used freely to prepare tertiary phosphines, it appears to have been used but once for the synthesis of a definite acidic compound. Michaelis[4] treated diethylamidophosphonyl chloride with 2 moles of bromobenzene and 4 atoms of sodium in ether solution, obtaining the N-diethylamide of diphenylphosphinic acid. Hydrolysis with hydrochloric acid gave the free acid. No yields were given. The reaction sequence may be shown by the equations below.

$(C_2H_5)_2NPOCl_2 + 2C_6H_5Br + 4Na \rightarrow$

$(C_2H_5)_2NP(O)(C_6H_5)_2 + 2NaBr + 2NaCl$

$(C_2H_5)_2NP(O)(C_6H_5)_2 + H_2O(HCl) \rightarrow$

$(C_6H_5)_2PO(OH) + (C_2H_5)_2NH \cdot HCl$

It is possible to visualize the extension of this reaction to many other compounds that have been prepared by means of the Grignard reaction in the past.

Direct Phosphonation of a Nitrogen Heterocycle

One instance of direct phosphonation by the reaction of phosphorus oxychloride with a pyrazole has been recorded. Michaelis and Pasternack[122] heated phosphorus oxychloride with antipyrine (1-phenyl-

---

[120] Willstätter and Sonnenfeld, *Ber.*, **47**, 2801 (1914).
[121] Montignie, *Bull. soc. chim. France*, (4), **49**, 73 (1931).
[122] Michaelis and Pasternack, *Ber.*, **32**, 2411 (1899).

3-methyl-5-chloropyrazole) or its methochloride for twelve hours in a sealed tube to 200°. On treatment with water a white solid was obtained which, after washing with ether and recrystallization from water, melted at 191°. It was given the structure of 1-phenyl-3-methyl-5-chloropyrazole-4-phosphonic acid. No yields were given. It is impossible to judge whether the particular pyrazole used possessed a unique configuration which made this reaction possible.

The overall reaction scheme is shown below.

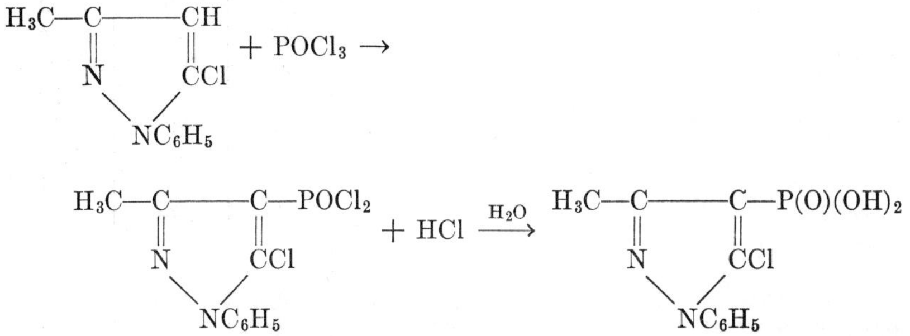

$$\underset{\underset{NC_6H_5}{\diagdown\;\;\diagup}}{\underset{N\;\;\;\;\;\;\;\;\;CCl}{H_3C-\underset{\|}{C}\!-\!-\!-\!-\!-\underset{\|}{CH}}} + POCl_3 \rightarrow$$

$$\underset{\underset{NC_6H_5}{\diagdown\;\;\diagup}}{\underset{N\;\;\;\;\;\;\;\;\;CCl}{H_3C-\underset{\|}{C}\!-\!-\!-\!-\!-\underset{\|}{C}\!-\!POCl_2}} + HCl \xrightarrow{H_2O} \underset{\underset{NC_6H_5}{\diagdown\;\;\diagup}}{\underset{N\;\;\;\;\;\;\;\;\;CCl}{H_3C-\underset{\|}{C}\!-\!-\!-\!-\!-\underset{\|}{C}\!-\!P(O)(OH)_2}}$$

TABLES OF COMPOUNDS REPORTED BEFORE MAY, 1948

The following tables summarize the syntheses of the compounds covered by this chapter which had been reported in the literature before May, 1948. The compounds prepared by the primary reactions which introduce the phosphorus atom into the molecule are listed; the tables of compounds prepared by the organomercury derivatives and by thermal decomposition of phosphonium-type compounds list the chlorophosphines which can be converted to the acids by hydrolysis.

TABLE I
Derivatives of Phosphonic and Phosphinic Acids Prepared by Alkylation of Phosphites or Other Trivalent Esters

Compound	Method*	Yield %	Reference
$CH_3PO(OCH_3)_2$	A	100	13
	D	—	6
$CH_3PO(OC_2H_5)_2$	A	100, 95	1, 13, 14
	B	45	10
$CH_3PO(OC_3H_7\text{-}iso)_2$	A	95	14
$CH_3PO(OC_6H_5)_2$	A	—	1, 13
$CH_3PO(OC_6H_4CH_3\text{-}p)_2$	A	—	1
$CH_3PO(OC_6H_4CH_3\text{-}m)_2$	A	—	1
$CH_3PO(OC_6H_4Cl\text{-}p)_2$	A	—	1
$CH_3PO[OC_6H_2(CH_3)_3]_2$	A	—	1
$CH_3PO(O_2C_6H_4\text{-}o)$	A	100	123
$CH_3PO[N(C_2H_5)_2]_2$	A	—	4
$C_2H_5PO(OC_2H_5)_2$	A	95	14
	B	35	3, 10
$C_2H_5PO(O_2C_6H_4\text{-}o)$	A	100	123
$CH_3(CH_2)_2PO(OC_2H_5)_2$	A	100	13
	B	67	10
$CH_3(CH_2)_2PO(OC_3H_7\text{-}n)_2$	A	—	101
	D	—	6
$CH_3CH\!=\!CHPO(OC_2H_5)_2$	A	—	14
$CH_3(CH_2)_3PO(OC_2H_5)_2$	A	—	14
$CH_3(CH_2)_3PO(OC_4H_9\text{-}n)_2$	A	—	124
	B	90	5
$iso\text{-}C_4H_9PO(OC_4H_9\text{-}iso)_2$	A	—	125
$CH_3(CH_2)_4PO(OC_2H_5)_2$	A	—	14
$CH_3(CH_2)_4PO(OC_4H_9\text{-}n)_2$	B	85	5
$iso\text{-}C_5H_{11}PO(OC_2H_5)_2$	A	—	14
$CH_3(CH_2)_5PO(OC_2H_5)_2$	A	65	5, 14
$CH_3(CH_2)_5PO(OC_4H_9\text{-}n)_2$	B	93	5
$CH_3(CH_2)_6PO(OC_2H_5)_2$	A	—	14
$CH_3(CH_2)_6PO(OC_4H_9\text{-}n)_2$	B	96	5
$CH_3(CH_2)_7PO(OC_2H_5)_2$	A	—	14
$CH_3(CH_2)_7PO(OC_4H_9\text{-}n)_2$	B	88	5
$CH_3(CH_2)_8PO(OC_2H_5)_2$	A	—	5
$CH_3(CH_2)_8PO(OC_4H_9\text{-}n)_2$	B	86	5
$CH_3(CH_2)_9PO(OC_2H_5)_2$	A	—	5
$CH_3(CH_2)_9PO(OC_4H_9\text{-}n)_2$	B	90	5

* A = ester procedure; B = sodium salt procedure; C = triarylcarbinol-phosphorus trichloride procedure; and D = special methods.

TABLE I—Continued

DERIVATIVES OF PHOSPHONIC AND PHOSPHINIC ACIDS PREPARED BY ALKYLATION OF PHOSPHITES OR OTHER TRIVALENT ESTERS

Compound	Method*	Yield %	Reference
$CH_3(CH_2)_{11}PO(OC_2H_5)_2$	A	63, 24	14, 126, 5
$CH_3(CH_2)_{11}PO(OC_4H_9\text{-}n)_2$	B	91	5
$CH_3(CH_2)_{13}PO(OC_2H_5)_2$	A	—	5
$CH_3(CH_2)_{13}PO(OC_4H_9\text{-}n)_2$	B	84	5
$CH_3(CH_2)_{15}PO(OC_4H_9\text{-}n)_2$	B	88	5
$CH_3(CH_2)_{17}PO(OC_4H_9\text{-}n)_2$	B	84	5
$HOCH_2PO(OC_2H_5)_2$	B	100	17
$HOCH_2CH_2PO(OC_2H_5)_2$	D	40	24
$ICH_2PO(OC_2H_5)_2$	A	60	14, 17
$Cl_3CPO(OCH_3)_2$	A	—	19, 127
$Cl_3CPO(OC_2H_5)_2$	A	93	18, 19, 127
$Cl_3CPO(OCH_2CH\!=\!CH_2)_2$	A	—	19, 127
$Cl_3CPO(OC_3H_7\text{-}n)_2$	A	—	19, 127
$Cl_3CPO(OC_3H_7\text{-}iso)_2$	A	60	19, 127
$Cl_3CPO(OC_4H_9\text{-}n)_2$	A	25	18, 19, 127
$Cl_3CPO(OC_4H_9\text{-}iso)_2$	A	60	19, 127
$ClCH_2CH_2PO(OC_2H_5)_2$	A	25	128
$ClCH_2CH_2PO(OCH_2CH_2Cl)_2$	A	40	33, 129
$BrCH_2CH_2PO(OC_2H_5)_2$	A	39	14
		61	130
$BrCH_2CH_2PO(OCH_2CH_2Br)_2$	A	32	131
$Br(CH_2)_3PO(OC_2H_5)_2$	A	90	32, 132
$NC(CH_2)_3PO(OC_2H_5)_2$	B	35–40	133
$C_2H_5O_2CPO(OC_2H_5)_2$	A	—	134
	B	50	10, 135
$CH_3O_2CCH_2PO(OC_2H_5)_2$	B	—	135
$C_2H_5O_2CCH_2PO(OC_2H_5)_2$	A	50	13, 134
	B	50, 58, 95	10, 134–137
$C_2H_5O_2CCH_2PO(OC_4H_9\text{-}iso)_2$	A	50	138
	B	32	139
$C_4H_9O_2CCH_2PO(OC_4H_9\text{-}n)_2$	B	69	137
$C_6H_5O_2CCH_2PO(OC_2H_5)_2$	A	Poor	138
	B	10	138
$C_2H_5O_2CCH(CH_3)PO(OC_2H_5)_2$	A	Poor	10, 134

* A = ester procedure; B = sodium salt procedure; C = triarylcarbinol-phosphorus trichloride procedure; and D = special methods.

TABLE I—Continued

DERIVATIVES OF PHOSPHONIC AND PHOSPHINIC ACIDS PREPARED BY ALKYLATION OF PHOSPHITES OR OTHER TRIVALENT ESTERS

Compound	Method*	Yield %	Reference
$C_2H_5O_2CCH_2CH_2PO(OC_2H_5)_2$	A	35	10, 134
	B	35, 78	35, 133
$C_2H_5O_2CCH(C_2H_5)PO(OC_2H_5)_2$	A	—	134
$C_2H_5O_2C(CH_2)_{10}PO(OC_4H_9\text{-}n)_2$	B	—	22
$(C_2H_5O_2C)_2CHPO(OC_2H_5)_2$	A	30	137
$(C_2H_5O_2C)_2CHPO(OC_4H_9\text{-}n)_2$	A	50	137
$(C_2H_5O)(NaO)P(O)CH_2PO(ONa)(OC_2H_5)$	B	—	10
$(C_2H_5O)_2P(O)CH_2PO(OC_2H_5)_2$	A	—	14
$(C_2H_5O)_2P(O)CH_2CH_2PO(OC_2H_5)_2$	A	26	14
$(C_6H_5O)_2P(O)CH_2CH_2PO(OC_6H_5)_2$	A	60	16
$(o\text{-}C_6H_4O_2)P(O)CH_2CH_2PO(O_2C_6H_4\text{-}o)$	A	60	16
$(C_2H_5O)_2P(O)CH_2CH_2CH_2PO(OC_2H_5)_2$	A	75	14, 32
	B	—	10
$(C_2H_5O)_2P(O)CH_2OCH_2PO(OC_2H_5)_2$	A	63	140
	B	85	28, 140
$C_6H_5CH_2OCH_2PO(OC_2H_5)_2$	A	48	28
	B	26	28
$C_6H_5CH_2OCH_2PO(OC_4H_9\text{-}n)_2$	B	33	28
$C_6H_5OCH_2CH_2PO(OC_2H_5)_2$	A	45	30
$CH_3OCH_2CH_2CH{=}CHCH_2PO(OCH_3)_2$	A, B	70	20
$CH_3OCH_2CH_2CH{=}CHCH_2PO(OC_2H_5)_2$	A, B	70	20
$CH_3OCH_2CH_2CH{=}CHCH_2PO(OC_4H_9\text{-}iso)_2$	A, B	70	20
$CH_2{=}CHCH{=}CHPO(OC_2H_5)_2$	A, B	—	42
$(C_6H_5)_2NCOCH_2PO(OC_2H_5)_2$	A	43	141
$CH_3COPO(OCH_3)_2$	A	58, 80	21, 142
$CH_3COPO(OC_2H_5)_2$	A	12, 50	21, 142
$CH_3COPO(OC_4H_9\text{-}n)_2$	A	50	142
$C_6H_5COPO(OCH_3)_2$	A	72	21
$C_6H_5COPO(OC_2H_5)_2$	A	62	21
$CH_3COCH_2PO(OC_2H_5)_2(?)$	B	—	10
$C_6H_5COCH_2PO(OC_2H_5)_2$	A	30	138
	B	37	138
$C_6H_5CH_2PO(OC_4H_9\text{-}n)_2$	B	85	31
$C_6H_5CH_2PO(OC_6H_5)_2$	A	—	1
$C_6H_5CH_2PO(O_2C_6H_4\text{-}o)$	A	70	123
$4\text{-}CH_3C_6H_4CH_2PO(OC_2H_5)_2$	A	78	31
$4\text{-}CH_3C_6H_4CH_2PO(OC_4H_9\text{-}n)_2$	B	85	31

* A = ester procedure; B = sodium salt procedure; C = triarylcarbinol-phosphorus trichloride procedure; and D = special methods.

TABLE I—Continued

Derivatives of Phosphonic and Phosphinic Acids Prepared by Alkylation of Phosphites or Other Trivalent Esters

Compound	Method*	Yield %	Reference
4-$C_2H_5C_6H_4CH_2PO(OC_2H_5)_2$	A	78	31
4-$C_2H_5C_6H_4CH_2PO(OC_4H_9$-$n)_2$	B	88	31
4-$C_4H_9C_6H_4CH_2PO(OC_4H_9$-$n)$	B	70	31
1-$C_{10}H_7CH_2PO(OC_2H_5)_2$	A	87	31
4-$C_6H_5C_6H_4CH_2PO(OC_4H_9$-$n)_2$	B	60	31
9-Phenanthrylmethanephosphonic acid	B	50	31
1,3,5-$(CH_3)_3C_6H[CH_2PO(OC_4H_9$-$n)_2]_2$-2,4	B	70	31
Bis-9,10-anthracenylenemethanephosphonic acid	B	75	31
$(C_6H_5)_3CPO(OCH_3)_2$	A	60	8
$(C_6H_5)_3CPO(OC_2H_5)_2$	A	100	8
$(C_6H_5)_3CPO(OC_3H_7$-$n)_2$	A	80	8
$(C_6H_5)_3CPO(OC_3H_7$-$iso)_2$	A	80	8
$(C_6H_5)_3CPO(OC_4H_9$-$n)_2$	A	—	8
4-$ClC_6H_4(C_6H_5)_2CPO(OH)_2$	C	93	9
4-$BrC_6H_4(C_6H_5)_2CPO(OH)_2$	C	90	9
3-$HOC_6H_4(C_6H_5)_2CPO(OH)_2$	C	—	9
1-$C_{10}H_7(C_6H_5)_2CPO(OH)_2$	C	—	9
2-$C_{10}H_7(C_6H_5)_2CPO(OH)_2$	C	—	9
4-$CH_3C_6H_4(C_6H_5)_2CPO(OH)_2$	C	75	9
$H_2NCH_2CH_2PO(OH)_2$	A	50	143
	B	—	22
$H_2NCH_2CH_2CH_2PO(OH)_2$	B	—	22
9-Acridinephosphonic acid	A	60	18
2-Thienylmethanephosphonic acid	B	71	144
$C_6H_5(CH_3)PO(OCH_3)$	A	34, 92	25, 26
$C_6H_5(C_2H_5)PO(OC_2H_5)$	A	90	2
$C_6H_5(iso$-$C_4H_9)PO(OC_4H_9$-$iso)$	A	72	27
(Isolated as the free acid)			
$C_6H_5(n$-$C_3H_7)PO(OC_3H_7$-$n)$	A	90	26
$C_6H_5(iso$-$C_3H_7)PO(OC_3H_7$-$iso)$	A	95	2, 11
$(C_6H_5)_3C(C_6H_5)PO(OC_4H_9$-$iso)$	A	58	27
$C_6H_5(C_2H_5OCH_2)PO(OC_2H_5)$	A	84	26
$C_6H_5(CH_3OCH_2)PO(OC_2H_5)$	A	80	26
$C_6H_5(C_2H_5O_2CCH_2)PO(OC_4H_9$-$iso)$	A	64	12
$C_6H_5[C_2H_5O_2CCH(CH_3)]PO(OC_4H_9$-$iso)$	A	76	12
$C_6H_5(Cl_3C)PO(OCH_3)$	A	—	127
$C_6H_5(Cl_3C)PO(OC_2H_5)$	A	—	127

* A = ester procedure; B = sodium salt procedure; C = triarylcarbinol-phosphorus trichloride procedure; and D = special methods.

TABLE I—Continued

Derivatives of Phosphonic and Phosphinic Acids Prepared by Alkylation of Phosphites or Other Trivalent Esters

Compound	Method*	Yield %	Reference
$C_6H_5(Cl_3C)PO(OC_3H_7\text{-}n)$	A	—	127
$C_6H_5(Cl_3C)PO(OC_4H_9\text{-}iso)$	A	—	127
$n\text{-}C_{17}H_{35}CONHCH_2PO(OH)_2$	D	60	7
$C_6H_5CONHCH_2PO(OH)_2$	D	—	7
$n\text{-}C_{17}H_{35}CO(CH_3)NCH_2PO(OH)_2$	D	—	145

* A = ester procedure; B = sodium salt procedure; C = triarylcarbinol-phosphorus trichloride procedure; and D = special methods.

[123] Arbuzov and Valitova, *Bull. acad. sci. U.R.S.S., classe sci. chim.*, **1940**, 529 [*C. A.*, **35**, 3990 (1941)].

[124] Arbuzov and Arbuzova, *J. Russ. Phys. Chem. Soc.*, **62**, 1533 (1930) [*C. A.*, **25**, 2414 (1931)].

[125] Arbuzov and Ivanov, *J. Russ. Phys. Chem. Soc.*, **45**, 690 (1913) [*C. A.*, **7**, 3599 (1913)].

[126] Goebel, U. S. pat. 2,436,141 [*C. A.*, **42**, 3425 (1948)].

[127] Kamai, *Compt. rend. acad. sci. U.R.S.S.*, **55**, 219 (1947) [*C. A.*, **41**, 5863 (1947)].

[128] Kabachnik and Rossiiskaya, *Bull. acad. sci. U.R.S.S., classe sci. chim.*, **1948**, 95 [*C. A.*, **42**, 5846 (1948)].

[129] Kabachnik and Rossiiskaya, *Bull. acad. sci. U.R.S.S., classe sci. chim.*, **1946**, 295 [*C. A.*, **42**, 7241 (1948)].

[130] Kosolapoff, *J. Am. Chem. Soc.*, **70**, 1971 (1948).

[131] Kabachnik and Rossiiskaya, *Bull. acad. sci. U.R.S.S., classe sci. chim.*, **1947**, 389.

[132] Parfent'ev and Shafiev, *Trudy Uzbek. Gosudarst. Univ.*, **15**, 87 (1939) [*C. A.*, **35**, 3963 (1941)].

[133] Nylen, *Ber.*, **59**, 1119 (1926).

[134] Arbuzov and Dunin, *J. Russ. Phys. Chem. Soc.*, **46**, 295 (1914) [*C. A.*, **8**, 2551 (1914)].

[135] Nylen, *Ber.*, **57**, 1023 (1924).

[136] Arbuzov and Kamai, *J. Russ. Phys. Chem. Soc.*, **61**, 619 (1929) [*C. A.*, **23**, 4443 (1929)].

[137] Kosolapoff, *J. Am. Chem. Soc.*, **68**, 1103 (1946).

[138] Arbuzov and Razumov, *J. Gen. Chem. U.S.S.R.*, **4**, 834 (1934) [*C. A.*, **29**, 2145 (1935)].

[139] Kamai, *Trudy Kirovsk. Inst. Khim. Tekhnol. (Kazan)*, **8**, 33 (1940) [*C. A.*, **35**, 2856 (1941)].

[140] Abramov and Azanovskaya, *J. Gen. Chem. U.S.S.R.*, **12**, 270 (1942) [*C. A.*, **37**, 3048 (1943)].

[141] Razumov and Kashurina, *Trudy Kirovsk. Inst. Khim. Tekhnol. (Kazan)*, **8**, 45 (1940) [*C. A.*, **35**, 2474 (1941)].

[142] Kabachnik, Rossiiskaya, and Shepeleva, *Bull. acad. sci. U.R.S.S., classe sci. chim.*, **1947**, 163 [*C. A.*, **42**, 4132 (1948)].

[143] Kosolapoff, *J. Am. Chem. Soc.*, **69**, 2112 (1947).

[144] Kosolapoff, *J. Am. Chem. Soc.*, **69**, 2248 (1947).

[145] Pikl, U. S. pat. 2,328,358 [*C. A.*, **38**, 754 (1944)].

TABLE II

Phosphonic Acids Prepared by Addition of Phosphorus Pentachloride to Unsaturated Compounds

Compound	Yield %	Reference
$C_6H_5CH{=}CHPO_3H_2$	55, 36	38, 39, 36
$C_6H_5(CH_3)C{=}CHPO_3H_2$	10	38
$(C_6H_5)_2C{=}CHPO_3H_2$	45	38
$C_6H_5(4\text{-}ClC_6H_4)C{=}CHPO_3H_2$	15	41
$C_6H_5(2\text{-}CH_3C_6H_4)C{=}CHPO_3H_2$	35	40
$(4\text{-}ClC_6H_4)_2C{=}CHPO_3H_2$	60	41
$(4\text{-}CH_3OC_6H_4)(C_6H_5)C{=}CHPO_3H_2$	40	41
$(4\text{-}ClC_6H_4)(4\text{-}CH_3OC_6H_4)C{=}CHPO_3H_2$	30	41
$C_6H_5(2\text{-}FC_6H_4)C{=}CHPO_3H_2$	30	41
$C_6H_5(4\text{-}C_6H_5C_6H_4)C{=}CHPO_3H_2$	33	41
$(4\text{-}C_6H_5C_6H_4)(4\text{-}CH_3C_6H_4)C{=}CHPO_3H_2$	30	41
$C_6H_5(3\text{-}ClC_6H_4)C{=}CHPO_3H_2$	25	41
$C_6H_5(1\text{-}C_{10}H_7)C{=}CHPO_3H_2$	—	40
$C_6H_5(2\text{-}C_{10}H_7)C{=}CHPO_3H_2$	—	40
$2,4\text{-}(CH_3)_2C_6H_3CH{=}CHPO_3H_2$	45	39
$2,4,6\text{-}(CH_3)_3C_6H_2CH{=}CHPO_3H_2$	55	39
$4\text{-}C_2H_5C_6H_4CH{=}CHPO_3H_2$	40	39
$2\text{-}tert\text{-}C_4H_9C_6H_4CH{=}CHPO_3H_2$	35	39
$2\text{-}C_6H_5C_6H_4CH{=}CHPO_3H_2$	45	39
$3\text{-}C_6H_5C_6H_4CH{=}CHPO_3H_2$	30	39
$4\text{-}C_6H_5C_6H_4CH{=}CHPO_3H_2$	45	39
$2\text{-}C_{10}H_7CH{=}CHPO_3H_2$	40	39
2-Indenephosphonic acid	50	36, 38
2-Vinylfluorene-2′-phosphonic acid	30	39
$CH_2{=}CHCH{=}CHPO_3H_2$	—	42
$C_6H_5CH{=}CHCH{=}CHPO_3H_2$	55, 88	41, 39
$1,4\text{-}H_2O_3PCH{=}C(C_6H_5)C_6H_4C(C_6H_5){=}CHPO_3H_2$	—	41
$(CH_3)_3CCH_2(CH_3)C{=}CHPO_3H_2$	50	39
$C_6H_5CCl{=}CHPO_3H_2$	3	37
$(CH_3)_2CClCH_2PO_3H_2$	Low	40
$(2\text{-}ClC_6H_4)CCl{=}CHPO_3H_2$	50	37
$(2\text{-}CH_3OC_6H_4)CCl{=}CHPO_3H_2$	100	37
$(4\text{-}CH_3OC_6H_4)CCl{=}CHPO_3H_2$	65	37
$(C_6H_5)_2C(C_6H_4\text{-}o)_2C{=}CHPO_3H_2$	—	40
$C_6H_5CH_2CCl{=}CHPO_3H_2$	15	37
$CH_3(CH_2)_4CCl{=}CHPO_3H_2$	—	37

TABLE III

Phosphonic and Phosphinic Acids Prepared by the Friedel-Crafts Reaction

Aromatic Compound	Phosphonic Acids		Phosphinic Acids	
	Yield %	Reference	Yield %	Reference
Benzene	5, 80	50, 55, 57	40	55
Chlorobenzene	25, 82	52, 55	33	55
Bromobenzene	10	52		
Toluene	25, 57	50, 54, 55, 59	22	55, 51
Ethylbenzene	15	52	—	52
Cymene	5	52, 59		
Cumene	—	52	—	52
2-Chlorotoluene	10	58		
1,2-Dichlorobenzene	36	55	2	55
1,4-Dichlorobenzene	3	55		
1,2,4-Trimethylbenzene	25	52	10	52
1,3,5-Trimethylbenzene	5	52, 64		
Biphenyl	5	51, 52, 63		
sym-Diphenylethane	—	52, 51		
Diphenylmethane	—	52, 51		
Naphthalene	15	54		
Anisole	20, 26	52, 60		
Phenetole	—	52		
Diphenyl ether	10	64		
Thiophene	5	53		
N,N-Dimethylaniline	30	65		

TABLE IV

Phosphonic and Phosphinic Acids Prepared by the Addition to Carbonyl Compounds

A. By Addition of P-H Linked Compounds

Products	Reference
$CH_3CH(OH)PO_3H_2$	79
$(CH_3)_2C(OH)PO_3H_2$	72, 73, 76, 77
$iso\text{-}C_4H_9CH(OH)PO_3H_2$	79, 84
$(CH_3)(C_2H_5)C(OH)PO_3H_2$	80
$(C_2H_5)_2C(OH)PO_3H_2$	83
$(CH_3)(n\text{-}C_3H_7)C(OH)PO_3H_2$	78
$C_6H_5CH(OH)PO_3H_2$	74
$(C_6H_5)_2C(OH)PO_3H_2$	78
$C_6H_5CH_2PO_3H_2$	66
$[(CH_3)_2C(OH)]_2PO_2H$	72, 76
$[(CH_3)_2C(OH)](CH_3CHOH)PO_2H$	81
$(iso\text{-}C_4H_9CHOH)_2PO_2H$	68, 71, 82
$(n\text{-}C_6H_{13}CHOH)_2PO_2H$	68, 71
$(n\text{-}C_6H_{13}CHOH)(CH_3CHOH)PO_2H$	81
$(n\text{-}C_6H_{13}CHOH)(iso\text{-}C_3H_7CHOH)PO_2H$	81
$[(CH_3)(C_2H_5)C(OH)](n\text{-}C_6H_{13}CHOH)PO_2H$	82
$(CH_3CHOH)(C_6H_5CHOH)PO_2H$	80
$[(CH_3)_2C(OH)](n\text{-}C_6H_{13}CHOH)PO_2H$	82
$[(CH_3)_2C(OH)](C_6H_5CHOH)PO_2H$	82
$[(C_2H_5)_2C(OH)](C_6H_5CHOH)PO_2H$	82
$[(CH_3)(n\text{-}C_3H_7)C(OH)](C_6H_5CHOH)PO_2H$	82
$[(C_6H_5)(CH_3)C(OH)](CH_3CHOH)PO_2H$	82
$(C_6H_5CHOH)_2PO_2H$	69
$(C_6H_5CH_2)_2PO_2H$	66

B. By Addition of Phosphorus Chlorides

Products	Yield %	Reference
$HOCH_2PO_3H_2$	93	84
$CH_3CH(OH)PO_3H_2$	—	67
$C_2H_5CH(OH)PO_3H_2$	—	67
$iso\text{-}C_3H_7CH(OH)PO_3H_2$	—	67
$iso\text{-}C_4H_9CH(OH)PO_3H_2$	65	67, 84
$n\text{-}C_6H_{13}CH(OH)PO_3H_2$	—	67
$(CH_3)_2C(OH)PO_3H_2$	91	92
$(CH_3)(C_2H_5)C(OH)PO_3H_2$	76	92

TABLE IV—Continued

Phosphonic and Phosphinic Acids Prepared by the Addition to Carbonyl Compounds

B. By Addition of Phosphorus Chlorides—Continued

Products	Yield %	Reference
$(CH_3)_3CC(OH)(CH_3)PO_3H_2$	56	92
$(C_2H_5)(n\text{-}C_3H_7)C(OH)PO_3H_2$	50	92
$CH_3(CH_2)_5C(OH)(H)PO_3H_2$	81	97
$(CH_3)_2C(OH)PO(OC_6H_5)_2$	50	88
$(CH_3)(C_2H_5)C(OH)PO(OC_6H_5)_2$	50	88
$(CH_3)(ClCH_2)C(OH)PO(OC_6H_5)_2$	10	88
$(CH_3)_2C(CH_2COCH_3)PO(OH)C_4H_9\text{-}n$	—	89
$(CH_3)_2C(CH_2COCH_3)PO(OC_6H_5)_2$	41	89
$CH_3COCH_2CH_2PO(OC_6H_5)_2$	14	89
$CH_3(CH_2)_3CH(C_2H_5)CH(PO_3H_2)CH_2COCH_3$	20	89
$(CH_3)_2C(PO_3H_2)CH_2COCH_3$	33	89
$CH_3C(PO_3H_2)(OH)CO_2H$	40	94
$C_6H_5CH(OH)PO_3H_2$	84, 72	67, 84, 93
$(C_6H_5)_2C(OH)PO_3H_2$	50	92
$(C_6H_5CH_2CH_2)_2C(OH)PO_3H_2$	56	92
$C_6H_5C(PO_3H_2)(Cl)CH_3$	82	97
$C_6H_5C(PO_3H_2)(OH)CH_3$	81	97
$CH_2\!=\!C(PO_3H_2)C_6H_5$	63, 90	92, 97
$C_6H_5COCH_2CH(PO_3H_2)COC_6H_5$	81	89
$C_6H_5CH(PO_3H_2)CH_2COC_6H_5$	78	90, 95
$4\text{-}CH_3OC_6H_4CH(PO_3H_2)CH_2COC_6H_5$	89	95
$C_6H_5CH(PO_3H_2)CH_2COC_6H_4Cl\text{-}4$	91	96
$C_6H_5CH(PO_3H_2)CH_2COCH\!=\!CHC_6H_5$	90	95
$C_6H_5P(O)(OH)[CH(C_6H_5)CH_2COC_6H_5]$	90	85
$C_6H_5CH\!=\!CHCH(PO_3H_2)CH_2COC_6H_5$	—	86
$C_6H_5CH_2C(PO_3H_2)(OH)CH_2C_6H_5$	50	92
$C_6H_5C(PO_3H_2)(OH)CH_2CH_2C_6H_5$	48	92
$C_6H_5CH[P(C_6H_5)O_2H]CH_2COCH\!=\!CHC_6H_5$	70	86
$C_6H_5CH\!=\!CHCH[P(C_6H_5)O_2H]CH_2COC_6H_5$	64	86
$C_6H_5CH(CH_2COC_6H_5)P(O)(OH)C_4H_9$	50	89
$C_6H_5CH(OH)P(O)(OH)OC_6H_5$	90	88
$C_6H_5CH(OH)P(O)(OH)OCH_3$	50	88
$C_6H_5CH(OH)P(O)(OH)OC_2H_5$	50	88
$C_6H_5CH(CH_2COC_6H_5)P(O)(OC_6H_5)_2$	30	88
$(CH_3)(C_6H_5)C(OH)P(O)(OC_6H_5)_2$	60	88
$C_6H_5CH(OH)P(O)(OC_6H_5)_2$	40	88
$C_6H_5P(O)(OH)(CHOHCH_3)$	—	52
$C_6H_5P(O)(OH)(CHOHC_6H_5)$	—	52
9-Keto-10-hydroxyphenanthrene-10-phosphonic acid	—	67

TABLE V

CHLOROPHOSPHINES PREPARED FROM ORGANOMERCURY INTERMEDIATES

Products	Yield %	Reference
$C_2H_5PCl_2$	—	106, 107
n-$C_3H_7PCl_2$	—	107
iso-$C_3H_7PCl_2$	—	107
n-$C_4H_9PCl_2$	61	89
iso-$C_4H_9PCl_2$	—	107
iso-$C_5H_{11}PCl_2$	—	107
$C_6H_5PCl_2$	100	100
$(C_6H_5)_2PCl$	64	146, 147
2-$CH_3C_6H_4PCl_2$	78	52, 59, 148
3-$CH_3C_6H_4PCl_2$	50	52
4-$CH_3C_6H_4PCl_2$	100	59
4-$CH_3OC_6H_4PCl_2$	Poor	52
4-$C_2H_5OC_6H_4PCl_2$	Poor	52
2,4-$(CH_3)_2C_6H_3PCl_2$	20	61
2,4,5-$(CH_3)_3C_6H_2PCl_2$	20	52
1-$C_{10}H_7PCl_2$	—	149
2-$C_{10}H_7PCl_2$	—	54
4-$(CH_3)_2NC_6H_4PCl_2$	45	65
$C_6H_5(4$-$CH_3C_6H_4)PCl$	64	51, 108, 150
$C_6H_5(4$-$BrC_6H_4)PCl$	47–53	109
$C_6H_5(4$-$CH_3OC_6H_4)PCl$	35	109
$C_6H_5[2,4,5$-$(CH_3)_3C_6H_2]PCl$	30	51
$(4$-$CH_3C_6H_4)_2PCl$	35	51

[146] Michaelis, *Ber.*, **10**, 627 (1877).
[147] Michaelis and Link, *Ann.*, **207**, 193 (1881).
[148] Michaelis and Panek, *Ber.*, **13**, 653 (1880).
[149] Kelbe, *Ber.*, **9**, 1051 (1876); **11**, 1499 (1878).
[150] Wedekind, *Ber.*, **45**, 2933 (1912).

TABLE VI

Chlorophosphines Prepared by Thermal Decomposition of Phosphonium Compounds

Starting Material	Product	Yield %	Reference
$(C_6H_5)_3PCl_2$	$(C_6H_5)_2PCl$	40	110, 112
$(2\text{-}CH_3C_6H_4)_3PCl_2$	$(2\text{-}CH_3C_6H_4)_2PCl$	50	112
$(4\text{-}CH_3C_6H_4)_3PCl_2$	$(4\text{-}CH_3C_6H_4)_2PCl$	50	112
$(1\text{-}C_{10}H_7)_3PCl_2$	$(1\text{-}C_{10}H_7)_2PCl$	30	112
$(4\text{-}CH_3C_6H_4)_2(C_6H_5)PCl_2$	$(4\text{-}CH_3C_6H_4)(C_6H_5)PCl$	50	112
$(2\text{-}ClC_6H_4)_3PCl_2$	$(2\text{-}ClC_6H_4)_2PCl$	60	112
$(4\text{-}ClC_6H_4)_3PCl_2$	$(4\text{-}ClC_6H_4)_2PCl$	55	112
$(4\text{-}O_2NC_6H_4)_3PCl_2$	$(4\text{-}O_2NC_6H_4)_2PCl$	60	112
$[4\text{-}(CH_3)_2NC_6H_4]_3PCl_2$	$[4\text{-}(CH_3)_2NC_6H_4]_2PCl$	50	112
$(CH_3)_2(C_6H_5)PCl_2$	$(CH_3)(C_6H_5)PCl$	60	112
$(C_2H_5)_2(C_6H_5)PCl_2$	$(C_2H_5)(C_6H_5)PCl$	50	112
$(C_2H_5)_3PCl_2$	$(C_2H_5)_2PCl$	70	111, 112
$(n\text{-}C_3H_7)_3PCl_2$	$(n\text{-}C_3H_7)_2PCl$	60	112
$(n\text{-}C_4H_9)_3PCl_2$	$(n\text{-}C_4H_9)_2PCl$	55	112
$(CH_3)(C_2H_5)_2PCl_2$	$(CH_3)(C_2H_5)PCl$	45	112
$(CH_3)(4\text{-}CH_3C_6H_4)(C_5H_{10}N)_2POH$	$(CH_3)(4\text{-}CH_3C_6H_4)PO_2H$	75	108, 114
$(CH_3)(C_6H_5)(C_5H_{10}N)_2POH$	$(CH_3)(C_6H_5)PO_2H$	75	52, 108, 114

CHAPTER 7

THE HALOGEN-METAL INTERCONVERSION REACTION WITH ORGANOLITHIUM COMPOUNDS

REUBEN G. JONES

Eli Lilly and Company

AND

HENRY GILMAN

Iowa State College

CONTENTS

	PAGE
INTRODUCTION	340
SCOPE OF THE REACTION	342
Nature of the Halogen Atoms	342
Nature of the Organolithium Compounds	343
Solvents	344
Types of Halogen Compounds that Undergo Interconversion	344
SIDE REACTIONS	348
EXPERIMENTAL CONDITIONS	351
EXPERIMENTAL PROCEDURES	352
Preparation of n-Butyllithium	352
Preparation of Phenyllithium	353
2-(m-Trifluoromethylphenyl)quinoline	354
2,4,5-Triphenyl-3-furancarboxylic Acid	354
Phenyl-di-(p-aminophenyl)arsenic	354
Triphenyl-o-hydroxymethylphenyllead	355
2,5-Thiophenedicarboxylic Acid	355
p-Bromophenyltrimethylsilane	355
Diphenyl-2,4-dimethoxy-5-bromophenylcarbinol	356
TABULAR SURVEY OF HALOGEN-LITHIUM INTERCONVERSION REACTIONS	356
Table I. Interconversions with n-Butyllithium Followed by Carbonation to Yield Carboxylic Acids	358
Table II. Miscellaneous Halogen-Metal Interconversion Reactions	362

INTRODUCTION

The reaction of an organic halide with an organometallic compound in which the metal and the halogen atoms exchange places is known as the halogen-metal interconversion reaction. This interconversion was independently discovered by Gilman and co-workers,[1,2] who found that o-bromoanisole reacted with n-butyllithium to yield o-anisyllithium and n-butyl bromide,

$$n\text{-}C_4H_9Li + \underset{OCH_3}{\underset{|}{C_6H_4}}\text{-}Br \rightarrow \underset{OCH_3}{\underset{|}{C_6H_4}}\text{-}Li + n\text{-}C_4H_9Br$$

and by Wittig, Pockels, and Dröge,[3] who observed that 4,6-dibromoresorcinol dimethyl ether and phenyllithium reacted to yield 2,4-dimethoxy-5-bromophenyllithium and bromobenzene.

$$C_6H_5Li + \underset{Br\ \ \ Br}{\underset{|\ \ \ \ |}{C_6H_2(OCH_3)_2}} \rightarrow \underset{Br\ \ \ Li}{\underset{|\ \ \ \ |}{C_6H_2(OCH_3)_2}} + C_6H_5Br$$

Numerous studies have established the fact that the halogen-lithium interconversion is a general and widely applicable reaction.

Ordinarily organolithium (RLi) compounds are employed as synthetic intermediates, and immediately after their preparation they are used in further reactions. In general, organolithium compounds undergo all the reactions that are characteristic of the well-known Grignard reagents (RMgX). There is usually no advantage in using the more expensive organolithium compounds for preparations that can be carried out successfully with Grignard reagents. However, because of their greater reactivity, organolithium compounds often may be successful in reactions where Grignard reagents fail. For example, addition of organolithium compounds to the azomethine linkage in pyridines or quinolines is a valuable method for preparing the 2-substituted compounds.

$$\underset{N}{C_5H_5N} + RLi \rightarrow \left[\underset{\underset{Li}{N}}{R\text{-}C_5H_5N}\right] \xrightarrow[(O)]{H_2O} R\text{-}C_5H_4N + LiOH$$

Grignard reagents add very slowly if at all to such azomethine linkages. Organothallium compounds of the type R_3Tl are prepared readily by the reaction of an R_2TlX compound with RLi. Grignard reagents do not react with R_2TlX compounds to give R_3Tl derivatives.

---

[1] Gilman, Langham, and Jacoby, *J. Am. Chem. Soc.*, **61**, 106 (1939).

[2] Gilman and Jacoby, *J. Org. Chem.*, **3**, 108 (1938).

[3] Wittig, Pockels, and Dröge, *Ber.*, **71**, 1903 (1938).

The simple reagents such as phenyllithium and *n*-butyllithium are most easily and economically prepared by the reaction of an organic halide with metallic lithium.

$$\text{RX} + 2\text{Li} \rightarrow \text{RLi} + \text{LiX}$$

Many organic halides do not react satisfactorily with metallic lithium to form RLi compounds, or with metallic magnesium to form Grignard reagents. However, the desired organolithium compound often can be obtained by a halogen-metal interconversion reaction. Thus the halogen-metal interconversion greatly extends the utility of organolithium and Grignard-type reactions. For example, *o*-hydroxyphenyllithium cannot be obtained by a reaction of *o*-bromophenol with metallic lithium, but, under the proper conditions, *o*-bromophenol reacts with *n*-butyllithium to give a high yield of the lithium salt of *o*-hydroxyphenyllithium.[4]

Halogen-metal interconversion reactions also have been observed with organometallic compounds of sodium,[5] magnesium,[6,7] barium,[8] and aluminum.[7] Dialkylmercury compounds undergo halogen-metal interconversion with aryl iodides under the influence of an organolithium compound as catalyst.[9] The present discussion, however, will be concerned only with halogen-metal interconversion reactions involving organolithium compounds.

The mechanism of the halogen-metal interconversion reaction has not been thoroughly investigated. However, it has been established that the reaction is reversible [10] and rapid even at low temperatures.[4,10] The interconversion has been pictured as an exchange between lithium and an electropositive halogen atom, and it is analogous to other halogen exchange reactions.[11]

The ease of interconversion is proportional to the degree of positive polarization of the halogen atoms. The relatively positive iodine and

---

[4] Gilman and Arntzen, *J. Am. Chem. Soc.*, **69**, 1537 (1947).
[5] Gilman, Moore, and Baine, *J. Am. Chem. Soc.*, **63**, 2479 (1941).
[6] Gilman and Spatz, *J. Am. Chem. Soc.*, **63**, 1553 (1941).
[7] Gilman and Haubein, *J. Am. Chem. Soc.*, **67**, 1033 (1945).
[8] Gilman, Haubein, O'Donnell, and Woods, *J. Am. Chem. Soc.*, **67**, 922 (1945).
[9] Gilman and Jones, *J. Am. Chem. Soc.*, **63**, 1443 (1941).
[10] Gilman and Jones, *J. Am. Chem. Soc.*, **63**, 1441 (1941).
[11] Meerwein, Hofmann, and Schill, *J. prakt. Chem.*, **154**, 266 (1940).

bromine atoms exchange readily with lithium, the less positive chlorine exchanges less readily, and the negative fluorine atom does not undergo halogen-metal interconversion at all.[12] The removal of the positive halogen atom is probably brought about through a nucleophilic attack by the anion of the organolithium compound.[12a] In this respect the halogen-metal interconversion resembles the hydrogen-metal interconversion or metalation reaction, the mechanism of which is probably a nucleophilic attack of the carbanion of the organometallic compound on a proton. Resonance and inductive forces of substituents in aromatic halogen compounds may have important effects on the halogen-metal interconversion reaction.[12a]

The position of equilibrium in the accompanying reaction is largely dependent upon the relative electronegativities of the radicals R and R'.

$$RLi + R'X \rightleftarrows R'Li + RX$$

Lithium tends to become attached to the more electronegative R group. Thus, n-propyllithium reacts with an equimolecular quantity of α-bromonaphthalene to give n-propyl bromide and α-naphthyllithium in 95% yield.[13] If the two radicals R and R' are of approximately equal electronegativity the yield of R'Li will be in the neighborhood of 50%. For example, the equilibrium mixture obtained from equal molecular quantities of phenyllithium and p-iodotoluene or from p-tolyllithium and iodobenzene contains about equal quantities of the four substances phenyllithium, p-tolyllithium, iodobenzene, and p-iodotoluene.[10] With a proper choice of organolithium compound, RLi, it is generally possible to convert the halide R'X to the desired R'Li in high yield.

SCOPE OF THE REACTION

Nature of the Halogen Atoms. For practical purposes the halogen-lithium interconversion is confined almost entirely to bromides and iodides. A few special examples of interconversions involving chlorides have been found,[14,15,16] but for the most part chlorides do not undergo the reaction.[6,13] No organic fluorides have been observed to enter into halogen-metal interconversion reactions.[12,17] Usually the more readily

---

[12] Wittig, *Naturwissenschaften*, **30**, 696 (1942).
[12a] Sunthankar and Gilman, *J. Org. Chem.*, **16**, 8 (1951).
[13] Gilman and Moore, *J. Am. Chem. Soc.*, **62**, 1843 (1940).
[14] Gilman and Haubein, *J. Am. Chem. Soc.*, **67**, 1420 (1945).
[15] Gilman and Melstrom, *J. Am. Chem. Soc.*, **68**, 103 (1946).
[16] Wittig and Witt, *Ber.*, **74**, 1474 (1941).
[17] Wittig and Fuhrmann, *Ber.*, **73**, 1197 (1940).

available bromides have been used instead of iodides, although in general iodides are more reactive [6] and give higher yields.[18,19]

Nature of the Organolithium Compounds. In the accompanying reaction with α-bromonaphthalene the order of decreasing effectiveness of some RLi compounds in diethyl ether solution is n-C_3H_7Li, C_2H_5Li, n-C_4H_9Li, C_6H_5Li, and CH_3Li.[13]

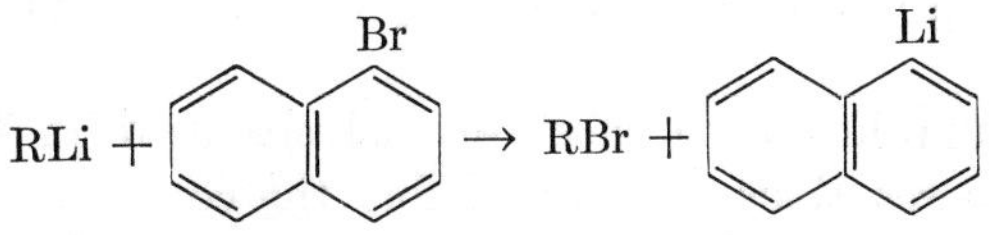

$$RLi + \underset{}{\text{(naphthyl-Br)}} \rightarrow RBr + \underset{}{\text{(naphthyl-Li)}}$$

In an analogous study of the interconversion of halogenated phenyl ethers [19] with various organolithium compounds it was found that n-propyllithium and n-butyllithium gave higher yields of interconversion products than did phenyllithium. Methyllithium gave no interconversion products under the same conditions.

A color test for the more reactive organolithium compounds is based upon an interconversion reaction.[20] The solution to be tested is treated with p-bromodimethylaniline followed by benzophenone, and then the mixture is hydrolyzed and acidified. The appearance of a red color indicates that halogen-metal interconversion has taken place as shown by the accompanying reactions.

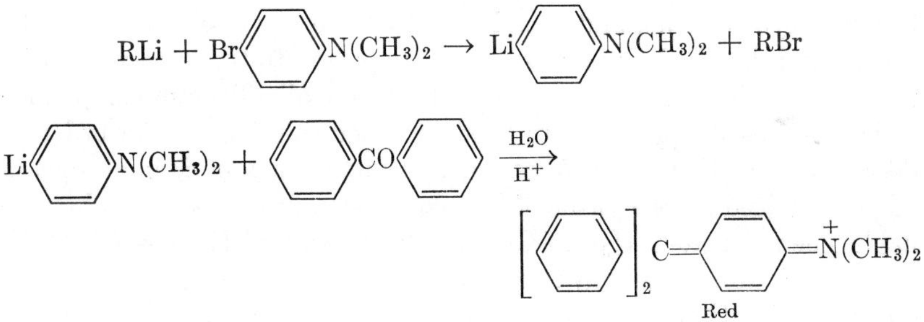

$$RLi + Br\text{-}C_6H_4\text{-}N(CH_3)_2 \rightarrow Li\text{-}C_6H_4\text{-}N(CH_3)_2 + RBr$$

$$Li\text{-}C_6H_4\text{-}N(CH_3)_2 + (C_6H_5)_2CO \xrightarrow[H^+]{H_2O} [C_6H_5]_2C\text{=}C_6H_4\text{=}\overset{+}{N}(CH_3)_2$$

Red

With this test it can be demonstrated that aliphatic organolithium compounds, with the exception of methyllithium, readily undergo interconversion with p-bromodimethylaniline, but aryl organolithium types such as phenyllithium do not.

Methyllithium slowly undergoes interconversion with some of the most reactive halides like o-bromoanisole and p-iodoanisole to give low yields of the expected products.[19] In general, methyllithium is of no value for interconversion reactions. Halogen-metal interconversion

---

[18] Gilman, Langham, and Moore, *J. Am. Chem. Soc.*, **62**, 2327 (1940).
[19] Langham, Brewster, and Gilman, *J. Am. Chem. Soc.*, **63**, 545 (1941).
[20] Gilman and Swiss, *J. Am. Chem. Soc.*, **62**, 1847 (1940).

between phenyllithium and certain aryl bromides and iodides such as
o-bromo- and o-iodo-anisole [16] and 2,5-diiodothiophene [21] takes place
quite satisfactorily. However, in addition to being less reactive than
the alkyllithium compounds, phenyllithium suffers from other disadvantages. For example, in contrast with n-butyllithium, which reacts
normally with p-bromoanisole to yield the expected p-anisyllithium,
phenyllithium gives largely 2-methoxy-5-bromophenyllithium [17] (see
p. 349).

n-Butyllithium has been most extensively used in halogen-metal
interconversion reactions. It is probably the compound of choice for
this purpose when diethyl ether is employed as the solvent. As stated
above, ethyllithium and n-propyllithium appear to be slightly more
effective in the reaction with α-bromonaphthalene than is n-butyllithium. n-Butyllithium, however, possesses the advantage of greater
stability, and it can be prepared in better yields than ethyl- or propyllithium. There appears to be no advantage in the use of higher alkyllithium compounds, for n-amyllithium is less effective than n-butyllithium.[5]

Solvents. In addition to the commonly used diethyl ether a number
of other solvents have been investigated as mediums for halogen-metal
interconversion reactions. Reaction takes place most rapidly in diethyl
ether; di-n-butyl ether is a satisfactory solvent, but dimethylaniline,
benzene, cyclohexane, and petroleum ether are less effective.[13] On the
other hand, low-boiling petroleum ether appears to be especially useful
for reactions involving secondary and tertiary alkyllithium compounds
which cannot be prepared or used in ether solution.[5] The order of
decreasing effectiveness of some organolithium compounds in petroleum
ether as judged by the yields of interconversion products with α-bromonaphthalene is: sec-C_4H_9Li, i-C_3H_7Li, t-C_4H_9Li, n-C_4H_9Li = i-C_4H_9Li,
n-C_3H_7Li.[5] Reactions may take an entirely different course in petroleum
ether than in ether. An example is the reaction of β-bromostyrene with
n-butyllithium. In low-boiling petroleum ether the product is β-styryllithium, but in diethyl ether solution the product is phenylethynyllithium.[18] This reaction will be discussed in more detail later.

Mixtures of ether and benzene often have been used in halogen-metal
interconversions. Such mixtures allow the reactions to be carried out
at reflux temperatures of 60° to 65°. These higher temperatures are
sometimes desirable, but usually the reactions are best conducted at
lower temperatures.

Types of Halogen Compounds that Undergo Interconversion. Generally the monobromides and iodides of simple aromatic compounds

[21] Campaigne and Foye, *J. Am. Chem. Soc.*, **70**, 3941 (1948).

such as benzene, naphthalene, anisole, and dimethylaniline readily enter into halogen-metal exchange with n-butyllithium to give high yields of aromatic lithium compounds. Much of the exploratory work on the interconversion reaction has been done with these simple types. As stated before, the interconversion reaction usually offers no advantage for the preparation of many of these organolithium compounds which can be obtained directly from the organic halide and metallic lithium. On the other hand, the preparation of certain aromatic lithium compounds directly from metallic lithium is accompanied by the formation of undesirable by-products. For example, perylene appears as a contaminant in preparations of α-naphthyllithium.[22] Other as yet unidentified substances are formed in β-naphthyllithium and p-biphenyllithium preparations.[23] These troublesome by-products, which often cause difficulty in the isolation and purification of the final reaction products, can be largely avoided when the organolithium compound is prepared by the halogen-lithium interconversion method.[23] Some halides that do not react at all successfully with lithium metal readily undergo halogen-metal interconversion to yield the desired organolithium compounds. Among this group may be mentioned m-bromobenzotrifluoride,[24] 2-bromobenzofuran,[25] various halides of pyridine [15, 26, 27] and quinoline,[6, 26] 1-bromo-3,4-dimethoxydibenzofuran,[28] a number of carbazole halides,[6] 9-bromophenanthrene,[29] and many others.

An aromatic compound containing two or more halogen atoms, like 4,6-dibromoresorcinol dimethyl ether [30] or 4,4'-dibromodiphenyl ether,[18] may react with one equivalent of an organolithium compound under controlled conditions so that only one of the halogen atoms is replaced by lithium. Unsymmetrical polyhalides such as 2,5-dibromotoluene may yield a mixture of isomeric mono-interconversion products.[18]

$$n\text{-}C_4H_9Li + \underset{Br}{\underset{|}{\bigcirc}}\text{-CH}_3\text{-Br} \xrightarrow[2\text{ min.}]{\text{Ether}} \xrightarrow[H^+]{CO_2} \underset{HO_2C}{\underset{|}{\bigcirc}}\text{-CH}_3\text{-Br} + \underset{Br}{\underset{|}{\bigcirc}}\text{-CH}_3\text{-CO}_2H$$
$$36\% \qquad\qquad 22\%$$

When an excess of organolithium compound is used, both halogen atoms of an aromatic dihalide may be replaced by lithium.[18, 21, 28] For example,

---

[22] Gilman and Brannen, *J. Am. Chem. Soc.*, **71**, 657 (1949).
[23] Gilman, Dunn, and Brannen, unpublished results.
[24] Gilman and Woods, *J. Am. Chem. Soc.*, **66**, 1981 (1944).
[25] Gilman and Melstrom, *J. Am. Chem. Soc.*, **70**, 1655 (1948).
[26] Gilman and Spatz, *J. Am. Chem. Soc.*, **62**, 446 (1940).
[27] Murray, Foreman, and Langham, *J. Am. Chem. Soc.*, **70**, 1037 (1948).
[28] Gilman, Swislowsky, and Brown, *J. Am. Chem. Soc.*, **62**, 348 (1940).
[29] Gilman and Cook, *J. Am. Chem. Soc.*, **62**, 2813 (1940).
[30] Wittig and Pockels, *Ber.*, **72**, 89 (1939).

4,6-dibromoresorcinol dimethyl ether reacts with two equivalents of phenyllithium to form the corresponding 4,6-dilithium compound.[30]

$$2C_6H_5Li + \underset{Br}{\underset{}{CH_3O}}\text{─}\underset{Br}{\underset{}{OCH_3}} \rightarrow \underset{Li}{\underset{}{CH_3O}}\text{─}\underset{Li}{\underset{}{OCH_3}}$$

The reaction of 2,8-dibromodibenzofuran with two equivalents of n-butyllithium followed by carbonation yields 2,8-dibenzofurandicarboxylic acid.[31]

$$2n\text{-}C_4H_9Li + \text{Br—[dibenzofuran]—Br} \xrightarrow[H^+]{CO_2} HO_2C\text{—[dibenzofuran]—}CO_2H$$

An ether linkage in the *ortho* position has a strong activating effect upon aryl halides, causing them to undergo interconversion more readily. In the reaction of 2,4,6-tribromoanisole with excess n-butyllithium only the two *ortho* bromine atoms are replaced.[18]

$$2n\text{-}C_4H_9Li + \underset{Br}{\underset{}{Br\text{—[}OCH_3\text{]—}Br}} \rightarrow \underset{Br}{\underset{}{Li\text{—[}OCH_3\text{]—}Li}}$$

Similarly, the reaction of 3,3′,5,5′-tetrabromo-2,2′-dimethoxybiphenyl with phenyllithium followed by carbonation yields 5,5′-dibromo-2,2′-dimethoxy-3,3′-biphenyldicarboxylic acid.[32]

$$2C_6H_5Li + \text{Br—[OCH}_3\text{]—[OCH}_3\text{]—Br (Br,Br)} \xrightarrow[H^+]{CO_2} HO_2C\text{—[OCH}_3\text{]—[OCH}_3\text{]—}CO_2H \text{ (Br,Br)}$$

Halogenated phenols [4, 33] and thiophenols [34] are easily converted to the corresponding lithium compounds by reaction with n-butyllithium.

$$2n\text{-}C_4H_9Li + \underset{}{\text{HO—[}C_6H_4\text{]—Br}} \rightarrow \underset{}{\text{LiO—[}C_6H_4\text{]—Li}} + n\text{-}C_4H_{10} + n\text{-}C_4H_9Br$$

$$2n\text{-}C_4H_9Li + \underset{}{\text{HS—[}C_6H_4\text{]—Br}} \rightarrow \underset{}{\text{LiS—[}C_6H_4\text{]—Li}} + n\text{-}C_4H_{10} + n\text{-}C_4H_9Br$$

---

[31] Gilman, Willis, and Swislowsky, *J. Am. Chem. Soc.*, **61**, 1371 (1939).
[32] Gilman, Swiss, and Cheney, *J. Am. Chem. Soc.*, **62**, 1963 (1940).
[33] Gilman, Arntzen, and Webb, *J. Org. Chem.*, **10**, 374 (1945).
[34] Gilman and Gainer, *J. Am. Chem. Soc.*, **69**, 1946 (1947).

In these transformations one equivalent of butyllithium is consumed by the active hydrogen atom. It is necessary therefore to use at least two moles of butyllithium per mole of halide. Other aryl halides containing hydroxyl groups, such as the bromobenzyl alcohols and bromophenylethyl alcohols,[35] also are subject to interconversion reactions.

$$2n\text{-}C_4H_9Li + HOH_2C\text{-}C_6H_4\text{-}Br \rightarrow LiOH_2C\text{-}C_6H_4\text{-}Li + n\text{-}C_4H_{10} + n\text{-}C_4H_9Br$$

p-Bromoaniline reacts with three equivalents of n-butyllithium, yielding p-dilithiumaminophenyllithium.[34,36,37,38]

$$3n\text{-}C_4H_9Li + H_2N\text{-}C_6H_4\text{-}Br \rightarrow Li_2N\text{-}C_6H_4\text{-}Li + 2n\text{-}C_4H_{10} + n\text{-}C_4H_9Br$$

This appears to be a valuable intermediate for a number of syntheses. Among the uses it has found is the micro-scale preparation of p-aminobenzoic acid containing radioactive carbon.[27]

$$Li_2N\text{-}C_6H_4\text{-}Li + C^{14}O_2 \xrightarrow{H^+} H_2N\text{-}C_6H_4\text{-}C^{14}O_2H$$

At low temperatures it is possible to bring about certain interconversions which cannot be realized at room temperature because of interfering side reactions. The reactions of n-butyllithium with bromo- and iodo-benzoic acid at $-70°$ lead to the formation of the corresponding carboxyphenyllithium derivatives.[4,26]

$$2n\text{-}C_4H_9Li + HO_2C\text{-}C_6H_4\text{-}I \rightarrow LiO_2C\text{-}C_6H_4\text{-}Li + n\text{-}C_4H_{10} + n\text{-}C_4H_9I$$

There is evidence [35] that p-cyanophenyllithium can be obtained from p-cyanobromobenzene and n-butyllithium at $-70°$. At ordinary temperatures organolithium compounds like n-butyllithium react rapidly with carboxyl and cyano groups to yield carbinols and ketones.

The halogen-metal interconversion method has been used to obtain organolithium compounds of pyridine and quinoline which are inaccessi-

---

[35] Gilman and Melstrom, *J. Am. Chem. Soc.*, **70**, 4177 (1948).
[36] Gilman and Stuckwisch, *J. Am. Chem. Soc.*, **63**, 2844 (1941).
[37] Gilman and Stuckwisch, *J. Am. Chem. Soc.*, **64**, 1007 (1942).
[38] Gilman and Stuckwisch, unpublished results.

ble otherwise. At $-35°$, 3-bromopyridine reacts with n-butyllithium to yield 3-pyridyllithium.[26, 27]

$$n\text{-}C_4H_9Li + \underset{N}{\underset{|}{\bigcirc}}\!\!\text{Br} \rightarrow \underset{N}{\underset{|}{\bigcirc}}\!\!\text{Li}$$

Likewise bromo- and iodo-quinolines [6, 26] may be converted to the quinolyllithium analogs by reaction with n-butyllithium at $-35°$. In these reactions it is essential to use low temperatures and short reaction periods. At higher temperatures secondary reactions predominate, particularly addition of the organolithium agent to the azomethine linkage.

Halogen-metal interconversion is not restricted to aryl halides. Aliphatic iodides and bromides and even some chlorides are subject to the reaction. At low temperatures either combination, n-butyllithium–ethyl iodide or ethyllithium–n-butyl iodide, undergoes reaction to form an equilibrium mixture containing all four components.[10]

$$n\text{-}C_4H_9Li + C_2H_5I \rightleftarrows n\text{-}C_4H_9I + C_2H_5Li$$

Phenylethynyl bromide and chloride react with n-butyllithium to yield phenylpropiolic acid after carbonation.[14]

$$n\text{-}C_4H_9Li + C_6H_5C{\equiv}CCl \xrightarrow[H_2O]{CO_2} C_6H_5C{\equiv}CCO_2H$$

The reaction of an organolithium compound with benzyl bromide [16] or benzyl chloride [39] appears to yield benzyllithium as an intermediate product, but this rapidly decomposes under the reaction conditions (see p. 350). In petroleum ether solution the reaction of β-bromostyrene with n-butyllithium followed by carbonation gives cinnamic acid. In diethyl ether the reaction takes a different course, and the only product isolated is phenylpropiolic acid,[18] which probably arises as indicated in the following sequence of reactions.

$$n\text{-}C_4H_9Li + C_6H_5CH{=}CHBr \begin{cases} \xrightarrow[\text{ether}]{\text{Petroleum}} C_6H_5CH{=}CHLi \xrightarrow[H^+]{CO_2} C_6H_5CH{=}CHCO_2H \\ \xrightarrow{\text{Ether}} [C_6H_5C{\equiv}CH] \xrightarrow{n\text{-}C_4H_9Li} C_6H_5C{\equiv}CLi \xrightarrow[H^+]{CO_2} C_6H_5C{\equiv}CCO_2H \end{cases}$$

SIDE REACTIONS

As previously pointed out it is often possible to avoid or at least minimize interfering side reactions by conducting certain halogen-metal interconversions at low temperatures and for short periods of time.

---

[39] Gilman and Haubein, *J. Am. Chem. Soc.*, **66**, 1515 (1944).

In other cases a relatively long reaction period may be desirable. For example, the best yields of p-hydroxyphenyllithium are obtained by heating n-butyllithium and p-bromophenol in ether solution under reflux for one and one-half hours or longer.[33] Even with a careful selection of experimental conditions it is occasionally difficult to avoid certain side reactions.

The *meta* and *para* brominated anisoles have been observed to undergo hydrogen-lithium exchange in some cases in preference to halogen-metal interconversion.[1, 3, 17, 18]

$$C_6H_5Li + \underset{Br}{\underset{}{\overset{OCH_3}{\bigcirc}}} \rightarrow \underset{Br}{\underset{}{\overset{OCH_3}{\bigcirc}}}{Li} + C_6H_6$$

Similar hydrogen-lithium exchange reactions with 2-bromodibenzofuran [40] and 3-bromodibenzofuran [31] are on record. It is significant that side reactions of this type are favored when the metalating agent is an aromatic lithium compound like phenyllithium and when the reaction is allowed to proceed for a long period of time. By using excess n-butyllithium and a short reaction period it is possible to obtain good yields of the expected halogen-metal interconversion products from the halogenated anisoles and dibenzofurans.[41]

The organometallic compounds of tin,[42] lead,[42] mercury,[9, 42] thallium,[43] bismuth,[44] and certain other metals react with organolithium compounds in such a way that a metal-metal exchange takes place. Generally this metal-metal interconversion is more rapid than the halogen-metal interconversion.[9] The reaction of n-butyllithium with di-p-bromophenylmercury, for example, yields only di-n-butylmercury and p-bromophenyllithium.

$$2n\text{-}C_4H_9Li + \underset{Br}{\bigcirc}\text{-Hg-}\underset{Br}{\bigcirc} \rightarrow (n\text{-}C_4H_9)_2Hg + 2\underset{Br}{\bigcirc}{Li}$$

It has not been possible to use the halogen-metal interconversion method to prepare an organolithium compound which also contains mercury, lead, tin, or other such metal.

The products of many halogen-metal interconversion reactions are unstable. Phenyllithium reacts with ethylene dibromide to yield

[40] Gilman, Cheney, and Willis, *J. Am. Chem. Soc.*, **61**, 951 (1939).
[41] Gilman, Langham, and Willis, *J. Am. Chem. Soc.*, **62**, 346 (1940).
[42] Gilman, Moore, and Jones, *J. Am. Chem. Soc.*, **63**, 2482 (1941).
[43] Gilman and Jones, *J. Am. Chem. Soc.*, **62**, 2357 (1940).
[44] Gilman, Yablunky, and Svigoon, *J. Am. Chem. Soc.*, **61**, 1170 (1939).

bromobenzene, ethylene, and lithium bromide.[45] The β-bromoethyllithium that may be formed probably decomposes in accordance with the accompanying reactions.

$$C_6H_5Li + BrCH_2CH_2Br \rightarrow C_6H_5Br + [BrCH_2CH_2Li] \rightarrow C_2H_4 + LiBr$$

Somewhat similar reactions take place between phenyllithium and ethylene iodide, ethylene chlorobromide, β-iodoethyl methyl ether, and 1,2-dibromocyclohexane.[45] In each reaction bromo- or iodo-benzene is formed, but the new organolithium compound which also may be formed promptly decomposes.

An interesting rearrangement occurs during the reaction of n-butyllithium with 3-bromobenzofuran at room temperature. Interconversion is followed by opening of the furan ring to yield, subsequent to hydrolysis, o-hydroxyphenylacetylene.[25]

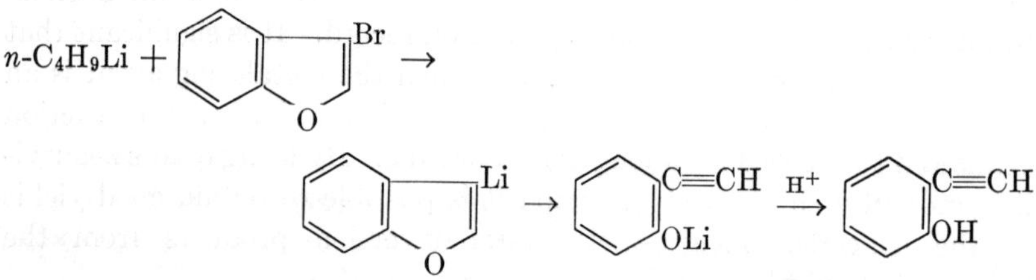

By conducting the reaction at a low temperature and for a short period of time followed by carbonation it has been possible to isolate the expected 3-benzofurancarboxylic acid.[25]

The reaction of an organolithium compound with an organic halide to form a coupling product, RR′, and lithium halide sometimes takes place in preference to halogen-metal interconversion. An example is the reaction of phenyllithium with o-chloroanisole to yield o-methoxybiphenyl and lithium chloride.[17]

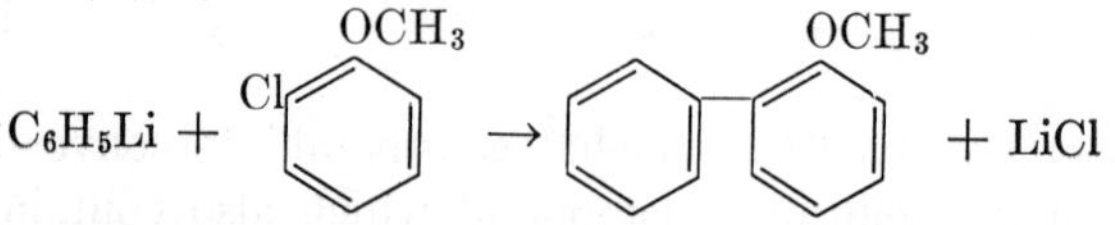

During the reactions of organolithium compounds with benzyl chloride [39] or benzyl bromide [16] there is good evidence that benzyllithium is formed intermediately, but it rapidly undergoes coupling so that the final products are lithium halides and bibenzyl or R-benzyl. These secondary coupling reactions may take place to some extent in many if not all halogen-metal interconversion systems. When the last stage of the

[45] Wittig and Harborth, *Ber.*, **77**, 306 (1944).

reaction, i.e., the reaction of R'Li with the final reactant, requires a relatively long time, a significant part of the R'Li may be used up in coupling with RX. This may constitute a serious drawback in certain interconversion reactions.

EXPERIMENTAL CONDITIONS

Three steps are involved in the halogen-metal interconversion reaction: (1) the preparation of an organolithium compound, usually n-butyllithium, from lithium metal and an organic halide; (2) the interaction of the RLi compound with a second organic halide R'X to yield R'Li; and (3) the reaction of the newly formed R'Li with the final reactant to yield the desired end product. Generally these three steps can be carried out in succession in the same apparatus. Organolithium compounds are prepared and handled in much the same way as the well-known Grignard reagents. They are extremely sensitive to air and to moisture. Therefore, the apparatus must be thoroughly dry, and an inert atmosphere must be maintained at all times. The conventional three-necked flask is ordinarily used as the reaction vessel. It is provided with a reflux condenser, a gas-tight mechanical stirrer, and a dropping funnel. Dry nitrogen gas serves as a most satisfactory inert atmosphere. Either a slow stream of the gas may be passed through the apparatus, or the outlets of the apparatus may be connected to a reservoir of the gas under a slight positive pressure.

The most commonly used solvent for reactions involving organolithium compounds is diethyl ether. It must be dry and free of ethanol. A satisfactory product is obtained by allowing commercial anhydrous ether to stand over metallic sodium. Petroleum ether (boiling range 30–40°) is particularly useful for certain reactions. It is best purified by shaking with sulfuric acid to remove unsaturated materials and then drying first with calcium chloride and finally with metallic sodium.

Alkyllithium compounds, with the exception of methyllithium, react slowly with ether to form hydrocarbons and lithium ethoxide.

$$RLi + C_2H_5OC_2H_5 \rightarrow RC_2H_5 \text{ (or } RH + C_2H_4\text{)} + LiOC_2H_5$$

Therefore, ether solutions of these compounds should be used soon after they are prepared. They cannot be stored for more than a few hours at room temperature without undergoing serious deterioration. The rate of this reaction of organolithium compounds with ether is markedly influenced by temperature. Ether solutions of n-butyllithium apparently can be kept for four days or longer [46] without significant decomposi-

---

[46] Gilman, Beel, Brannen, Bullock, Dunn, and Miller, *J. Am. Chem. Soc.*, **71**, 1499 (1949).

tion if the temperature is maintained at or below 10°. The yield of an approximately 0.5 M preparation of n-butyllithium freshly prepared from n-butyl bromide and lithium metal in ether may vary from about 40% to 90%, depending upon the *temperature* at which the reaction is conducted. The higher yields are obtained at lower temperatures ($-10°$ or below).[46] In low-boiling petroleum ether the alkyllithium compounds have been prepared in yields up to 90%.[5] Furthermore, these preparations are stable, and they can be stored for reasonably long periods of time. The aryllithium compounds like phenyllithium are obtained in yields of 90% or more from metallic lithium and aryl bromides. They are relatively stable in ether solution.

In carrying out the interconversion step it is often desirable to add the RLi compound to the organic halide, R′X. This order of addition is especially recommended when the halide, R′X, also contains an active hydrogen atom; otherwise part of the newly formed R′Li may be consumed in secondary reactions. This is illustrated by the accompanying sequence of reactions.[4] Reaction I generally proceeds more rapidly

$$n\text{-}C_4H_9Li + \underset{\text{OH}}{\underset{\text{Br}}{C_6H_4}} \rightarrow \underset{\text{OLi}}{\underset{\text{Br}}{C_6H_4}} + n\text{-}C_4H_{10} \quad (I)$$

$$n\text{-}C_4H_9Li + \underset{\text{OLi}}{\underset{\text{Br}}{C_6H_4}} \rightarrow \underset{\text{OLi}}{\underset{\text{Li}}{C_6H_4}} + n\text{-}C_4H_9Br \quad (II)$$

$$\underset{\text{OLi}}{\underset{\text{Li}}{C_6H_4}} + \underset{\text{OH}}{\underset{\text{Br}}{C_6H_4}} \rightarrow \underset{\text{OLi}}{\underset{\text{}}{C_6H_5}} + \underset{\text{OLi}}{\underset{\text{Br}}{C_6H_4}} \quad (III)$$

than reaction II. When II is under way there are contained in the mixture two organolithium compounds, $n\text{-}C_4H_9Li$ and $o\text{-}LiC_6H_4OLi$. Part of the latter may then be destroyed by entering into reaction III. Accordingly, the n-butyllithium is added to the o-bromophenol and reaction I goes to completion before reaction II begins.

Solutions of organolithium compounds may be measured and transferred from one reaction vessel to another with large pipets. These are conveniently filled by means of rubber aspirator bulbs.

EXPERIMENTAL PROCEDURES

Preparation of n-Butyllithium.[46] In a 500-ml. three-necked flask equipped with a stirrer, a low-temperature thermometer, and a dropping funnel is placed 200 ml. of anhydrous ether. After the apparatus has been swept with dry, oxygen-free nitrogen, 8.6 g. (1.25 gram atoms) of lithium wire (or any other convenient form of lithium metal) [18] is cut into small pieces which are allowed to fall directly into the reaction flask in a stream of nitrogen. With the stirrer started, about 30 drops of a solution of 68.5 g. (0.50 mole) of n-butyl bromide in 100 ml. of anhydrous ether is added from the dropping funnel. The reaction mixture is then cooled to $-10°$ by immersing the flask in a Dry Ice-acetone bath kept at about $-30°$ to $-40°$. The solution becomes slightly cloudy and bright spots appear on the lithium when the reaction has started. The remainder of the n-butyl bromide solution is then added at an even rate over a period of thirty minutes while the internal temperature is maintained at $-10°$. After addition is complete the reaction mixture is allowed to warm up to $0°$ to $10°$ with stirring during one to two hours. The reaction mixture is then filtered under an atmosphere of nitrogen by decantation through a narrow tube loosely plugged with glass wool into a graduated dropping funnel previously flushed with nitrogen.

The yield is 80% to 90%, and this is determined as follows:[39] A 5- or 10-ml. aliquot of the solution is withdrawn by means of a pipet connected to a rubber suction bulb, and hydrolyzed by adding to 10 ml. of distilled water. This is titrated with standard acid to determine the total alkali, using phenolphthalein as indicator. A second 5- or 10-ml. aliquot is withdrawn and run into a solution of 10 ml. of anhydrous ether containing 1 ml. of benzyl chloride. The mixture is allowed to stand for one minute after the addition and is then hydrolyzed with 10 ml. of water and titrated with standard acid. Care must be taken not to overstep the end point since the aqueous layer becomes decolorized before the ether layer. To overcome this the mixture should be shaken vigorously near the end point. The second titration determines the alkali present in the form of compounds other than n-butyllithium. The difference between the two titration values represents the concentration of n-butyllithium.

Preparation of Phenyllithium.[47] A 2-l. three-necked flask is provided with a gas-tight stirrer, a dropping funnel, and a reflux condenser. In the flask is placed 500 ml. of anhydrous ether. The apparatus is flushed with dry nitrogen gas, and 29.4 g. (4.2 gram atoms) of lithium metal (conveniently in the form of wire) is cut into small pieces [18] and allowed to fall directly into the flask. About 40 drops of a solution of 314 g. (2.0 moles) of bromobenzene in 1 l. of anhydrous ether is then added at room temperature. A slight cloudiness appearing in the ether solution after about three minutes indicates that the reaction has started. The addition of the bromobenzene solution is continued at a moderate rate until vigorous refluxing begins, and then the reaction flask is gradually

[47] Gilman and Miller, unpublished results.

immersed in an ice bath while the rate of addition of the bromobenzene is regulated so that refluxing is maintained. The reaction mixture must not be allowed to cool below the refluxing temperature at any time, and for small preparations a cooling bath should not be used. Toward the end of the addition the cooling bath is removed and stirring is continued until refluxing stops. The preparation requires about two hours. The solution is decanted under nitrogen through an L-shaped glass tube loosely plugged with glass wool into a graduated dropping funnel which has been previously flushed with nitrogen. To determine the concentration a small measured aliquot is hydrolyzed with distilled water and titrated with standard acid, using phenolphthalein as indicator. The yield by this procedure and by this analysis is 95–99%.

2-(m-Trifluoromethylphenyl)quinoline.[24] A solution of 62.5 g. (0.29 mole) of m-trifluoromethylphenyl bromide in 100 ml. of anhydrous ether under a nitrogen atmosphere is cooled in an ice bath. With stirring, a solution of 0.30 mole of n-butyllithium in 380 ml. of ether is added during one hour. The resulting solution of m-trifluoromethylphenyllithium is added during one hour to a stirred solution of 25.8 g. (0.2 mole) of quinoline in 50 ml. of ether. After being heated under reflux for two hours, the mixture is poured upon 200 g. of ice. The ether layer is separated and mixed with 25 ml. of nitrobenzene. After removal of the ether by distillation, the residual liquid is heated under reflux for twenty minutes and then distilled under reduced pressure to yield 37.2 g. (68%) of 2-(m-trifluoromethylphenyl)quinoline which boils at 142–144°/1–2 mm.

Heating with nitrobenzene serves to oxidize the intermediate 2-(m-trifluoromethylphenyl)-1,2-dihydroquinoline to the desired 2-(m-trifluoromethylphenyl)quinoline.

2,4,5-Triphenyl-3-furancarboxylic Acid.[15] A solution of 3.75 g. (0.01 mole) of 2,4,5-triphenyl-3-bromofuran in 50 ml. of warm ether is added rapidly to a solution of 0.02 mole of n-butyllithium in 50 ml. of ether. The mixture is stirred for thirty minutes at room temperature and then poured on about 50 g. of crushed, solid carbon dioxide. After evaporation of the excess carbon dioxide, the mixture is extracted with 50 ml. of dilute aqueous potassium hydroxide solution. The aqueous solution is acidified with hydrochloric acid to precipitate 2.7 g. of crude 2,4,5-triphenyl-3-furancarboxylic acid. The product is recrystallized from glacial acetic acid. There is obtained about 2.2 g. (65% yield) of pure acid melting at 257–258°.

Phenyl-di-(p-aminophenyl)arsenic.[36] To a solution of 0.3 mole of n-butyllithium in 500 ml. of ether is added 17.2 g. (0.1 mole) of p-bromoaniline. After nine minutes a solution of 11.1 g. (0.05 mole) of phenylarsenic dichloride in 50 ml. of ether is added dropwise during a period of five minutes. The mixture is heated under reflux for one hour and then hydrolyzed by the dropwise addition of 10% hydrochloric acid solution. The aqueous layer is separated and treated with sodium hydroxide solution to precipitate the phenyl-di-(p-aminophenyl)-arsenic. After crystallization from 50% ethanol, the product melts at 69°. The yield is about 11 g. (65%).

Triphenyl-o-hydroxymethylphenyllead.[48] A solution of 0.30 mole of n-butyllithium in 415 ml. of ether is added during fifteen minutes to a solution of 28.1 g. (0.15 mole) of o-bromobenzyl alcohol in 75 ml. of ether. After the resulting solution has been stirred for one-half hour, 56.9 g. (0.12 mole) of solid triphenyllead chloride is added as rapidly as possible with vigorous stirring. The reaction mixture is then immediately hydrolyzed by pouring onto iced ammonium chloride solution. The solutions are filtered from a little impure tetraphenyllead. The ether layer of the filtrate is separated and dried over sodium sulfate, and the ether is distilled. The last traces of ether and some octane (formed by coupling during preparation of the n-butyllithium) are removed by heating on the steam bath under vacuum. The partially solidified residue is boiled with 200 ml. of absolute ethanol, and the solution is filtered from a little insoluble material (impure triphenyllead chloride). The filtrate is cooled to give 26 g. (40%) of triphenyl-o-hydroxymethylphenyllead, melting at 133–136°. A further 20 g. (30%) of less pure product is obtained by distilling some of the ethanol from the mother liquor and diluting the remainder with water. The melting point of the pure product, obtained by crystallization of some of the first fraction from a mixture of benzene and petroleum ether (b.p. 60–68°), is 134–136°.

2,5-Thiophenedicarboxylic Acid.[21] To a solution of phenyllithium prepared from 0.05 mole of bromobenzene is added with stirring 3.5 g. (0.01 mole) of 2,5-diiodothiophene in 20 ml. of ether during a period of ten minutes. After stirring for an additional ten minutes the mixture is poured onto 40 g. of crushed, solid carbon dioxide. After evaporation of the carbon dioxide the mixture is extracted with dilute sodium hydroxide solution and the aqueous layer is acidified to yield 0.9 g. (53%) of 2,5-thiophenedicarboxylic acid. The acid sublimes at 150–300°; its dimethyl ester melts at 145–146°.

p-Bromophenyltrimethylsilane.[49] p-Bromophenyllithium is prepared by adding a solution of 0.132 mole of n-butyllithium in 275 ml. of ether to 33 g. (0.14 mole) of p-dibromobenzene dissolved in 60 ml. of dry

---

[48] Melstrom, doctoral dissertation, Iowa State College, 1943.
[49] Gilman and Melvin, unpublished results.

ether. After this mixture has stood at room temperature for about thirty minutes, a solution of 13 g. (0.12 mole) of trimethylchlorosilane in 30 ml. of ether is added at such a rate that gentle refluxing is maintained. The solution is heated under reflux for two hours and is then poured into cold 2 N hydrochloric acid. The ether layer is separated, dried, and evaporated. Distillation of the residual liquid in vacuum yields 15–17 g. (55–61%) of p-bromophenyltrimethylsilane which boils at 74–76°/2.5 mm.

Diphenyl-2,4-dimethoxy-5-bromophenylcarbinol.[3] To 17.8 g. (0.06 mole) of 4,6-dibromoresorcinol dimethyl ether under nitrogen is added 0.06 mole of phenyllithium in 60 ml. of ether solution. After ten minutes a solution of 10.9 g. (0.06 mole) of benzophenone in 20 ml. of anhydrous ether is added dropwise. The thick mixture is well stirred and then hydrolyzed by shaking vigorously with water. The white crystalline carbinol is collected on a filter and recrystallized from glacial acetic acid; m.p. 193–194°. The yield is 22.3 g. (96%).

TABULAR SURVEY OF HALOGEN-LITHIUM INTERCONVERSION REACTIONS

An effort has been made to include in Tables I and II all the examples of halogen-lithium interconversion reactions that had been investigated up to 1950. In Table I are presented the reactions in which the organolithium compound formed from a halide and n-butyllithium has been carbonated to yield a carboxylic acid. In Table II are listed the other halogen-lithium interconversion reactions. The yields of products given in the tables are usually based upon one or two experiments, and therefore it might be expected that the yields could be improved in some reactions.

TABLE I
INTERCONVERSIONS WITH n-BUTYLLITHIUM FOLLOWED BY CARBONATION TO YIELD CARBOXYLIC ACIDS
$$RX + n\text{-}C_4H_9Li \rightarrow RLi \rightarrow RCO_2H$$

	Organic Halide	Acid	Yield %	Reference
C_2H_5I	Ethyl iodide	Propionic	43	10
C_5H_4BrN	3-Bromopyridine	Nicotinic	30–62	26, 27
$C_6H_3Br_3$	1,3,5-Tribromobenzene	3,5-Dibromobenzoic	71	18
$C_6H_4Br_2$	p-Dibromobenzene	p-Bromobenzoic	90	9, 18
$C_6H_4Br_2$	p-Dibromobenzene	Terephthalic	89	9, 18
C_6H_5Br	Bromobenzene	Benzoic	51	18
C_6H_5I	Iodobenzene	Benzoic	51	18
C_6H_4BrCl	m-Bromochlorobenzene	m-Chlorobenzoic	70	50
C_6H_4BrCl	p-Bromochlorobenzene	p-Chlorobenzoic	90	18
C_6H_4ClI	m-Iodochlorobenzene	m-Chlorobenzoic	42	18
C_6H_5BrO	o-Bromophenol	Salicylic	67	4, 33
C_6H_5BrO	p-Bromophenol	p-Hydroxybenzoic	41	4, 33
C_6H_5IO	p-Iodophenol	p-Hydroxybenzoic	50	4
C_6H_5BrS	p-Bromothiophenol	p-Mercaptobenzoic	75	34
C_6H_6BrN	o-Bromoaniline	Anthranilic	40	38
C_6H_6BrN	p-Bromoaniline	p-Aminobenzoic	68	27, 36, 38
$C_6H_6BrNO_2S$	p-Bromobenzenesulfonamide	p-Sulfonamidobenzoic	14	35
$C_7H_5Br_3O$	2,4,6-Tribromoanisole	3,5-Dibromo-4-methoxybenzoic	16–19	18
$C_7H_5Br_3O$	2,4,6-Tribromoanisole	4-Bromo-2,6-anisoledicarboxylic	88	18
$C_7H_6Br_2$	2,5-Dibromotoluene	2-Methyl-4-bromobenzoic	22	18
$C_7H_6Br_2$	2,5-Dibromotoluene	3-Methyl-4-bromobenzoic	36	18
$C_7H_6Br_2$	2,5-Dibromotoluene	2,5-Toluenedicarboxylic	64	18
C_7H_7Br	o-Bromotoluene	o-Toluic	84	18

Formula	Starting material	Product	Yield	Refs.
C_7H_7Br	m-Bromotoluene	m-Toluic	65	18
C_7H_7Br	p-Bromotoluene	p-Toluic	86	18
C_7H_7Cl	Benzyl chloride	Phenylacetic	Trace	39
C_7H_7I	p-Iodotoluene	p-Toluic	72	18
$C_7H_4BrF_3$	m-Bromobenzotrifluoride	m-Trifluoromethylbenzoic	62	24
C_7H_4BrN	p-Bromobenzonitrile	Terephthalic	17	35
$C_7H_5BrO_2$	o-Bromobenzoic acid	Phthalic	35	4, 26
$C_7H_5IO_2$	o-Iodobenzoic acid	Phthalic	12	4
$C_7H_5IO_2$	p-Iodobenzoic acid	Terephthalic	62	4
C_7H_7BrO	o-Bromoanisole	o-Anisic	72	1, 9, 18
C_7H_7BrO	p-Bromoanisole	p-Anisic	52	1, 18, 41
C_7H_7BrO	m-Bromobenzyl alcohol	m-Hydroxymethylbenzoic	32	35
C_7H_7BrO	p-Bromobenzyl alcohol	p-Hydroxymethylbenzoic	18	35
C_7H_7IO	o-Iodoanisole	o-Anisic		9
C_7H_7IO	p-Iodoanisole	p-Anisic	78	18, 41
C_7H_8BrN	p-Bromo-N-methylaniline	p-Methylaminobenzoic	27	38
C_8H_5Br	Phenylethynyl bromide	Phenylpropiolic	87	14
C_8H_5Cl	Phenylethynyl chloride	Phenylpropiolic	23	14
C_8H_7Br	β-Bromostyrene	Cinnamic	23	18
C_8H_5BrO	2-Bromobenzofuran	2-Benzofurancarboxylic	62	25
C_8H_5BrO	3-Bromobenzofuran	3-Benzofurancarboxylic	12	25
C_8H_9BrO	p-Bromophenethyl alcohol	p-(β-Hydroxyethyl)benzoic	52	35
C_8H_9BrO	p-Bromo-α-methylbenzyl alcohol	p-(α-Hydroxyethyl)benzoic	45	35
$C_8H_9BrO_2$	4-Bromoveratrole	Veratric	90	51
$C_8H_{10}BrN$	m-Bromodimethylaniline	m-Dimethylaminobenzoic	32	52
$C_8H_{10}BrN$	p-Bromodimethylaniline	p-Dimethylaminobenzoic	56	52
C_8H_8BrNO	o-Bromoacetanilide	o-Acetylaminobenzoic	52	53
C_9H_6BrN	3-Bromoquinoline	3-Quinolinecarboxylic	52	6, 26
$C_{10}H_7Br$	α-Bromonaphthalene	α-Naphthoic	90	5, 13
$C_{10}H_7Br$	β-Bromonaphthalene	β-Naphthoic	61, 77	13, 23
$C_{10}H_7BrO$	1-Bromo-2-hydroxynaphthalene	2-Hydroxy-1-naphthoic	60	54

TABLE I—*Continued*

INTERCONVERSIONS WITH *n*-BUTYLLITHIUM FOLLOWED BY CARBONATION TO YIELD CARBOXYLIC ACIDS

$$RX + n\text{-}C_4H_9Li \rightarrow RLi \rightarrow RCO_2H$$

	Organic Halide	Acid	Yield %	Reference
$C_{10}H_7BrO$	2-Bromo-6-hydroxynaphthalene	6-Hydroxy-2-naphthoic	68	54
$C_{10}H_8IN$	2-Iodo-4-methylquinoline	4-Methyl-2-quinolinecarboxylic	53	6
$C_{10}H_{14}INO_2S$	p-Iodo-N,N-diethylbenzenesulfonamide	p-(N,N-Diethylsulfonamido)benzoic	78	4
$C_{11}H_9BrO$	1-Bromo-2-methoxynaphthalene	2-Methoxy-1-naphthoic	70	54
$C_{12}H_8Br_2$	3,3′-Dibromobiphenyl	3,3′-Biphenyldicarboxylic	46	55
$C_{12}H_8Br_2$	4,4′-Dibromobiphenyl	4′-Bromo-4-biphenylcarboxylic	67	18
$C_{12}H_8Br_2$	4,4′-Dibromobiphenyl	4,4′-Biphenyldicarboxylic	91	18
$C_{12}H_9Br$	3-Bromoacenaphthene	3-Acenaphthenecarboxylic	84	18
$C_{12}H_9Br$	4-Bromobiphenyl	4-Biphenylcarboxylic	62	18, 23
$C_{12}H_6Br_2O$	2,8-Dibromodibenzofuran	2,8-Dibenzofurandicarboxylic	72	31
$C_{12}H_7BrO$	2-Bromodibenzofuran	2-Dibenzofurancarboxylic	87	41
$C_{12}H_7BrO$	3-Bromodibenzofuran	3-Dibenzofurancarboxylic	81	31
$C_{12}H_7BrO$	4-Bromodibenzofuran	4-Dibenzofurancarboxylic	58	31
$C_{12}H_8BrN$	2-Bromocarbazole	2-Carbazolecarboxylic	56	6
$C_{12}H_8Br_2O$	4,4′-Dibromodiphenyl ether	p-(4′-Bromophenoxy)benzoic	66	18
$C_{12}H_8Br_2O$	4,4′-Dibromodiphenyl ether	4,4′-Diphenyl ether dicarboxylic	65	18
$C_{12}H_9BrO$	2-Bromodiphenyl ether	o-Phenoxybenzoic	70	19
$C_{12}H_9BrO$	4-Bromodiphenyl ether	p-Phenoxybenzoic	79	18, 19
$C_{12}H_9IO$	2-Iododiphenyl ether	o-Phenoxybenzoic	75	19
$C_{12}H_9IO$	3-Iododiphenyl ether	m-Phenoxybenzoic	93	19
$C_{12}H_9IO$	4-Iododiphenyl ether	p-Phenoxybenzoic	47	19
$C_{13}H_9BrO_2$	1-Bromo-2-methoxydibenzofuran	2-Methoxy-1-dibenzofurancarboxylic		28

Formula	Compound			
$C_{13}H_9BrO_2$	1-Bromo-4-methoxydibenzofuran	4-Methoxy-1-dibenzofurancarboxylic	69	28
$C_{13}H_9BrO_2$	3-Bromo-2-methoxydibenzofuran	2-Methoxy-3-dibenzofurancarboxylic	59	28
$C_{13}H_9BrO_2$	6-Bromo-4-methoxydibenzofuran	4-Methoxy-6-dibenzofurancarboxylic	57	28
$C_{13}H_{11}BrO_2$	3-Bromo-4-methoxydiphenyl ether	2-Methoxy-5-phenoxybenzoic	66	19
$C_{13}H_{11}IO_2$	4-Iodo-4'-methoxydiphenyl ether	p-(4-Methoxyphenoxy)benzoic	61	19
$C_{14}H_9Br$	2-Bromophenanthrene	2-Phenanthrenecarboxylic	37	29
$C_{14}H_9Br$	3-Bromophenanthrene	3-Phenanthrenecarboxylic	32	29
$C_{14}H_9Br$	9-Bromophenanthrene	9-Phenanthrenecarboxylic	51	29
$C_{14}H_{10}Br_2O_3$	1,9-Dibromo-2,8-dimethoxydibenzofuran	2,8-Dimethoxy-1,9-dibenzofurandicarboxylic	67	28
$C_{14}H_{10}Br_2O_3$	3,7-Dibromo-2,8-dimethoxydibenzofuran	2,8-Dimethoxy-3,7-dibenzofurandicarboxylic	18	28
$C_{14}H_{11}BrO_3$	1-Bromo-3,4-dimethoxydibenzofuran	3,4-Dimethoxy-1-dibenzofurancarboxylic	54	28
$C_{14}H_{11}BrO_3$	1-Bromo-4,6-dimethoxydibenzofuran	4,6-Dimethoxy-1-dibenzofurancarboxylic	60	28
$C_{14}H_{11}Br_2N$	2,8-Dibromo-5-ethylcarbazole	5-Ethyl-2,8-carbazoledicarboxylic	84	6
$C_{14}H_{11}I_2N$	2,8-Diiodo-5-ethylcarbazole	5-Ethyl-2,8-carbazoledicarboxylic	79	6
$C_{14}H_{12}BrN$	2-Bromo-5-ethylcarbazole	5-Ethyl-2-carbazolecarboxylic	71	6
$C_{14}H_{12}IN$	2-Iodo-5-ethylcarbazole	5-Ethyl-2-carbazolecarboxylic	67	6
$C_{14}H_{12}Br_2O_2$	5,5'-Dibromo-2,2'-dimethoxybiphenyl	2,2'-Dimethoxy-5,5'-biphenyldicarboxylic	63	32
$C_{14}H_8INO_2$	p-Iodophthalimidobenzene	p-Phthalimidobenzoic	37	38
$C_{18}H_{14}BrP$	3-Bromophenyldiphenylphosphine	m-Diphenylphosphinobenzoic	11	56
$C_{18}H_{14}BrP$	4-Bromophenyldiphenylphosphine	p-Diphenylphosphinobenzoic	57	56
$C_{22}H_{15}BrO$	3-Bromo-2,4,5-triphenylfuran	2,4,5-Triphenyl-3-furancarboxylic	78	15
$C_{22}H_{15}ClO$	3-Chloro-2,4,5-triphenylfuran	2,4,5-Triphenyl-3-furancarboxylic	14	15
$C_{23}H_{16}BrN$	2-Bromo-3,4,6-triphenylpyridine	3,4,6-Triphenyl-2-pyridinecarboxylic	67	15

[50] Gilman and Spatz, *J. Am. Chem. Soc.*, **66**, 621 (1944).
[51] Calvin, Heidelberger, Reid, Tolbert, and Yankwich, *Isotopic Carbon*, p. 183, John Wiley & Sons, New York, 1949.
[52] Gilman and Banner, *J. Am. Chem. Soc.*, **62**, 344 (1940).
[53] Murray and Ronzio, unpublished results.
[54] Gilman and Sunthankar, unpublished results.
[55] Snyder, Weaver, and Marshall, *J. Am. Chem. Soc.*, **71**, 289 (1949).
[56] Gilman and Brown, *J. Am. Chem. Soc.*, **67**, 824 (1945).

TABLE II
Miscellaneous Halogen-Metal Interconversion Reactions

	Organic Halide	Organolithium Compound	Reactant	Final Product	Yield %	Reference
C_4H_9I	n-Butyl iodide	Ethyl	Carbon dioxide	n-Valeric acid	36	10
$C_4H_2I_2S$	2,5-Diiodothiophene	Phenyl	Carbon dioxide	2,5-Thiophenedicarboxylic acid	53	21
C_4H_3IS	2-Iodothiophene	Phenyl	Carbon dioxide	2-Thiophenecarboxylic acid	58	21
$C_6H_4Br_2$	p-Dibromobenzene	n-Butyl	Trimethylchlorosilane	Trimethyl-p-bromophenylsilane	55–61	49
$C_6H_4Br_2$	p-Dibromobenzene	n-Butyl	Triphenylchlorosilane	Triphenyl-p-bromophenylsilane	68–70	49
C_6H_4BrCl	m-Bromochlorobenzene	n-Butyl	6-Methylquinoline	2-(3'-Chlorophenyl)-6-methylquinoline	57	57
C_6H_4BrCl	m-Bromochlorobenzene	n-Butyl	6-Methoxyquinoline	2-(3'-Chlorophenyl)-6-methoxyquinoline	53	50
C_6H_4BrCl	p-Bromochlorobenzene	n-Butyl	6-Methoxyquinoline	2-(4'-Chlorophenyl)-6-methoxyquinoline	50	50
C_6H_4BrCl	p-Bromochlorobenzene	n-Butyl	6-Methylquinoline	2-(4'-Chlorophenyl)-6-methylquinoline	30	57
C_6H_4BrCl	p-Bromochlorobenzene	n-Butyl	7-Methylquinoline	2-(4'-Chlorophenyl)-7-methylquinoline	25	57
C_6H_4BrCl	p-Bromochlorobenzene	n-Butyl	8-Methylquinoline	2-(4'-Chlorophenyl)-8-methylquinoline	54	57
C_6H_5I	Iodobenzene	p-Tolyl	Carbon dioxide	Benzoic acid	47	10
C_6H_5BrO	o-Bromophenol	n-Butyl	$MgBr_2$, then triethyltin bromide	Triethyl-o-hydroxyphenyltin	54	58
C_6H_5BrO	o-Bromophenol	n-Butyl	$MgBr_2$, then diphenyltin dichloride	Di-o-hydroxyphenyldiphenyltin	68	58

HALOGEN-METAL INTERCONVERSION

C_6H_5BrO	o-Bromophenol	n-Butyl	$MgBr_2$, then triphenyltin chloride	Triphenyl-o-hydroxyphenyltin	57	58
C_6H_5BrO	o-Bromophenol	n-Butyl	Triphenylchlorosilane	Triphenyl-o-hydroxyphenylsilane	55–65	49
C_6H_5BrO	p-Bromophenol	n-Butyl	$MgBr_2$, then triphenyltin chloride	Triphenyl-p-hydroxyphenyltin	10	58
C_6H_5BrO	p-Bromophenol	n-Butyl	Triphenylchlorosilane	Triphenyl-p-hydroxyphenylsilane	78–82	49
C_6H_5BrS	p-Bromothiophenol	n-Butyl	Isoquinoline	1-(p-Mercaptophenyl)isoquinoline	20	34
C_6H_6BrN	m-Bromoaniline	n-Butyl	Trimethylchlorosilane	Trimethyl-m-aminophenylsilane	2	59
C_6H_6BrN	m-Bromoaniline	n-Butyl	Triphenylchlorosilane	Triphenyl-m-aminophenylsilane	2	59
C_6H_6BrN	p-Bromoaniline	n-Butyl	Isoquinoline	1-(p-Aminophenyl)isoquinoline	70	34
C_6H_6BrN	p-Bromoaniline	n-Butyl	$MgBr_2$, then triphenyllead chloride	p-Aminophenyltriphenyllead	66	37
C_6H_6BrN	p-Bromoaniline	n-Butyl	Phenylarsenic dichloride	Phenyl-di-(p-aminophenyl)arsenic	91	36
C_6H_6BrN	p-Bromoaniline	n-Butyl	Phenylphosphorus dichloride	Phenyl-di-(p-aminophenyl)phosphorus	93	36
$C_7H_4BrF_3$	m-Bromobenzotrifluoride	n-Butyl	Quinoline	2-(m-Trifluoromethylphenyl)-quinoline	68	24
$C_7H_4BrF_3$	m-Bromobenzotrifluoride	n-Butyl	8-Methylquinoline	2-(m-Trifluoromethylphenyl)-8-methylquinoline	72	24
C_7H_7I	p-Iodotoluene	Phenyl	Carbon dioxide	p-Toluic acid	34	10
C_7H_7BrO	o-Bromoanisole	n-Butyl	Quinoline	2-(2'-Methoxyphenyl)quinoline	41	50
C_7H_7BrO	o-Bromoanisole	Methyl	Carbon dioxide	o-Anisic acid	17	19
C_7H_7BrO	o-Bromoanisole	Phenyl	Water	Anisole	90	17
C_7H_7BrO	o-Bromoanisole	Phenyl	Benzophenone	Diphenyl-2-methoxyphenylcarbinol	88	17
C_7H_7BrO	p-Bromoanisole	n-Butyl	6-Methylquinoline	2-(p-Methoxyphenyl)-6-methylquinoline	46	60
C_7H_7BrO	p-Bromoanisole	n-Butyl	7-Methylquinoline	2-(p-Methoxyphenyl)-7-methylquinoline	61	60

TABLE II—Continued

MISCELLANEOUS HALOGEN-METAL INTERCONVERSION REACTIONS

	Organic Halide	Organo-lithium Compound	Reactant	Final Product	Yield %	Reference
C_7H_7BrO	p-Bromoanisole	n-Butyl	8-Methylquinoline	2-(p-Methoxyphenyl)-8-methylquinoline	31	60
C_7H_7BrO	o-Bromobenzyl alcohol	n-Butyl	Triphenyllead chloride	Triphenyl-o-hydroxymethylphenyllead	70	48
C_7H_7BrO	o-Bromobenzyl alcohol	n-Butyl	$MgBr_2$, then triphenyltin chloride	Triphenyl-o-hydroxymethylphenyltin	64	58
C_7H_7BrO	m-Bromobenzyl alcohol	n-Butyl	Triphenyllead chloride	Triphenyl-m-hydroxymethylphenyllead	41	48
C_7H_7BrO	p-Bromobenzyl alcohol	n-Butyl	Triphenyllead chloride	Triphenyl-p-hydroxymethylphenyllead	63	48
C_7H_7BrO	p-Bromobenzyl alcohol	n-Butyl	$MgBr_2$, then triphenyltin chloride	Triphenyl-p-hydroxymethylphenyltin	66	58
C_7H_7IO	o-Iodoanisole	Methyl	Carbon dioxide	o-Anisic acid	13	19
C_7H_7IO	o-Iodoanisole	Phenyl	Water	Anisole	90	17
C_7H_7IO	o-Iodoanisole	Phenyl	Benzophenone	Diphenyl-2-methoxyphenylcarbinol	92	17
C_7H_7IO	p-Iodoanisole	Methyl	Carbon dioxide	p-Anisic acid	13	19
$C_8H_8Br_2O_2$	4,6-Dibromoresorcinol dimethyl ether	Phenyl	Water	Resorcinol dimethyl ether	72	30
$C_8H_8Br_2O_2$	4,6-Dibromoresorcinol dimethyl ether	Phenyl	Water	4-Bromoresorcinol dimethyl ether	95	3, 30

HALOGEN-METAL INTERCONVERSION

$C_8H_8Br_2O_2$	4,6-Dibromoresorcinol dimethyl ether	Phenyl	Carbon dioxide	2,4-Dimethoxy-5-bromobenzoic acid		3
$C_8H_8Br_2O_2$	4,6-Dibromoresorcinol dimethyl ether	Phenyl	Benzophenone	Diphenyl-2,4-dimethoxy-5-bromophenylcarbinol	96	3
C_8H_9BrO	p-Bromophenethyl alcohol	n-Butyl	Triphenyllead chloride	Triphenyl-p-(β-hydroxyethyl)-phenyllead	57	48
C_8H_9BrO	p-Bromo-α-methylbenzyl alcohol	n-Butyl	Triphenyllead chloride	Triphenyl-p-(α-hydroxyethyl)-phenyllead	52	48
C_8H_9BrO	o-Bromobenzylmethyl ether	n-Butyl	MgBr$_2$, then triphenyltin chloride	Triphenyl-o-methoxymethylphenyltin	38	58
$C_8H_9BrO_2$	4-Bromoresorcinol dimethyl ether	Phenyl	Carbon dioxide	2,4-Dimethoxybenzoic acid		30
$C_8H_9BrO_2$	4-Bromoresorcinol dimethyl ether	Phenyl	Benzophenone	Diphenyl-2,4-dimethoxyphenylcarbinol	95	30
$C_9H_{10}BrN$	1-Bromo-5,6,7,8-tetrahydroisoquinoline	n-Butyl	Benzaldehyde	1-α-Hydroxybenzyl-5,6,7,8-tetrahydroisoquinoline	59	61
$C_{10}H_7Br$	α-Bromonaphthalene	Ethyl	Carbon dioxide	α-Naphthoic acid	90	5, 13
$C_{10}H_7Br$	α-Bromonaphthalene	n-Propyl	Carbon dioxide	α-Naphthoic acid	95	5, 13
$C_{10}H_7Br$	α-Bromonaphthalene	i-Propyl	Carbon dioxide	α-Naphthoic acid	78	5
$C_{10}H_7Br$	α-Bromonaphthalene	i-Butyl	Carbon dioxide	α-Naphthoic acid	35	5
$C_{10}H_7Br$	α-Bromonaphthalene	sec-Butyl	Carbon dioxide	α-Naphthoic acid	80	5
$C_{10}H_7Br$	α-Bromonaphthalene	t-Butyl	Carbon dioxide	α-Naphthoic acid	34	5
$C_{10}H_7Br$	α-Bromonaphthalene	n-Amyl	Carbon dioxide	α-Naphthoic acid	58	5
$C_{10}H_7Br$	α-Bromonaphthalene	Phenyl	Carbon dioxide	α-Naphthoic acid	38	13
$C_{12}H_7BrS$	2-Bromodibenzothiophene	n-Butyl	Trimethylchlorosilane	Trimethyl-2-dibenzothienylsilane	50	62
$C_{12}H_7BrS$	3-Bromodibenzothiophene	n-Butyl	Trimethylchlorosilane	Trimethyl-3-dibenzothienylsilane	80	62
$C_{12}H_7BrS$	4-Bromodibenzothiophene	n-Butyl	Trimethylchlorosilane	Trimethyl-4-dibenzothienylsilane	51	63

TABLE II—Continued

MISCELLANEOUS HALOGEN-METAL INTERCONVERSION REACTIONS

	Organic Halide	Organo-lithium Compound	Reactant	Final Product	Yield %	Reference
$C_{12}H_9BrO$	Phenyl 4-bromophenyl ether	n-Propyl	Carbon dioxide	p-Phenoxybenzoic acid	80	19
$C_{12}H_9BrO$	Phenyl 4-bromophenyl ether	Methyl	Carbon dioxide	p-Phenoxybenzoic acid	14	19
$C_{12}H_9IO$	Phenyl 2-iodophenyl ether	Methyl	Carbon dioxide	o-Phenoxybenzoic acid	14	19
$C_{12}H_9IO$	Phenyl 2-iodophenyl ether	Phenyl	Carbon dioxide	o-Phenoxybenzoic acid	79	19
$C_{12}H_9IO$	Phenyl 3-iodophenyl ether	Phenyl	Carbon dioxide	m-Phenoxybenzoic acid	37	19
$C_{12}H_9IO$	Phenyl 4-iodophenyl ether	Phenyl	Carbon dioxide	p-Phenoxybenzoic acid	33	19
$C_{13}H_{11}IO_2$	4-Iodophenyl 4'-methoxyphenyl ether	Phenyl	Carbon dioxide	p-(4'-Methoxyphenoxy)benzoic acid	47	19
$C_{14}H_{10}Br_4O_2$	3,3',5,5'-Tetrabromo-2,2'-dimethoxybiphenyl	Phenyl	Carbon dioxide	5,5'-Dibromo-2,2'-dimethoxy-3,3'-biphenyldicarboxylic acid	13	32

[57] Gilman, Christian, and Spatz, J. Am. Chem. Soc., **68**, 979 (1946).
[58] Arntzen, doctoral dissertation, Iowa State College, 1942.
[59] Gilman and Summers, unpublished results.
[60] Gilman, Towle, and Spatz, J. Am. Chem. Soc., **68**, 2017 (1946).
[61] Grewe, Mondon, and Nolte, Ann., **564**, 161 (1949).
[62] Illuminati, Nobis and Gilman, unpublished results.
[63] Gilman and Nobis, unpublished results.

CHAPTER 8

THE PREPARATION OF THIAZOLES

RICHARD H. WILEY

University of Louisville

D. C. ENGLAND

E. I. du Pont de Nemours and Company

AND

LYELL C. BEHR

Mississippi State College

CONTENTS

	PAGE
INTRODUCTION	368
THE REACTION OF THIOAMIDES AND α-HALO CARBONYL COMPOUNDS	369
Scope and Limitations	369
Table I. Preparation of Thiazoles Substituted by Hydrocarbon Groups	370
Mechanism	373
THE REACTION OF AMMONIUM DITHIOCARBAMATE AND α-HALO CARBONYL COMPOUNDS	374
Scope and Limitations	374
Mechanism	375
THE REACTION OF α-ACYLAMINO CARBONYL COMPOUNDS AND PHOSPHORUS PENTASULFIDE	376
Scope and Limitations	376
Mechanism	376
THE REARRANGEMENT OF α-THIOCYANO KETONES	377
EXPERIMENTAL PROCEDURES	378
The Reaction of Thioamides with α-Halo Carbonyl Compounds	378
2,4-Dimethylthiazole	379
4-Methyl-5-(β-hydroxyethyl)thiazole	379
2-Phenyl-4,5-dimethylthiazole	379
2-Aminothiazole	379
2-Amino-4-methylthiazole	380
2-Amino-4-phenylthiazole	380

	PAGE
The Reaction of Ammonium Dithiocarbamate and α-Halo Ketones	380
2-Mercapto-4-methylthiazole	381
The Reaction of α-Acylamino Carbonyl Compounds and Phosphorus Pentasulfide	381
5-Phenylthiazole	381
The Rearrangement of α-Thiocyano Ketones	381
2-Hydroxy-4-methylthiazole	382
TABULAR SURVEY OF THIAZOLES	382
Table II. Thiazoles in Which Substituents Are Linked through Carbon	384
Table III. Quaternary Thiazolium Salts	391
Table IV. Hydroxythiazoles and Ethers	395
Table V. 2-Mercaptothiazoles from Ammonium Dithiocarbamate	396
Table VI. Thiazyl Ketones	396
Table VII. Thiazole Carboxylic Acids and Esters	397
Table VIII. 2-Aminothiazoles from Thiourea	398
Table IX. Alkylamino- and Arylamino-thiazoles	400
Table X. Dialkyl (or Aryl) Aminothiazoles	402
Table XI. Compounds Containing More Than One Thiazole Ring	403

INTRODUCTION

Thiazoles have become increasingly important in pharmaceutical, biochemical, and technical fields. Commercially important compounds that contain the thiazole ring are the mercaptothiazoles, which are valuable rubber accelerators, various "sulfa" and antitubercular drugs, the penicillins, and thiamin. Certain thiazole derivatives show great promise as intermediates in the synthesis of amino acids, peptides, and purines. This application has been discussed by Heilbron.[1]

As a consequence of the varied interest in the thiazoles, an extensive body of literature dealing with their syntheses and properties is available. This chapter summarizes information on the methods of preparation of the thiazoles but is restricted to those in which the thiazole ring is not part of a condensed system. Various reduced rings such as thiazolidines and thiazolones are also omitted from consideration.

In general, the methods of preparation of the thiazoles involve the use of substituted carbonyl compounds. The most valuable method is the reaction of thioamides with α-halo carbonyl compounds. It finds its greatest application in the synthesis of thiazoles containing alkyl, aryl, or heterocyclic substituents. Mono-, di-, and tri-substituted thiazoles in any combination can be prepared. Of great importance is the synthesis of a variety of 2-aminothiazoles, using thiourea and its N-substituted derivatives. A closely allied reaction (which will be

---

[1] Heilbron, *J. Chem. Soc.*, **1949**, 2099.

considered separately) is that between ammonium dithiocarbamate and α-halo ketones, which constitutes the best method for the preparation of the 2-mercaptothiazoles.

A third preparative method for the thiazoles is the reaction of α-acylamino carbonyl compounds with phosphorus pentasulfide. This reaction, which is formally similar to the preparation of thiophene derivatives from 1,4-diketones (see Chapter 9), has not received the attention it appears to deserve. It is suitable for di- and tri-alkyl (or aryl) thiazoles as well as 5-alkoxy derivatives.

Last will be considered the rearrangement of α-thiocyano ketones in aqueous solution which produces 2-hydroxythiazoles substituted in the 4 or 4,5 positions. There is little information as to the scope of this reaction.

Mention should be made of the preparation of "chrysean" from hydrogen sulfide and potassium cyanide, first carried out by Wallach.[2] The reaction was studied by Hellsing,[3,4,5] who suggested without rigorous proof that "chrysean" was 5-amino-2-thiocarbamylthiazole (I). It has

$$H_2N-\underset{S}{\underset{|}{\diagdown}}\overset{N}{\underset{|}{\diagup}}-CSNH_2$$
I

since been shown that chrysean does indeed possess this structure.[6] The mechanism by which it is produced is unknown. Attempts to obtain greater than a 15–20% yield have been fruitless.[7,8,9]

THE REACTION OF THIOAMIDES AND α-HALO CARBONYL COMPOUNDS

Scope and Limitations

In the simplest general sense, the reaction of α-halo carbonyl compounds and thioamides produces 3,4,5-trisubstituted thiazoles, as shown in the accompanying equation. When one or more of the R's is hydro-

$$\begin{array}{c} RCHX \\ | \\ R'C=O \end{array} + \begin{array}{c} S \\ \diagdown \\ H_2N \end{array}CR'' \longrightarrow \begin{array}{c} R \\ \| \\ R'-N \end{array}\begin{array}{c} S \\ \diagdown \\ \end{array}CR'' + HX + H_2O$$

[2] Wallach, *Ber.*, **7**, 902 (1874).
[3] Hellsing, *Ber.*, **32**, 1497 (1899).
[4] Hellsing, *Ber.*, **33**, 1774 (1900).
[5] Hellsing, *Ber.*, **36**, 3546 (1903).
[6] Erlenmeyer, Mengisen, and Prijs, *Helv. Chim. Acta*, **30**, 1865 (1947).
[7] Arnold and Scaife, *J. Chem. Soc.*, **1944**, 103.
[8] Arnold and Scaife, Brit. pat. 569,221 [*C. A.*, **41**, 5149 (1947)].
[9] Arnold, Scaife, and Starr, Brit. pat. 569,220 [*C. A.*, **41**, 5149 (1947)].

gen, mono- or di-substituted derivatives or thiazole itself can be obtained. The possibilities are summarized in Table I.

TABLE I

PREPARATION OF THIAZOLES SUBSTITUTED BY HYDROCARBON GROUPS

Thiazole Position			Reactants	
2	4	5	Thioamide	Halogen Compound
H	H	H	HCSNH$_2$	XCH$_2$CHO or derivatives
H	R	H	HCSNH$_2$	RCOCH$_2$X
H	H	R	HCSNH$_2$	RCHXCHO
H	R	R'	HCSNH$_2$	RCOCHXR'
R	H	H	RCSNH$_2$	XCH$_2$CHO or derivatives
R	H	R'	RCSNH$_2$	R'CHXCHO
R	R'	H	RCSNH$_2$	R'COCH$_2$X
R	R'	R''	RCSNH$_2$	R'COCHXR''

Thioformamide produces thiazoles unsubstituted in the 2 position. Thus chloroacetaldehyde and thioformamide yield thiazole itself,[10,11] and halogen derivatives of the higher aldehydes produce 5-substituted thiazoles. In this way, 5-methylthiazole has been obtained from α-bromopropionaldehyde [12] and 5-phenylthiazole from α-bromophenylacetaldehyde.[10] Similarly, 4-substituted thiazoles result when halomethyl ketones are used. Chloroacetone and thioformamide furnish 4-methylthiazole,[10,13] and phenacyl bromide gives 4-phenylthiazole in 40% yield. If thioformamide is condensed with higher α-halo ketones, 4,5-disubstituted compounds result. In this way, 4-phenyl-5-methylthiazole is produced from α-bromopropiophenone [14] and 4-methyl-5-(β-hydroxyethyl)thiazole from 3-chloropentan-1-ol-4-one [10,15,16,17] or

[10] Hromatka, U. S. pat. 2,160,867 [C. A., **33**, 7320 (1939)].
[11] Willstätter and Wirth, Ber., **42**, 1908 (1909).
[12] Erlenmeyer and Schmidt, Helv. Chim. Acta, **29**, 1957 (1946).
[13] Clarke and Gurin, J. Am. Chem. Soc., **57**, 1876 (1935).
[14] Ochiai, Kakuda, Nakayama, and Masuda, J. Pharm. Soc. Japan, **59**, 462 (1939) [C. A., **34**, 101 (1940)].
[15] Buchman, J. Am. Chem. Soc., **58**, 1803 (1936).
[16] Buchman, Ger. pat. 673,174 [C. A., **33**, 4271 (1939)].
[17] Pesina, J. Gen. Chem. U.S.S.R., **9**, 804 (1939) [C. A., **34**, 425 (1940)].

the corresponding bromo compound.[16-19] In another variation, halogen derivatives of active methylene compounds are used. For example, 3-chloro-2,4-pentanedione produces 4-methyl-5-acetylthiazole in 55% yield,[20,21] and halogen derivatives of acetoacetic esters furnish 4-methyl-5-thiazolecarboxylic acid esters in 50–75% yields.[13,22-29]

The use of simple thioamides other than formamide leads to 2-substituted thiazoles. Reported examples are numerous. The reaction of a thioamide with chloroacetaldehyde, or substances readily yielding the aldehyde, produces 2-alkyl (or aryl) thiazoles. For example, 2-methylthiazole can be obtained by reaction of thioacetamide with ethyl α,β-dichloroethyl ether [30], and 2-phenylthiazole from thiobenzamide and the same dichloro compound.[31] If the aldehyde is replaced by a chloromethyl ketone, 2,4-disubstituted thiazoles result. Thus, 2,4-dimethylthiazole is the product of the condensation of chloroacetone and thioacetamide,[10,13,30,32] and 2-phenyl-4-ethylthiazole results from the reaction of ethyl chloromethyl ketone and thiobenzamide.[33] Trisubstituted thiazoles are obtained by the reaction of a thioamide with a higher α-halo ketone. This modification has been very widely used. As examples may be cited 2-phenyl-4,5-dimethylthiazole, obtained from thiobenzamide and methyl α-chloroethyl ketone in 65% yield,[33] and 2,4,5-trimethylthiazole, prepared from thioacetamide and methyl α-chloroethyl ketone.[34,35] The reaction of thioamides with α-chloroaldehydes (other than acetaldehyde) yields 2,5-disubstituted thiazoles. Thioacetamide and α-chloropropionaldehyde thus yield 2,5-dimethylthiazole.[31,36]

[18] Research Corporation, Brit. pat. 472,459 [*C. A.*, **32**, 1408 (1938)]; French pat. 803,495 [*C. A.*, **31**, 2616 (1937)].
[19] Slobodin and Hel'ms, *Compt. rend. acad. sci. U.R.S.S.*, **39**, 145 (1943) [*C. A.*, **38**, 1239 (1944)].
[20] Baumgarten, Dornow, Gutschmidt, and Krehl, *Ber.*, **75B**, 442 (1942).
[21] Buchman and Richardson, *J. Am. Chem. Soc.*, **67**, 395 (1945).
[22] Buchman and Richardson, *J. Am. Chem. Soc.*, **61**, 891 (1939).
[23] Erlenmeyer, Epprecht, and Meyenburg, *Helv. Chim. Acta*, **20**, 310 (1937).
[24] Harington and Moggridge, *J. Chem. Soc.*, **1939**, 443.
[25] Soc. pour l'ind. chim. à Bâle, Ger. pat. 658,353 [*C. A.*, **32**, 4727 (1938)].
[26] Tomlinson, *J. Chem. Soc.*, **1935**, 1030.
[27] Cerecedo and Tolpin, *J. Am. Chem. Soc.*, **59**, 1660 (1937).
[28] Buchman and Sargent, *J. Am. Chem. Soc.*, **67**, 400 (1945).
[29] Price and Pickel, U. S. pat. 2,209,092 [*C. A.*, **35**, 141 (1941)].
[30] Hantzsch, *Ber.*, **21**, 942 (1888).
[31] Hubacher, *Ann.*, **259**, 228 (1890).
[32] Hantzsch, *Ann.*, **250**, 257 (1889).
[33] Friedman, Sparks, Meredith, and Adams, *J. Am. Chem. Soc.*, **59**, 2262 (1937).
[34] Roubleff, *Ann.*, **259**, 253 (1890).
[35] Hantzsch, *Ber.*, **23**, 2339 (1890).
[36] McLean and Muir, *J. Chem. Soc.*, **1942**, 383.

Thioamides of dibasic acids produce symmetrically substituted 2,2'-bithiazoles, and the two heterocyclic rings are connected by the same group originally connecting the thioamide functions. Dithioöxamide,

$$(CH_2)_n \begin{matrix} CSNH_2 \\ \\ CSNH_2 \end{matrix} + 2RCOCHXR' \longrightarrow \underset{R'}{\overset{R}{\underset{S}{\boxed{}}}}(CH_2)_n \underset{S}{\overset{R}{\underset{R'}{\boxed{}}}} + 2HX + 2H_2O$$

the simplest dithioamide, yields 4,4'-dimethyl-2,2'-bithiazole when treated with chloroacetone.[31,37] A polymer results if an α,α'-dihalo carbonyl compound is used.

Thionurethans, ROCSNH$_2$, produce 2-alkoxythiazoles when treated with α-halo carbonyl compounds.[31,38,39] Neither this reaction nor the similar one in which thioöxamates react to produce esters of 2-thiazolecarboxylic acids [12,40] has been extensively studied.

If an N-substituted thioamide is allowed to undergo reaction with an α-halo carbonyl compound, quaternary thiazolium salts result, sometimes in quantitative yield. Thus, N-methylthioacetamide condenses

$$\begin{matrix} RCHX \\ | \\ R'C{=}O \end{matrix} + \begin{matrix} S \\ \diagdown \\ \diagup \\ HN \\ | \\ R''' \end{matrix} CR'' \longrightarrow \begin{matrix} RC{-\!-}S \\ \| \qquad \diagdown \\ \qquad \quad CR'' \\ \diagup \\ R'C{-\!-}N \\ | \\ R''' \end{matrix} + H_2O \\ \quad X^-$$

with chloroacetone to furnish 2,3,4-trimethylthiazolium chloride in 100% yield,[41] and various thioformamidomethylpyrimidine derivatives condense with α-halo carbonyl compounds to give thiamin-like substances.

One of the most valuable of the modifications of the thioamide reaction is that using thiourea or its substitution products which affords 2-aminothiazoles, usually in very good yields. Chloroacetaldehyde or various of its derivatives with thiourea gives 2-aminothiazole itself. The yield

[37] Karrer, Leiser, and Graf, *Helv. Chim. Acta*, **27**, 624 (1944).
[38] Schenkel, Marbet, and Erlenmeyer, *Helv. Chim. Acta*, **27**, 1437 (1944).
[39] Schwaneberg, inaugural dissertation, University of Leipzig, Leipzig, 1930 [*C. A.*, **25**, 3664 (1931)].
[40] Boon, Brit. pat. 546,994 [*C. A.*, **37**, 5556 (1943)]; U. S. pat. 2,341,687 [*C. A.*, **38** 4385 (1944)].
[41] Todd, Bergel, and Karimullah, *Ber.*, **69**, 217 (1936).

using chloroacetaldehyde diethyl acetal may reach 92%.[42-46] Excellent yields of 2-aminothiazole can also be obtained from α,β-dihaloethyl acetates and thiourea.[44,47] The halo esters are readily obtained by the reaction of the free halogen and vinyl acetate and can be used without purification. Phenacyl chloride and thiourea furnish 2-amino-4-phenylthiazole in 90% yield.[48] Replacement of thiourea by a N-substituted derivative yields 2-alkylamino- (or arylamino)-thiazoles, and N,N-disubstituted thioureas furnish 2-dialkylaminothiazoles.

Mechanism

Although no exhaustive studies of the reaction of a thioamide with an α-halo carbonyl compound have been made, it appears that the first stage involves formation of a carbon-sulfur link by elimination of a molecule of hydrogen halide. In the second step, ring closure takes

$$\begin{array}{c} \text{RCHX} \\ | \\ \text{R'C}=\text{O} \end{array} + \begin{array}{c} \text{HS} \\ \diagdown \\ \diagup \text{CR''} \\ \text{HN} \end{array} \rightarrow \left[\begin{array}{cc} \text{RCH}\!-\!\!-\!\!-\!\!-\!\!-\!\text{S} \\ | & | \\ \text{R'C}=\text{O} & \text{CR''} \\ & \diagdown \\ & \text{N} \\ & | \\ & \text{H} \end{array} \right] + \text{HX}$$

place with the enolic form of the ketone, and a molecule of water is eliminated. The reaction as ordinarily carried out is exothermic, and there is no real evidence of a stepwise process.

$$\left[\begin{array}{cc} \text{RC}\!-\!\!-\!\!-\!\!-\!\!-\!\text{S} \\ \| & | \\ \text{R'C} & \text{CR''} \\ \diagdown & \diagup \\ \text{OH} & \text{HN} \end{array} \right] \rightarrow \begin{array}{c} \text{R'}\!-\!\!-\!\text{N} \\ \text{R}\!\diagup\!\!\diagdown\!\text{R''} \\ \text{S} \end{array} + \text{H}_2\text{O}$$

There has been no report of the alternative possibility: loss of hydrogen halide by reaction of the halogen with a hydrogen atom bound to nitrogen. If the reaction were to take such a course, the product obtained would be different from that actually isolated. The product of the

---

[42] Postovskiĭ, Khmelevskiĭ, and Bednyagina, *J. Applied Chem. U.S.S.R.*, **17**, 65 (1944) [*C. A.*, **39**, 1410 (1945)].
[43] Skrimshire, Brit. pat. 540,032 [*C. A.*, **36**, 4138 (1942)].
[44] Christiansen, U. S. pat. 2,242,237 [*C. A.*, **35**, 5518 (1941)]; Brit. pat. 549,846 [*C. A.*, **38**, 1078 (1944)].
[45] Khmelevskiĭ, Postovskiĭ, and Bednyagina, U. S. S. R. pat. 64,732 [*C. A.*, **40**, 5776 (1946)].
[46] Kyrides, U. S. pat. 2,330,223 [*C. A.*, **38**, 1250 (1944)].
[47] Morren and DuPont, *J. Pharm. Belg.*, **1**, 126 (1942) [*C. A.*, **38**, 3284 (1944)].
[48] Dodson and King, *J. Am. Chem. Soc.*, **67**, 2242 (1945).

reaction of an α-halo aldehyde with a thioamide is a 2,5-disubstituted thiazole, not the 2,4 isomer. The product from an α-halo ketone is a 2,4- or 2,4,5-trisubstituted thiazole. Considerations similar to these arise in the preparation of the oxazoles by reaction between amides and α-halo carbonyl compounds.[49]

An interesting fact is that the reaction may be carried out without isolating the halo carbonyl compound by merely heating the thioamide with the ketone and a halogen.[48] The halogen may be dispensed with by substituting oxidizing agents such as sulfur trioxide, sulfuric acid, or nitric acid, which seems to indicate that the halogenated ketone is not necessary.[50] It should be noted, however, that the yields are much poorer.

THE REACTION OF AMMONIUM DITHIOCARBAMATE AND α-HALO CARBONYL COMPOUNDS

Scope and Limitations

That 2-mercaptothiazoles can be prepared from ammonium dithiocarbamate and α-halo ketones was first reported in 1893 by Miolati.[51] The general overall reaction may be written as in the accompanying equation. It is possible to prepare a wide variety of 2-mercaptothiazoles

$$\begin{matrix} RCHX \\ | \\ R'CO \end{matrix} + H_2NCSNH_4 \longrightarrow \underset{R}{\overset{R'}{\diagdown}}\!\!\underset{S}{\boxed{}}\!\!\overset{N}{\diagup}SH + NH_4X$$

by using various types of halogenated ketones. Chloroacetone yields 2-mercapto-4-methylthiazole,[51-55] and 2-chloro-3-butanone produces 2-mercapto-4,5-dimethylthiazole.[52, 55] The reaction is also applicable in the aromatic series; phenacyl chloride or bromide yields 2-mercapto-4-phenylthiazole.[51, 55, 56, 57] Complex groups can also be introduced. For example, 2-mercapto-4-methyl-5-(β-acetoxyethyl)thiazole, a useful intermediate in the preparation of thiamin, can be obtained by reaction of ammonium dithiocarbamate and 3-chloro-5-acetoxy-2-pentanone.[58, 59]

[49] Wiley, *Chem. Revs.*, **37**, 401 (1945).
[50] Dodson and King, *J. Am. Chem. Soc.*, **68**, 871 (1946).
[51] Miolati, *Gazz. chim. ital.*, **23I**, 575 (1893).
[52] Buchman, Reims, and Sargent, *J. Org. Chem.*, **6**, 764 (1941).
[53] Gibbs and Robinson, *J. Chem. Soc.*, **1945**, 925.
[54] Levi, *Gazz. chim. ital.*, **61**, 719 (1931).
[55] Mathes, U. S. pat. 2,186,419 [*C. A.*, **34**, 3537 (1940)].
[56] Ubaldini and Fiorenza, *Gazz. chim. ital.*, **73**, 169 (1943).
[57] Mathes, U. S. pat. 2,186,421 [*C. A.*, **34**, 3537 (1940)].
[58] Hoffmann, LaRoche and Co. A.-G., Ger. pat. 678,153 [*C. A.*, **33**, 7819 (1939)]; Brit. pat. 492,637 [*C. A.*, **33**, 1760 (1939)]; Swiss pat. 196,649 [*C. A.*, **33**, 1883 (1939)].
[59] Gravin, *J. Applied Chem. U.S.S.R.*, **16**, 105 (1943) [*C. A.*, **38**, 1239 (1944)].

No reaction using an α-halo aldehyde has been reported. The expected product would be a 2-mercaptothiazole unsubstituted in the 4 position.

The value of the dithiocarbamate reaction as a synthetic tool is enhanced by the fact that the thiol group can be replaced by hydrogen with hydrogen peroxide in the presence of a strong acid.[58, 60] Thus 2-mercapto-4-(β-hydroxyethyl)-5-methylthiazole treated with hydrogen peroxide and hydrochloric acid yields chiefly 4-(β-hydroxyethyl)-5-methylthiazole and a small amount of 2-chloro-4-(β-hydroxyethyl)-5-methylthiazole.

$$\underset{H_3C}{\overset{HOH_2CH_2C}{\diagup}}\underset{S}{\overset{N}{\diagdown}}SH \xrightarrow{\underset{HCl}{H_2O_2}} \underset{H_3C}{\overset{HOH_2CH_2C}{\diagup}}\underset{S}{\overset{N}{\diagdown}} + \underset{H_3C}{\overset{HOH_2CH_2C}{\diagup}}\underset{S}{\overset{N}{\diagdown}}Cl$$

Mechanism

The first product in the reaction of ammonium dithiocarbamate and the α-halo ketone is a substituted dithiocarbamate, which is formed with the elimination of a molecule of ammonium halide. Several such inter-

$$H_2NCSNH_4 + RCHXCOR' \rightarrow H_2NCSCHCOR' + NH_4X$$
$$\quad \| \qquad \qquad \qquad \qquad \quad \| \; |$$
$$\quad S \qquad \qquad \qquad \qquad \qquad S \; R$$

mediates have been isolated by allowing the reaction to proceed for only a few minutes in ether. Acetyl and phenacyl dithiocarbamate have been isolated in this way by Levi [54] and by Ubaldini and Fiorenza.[56] The dithiocarbamate can be cyclized merely by heating, a transformation which can be looked upon as a double enolization followed by loss of water.

$$\underset{RCH \quad \; C=S}{\overset{R'C=O \; \; NH_2}{\diagdown S \diagup}} \rightarrow \underset{RC \quad \; CSH}{\overset{R'COH \; \; HN}{\diagdown S \diagup}} \rightarrow \underset{R}{\overset{R'}{\diagup}}\underset{S}{\overset{N}{\diagdown}}SH + H_2O$$

The reaction between methyl dithiocarbamate and chloroacetone to give 2-methylthio-4-methylthiazole [52] must proceed by a different mechanism, since ammonium chloride cannot be split out in the way prescribed above. This conclusion is supported by the fact that, by contrast, the reaction is slow and the yield low.

[60] Karrer and Sanz, *Helv. Chim. Acta*, **26**, 1778 (1943).

THE REACTION OF α-ACYLAMINO CARBONYL COMPOUNDS AND PHOSPHORUS PENTASULFIDE

Scope and Limitations

The reaction of 1,4-dicarbonyl compounds with phosphorus pentasulfide to produce thiophene derivatives is well known. If one of the

$$RCOCH_2CH_2COR' \xrightarrow{P_2S_5} R\underset{S}{\underset{}{\boxed{}}}R'$$

methylene groups between the carbonyl functions is replaced by an imino group, as in the acylamino carbonyl compounds, thiazoles result.

$$\underset{II}{RCOCH_2NHCOR'} \xrightarrow{P_2S_5} R\underset{S}{\underset{}{\boxed{}}}\overset{N}{}R'$$

As the equation is written, the product is a 2,5-disubstituted thiazole. Acetamidoacetone, for example, yields 2,5-dimethylthiazole.[61] If R or R' is hydrogen, a monosubstituted product results; thus formaminoacetophenone furnishes 5-phenylthiazole in 70% yield.[62] 2,4,5-Trisubstituted thiazoles can be prepared by using a derivative of an acylamino carbonyl compound in which a methylene hydrogen atom is replaced by an alkyl group. N-(α-Benzoylethyl)acetamide thus affords a 50% yield

$$RCOC\overset{R''}{\underset{|}{H}}NHCOR' \xrightarrow{P_2S_5} R\underset{S}{\underset{}{\boxed{}}}\overset{R''N}{}R'$$

of 5-phenyl-2,4-dimethylthiazole.[62] Replacement of both hydrogen atoms eliminates the ability to form a thiazole.[62] The use of ethyl esters (II, $R = OC_2H_5$) has been described.[63] The products are 5-ethoxythiazoles.

Although few yields have been reported, this reaction appears to be promising, especially for the preparation of thiazoles substituted with hydrocarbon groups. It is possible that further investigation will result in an extended usefulness.

Mechanism

There are two plausible routes by which a thiazole could be formed from an acylamino carbonyl compound and phosphorus pentasulfide. In the first, the oxygen atoms of the carbonyl groups are replaced by

---

[61] Gabriel, *Ber.*, **43**, 1283 (1910).
[62] Bachstez, *Ber.*, **47**, 3163 (1914).
[63] Miyamichi, *J. Pharm. Soc. Japan*, No. **528**, 103 (1926) [*C. A.*, **20**, 2679 (1926)].

sulfur, and the product undergoes cyclization by loss of hydrogen sulfide (A). In the second route, cyclization (by dehydration) occurs first; the

(A) $\text{RCOCH}_2\text{NHCOR}' \xrightarrow{P_2S_5} \text{RCSCH}_2\text{NHCSR}' \rightleftarrows \begin{bmatrix} \text{RC}=\text{CH}-\text{N}=\text{CR}' \\ | \quad\quad\quad\quad\quad | \\ \text{SH} \quad\quad\quad \text{HS} \end{bmatrix}$

(B) $\text{RCOCH}_2\text{NHCOR}' \longrightarrow \underset{R\quad O\quad R'}{\text{[oxazole]}} \xrightarrow{P_2S_5} \underset{R\quad S\quad R'}{\text{[thiazole]}}$

pentasulfide then reacts with the oxazole thus produced. The sequence A is preferable to B for two reasons. Phosphorus pentasulfide reacts but slowly with water [64] and would therefore not be expected to be a sufficiently strong dehydrating agent to form the oxazole directly. Second, in other reactions where oxazole formation could precede thiazole formation, no oxazole has been detected. A case in point is the reaction of phosphorus pentasulfide with acetamide and chloroacetone.[65] Support for A can be found in the known conversion of amides or ketones to their thio analogs by phosphorus pentasulfide.

THE REARRANGEMENT OF α-THIOCYANO KETONES

The α-thiocyano ketones are sensitive substances which isomerize to 2-hydroxythiazoles under a variety of conditions.

The mechanism is believed to be that shown in the accompanying equation.

$\text{RCOCH}_2\text{SCN} \xrightarrow{H_2O} \text{RCOCH}_2\text{SCONH}_2 \rightleftarrows \begin{bmatrix} \text{RC}=\text{CHSC}-\text{OH} \\ | \quad\quad\quad\quad || \\ \text{OH} \quad\quad \text{HN} \end{bmatrix}$

$\underset{S}{\text{RC}—\text{NH}} \quad \rightleftarrows \quad \underset{S}{\text{RC}—\text{N}}$
(with HC=C—O → HC=C—OH ring forms)

The rearrangement is carried out in an aqueous solution and is strongly influenced by the presence of acids or alkalies. Choice of the medium is of great importance, for often several products may be formed. In

---

[64] Yost and Russell, *Systematic Inorganic Chemistry*, p. 183, Prentice-Hall, New York, 1944.

[65] Schwarz, *Org. Syntheses*, **25**, 35 (1945).

some instances excellent yields have been obtained under the proper conditions.

The paucity of examples in the literature makes it difficult to draw a conclusion as to the generality of the reaction. It has been intensively studied, however, using thiocyanoacetone and thiocyanoacetophenone. With the latter compound, concentrated hydrochloric acid as the hydrolytic medium enables one to isolate the intermediate phenacylthiolcarbamate, $C_6H_5COCH_2SCONH_2$. With dilute hydrochloric acid, a quantitative yield of 2-hydroxy-4-phenylthiazole has been attained.[66]

EXPERIMENTAL PROCEDURES

The Reaction of Thioamides with α-Halo Carbonyl Compounds

The common method of carrying out this synthesis is to warm the reactants without solvent for a short time to initiate the reaction; it thereupon proceeds spontaneously. External cooling may be necessary, for the reaction is exothermic. Often there is a considerable amount of frothing which necessitates a larger flask than might normally be used. A preferred process uses an inert solvent to aid in controlling the reaction. Water or ethanol is most frequently employed, but the identity of the medium is not important as long as it is inert. The aqueous suspension or ethanolic solution of the reactants is heated under reflux for several hours.

Since the heterocyclic nitrogen atom of the thiazole ring is basic, the product is often obtained as the hydrohalide. The free base is readily produced, however, by use of alkali. The crude thiazoles are purified by distillation under reduced pressure or by crystallization. The heterocyclic ring is quite stable thermally, and many high-boiling thiazoles may be distilled with safety.

Preparation and purification of the thioamide (from the amide and phosphorus pentasulfide) is sometimes difficult. A modification introduced by Hromatka avoids isolation of the thioamide. It consists in heating a mixture of the amide, phosphorus pentasulfide, and the α-halo carbonyl compound.[10, 67] Presumably the thioamide is formed first and then reacts.

Because of the instability of the α-halo aldehydes, particularly the haloacetaldehydes, it is preferable to use some more stable derivative. Among these are the acetals,[32, 34, 42, 43, 68, 69, 70] ethyl α,β-dichloroethyl

---

[66] Arapides, *Ann.*, **249**, 7 (1888).

[67] Hromatka, Ger. pat. 670,131 [*C. A.*, **33**, 2909 (1939)].

[68] Suter and Johnson, *J. Am. Chem. Soc.*, **52**, 1585 (1930); U. S. pat. 1,970,656 [*C. A.*, **28**, 6250 (1934)].

[69] Short and Kelly, Brit. pat. 558,956 [*C. A.*, **39**, 4632 (1945)].

[70] Leitch and Brickman, U. S. pat. 2,230,962 [*C. A.*, **35**, 3270 (1941)].

ether,[30, 31, 33, 34, 71, 72] α,β-dichloroethyl acetate,[44, 47] and tribromoparaldehyde.[70]

2,4-Dimethylthiazole. Preparation of this compound by reaction of acetamide, phosphorus pentasulfide, and chloroacetone in benzene is described in *Organic Syntheses* by Schwarz.[65] The yield of thiazole, boiling at 143–145°, is 41–45% based on the phosphorus pentasulfide used.

4-Methyl-5-(β-hydroxyethyl)thiazole.[15] A mixture of 9.5 g. of γ-chloro-γ-acetopropyl alcohol, 6.7 g. of crude thioformamide, and 3 ml. of ethanol is allowed to stand for three days at room temperature. An additional 3.0 g. of thioformamide is added portionwise during this period. The reaction mixture is then heated for one hour on the steam bath and after cooling is taken up in water. The aqueous solution of the thiazole salt is washed with ether and then treated with aqueous sodium hydroxide to liberate the free thiazole. The latter is taken up in ether, and the solution is dried over anhydrous magnesium or sodium sulfate. Filtration of the solution and evaporation of the filtrate produces the crude thiazole, which upon distillation under reduced pressure yields 4.9 g. (50%) of pure 4-methyl-5-(β-hydroxyethyl)thiazole boiling at 93–95°/2 mm. The thiazole forms a quaternary methiodide, m.p. 89°.

2-Phenyl-4,5-dimethylthiazole.[33] Equimolar amounts of methyl α-chloroethyl ketone and thiobenzamide are heated with ethanol (5 ml. for each gram of the thioamide) on a steam bath until the ethanol has evaporated. Sufficient water is added to dissolve the crude thiazole; the acid is neutralized, and the product is removed by solution in ether. The ether solution is dried and filtered, and the ether is evaporated. The crude product is distilled under reduced pressure, and the 2-phenyl-4,5-dimethylthiazole distils as a straw-colored oil at 126–128°/6 mm. The yield is 65%.

2-Aminothiazole.[71, 72] To a solution of 76 g. (1.0 mole) of thiourea in 140 ml. of water, 143 g. (1.0 mole) of ethyl α,β-dichloroethyl ether is added. The mixture is heated under reflux, and as the reaction proceeds the two layers gradually merge. The solution is heated for a short additional period and cooled, and sufficient alkali is added to free the thiazole from its salt. Ether is added to dissolve the product, and the solution thus obtained is dried and filtered. Evaporation of the ether affords the crude 2-aminothiazole, which is pink from the presence of some aldehyde resin. Recrystallization from ethanol furnishes a nearly quantitative yield of 2-aminothiazole. It crystallizes as yellow tablets, melting at 90°.

[71] Bogert and Chertcoff, *J. Am. Chem. Soc.*, **46**, 2864 (1924); *Proc. Natl. Acad. Sci. U. S.*, **10**, 418 (1924).

[72] Traumann, *Ann.*, **249**, 31 (1888).

2-Amino-4-methylthiazole. The preparation of this amine, m.p. 44–45°, is described by Byers and Dickey in *Organic Syntheses*.[73] It is obtained in 70–75% yield by reaction of thiourea and chloroacetone.

2-Amino-4-phenylthiazole.[48] To a slurry consisting of 24.0 g. (0.2 mole) of acetophenone and 30.4 g. (0.4 mole) of thiourea is added 50.8 g. (0.2 mole) of iodine. The mixture is heated overnight on the steam bath in a closed vessel, then diluted with water, and heated with water until solution occurs. A small amount of sulfur is removed by filtration, and the filtrate is cooled and made alkaline with aqueous ammonia. The insoluble free thiazole is removed by filtration and crystallized from ethanol. The yield of 2-amino-4-phenylthiazole melting at 147° is 94%. Poorer yields are obtained using chlorine or bromine in place of iodine.

The Reaction of Ammonium Dithiocarbamate and α-Halo Ketones

The reaction of ammonium dithiocarbamate and α-halo ketones is exothermic and is therefore almost universally carried out in a solvent, a variety of which has been used. Ethanol is most common, but water,[55] hydrocarbons,[74] ether,[53,54,56] and isopropyl acetate[57] have been used. In general, the reaction is carried out by stirring the mixture of the dithiocarbamate and the α-halo ketone in the solvent at room temperature or lower, sometimes with slight warming to initiate the reaction. Once started, the reaction may proceed very vigorously, and unless a large volume of solvent is used external cooling may be desirable. The yields vary considerably; the best reported (97%) has been obtained with water. From this solvent, the mercaptothiazole precipitates as a white solid or an oil which shortly solidifies. If ethanol is the solvent, the product remains in solution and may be obtained by evaporation of the ethanol or by addition of water. One recrystallization of crude material from ethanol or ethyl acetate is usually sufficient to produce a pure product.

Equimolar quantities of the reactants should preferably be used. Excess halo ketone leads to reaction with the thiol hydrogen of the product, and a thiazyl thioether results. To avoid this complication,

$$\text{H}_3\text{C-thiazole-SH} + \text{ClCH}_2\text{COCH}_3 \longrightarrow \text{H}_3\text{C-thiazole}\cdot\text{HCl-SCH}_2\text{COCH}_3$$
III

the halo ketone can be added gradually to the dithiocarbamate. If two moles of the halo ketone are used, the thioether results almost exclu-

---

[73] Byers and Dickey, *Org. Syntheses*, Coll. Vol. **2**, 31 (1943).
[74] Mathes, U. S. pat. 2,186,420 [*C. A.*, **34**, 3537 (1940)].

sively at the expense of the mercaptothiazole. Thus, ammonium dithiocarbamate and two moles of chloroacetone in ether afford a good yield of the thioether III.

2-Mercapto-4-methylthiazole.[52] To a flask surrounded by an ice bath and containing 71.5 g. (0.65 mole) of ammonium dithiocarbamate in 140 ml. of absolute ethanol is slowly added 60 g. (0.65 mole) of chloroacetone. During the addition the slurry is mechanically stirred or shaken vigorously. The flask is removed from the ice bath, allowed to stand at room temperature for twelve hours, and heated one hour on the water bath. The ethanol is then removed by distillation. Addition of water to the oily residue and shaking induce crystallization. The yield of substantially pure material is 51.5 g. (85%). Recrystallization from a diisopropyl ether-ethanol mixture yields a purer product, m.p. 88.0–88.5°.

The Reaction of α-Acylamino Carbonyl Compounds and Phosphorus Pentasulfide

The procedure consists in heating an intimate mixture of an excess of phosphorus pentasulfide and the acylamino carbonyl compound at 100–170° until foaming (evolution of hydrogen sulfide) ceases. Usually only a short time is required. The crude thiazole is treated with aqueous alkali or acid to remove excess pentasulfide. After neutralization of the acid present, isolation is accomplished by steam distillation or by filtration, if the product is a solid.

5-Phenylthiazole.[62] A mixture of 1 g. (0.0061 mole) of α-formaminoacetophenone and 1.5 g. (0.0060 mole) of phosphorus pentasulfide is warmed on a water bath for ten minutes, at which time foaming should have ceased. To the dark brown mass, water is added to destroy any excess pentasulfide. The mixture is then acidified with hydrochloric acid and filtered. The filtrate is made just alkaline with aqueous sodium hydroxide, and the thiazole is distilled in steam. It solidifies on cooling to small, iridescent leaflets, m.p. 45–46°. The yield of 5-phenylthiazole is 0.6 g. (70%).

The Rearrangement of α-Thiocyanoketones

The necessary starting materials are best prepared by reaction of an α-halo ketone with barium thiocyanate.[75–78] Alkali metal thiocyanates

[75] Hantzsch, *Ber.*, **60**, 2537 (1927).
[76] Hantzsch, *Ber.*, **61**, 1776 (1928).
[77] Tcherniac, *Ber.*, **61**, 574 (1928).
[78] Tcherniac and Hellon, *Ber.*, **16**, 348 (1883).

afford somewhat lower yields; ammonium thiocyanate should not be used, for it causes partial rearrangement of the ketone and formation of a 2-aminothiazole.

Rearrangement of the thiocyanoketone to the 2-hydroxythiazole is carried out in aqueous solution, either acidic or alkaline. Selection of the hydrolytic agent is of great importance in order to prevent the formation of undesirable by-products. Aqueous ammonia, for example, yields a considerable amount of the 2-aminothiazole, in addition to other products. Dilute hydrochloric acid or sodium bicarbonate solution appears to be a fairly trustworthy hydrolytic medium. Because of the solubility of the lower-molecular-weight 2-hydroxythiazoles in water, they are obtained by exhaustive extraction with ether.

Several 2-hydroxythiazoles have been prepared without isolating the thiocyanoketone.[79-82] Because of the lack of yield data, it is difficult to assess the relative merits of the two modifications.

2-Hydroxy-4-methylthiazole. (a) [76] A solution of 5.0 g. (0.043 mole) of thiocyanoacetone in 100 ml. of water is heated on the water bath with 15 ml. of 2 N hydrochloric acid for two hours. The cooled solution is extracted six times with ether. The ether solution is dried with anhydrous calcium chloride and filtered. Evaporation of the ether yields 3.6 g. (72%) of crude 2-hydroxy-4-methylthiazole. Additional extractions with ether enhance the yield. Recrystallization of the crude material from ligroin yields 3.0 g. (60%) of pure product, m.p. 102°.

(b) [81] To a suspension of 92.5 g. (1.0 mole) of chloroacetone in 1.5 l. of water are added 125 g. (1.29 moles) of potassium thiocyanate (or 104.5 g. of the sodium salt) and 30 g. of sodium bicarbonate. The mixture is shaken from time to time during a period of ten days. A brown resin is gradually deposited; it is removed by filtration, and the filtrate is heated to 45°. After addition of 20 g. of decolorizing charcoal the suspension is allowed to cool for two hours with frequent shaking. It is filtered and extracted exhaustively with ether in a liquid-liquid extractor. There is thus obtained 47 g. (41%) of 2-hydroxy-4-methylthiazole, m.p. 103–104°, some of which crystallizes and some of which is obtained by evaporating the ether solution.

TABULAR SURVEY OF THIAZOLES

The following tables list methods of preparation which have appeared in the literature through 1946. A few later references have also been

[79] Andersag and Westphal, U. S. pat. 2,139,570 [*C. A.*, **33**, 2287 (1939)].
[80] I.G. Farbenindustrie A.-G., Brit. pat. 456,751 [*C. A.*, **31**, 2232 (1937)].
[81] Tcherniac, *J. Chem. Soc.*, **115**, 1071 (1919).
[82] Hantzsch and Weber, *Ber.*, **20**, 3118, 3336 (1887).

included. No specific notation of method is presented, since it will be obvious from the list of reactants. A few instances are included where no thiazole was obtained; these are usually cases in which some other investigator has reported a successful synthesis. Attention is drawn to the fact that many individual thiazoles can best be prepared from other thiazoles rather than by direct cyclization. Therefore, the methods shown are not always the best preparative methods.

TABLE II
THIAZOLES IN WHICH SUBSTITUENTS ARE LINKED THROUGH CARBON

$$\underset{R''}{\overset{R'}{}}\!\!\diagdown\!\!\underset{S}{\overset{N}{\bigcirc}}\!\!-\!\!R$$

Product			Reactants		Yield %	Reference*
R	R'	R''				
H	H	H	$ClCH_2CHO$	$HCSNH_2$	—	10, 11
			$ClCH_2CHClOC_2H_5$	$HCSNH_2$	0	36
			$(C_2H_5O)_2CHCH_2NHCHO$	P_2S_5	62†	83
CH_3—	H	H	$ClCH_2CHClOC_2H_5$	CH_3CSNH_2	—	30
			$ClCH_2CHClOC_2H_5$	CH_3CSNH_2	0	36
			$ClCH_2CH(OC_2H_5)_2$	CH_3CSNH_2	—	32, 36
C_6H_5—	H	H	$ClCH_2CHClOC_2H_5$	$C_6H_5CSNH_2$	—	31
			$(C_2H_5O)_2CHCH_2NHCOC_6H_5$	P_2S_5	—	62
$CH_2\!=\!CH(CH_2)_9CH_2$—	H	H	$ClCH_2CH(OH)_2$	$CH_2\!=\!CH(CH_2)_9CHCSNH_2\;CO_2C_2H_5$	63	88
$p\text{-}CH_3OC_6H_4$—	H	H	$BrCH_2CH(OC_2H_5)_2$	$p\text{-}CH_3OC_6H_4CSNH_2$	—	68
$p\text{-}C_2H_5OC_6H_4$—	H	H	$ClCH_2CHClOC_2H_5$	$p\text{-}C_2H_5OC_6H_4CSNH_2$	—	33
$3,4\text{-}(CH_2O_2)C_6H_3$—	H	H	$BrCH_2CH(OC_2H_5)_2$	$3,4\text{-}(CH_2O_2)C_6H_3CSNH_2$	—	68
$3,4\text{-}(HO)_2C_6H_3$—	H	H	$BrCH_2CH(OC_2H_5)_2$	$3,4\text{-}(HO)_2C_6H_3CSNH_2$	—	68
![quinolinyl]	H	H	$BrCH_2CH(OC_2H_5)_2$	8-quinolinyl-$CSNH_2$	—	84
H	CH_3—	H	$ClCH_2COCH_3$	$HCSNH_2$	—	10, 13

					Yield %	Refs.
C₆H₅—		H	BrCH₂COC₆H₅	HCSNH₂	40	10
![structure: C6H5 / N-CH3 thiazole-like with H]		H	![structure with COCH2Br, CH3]	HCSNH₂	—	107
H		H	CH₃CHBrCHO	HCSNH₂	30	12
H		CH₃	C₆H₅CHBrCHO	HCSNH₂	—	10
		C₆H₅	C₆H₅COCH₂NHCHO	P₂S₅	70	62
CH₃—		H	ClCH₂COCH₃	CH₃CSNH₂	—	10, 13, 30, 32, 65
ClCH₂—		H	ClCH₂COCH₂Cl	CH₃CSNH₂	75	85
C₂H₅OCOCH₂—		H	C₂H₅OCOCH₂COCH₂Br	CH₃CSNH₂	—	35
CH₃CH— CO₂C₂H₅		H	CH₃	CH₃CSNH₂	—	34, 35
CH₃—		H	BrCH₂COCHCO₂C₂H₅	CH₃CSNH₂	—	30, 32
C₆H₅—		H	C₆H₅COCH₂Br	CH₃CSNH₂	—	86
p-FC₆H₄—		H	p-FC₆H₄COCH₂Br	CH₃CSNH₂	85–95	86
p-ClC₆H₄—		H	p-ClC₆H₄COCH₂Br	CH₃CSNH₂	85	86
p-BrC₆H₄—		H	p-BrC₆H₄COCH₂Br	CH₃CSNH₂	85	86
p-IC₆H₄—		H	p-IC₆H₄COCH₂Br	CH₃CSNH₂	—	87
β-C₁₀H₇—		H	β-C₁₀H₇COCH₂Br	CH₃CSNH₂	—	107
![N-CH3 thiazole structure]			![COCH2Br, CH3 structure]			
CH₃—		H	ClCH₂COCH₃	C₂H₅CSNH₂	—	31
C₆H₅—		H	C₆H₅COCH₂Br	C₂H₅CSNH₂	—	31
p-FC₆H₄—		H	p-FC₆H₄COCH₂Br	C₂H₅CSNH₂	85–95	86
p-ClC₆H₄—		H	p-ClC₆H₄COCH₂Br	C₂H₅CSNH₂	85–95	86
p-BrC₆H₄—		H	p-BrC₆H₄COCH₂Br	C₂H₅CSNH₂	85–95	86
p-IC₆H₄—		H	p-IC₆H₄COCH₂Br	C₂H₅CSNH₂	85–95	86
CH₃—		H	CH₃COCH₂Cl	C₂H₅CSNH₂	—	10
(CH₃)₂CHCH₂—		H	3,4-(HO)₂C₆H₃COCH₂Cl	(CH₃)₂CHCH₂CSNH₂	—	89
CH₃CONHCH₂—		H	3,4-(HO)₂C₆H₃COCH₂Cl	CH₃CONHCH₂CSNH₂	—	89
CH₃CONHC— CH₃				CH₃CONHCCSNH₂ CH₃		

* References 83–193 are on pp. 407–409.   † The thiazole was isolated as the mercuric chloride complex.

TABLE II—Continued
Thiazoles in Which Substituents Are Linked Through Carbon

Product			Reactants		Yield %	Reference*
R	R'	R''				
(phthalimido-NCH₂—)	ClCH₂—	H	ClCH₂COCH₂Cl	NCH₂CSNH₂ (phthalimido)	32	90
(phthalimido-NCH₂CH₂—)	3,4-(HO)₂C₆H₃—	H	3,4-(HO)₂C₆H₃COCH₂Cl	NCH₂CH₂CSNH₂ (phthalimido)	—	91
(phthalimido-N(CH₂)₃—)	3,4-(HO)₂C₆H₃—	H	3,4-(HO)₂C₆H₃COCH₂Cl	N(CH₂)₃CSNH₂ (phthalimido)	—	91
C₂H₅OCH₂—	C₆H₅—	H	C₆H₅COCH₂Br	C₂H₅OCH₂CSNH₂	—	92
C₆H₅CO₂CH₂—	C₆H₅—	H	C₆H₅COCH₂Br	C₆H₅CO₂CH₂CSNH₂	80	92
C₆H₅CO₂CH(CH₃)—	C₆H₅—	H	C₆H₅COCH₂Br	C₆H₅CO₂CH(CH₃)CSNH₂	35	93
C₆H₅CO₂CH(CH₃)—	3,4-(HO)₂C₆H₃—	H	3,4-(HO)₂C₆H₃COCH₂Cl	C₆H₅CO₂CH(CH₃)CSNH₂	83	94
C₆H₅CO₂CH(CH₃)—	3,4-(CH₃O)₂-C₆H₃—	H	3,4-(CH₃O)₂C₆H₃COCH₂Cl	C₆H₅CO₂CH(CH₃)CSNH₂	76	94
C₆H₅—	CH₃—	H	CH₃COCH₂Cl	C₆H₅CSNH₂	67	31
C₆H₅—	C₂H₅—	H	C₂H₅COCH₂Cl	C₆H₅CSNH₂	71	33
C₆H₅—	ClCH₂—	H	ClCH₂COCH₂Cl	C₆H₅CSNH₂	81	95
C₆H₅—	C₆H₅—	H	C₆H₅COCH₂Br	C₆H₅CSNH₂	—	96
C₆H₅—	3,4-(HO)₂C₆H₃—	H	3,4-(HO)₂C₆H₃COCH₂Cl	C₆H₅CSNH₂	— †	31
p-CH₃OC₆H₄—	CH₃—	H	CH₃COCH₂Cl	p-CH₃OC₆H₄CSNH₂	90	96a
p-CH₃OC₆H₄—	ClCH₂—	H	ClCH₂COCH₂Cl	p-CH₃OC₆H₄CSNH₂	72	68
p-CH₃OC₆H₄—	3,4-(HO)₂C₆H₃—	H	3,4-(HO)₂C₆H₃COCH₂Cl	p-CH₃OC₆H₄CSNH₂	—	96, 97
p-C₂H₅OC₆H₄—	CH₃—	H	CH₃COCH₂Cl	p-C₂H₅OC₆H₄CSNH₂	—	33

THE PREPARATION OF THIAZOLES

2-substituent	4-substituent	5-substituent	α-halo carbonyl	thioamide	yield	ref
$p\text{-}C_2H_5OC_6H_4-$	C_2H_5-	H	$CH_3CH_2COCH_2Cl$	$p\text{-}C_2H_5OC_6H_4CSNH_2$	—	33
$3,4\text{-}(CH_2O)_2C_6H_3-$	$ClCH_2-$	H	$ClCH_2COCH_2Cl$	$3,4\text{-}(CH_3O)_2C_6H_3CSNH_2$	—	98
$3,4\text{-}(CH_2O_2)C_6H_3-$	CH_3-	H	CH_3COCH_2Cl	$3,4\text{-}(CH_2O_2)C_6H_3CSNH_2$	91	68
$3,4\text{-}(CH_2O_2)C_6H_3-$	$ClCH_2-$	H	$ClCH_2COCH_2Cl$	$3,4\text{-}(CH_2O_2)C_6H_3CSNH_2$	60	96
$3,4\text{-}(CH_2O_2)C_6H_3-$	$3,4\text{-}(HO)_2C_6H_3-$	H	$3,4\text{-}(HO)_2C_6H_3COCH_2Cl$	$3,4\text{-}(CH_2O_2)C_6H_3CSNH_2$	90	68
$3,4\text{-}(HO)_2C_6H_3-$	CH_3-	H	CH_3COCH_2Cl	$3,4\text{-}(HO)_2C_6H_3CSNH_2$	—	68
$3,4\text{-}(HO)_2C_6H_3-$	$3,4\text{-}(HO)_2C_6H_3-$	H	$3,4\text{-}(HO)_2C_6H_3COCH_2Cl$	$3,4\text{-}(HO)_2C_6H_3CSNH_2$	80	68
$p\text{-}CH_3CONHC_6H_4-$	C_6H_5-	H	$C_6H_5COCH_2Cl$	$p\text{-}CH_3CONHC_6H_4CSNH_2$	—	94
	CH_3-	H	CH_3COCH_2Br	[quinoline-2-CSNH$_2$]	100	84
	CH_3-	H	CH_3COCH_2Br	[quinoline-2-CSNH$_2$]	—	84
	CH_3-	H	CH_3COCH_2Br	[quinoline-3-CSNH$_2$]	—	84
	CH_3-	H	CH_3COCH_2Br	[quinoline-4-CSNH$_2$]	—	84
		H	CH_3COCH_2Br	[quinoline-H$_2$NSC]	—	84
	H	CH_3-	$CH_3CHClCHO$	CH_3CSNH_2	—	31, 36
	H	CH_3-	$CH_3COCH_2NHCOCH_3$	P_2S_5	—	61
	H	C_6H_5-	$C_6H_5CHBrCHO$	CH_3CSNH_2	—	100
	H	CH_3-	$C_6H_5COCH_2NHCOCH_3$	P_2S_5	—	61
	H	$p\text{-}CH_3C_6H_4-$	$p\text{-}CH_3C_6H_4COCH_2NHCOCH_3$	P_2S_5	—	101

* References 83–193 are on pp. 407–409.
† The thiazole was isolated as the mercuric chloride complex.

TABLE II—*Continued*

THIAZOLES IN WHICH SUBSTITUENTS ARE LINKED THROUGH CARBON

Product			Reactants		Yield %	Reference *
R	R'	R''				
C_6H_5—	H	CH_3—	$CH_3COCH_2NHCOC_6H_5$	P_2S_5	—	61
C_6H_5—	H	C_6H_5—	$C_6H_5COCH_2NHCOC_6H_5$	P_2S_5	—	102
C_6H_5—	H	$p\text{-}CH_3C_6H_4$—	$p\text{-}CH_3C_6H_4COCH_2NHCOC_6H_5$	P_2S_5	—	101
H	CH_3—	CH_3CH_2—	$CH_3COCHClC_2H_5$	$HCSNH_2$	29	21
H	CH_3—	$HOCH_2$—	$CH_3COCHBrCH_2OH$	$HCSNH_2$	7	28
					28	17
H	CH_3—	$BrCH_2$—	$CH_3COCHBrCH_2Br$	$HCSNH_2$	80	28
					0	28
H	CH_3—	$CH_3CO_2CH_2$—	$CH_3COCHClCH_2OCOCH_3$	$HCSNH_2$	16	27, 28
H	CH_3—	$C_2H_5OCOCH_2$—	$CH_3COCHBrCH_2CO_2C_2H_5$	$HCSNH_2$	71	29
H	CH_3—	$HOCH_2CH_2$—	$CH_3COCHClCH_2CH_2OH$	$HCSNH_2$	50	10, 15, 16, 17
			$CH_3COCHBrCH_2CH_2OH$	$HCSNH_2$	28–47	16, 17, 18, 19
			$CH_3COCHBrCH_2CH_2OCOCH_3$	$HCSNH_2$	15.5	103, 104 105
			$\begin{array}{c}\text{Cl}\\ \text{CH}_2\text{—CCOCH}_3\\ \text{CH}_2\text{—O}\hspace{6pt}\text{CO}\\ \hspace{10pt}\text{CH}_3\end{array}$	$HCSNH_2$	—	106
			$\begin{array}{c}\text{CH}_2\text{—CHCl}\\ \text{CH}_2\text{—C}\hspace{6pt}\text{CH}_3\\ \hspace{10pt}\text{O}\hspace{6pt}\text{OCH}_2\text{CH}_3\end{array}$			
H	CH_3—	$BrCH_2CH_2$—	$CH_3COCHBrCH_2CH_2Br$	$HCSNH_2$	—	80
H	CH_3—	$CH_3CH_2OCH_2CH_2$—	$CH_3COCHClCH_2CH_2OC_2H_5$	$HCSNH_2$	—	13
H	CH_3—	$CH_3CHOHCH_2$—	$CH_3COCHClCH_2CHOHCH_3$	$HCSNH_2$	24	21

THE PREPARATION OF THIAZOLES

				Yield %	Ref.
H	CH₃—	HOCH₂CH₂CH₂—	CH₃COCHBrCH₂CH₂CH₂OH	HCSNH₂	48	21
H	CH₃—	BrCH₂CH₂CH₂—	CH₃COCHBrCH₂CH₂CH₂Br	HCSNH₂	—	80
H	CH₃—	C₆H₅—	CH₃COCHBrC₆H₅	HCSNH₂	—	108
H	CH₃—	BrCH₂CH₂—	CH₃COCHBrCH₂Br	HCSNH₂	—	109
H	C₆H₅—	CH₃—	C₆H₅COCHBrCH₃	HCSNH₂	—	14
CH₃—	CH₃—		CH₃COCHClCH₃	CH₃CSNH₂	11	34, 35
CH₃—	HOCH₂CH₂—		CH₃COCHBrCH₂CH₂OH	CH₃CSNH₂	10	17
			(epoxide: CH₂—CCOCH₃ / CO / CH₂—O)			105
CH₃—	C₆H₅OCH₂CH₂—		CH₃COCHClCH₂OC₆H₅	CH₃CSNH₂	—	104
CH₃—	CH₂—CHCH₂— (epoxide O)		ClCH₂CHClCH₂COCOCH₃	CH₃CSNH₂	—	110
CH₃—	C₆H₅—	CH₃—	C₆H₅COCHBrCH₃	CH₃CSNH₂	40	87
CH₃—	C₆H₅—	CH₃CH(CO₂H)—	C₆H₅COCHBrCH(CH₃)CO₂H	CH₃CSNH₂	10	111
CH₃—	C₆H₅—	HO₂CCH₂—	C₆H₅COCHBrCH₂CO₂H	CH₃CSNH₂	93	111
CH₃—	p-CH₃C₆H₄—	HO₂CCH₂—	p-CH₃C₆H₄COCHBrCH₂CO₂H	CH₃CSNH₂	90	111
CH₃—	p-CH₃CH₂C₆H₄—	HO₂CCH₂—	p-CH₃CH₂C₆H₄COCHBrCH₂CO₂H	CH₃CSNH₂	60	111
CH₃—	p-iso-C₃H₇C₆H₄—	HO₂CCH₂—	p-iso-C₃H₇C₆H₄COCHBrCH₂CO₂H	CH₃CSNH₂	74	111
CH₃—	2,4-(CH₃)₂C₆H₃—	HO₂CCH₂—	2,4-(CH₃)₂C₆H₃COCHBrCH₂CO₂H	CH₃CSNH₂	86	111
CH₃—	p-ClC₆H₄—	HO₂CCH₂—	p-ClC₆H₄COCHBrCH₂CO₂H	CH₃CSNH₂	91	111
CH₃—	α-C₁₀H₇—	HO₂CCH₂—	α-C₁₀H₇COCHBrCH₂CO₂H	CH₃CSNH₂	43	111
CH₃—	β-C₁₀H₇—	HO₂CCH₂—	β-C₁₀H₇COCHBrCH₂CO₂H	CH₃CSNH₂	68	111
CH₃—	p-CH₃OC₆H₄—	HO₂CCH₂—	p-CH₃OC₆H₄COCHBrCH₂CO₂H	CH₃CSNH₂	43	111
CH₃—	p-CH₃CH₂OC₆H₄—	HO₂CCH₂—	p-CH₃CH₂OC₆H₄COCHBrCH₂CO₂H	CH₃CSNH₂	12	111
CH₃—	(2-thienyl)—	HO₂CCH₂—	(2-thienyl)COCHBrCH₂CO₂H	CH₃CSNH₂	84	111
CH₃—	C₆H₅—		C₆H₅COCH₂NHCOCH₃	P₂S₅	50	62
CH₃—	C₆H₅—		C₆H₅COCHClC₆H₅	CH₃CSNH₂	—	31
CH₃CH₂—	HOCH₂CH₂—		CH₃COCHBrCH₂OH	CH₃CH₂CSNH₂	—	14, 17

* References 83–193 are on pp. 407–409.

TABLE II—Continued

THIAZOLES IN WHICH SUBSTITUENTS ARE LINKED THROUGH CARBON

Product			Reactants		Yield %	Reference*
R	R'	R''				
$CH_3CH_2CH_2$—			CH_2—$CCOCH_3$ | | CH_2—O—CO	$CH_3CH_2CSNH_2$	—	105
C_6H_5—	CH_3—	$HOCH_2CH_2$—	$CH_3COCHBrCH_2CH_2OH$	$CH_3CH_2CH_2CSNH_2$	—	17
C_6H_5—	CH_3—	CH_3—	CH_2—$CCOCH_3$ | | CH_2—O—CO	$CH_3CH_2CH_2CSNH_2$	—	105
		$HOCH_2CH_2$—	$CH_3CHClCOCH_3$	$C_6H_5CSNH_2$	65	33
C_6H_5—	CH_3—	$ClCH_2CHOHCH_2$—	CH_2—$CCOCH_3$ | | CH_2—O—CO	$C_6H_5CSNH_2$	—	105
			Cl—$CCOCH_3$ | | CH_2—CCH—O—CO C_6H_5	$C_6H_5CSNH_2$	—	110
C_6H_5—	C_6H_5—	C_6H_5—	$C_6H_5COCHBrC_6H_5$	$C_6H_5CSNH_2$	—	31
p-$C_2H_5OC_6H_4$—	CH_3—	CH_3—	$CH_3COCHClCH_3$	p-$C_2H_5OC_6H_4CSNH_2$	—	33

* References 83–193 are on pp. 407–409.

TABLE III
Quaternary Thiazolium Salts

$$\underset{R'''}{\overset{R''}{\Big|}}\!\!\!\underset{S}{\overbrace{}}\!\!\!\overset{+NR'}{\underset{R}{\Big|}}\ X^-$$

Product				Reactants		Yield %	Reference*
R	R'	R''	R'''				
H	CH$_3$CH$_2$—	CH$_3$—	H	CH$_3$COCH$_2$Cl	HCSNHCH$_2$CH$_3$	8	10
H	iso-C$_3$H$_7$—	CH$_3$—	H	CH$_3$COCH$_2$Cl	HCSNHCH(CH$_3$)$_2$	—	10
H	n-C$_4$H$_9$—	CH$_3$—	H	CH$_3$COCH$_2$Cl	HCSNHCH$_2$CH$_2$CH$_2$CH$_3$	—	10
H	C$_6$H$_{11}$CH$_2$—	CH$_3$—	H	CH$_3$COCH$_2$Cl	HCSNHCH$_2$C$_6$H$_{11}$	19	10
H	C$_6$H$_5$CH$_2$—	CH$_3$—	H	CH$_3$COCH$_2$Cl	HCSNHCH$_2$C$_6$H$_5$	11	10
H	C$_6$H$_5$—	CH$_3$—	H	CH$_3$COCH$_2$Cl	HCSNHC$_6$H$_5$	—	112
H	3,4-(CH$_2$O$_2$)C$_6$H$_3$—	CH$_3$—	H	CH$_3$COCH$_2$Cl	HCSNHC$_6$H$_3$(O$_2$CH$_2$)-3,4	54	13
H	o-O$_2$NC$_6$H$_4$—	CH$_3$—	H	CH$_3$COCH$_2$Cl	HCSNHC$_6$H$_4$NO$_2$-o	—	10
H	o-CH$_3$CONHC$_6$H$_4$—	CH$_3$—	H	CH$_3$COCH$_2$Cl	HCSNHC$_6$H$_4$NHCOCH$_3$-o	85	112
H	(CH$_3$)$_2$CHCH$_2$CH$_2$—	C$_6$H$_5$—	H	C$_6$H$_5$COCH$_2$Br	HCSNHCH$_2$CH$_2$CH(CH$_3$)$_2$	—	112
CH$_3$—	CH$_3$—	CH$_3$—	H	CH$_3$COCH$_2$Cl	CH$_3$CSNHCH$_3$	100	113
CH$_3$—	C$_6$H$_5$CH(CH$_3$)—	CH$_3$—	H	CH$_3$COCH$_2$Cl	CH$_3$CSNHCH(CH$_3$)C$_6$H$_5$	—	41
CH$_3$—	C$_6$H$_5$—	CH$_3$—	H	CH$_3$COCH$_2$Cl	CH$_3$CSNHC$_6$H$_5$	100	114
CH$_3$—	o-CH$_3$C$_6$H$_4$—	CH$_3$—	H	CH$_3$COCH$_2$Cl	CH$_3$CSNHC$_6$H$_4$CH$_3$-o	—	13
CH$_3$—	o-O$_2$NC$_6$H$_4$—	CH$_3$—	H	CH$_3$COCH$_2$Cl	CH$_3$CSNHC$_6$H$_4$NO$_2$-o	—	41
H	H$_3$C—(NH$_2$-pyrimidinyl)—CH$_2$—	CH$_3$—	H	CH$_3$COCH$_2$Cl	HCSHNH$_2$C—(pyrimidinyl)—CH$_3$	—	115

*References 83–193 are on pp. 407–409.

TABLE III—*Continued*

QUATERNARY THIAZOLIUM SALTS

Product			Reactants		Yield %	Reference*
R	R'	R''	R'''			
H	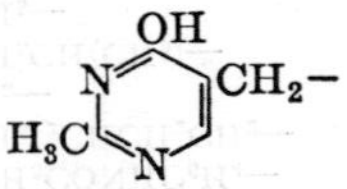 (4-methyl-2-hydroxypyrimidin-5-yl)CH$_2$–	CH$_3$–	H	CH$_3$COCH$_2$Cl	HCSHNH$_2$C-pyrimidine(OH,CH$_3$)	—	115
H	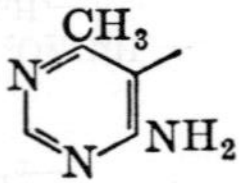 (4-methyl-2-aminopyrimidin-5-yl)	CH$_3$–	H	CH$_3$COCH$_2$Cl	HCSHN-pyrimidine(CH$_3$,NH$_2$)	—	116
H	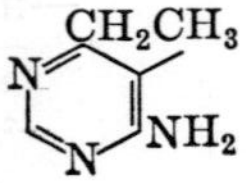 (2-amino-4-methylpyrimidin-5-yl)CH$_2$–	CH$_3$–	H	CH$_3$COCH$_2$Cl	HCSHN-pyrimidine(CH$_2$CH$_3$,NH$_2$)	—	116
H	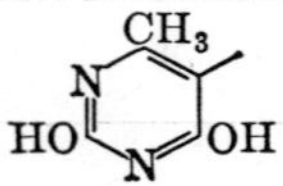 (4-methyl-2-hydroxypyrimidin-5-yl)	CH$_3$–	H	CH$_3$COCH$_2$Cl	HCSHN-pyrimidine(CH$_3$,OH)	—	116
H	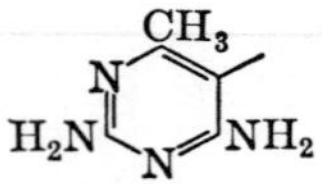 (2,4-diaminopyrimidin-5-yl)	CH$_3$–	H	CH$_3$COCH$_2$Cl	HCSHN-pyrimidine(CH$_3$,NH$_2$)	—	116
H	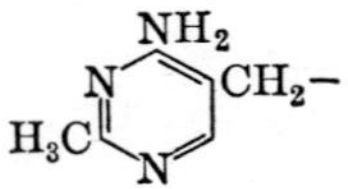 (2-amino-4-methylpyrimidin-5-yl)CH$_2$–	CH$_3$–	HOCH$_2$CH$_2$–	CH$_3$COCHClCH$_2$CH$_2$OH	HCSHNH$_2$C-pyrimidine(NH$_2$,CH$_3$)	—	117, 118

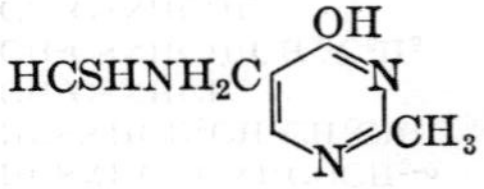

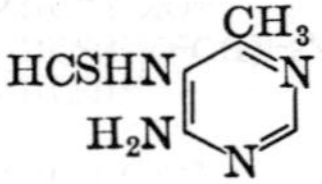

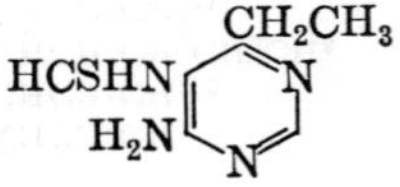

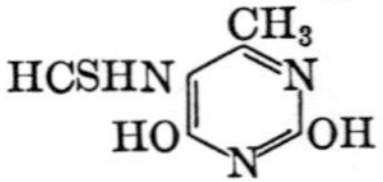

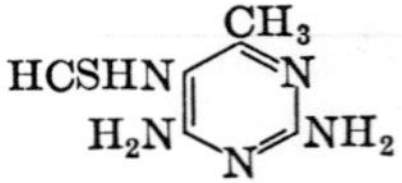

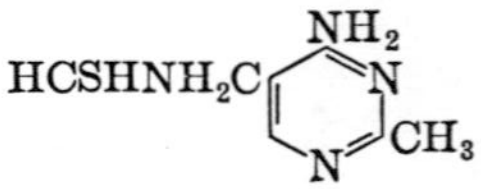

THE PREPARATION OF THIAZOLES

H	![pyrimidine with NH2, CH2-, H3C, N, N]	CH_3—	$CH_3O_2CCH_2$—	$CH_3COCHBrCH_2CO_2CH_3$	HCSHNH₂C (pyrimidine NH₂, CH₃)	—	119
H	![pyrimidine NH2, CH2-, H3C]	CH_3—	$HOCH_2CH_2$—	$CH_3COCHClCH_2CH_2OCOCH_3$	HCSHNH₂C (NH₂, CH₃)	—	10, 118, 120, 121
H	![pyrimidine NH2, CH2-, H3C]	CH_3—	$HOCH_2CH_2$—	$CH_3COCHBrCH_2CH_2OCOC_6H_5$	HCSHNH₂C (NH₂, CH₃)	—	122
H	![pyrimidine NH2, CH2-, H3C]	CH_3—	$HOCH_2CH_2$—	(cyclic structure with CHCl, CH₃, OH, O)	HCSHNH₂C (NH₂, CH₃)	—	117
H	![pyrimidine OH, CH2-, H3C]	CH_3—	$HOCH_2CH_2OH$	$CH_3COCHBrCH_2CH_2OH$	HCSHNH₂C (OH, CH₃)	—	123
H	![pyrimidine OH, CH2-, H3C]	CH_3—	$HOCH_2CH_2$—	$CH_3COCHClCH_2CH_2OH$	HCSHNH₂C (OH, CH₃)	—	115
H	![pyrimidine OH, CH2-, H3C]	CH_3—	$HOCH_2CH_2$—	$CH_3COCHBrCH_2CH_2OCOCH_3$	HCSHNH₂C (OH, CH₃)	—	115
				$CH_3COCHBrCH_2CH_2OCOCH_3$	HCSHNH₂C (OH, CH₃)	—	115

* References 83–193 are on pp. 407–409.

TABLE III—*Continued*

QUATERNARY THIAZOLIUM SALTS

Product				Reactants		Yield %	Reference*
R	R'	R''	R'''				
H	4-methyl-6-amino-pyrimidin-5-yl	CH$_3$—	HOCH$_2$CH$_2$—	CH$_3$COCHClCH$_2$OH	HCSHN + H$_2$N–pyrimidine(CH$_3$)	—	116
H	4-ethyl-6-amino-pyrimidin-5-yl (CH$_2$CH$_3$)	CH$_3$—	HOCH$_2$CH$_2$—	CH$_3$COCHClCH$_2$OH	HCSHN + H$_2$N–pyrimidine(CH$_2$CH$_3$)	—	116
H	4-amino-pyrimidin-5-ylmethyl (NH$_2$)	CH$_3$—	HOCH$_2$CH$_2$—	CH$_3$COCHClCH$_2$OCOCH$_3$	HCSHNH$_2$C–pyrimidine(NH$_2$, CH$_2$CH$_3$)	—	122
H	2,4-diamino-pyrimidin-5-ylmethyl	CH$_3$—	HOCH$_2$CH$_2$—	CH$_3$COCHBrCH$_2$OCOCH$_3$	HCSHNH$_2$C–pyrimidine(NH$_2$, NH$_2$)	80	124
CH$_3$	2-methyl-4-amino-pyrimidin-5-ylmethyl	CH$_3$—	HOCH$_2$CH$_2$—	CH$_3$COCHBrCH$_2$OCOCH$_3$	CH$_3$CSHNH$_2$C–pyrimidine(NH$_2$, CH$_3$)	—	115

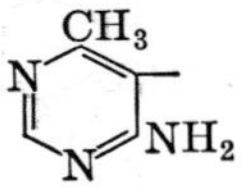

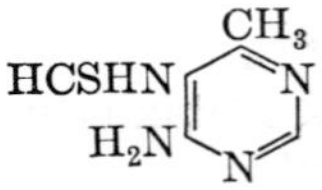

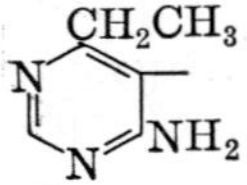

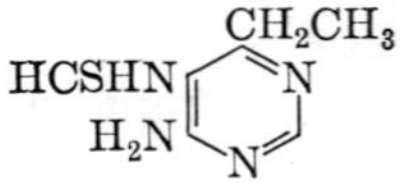

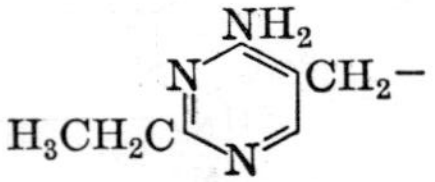

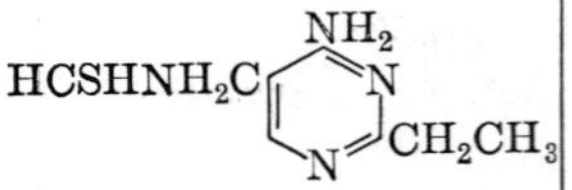

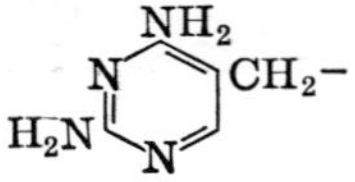

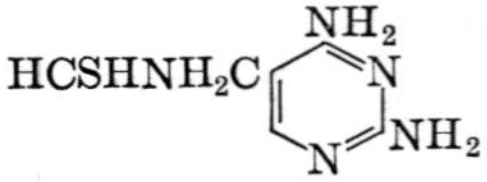

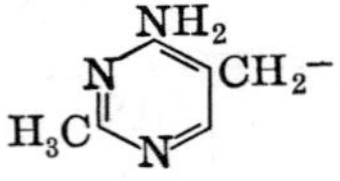

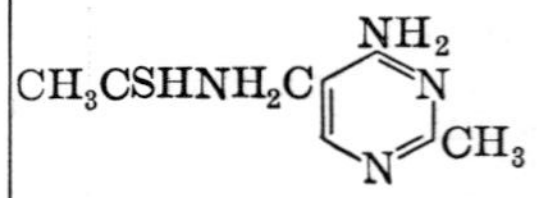

* References 83–193 are on pp. 407–409.

TABLE IV
Hydroxythiazoles and Ethers

$$\begin{array}{c} R' \underset{R''}{\overset{N}{\diagdown}} \!\!\!\diagup R \\ S \end{array}$$

Product			Reactants		Yield %	Reference *
R	R'	R''				
C_2H_5O—	H	H	$CH_3COCH_2OCHClCH_2Cl$	$C_2H_5OCSNH_2$	—	39
HO—	CH_3—	H	CH_3COCH_2Cl	NH_2CSONH_4	93	81
			CH_3COCH_2SCN	—	72	76
			CH_3COCH_2SCN	—	—	39, 66, 76–78, 125–127
$C_2H_5O_2C$—	CH_3—	H	CH_3COCH_2Cl	Metal thiocyanate	41	81, 82, 128
HO—	C_6H_5—	H	CH_3COCH_2Cl	$C_2H_5O_2CCSNH_2$	—	38
			$C_6H_5COCH_2SCN$	—	—	66, 76, 125
			$C_6H_5COCH_2Cl$	KSCN	—	129
C_2H_5O—	C_6H_5—	H	$C_6H_5COCH_2Br$	$C_2H_5OCSNH_2$	—	31
HO—	CH_3—	CH_3—	$CH_3COCH(CH_3)SCN$	—	—	39
CH_3O—	CH_3—	CH_3—	$CH_3CHBrCOCH_3$	CH_3OCSNH_2	—	39
HO—	CH_3—	$BrCH_2CH_2$—	$BrCH_2CH_2CHBrCOCH_3$	$Ba(SCN)_2$	—	79, 80
HO—	CH_3—	$C_2H_5OCOCH_2$—	$C_2H_5OCOCH_2CHBrCOCH_3$	$Ba(SCN)_2$	—	79, 80
HO—	CH_3—	$C_2H_5OCOCH_2CH_2$—	$C_2H_5OCOCH_2CH_2CHBrCOCH_3$	$Ba(SCN)_2$	—	79, 80
HO—	CH_3—	$HOCH_2CH_2$—	$HOCH_2CH_2CHBrCOCH_3$	$Ba(SCN)_2$	—	79, 80
HO—	CH_3—	$CH_3CO_2CH_2CH_2$—	$CH_3CO_2CH_2CH_2CHBrCOCH_3$	$Ba(SCN)_2$	—	79, 130
HO—	CH_3—	$C_6H_5CO_2CH_2CH_2$—	$C_6H_5CO_2CH_2CH_2CHBrCOCH_3$	$Ba(SCN)_2$	—	79, 80
CH_3—	H	C_2H_5O—	$C_2H_5OCOCH_2NHCOCH_3$	P_2S_5	—	63
C_6H_5—	H	C_2H_5O—	$C_2H_5OCOCH_2NHCOC_6H_5$	P_2S_5	—	63
C_6H_5—	CH_3—	C_2H_5O—	$C_2H_5OCOCH(CH_3)NHCOC_6H_5$	P_2S_5	—	63

* References 83–193 are on pp. 407–409.

TABLE V

2-Mercaptothiazoles from Ammonium Dithiocarbamate

$$\underset{S}{\underset{R'}{R}}\!\!\!\!\!\!\!\!\!\!\!\!\!\!\diagdown\!\!\!\!\!\diagup\!\!\!\!\!\!\!\!\!\!\!\overset{N}{}\text{—SH}$$

Product		Reactant	Medium	Yield %	References *
R	R'				
CH_3-	H	CH_3COCH_2Cl	Water	78	55
		CH_3COCH_2Cl	Ether	—	54
		CH_3COCH_2Cl	Ether	—	53
		CH_3COCH_2Cl	Ethanol	—	51
		CH_3COCH_2Cl	Ethanol	85	52
C_2H_5-	H	$CH_3CH_2COCH_2Cl$	Water	97	55
C_2H_5-	H	Brominated $CH_3COCH_2CH_3$ †	Ethanol	29	52
$CH_3(CH_2)_{10}CH_2-$	H	$CH_3(CH_2)_{10}CH_2COCH_2Cl$	Methanol	45	131
$CH_3(CH_2)_{12}CH_2-$	H	$CH_3(CH_2)_{12}CH_2COCH_2Cl$	Methanol	52	131
C_6H_5-	H	$C_6H_5COCH_2Cl$	Water	95	55
		$C_6H_5COCH_2Cl$	Isopropyl alcohol	91	59
		$C_6H_5COCH_2Br$	Ether	55 ‡	56
		$C_6H_5COCH_2Br$	Ethanol	—	51
CH_3-	CH_3-	$CH_3COCHBrCH_3$	—	—	58
		$CH_3COCHBrCH_3$	Ethanol	75	52
		Brominated $CH_3COCH_2CH_3$ †	Ethanol	43	52
		$CH_3COCHClCH_3$	Water	91	55
		$CH_3COCHClCH_3$	Ethanol	47	52
CH_3-	$HOCH_2CH_2-$	$CH_3COCHBrCH_2CH_2OH$	Ether	—	53
CH_3-	$CH_3CO_2CH_2CH_2-$	$CH_3COCHClCH_2CH_2OCOCH_3$	Ethanol	71	57
		$CH_3COCHClCH_2CH_2OCOCH_3$	—	—	58
		$CH_3COCHBrCH_2CH_2OCOCH_3$	Ether	—	53
CH_3-	$ClCH_2CH_2-$	$CH_3COCHClCH_2CH_2Cl$		—	58

* References 83–193 are on pp. 407–409.
† The mixture of monobromo derivatives obtained by brominating methyl ethyl ketone was used without separation.
‡ The phenacyl thio ether was also obtained.

TABLE VI

Thiazyl Ketones

$$CH_3CO\!\!-\!\!\underset{S}{\underset{}{\diagdown\!\!\!\!\!\diagup}}\!\!\!\!\!\!\overset{R'\!\!-\!\!N}{}\!\!-R$$

Product		Reactants		Yield %	Reference
R	R'				
H	CH_3-	$CH_3COCHClCOCH_3$	$HCSNH_2$	55	20, 21
CH_3-	CH_3-	$CH_3COCHBrCOCH_3$	CH_3CSNH_2	—	132

TABLE VII
Thiazole Carboxylic Acids and Esters

$$\underset{R''}{\overset{R'}{\diagdown}}\underset{S}{\overset{N}{\diagup}}R$$

Product			Reactants		Yield %	Reference*
R	R'	R''				
$C_2H_5O_2C-$	CH_3-	H	CH_3COCH_2Cl	$C_2H_5O_2CCSNH_2$	—	40
$C_2H_5O_2C-$	CH_3-	$HOCH_2CH_2-$	$CH_3COCHBrCH_2CH_2OH$	$C_2H_5O_2CCSNH_2$	—	40
$C_2H_5O_2C-$	CH_3-	$CH_3CO_2CH_2CH_2-$	$CH_3COCHBrCH_2CH_2OCOCH_3$	$C_2H_5O_2CCSNH_2$	—	40
$C_2H_5O_2C-$	H	CH_3-	$CH_3CHBrCHO$	$C_2H_5O_2CCSNH_2$	—	12
H	H	HO_2C-	$C_2H_5OCOCHClCHO$	$HCSNH_2$	—	23, 133–135
H	H	$C_2H_5O_2C-$	$C_2H_5OCOCHClCHO$	$HCSNH_2$	—	25, 133
C_6H_5-	$C_2H_5O_2C-$	H	$C_2H_5OCOCOCH_2Br$	$C_6H_5CSNH_2$	—	137
CH_3-	H	$C_2H_5O_2C-$	$C_2H_5OCOCHClCHO$	CH_3CSNH_2	37	136
C_6H_5-	H	$C_2H_5O_2C-$	$C_2H_5OCOCHClCHO$	$C_6H_5CSNH_2$	75	137, 138
H	H	HO_2C-	$C_2H_5OCOCHClCOCH_3$	$HCSNH_2$	58	13, 23, 24
H	CH_3-	CH_3O_2C-	$C_2H_5OCOCHClCO_2CH_3$	$HCSNH_2$	50	22
H	CH_3-	$C_2H_5O_2C-$	$CH_3COCHBrCO_2C_2H_5$	$HCSNH_2$	—	25, 26
H	CH_3-	$C_2H_5O_2C-$	$C_6H_5COCHBrCO_2C_2H_5$	$HCSNH_2$	—	25
H	C_6H_5-	HO_2C-	$CH_3COCHClCO_2C_2H_5$	CH_3CSNH_2	—	35, 100
CH_3-	CH_3-	$C_2H_5O_2C-$	$CH_3COCHClCO_2C_2H_5$	CH_3CSNH_2	—	35
CH_3-	CH_3-	$C_2H_5O_2C-$	$C_6H_5COCHClCO_2C_2H_5$	$C_6H_5CSNH_2$	—	31
C_6H_5-	CH_3-	$C_2H_5O_2C-$	$C_2H_5OCOCOCHClCO_2C_2H_5$	$HCSNH_2$	—	25, 133, 139
H	$C_2H_5O_2C-$	HO_2C-	$C_2H_5OCOCOCHClCO_2C_2H_5$	CH_3CSNH_2	83	41, 136
CH_3-	HO_2C-	$C_2H_5O_2C-$	$C_2H_5OCOCOCHClCO_2C_2H_5$	$C_6H_5CSNH_2$	—	137, 138
C_6H_5-	$C_2H_5O_2C-$	$C_2H_5O_2C-$	$C_2H_5OCOCOCHClCO_2C_2H_5$†‡	$Ba(SCN)_2$	—	82, 140
$HO-$	CH_3-	$C_2H_5O_2C-$	$CH_3COCHClCO_2C_2H_5$	NH_2CSSNH_4	—	51
$HS-$	CH_3-	$C_2H_5O_2C-$	$CH_3COCHClCO_2C_2H_5$	$C_2H_5O_2CCSNH_2$	43	99

* References 83–193 are on pp. 407–409.
† The reaction of $C_2H_5O_2CCH_2SCN$ in water yielded a product not clearly described. See ref. 141.
‡ The product of rearrangement of HO_2CCH_2SCN is 2,4-diketothiazolidine. See ref. 142.

TABLE VIII

2-Aminothiazoles from Thiourea

$$\underset{R'}{\overset{R}{\longrightarrow}}\underset{S}{\overset{N}{\bigg|}}NH_2$$

Product		Reactant	Yield %	Reference *
R	R'			
H	H	$ClCH_2CHO$	70	42
		$ClCH_2CHO$	80	143
		$ClCH_2CH(OCH_3)_2$	75	69
		$ClCH_2CH(OC_2H_5)_2$	68–92	42–46
		$(BrCH_2CHO)_3$	65	70
		$ClCH_2CHClOC_2H_5$	60–100	34, 71, 72
		$CH_3CH_2CH_2CH_2OCHClCH_2Cl$	80	42
		$(CH_3)_2CHCH_2CH_2OCHClCH_2Cl$	72	42
		$CH_3CO_2CHClCH_2Cl$	50–86	44, 47
		$CH_3CO_2CHBrCH_2Br$	—	44
		$CH_3CH_2CH_2CO_2CHClCH_2Cl$	—	44
		$ClCH_2CHClOCHClCH_2Cl$	Quant.	144
		$BrCH_2CHBrOCHBrCH_2Br$	86	144
CH_3—	H	$CH_3CO_2CHClCHClCH_3$	—	44
		CH_3COCH_2Cl	70–75	73
		CH_3COCH_2Cl	—	48, 71, 72, 145–147
		CH_3COCH_2Br	36	48
		CH_3COCH_2I	77	48
n-C_5H_{11}—	H	n-$C_5H_{11}COCH_2Cl$	—	148
n-C_7H_{15}—	H	n-$C_7H_{15}COCH_2Cl$	—	148
n-C_9H_{19}—	H	n-$C_9H_{19}COCH_2Cl$	—	148
n-$C_{11}H_{23}$—	H	n-$C_{11}H_{23}COCH_2Cl$	—	148
n-$C_{13}H_{27}$—	H	n-$C_{13}H_{27}COCH_2Cl$	—	148
n-$C_{15}H_{31}$—	H	n-$C_{15}H_{31}COCH_2Cl$	—	148
$ClCH_2$—	H	$ClCH_2COCH_2Cl$	58	149
$ClCH_2CH_2$—	H	$ClCH_2CH_2COCH_2Cl$	—	150
C_6H_5—	H	$CH_3CO_2CHBrCHBrC_6H_5$	—	44
C_6H_5—	H	$C_6H_5COCH_2Cl$	49	71
			90	48
		$C_6H_5COCH_2Br$	85	32, 48, 51, 72, 146
		$C_6H_5COCH_2I$	94	48
m-$O_2NC_6H_4$—	H	m-$O_2NC_6H_4COCH_2Cl$	75	48
		m-$O_2NC_6H_4COCH_2Br$	95	48
		m-$O_2NC_6H_4COCH_2I$	52	48
3,4-$(HO)_2C_6H_3$—	H	3,4-$(HO)_2C_6H_3COCH_2Cl$	—	96a, 152
$C_2H_5O_2CCH_2$—	H	$BrCH_2COCH_2CO_2C_2H_5$	86	35, 153–155
CH_3CH— | CO_2H	H	$BrCH_2COCH(CH_3)CO_2CH_3$	—	153
$CH_3(CH_2)_3CH$— | CO_2H	H	$BrCH_2COCH(CO_2C_2H_5)C_4H_9$-$n$	—	156
$(CH_3)_2NCH_2CH_2$—	H	$(CH_3)_2NCH_2CH_2COCH_2Br$ HBr	81	157
![pyrazole with CH3]	H	![pyrazole with CH3 and COCH2Br]	—	107

* References 83–193 are on pp. 407–409.

TABLE VIII—Continued

2-AMINOTHIAZOLES FROM THIOUREA

Product		Reactant	Yield %	Reference *
R	R'			
H	CH_3-	$CH_3CHClCHO$	—	31
		$CH_3CHBrCHO \cdot H_2O$	—	36
H	C_2H_5-	$C_2H_5CHClCHO$	—	158
H	$iso\text{-}C_3H_7-$	$(CH_3)_2CHCHClCHO$	—	158
H	$n\text{-}C_4H_9-$	$n\text{-}C_4H_9CHClCHO$	—	158
H	$n\text{-}C_5H_{11}-$	$n\text{-}C_5H_{11}CHClCHO$	—	158
CH_3-	CH_3-	$CH_3COCHClCH_3$	—	159
CH_3-	CH_3CH_2-	$CH_3COCHClCH_2CH_3$	50	158
CH_3-	$n\text{-}C_3H_7-$	$CH_3COCHClCH_2CH_2CH_3$	50	158
CH_3-	$n\text{-}C_4H_9-$	$CH_3COCHClCH_2(CH_2)_2CH_3$	50	158
CH_3-	$n\text{-}C_5H_{11}-$	$CH_3COCHClC_5H_{11}\text{-}n$	50	158
CH_3-	$(CH_3)_2CHCH_2CH_2-$	$CH_3COCHClCH_2CH_2CH(CH_3)_2$	50	158
CH_3-	$n\text{-}C_6H_{13}-$	$CH_3COCHClC_6H_{13}\text{-}n$	50	158
CH_3-	$HOCH_2CH_2-$	$CH_3COCHClCH_2CH_2OH$	—	151, 159, 160
		![structure: CH2—CH2—C(Cl)(COCH3)—CO—O ring]	83	105
		![structure: CH2—CH2—C(Br)(COCH3)—CO—O ring]	91	105
CH_3-	CH_2-CHCH_2- with epoxide O	$ClH_2CCH-CH_2$ with $C(Cl)(COCH_3)$—CO—O ring	—	110
C_2H_5-	CH_3-	$C_2H_5COCHBrCH_3$	—	161
C_6H_5-	CH_3-	$C_6H_5COCHClCH_3$	68	48
		$C_6H_5COCHBrCH_3$	80	48
		$C_6H_5COCHICH_3$	94	48
C_6H_5-	HO_2CCH_2-	$C_6H_5COCHBrCH_2CO_2H$	90	111
C_6H_5-	$HO_2CCH(CH_3)-$	$C_6H_5COCHBrCH(CH_3)CO_2H$	83	111
$p\text{-}CH_3C_6H_4-$	HO_2CCH_2-	$p\text{-}CH_3C_6H_4COCHBrCH_2CO_2H$	90	111
$\alpha\text{-}C_{10}H_7-$	HO_2CCH_2-	$\alpha\text{-}C_{10}H_7COCHBrCH_2CO_2H$	90	111
$\beta\text{-}C_{10}H_7-$	HO_2CCH_2-	$\beta\text{-}C_{10}H_7COCHBrCH_2CO_2H$	90	111
![thienyl]	HO_2CCH_2-	![thienyl]$COCHBrCH_2CO_2H$	90	111
HO_2C-	H	$BrCH_2COCO_2H$	56	162
		$Br_2CHCOCO_2H$	—	163
		Br_3CCOCO_2H	—	164
$C_2H_5O_2C-$	H	$BrCH_2COCO_2C_2H_5$	66	165
H	$C_2H_5O_2C-$	$HCOCHClCO_2C_2H_5$	39–60	137, 156, 166, 167
CH_3-	$C_2H_5O_2C-$	$CH_3COCHClCO_2C_2H_5$	60–100	35, 48, 140, 153
		$CH_3COCHBrCO_2C_2H_5$	82	35, 48, 153
		$CH_3COCHICO_2C_2H_5$	63	48
		$CH_3COCBr_2CO_2C_2H_5$ †	—	153
$C_2H_5O_2CCH_2-$	$C_2H_5O_2C-$	$C_2H_5O_2CCHClCOCH_2CO_2C_2H_5$	84	162
C_6H_5-	$C_2H_5O_2C-$	$C_6H_5COCHClCO_2C_2H_5$	—	168
$C_2H_5O_2C-$	$C_2H_5O_2C-$	$C_2H_5O_2CCOCHClCO_2C_2H_5$	—	137

* References 83–193 are on pp. 407–409.
† The corresponding dichloro compound gave no thiazole. See ref. 140.

TABLE IX
ALKYLAMINO- AND ARYLAMINO-THIAZOLES

$$\underset{R''}{\overset{R'}{}}\!\!\diagdown\!\!\!\underset{S}{\overset{N}{\bigcirc}}\!\!\!\diagup\!\! R$$

Product			Reactants		Yield %	Reference*
R	R'	R''				
CH_3NH-	H	H	$ClCH_2CHClOC_2H_5$	$CH_3NHCSNH_2$	—	169
CH_3NH-	CH_3-	H	CH_3COCH_2Cl	$CH_3NHCSNH_2$	—	72, 170
CH_3NH-	C_6H_5-	H	$C_6H_5COCH_2Br$	$CH_3NHCSNH_2$	—	72
CH_3NH-	$3,4-(HO)_2C_6H_3-$	H	$3,4-(HO)_2C_6H_3COCH_2Cl$	$CH_3NHCSNH_2$	—	96a
$CH_2=CHCH_2NH-$	CH_3-	H	CH_3COCH_2Cl	$CH_2=CHCH_2NHCSNH_2$	—	171
$CH_2=CHCH_2NH-$	C_6H_5-	H	$C_6H_5COCH_2Cl$	$CH_2=CHCH_2NHCSNH_2$	—	172
$CH_2=CHCH_2NH-$	$3,4-(HO)_2C_6H_3-$	H	$3,4-(HO)_2C_6H_3COCH_2Cl$	$CH_2=CHCH_2NHCSNH_2$	—	152
C_6H_5NH-	H	H	$ClCH_2CHClOC_2H_5$	$C_6H_5NHCSNH_2$	—	72, 169
C_6H_5NH-	CH_3-	H	CH_3COCH_2Cl	$C_6H_5NHCSNH_2$	—	72, 171
$p-CH_3C_6H_4NH-$	C_6H_5-	C_6H_5-	$C_6H_5COCHBrC_6H_5$	$p-CH_3C_6H_4NHCSNH_2$	—	172
$p-CH_3C_6H_4NH-$	CH_3-	H	CH_3COCH_2Cl	$p-CH_3C_6H_4NHCSNH_2$	—	174
$p-HOC_6H_4NH-$	CH_3-	H	CH_3COCH_2Cl	$p-HOC_6H_4NHCSNH_2$	—	173
$p-C_2H_5OC_6H_4NH-$	CH_3-	H	CH_3COCH_2Cl	$p-C_2H_5OC_6H_4NHCSNH_2$	—	173
$p-ClC_6H_4NH-$	CH_3-	H	CH_3COCH_2Cl	$p-ClC_6H_4NHCSNH_2$	33	173
$p-BrC_6H_4NH-$	CH_3-	H	CH_3COCH_2Cl	$p-BrC_6H_4NHCSNH_2$	—	173
$p-IC_6H_4NH-$	CH_3-	H	CH_3COCH_2Cl	$p-IC_6H_4NHCSNH_2$	—	173
$2,4-Br_2C_6H_3NH-$	CH_3-	H	CH_3COCH_2Cl	$2,4-Br_2C_6H_3NHCSNH_2$	—	173
$p-H_2NO_2SC_6H_4NH-$	H	H	$ClCH_2CHClOC_2H_5$	$p-H_2NO_2SC_6H_4NHCSNH_2$	—	175
$p-H_2NO_2SC_6H_4NH-$	CH_3-	H	CH_3COCH_2Cl	$p-H_2NO_2SC_6H_4NHCSNH_2$	92	176
$p-H_2NO_2SC_6H_4NH-$	CH_3-	$HOCH_2CH_2-$	$CH_3COCHBrCH_2CH_2OCOCH_3$	$p-H_2NO_2SC_6H_4NHCSNH_2$	80	176
$p-CH_3CONHC_6H_4O_2SNH-$	H	H	$ClCH_2CHClOCOCH_3$	$p-CH_3CONHC_6H_4SO_2NHCSNH_2$	79	177
$p-H_2NO_2SC_6H_4NH-$	C_6H_5-	H	$C_6H_5COCH_2Br$	$p-H_2NO_2SC_6H_4NHCSNH_2$	—	175

THE PREPARATION OF THIAZOLES

p-H₂NO₂SC₆H₄NH—	C₂H₅O₂CCH₂—	H	BrCH₂COCH₂CO₂C₂H₅	p-H₂NO₂SC₆H₄NHCSNH₂	—	175
p-CH₃CONHO₂SC₆H₄NH—	CH₃—	HOCH₂CH₂—	CH₃COCHClCH₂CH₂OCOCH₃	p-CH₃CONHO₂SC₆H₄NHCSNH₂	25	178
p-H₂NO₂SC₆H₄NH—	CH₃—	C₂H₅OCOCH₂—	CH₃COCHBrCH₂CO₂C₂H₅	p-H₂NO₂SC₆H₄NHCSNH₂	—	175
p-H₂NO₂SC₆H₄NH—	CH₃—	C₂H₅OCO—	CH₃COCHBrCO₂C₂H₅	p-H₂NO₂SC₆H₄NHCSNH₂	—	175
CH₃CONH—	4-HOC₆H₄—(HO)	H	3,4-(HO)₂C₆H₃COCH₂Cl	CH₃CONHCSNH₂	—	152
H₂NNH—	CH₃—	H	CH₃COCH₂Cl	H₂NNHCSNH₂	—	132
o-CH₃C₆H₄NHNH—	CH₃—	H	CH₃COCH₂Cl	o-CH₃C₆H₄NHNHCSNH₂	—	179
o-CH₃C₆H₄NHNH—	C₆H₅—	H	C₆H₅COCH₂Br	o-CH₃C₆H₄NHNHCSNH₂	—	179
o-CH₃C₆H₄NHNH—	p-CH₃C₆H₄—	H	p-CH₃C₆H₄COCH₂Br	o-CH₃C₆H₄NHNHCSNH₂	—	179
m-CH₃C₆H₄NHNH—	CH₃—	H	CH₃COCH₂Cl	m-CH₃C₆H₄NHNHCSNH₂	—	179
m-CH₃C₆H₄NHNH—	C₆H₅—	H	C₆H₅COCH₂Br	m-CH₃C₆H₄NHNHCSNH₂	—	179
m-CH₃C₆H₄NHNH—	p-CH₃C₆H₄—	H	p-CH₃C₆H₄COCH₂Br	m-CH₃C₆H₄NHNHCSNH₂	—	179
p-CH₃C₆H₄NHNH—	CH₃—	H	CH₃COCH₂Cl	p-CH₃C₆H₄NHNHCSNH₂	—	179
p-CH₃C₆H₄NHNH—	C₆H₅—	H	C₆H₅COCH₂Br	p-CH₃C₆H₄NHNHCSNH₂	—	179
p-CH₃C₆N₄NHNH—	p-CH₃C₆H₄—	H	p-CH₃C₆H₄COCH₂Br	p-CH₃C₆H₄NHNHCSNH₂	—	179
m-O₂NC₆H₄NHNH—	CH₃—	H	CH₃COCH₂Cl	m-O₂NC₆H₄NHNHCSNH₂	—	180
m-O₂NC₆H₄NHNH—	C₆H₅—	H	C₆H₅COCH₂Br	m-O₂NC₆H₄NHNHCSNH₂	—	180
m-O₂NC₆H₄NHNH—	p-CH₃C₆H₄—	H	p-CH₃C₆H₄COCH₂Br	m-O₂NC₆H₄NHNHCSNH₂	—	180
α-C₁₀H₇NHNH—	CH₃—	H	CH₃COCH₂Cl	α-C₁₀H₇NHNHCSNH₂	—	180
α-C₁₀H₇NHNH—	C₆H₅—	H	C₆H₅COCH₂Br	α-C₁₀H₇NHNHCSNH₂	—	180
α-C₁₀H₇NHNH—	p-CH₃C₆H₄—	H	p-CH₃C₆H₄COCH₂Br	α-C₁₀H₇NHNHCSNH₂	—	181
6-methoxyquinolin-8-yl-NHNH—		H	ClCH₂CHO	6-methoxyquinolin-8-yl-NHNHCSNH₂	—	183
(CH₃)₂C=NNH—	CH₃—	H	CH₃COCH₂Cl	(CH₃)₂C=NNHCSNH₂	—	183
(CH₃)₂C=NNH—	CH₃—	H	CH₃COCH₂Cl	(CH₃)₂C=NNHCSNH₂	—	183
C₆H₅CH=NNH—	CH₃—	H	CH₃COCH₂Cl	C₆H₅CH=NNHCSNH₂	—	183
C₆H₅C(CH₃)=NNH—	CH₃—	H	CH₃COCH₂Cl	C₆H₅C(CH₃)=NNHCSNH₂	—	183

* References 83–193 are on pp. 407–409.

TABLE X
Dialkyl (or Aryl) Aminothiazoles

$$\underset{R'''}{\overset{R''}{\underset{S}{\bigsqcup}}}\underset{N}{\overset{N}{\bigsqcup}}N\underset{R'}{\overset{R}{\diagdown}}$$

Product				Reactants		Yield %	Reference *
R	R'	R''	R'''				
CH$_3$—	C$_6$H$_5$—	CH$_3$—	H—	CH$_3$COCH$_2$Cl	C$_6$H$_5$$\overset{\mid}{\text{N}}$CSNH$_2$ †‡ $\underset{\mid}{\text{CH}_3}$	—	173
C$_6$H$_5$CH$_2$—	p-CH$_3$C$_6$H$_4$—	C$_6$H$_5$—	H—	C$_6$H$_5$COCH$_2$Br	C$_6$H$_5$CH$_2$NCSNH$_2$ $\underset{\mid}{\text{C}_6\text{H}_4\text{CH}_3\text{-}p}$	—	172

* References 83–193 are on pp. 407–409.

† The following unsymmetrical thioureas have been found not to yield thiazole derivatives with either chloroacetone or phenacyl chloride: (CH$_3$)$_2$NCSNH$_2$, [(CH$_3$)$_2$CHCH$_2$CH$_2$]$_2$NCSNH$_2$, (C$_6$H$_5$)$_2$-NCSNH$_2$. See ref. 182.

‡ 2-Phenylimino-3,4-dimethyl-2,3-dihydrothiazole resulted from the reaction of sym-methylphenylthiourea and chloroacetone. See ref. 173.

TABLE XI
Compounds Containing More Than One Thiazole Ring

Product	Reactants	Yield %	Reference*
Polymeric H₃C−[thiazole]−CH₃	BrCH₂COCOCH₂Br CH₃COCH₂Cl	— —	37 31, 37
Polymeric H₅C₆−[thiazole]−C₆H₅	CH₃COCHClCHClCOCH₃ C₆H₅COCH₂Br	— —	31 184
H₅C₆−[thiazole]−C₆H₅ (bis with C₆H₅)	C₆H₅CHClCOC₆H₅	—	185
p-C₆H₅C₆H₄−[thiazole]−C₆H₄C₆H₅-p	p-C₆H₅C₆H₄CHClCOC₆H₄C₆H₅-p	—	185
H₃C−[thiazole]−CH₂CH₂OH (bis)	CH₃COCHClCH₂OH	—	186
H₃C−[thiazole]−CH₂CH₂OCOCH₃ (bis)	CH₃COCHClCH₂OCOCH₃	—	186

Reactants column (H₂NCSNH₂ series): H₂NCSCSNH₂; H₂NCSCSNH₂; CH₃CSNH₂, H₂NCSCSNH₂; H₂NCSCSNH₂; H₂NCSCSNH₂; H₂NCSCSNH₂; H₂NCSCSNH₂.

* References 83–193 are on pp. 407–409.

TABLE XI—Continued

COMPOUNDS CONTAINING MORE THAN ONE THIAZOLE RING

Product	Reactants		Yield %	Reference *
2-hydroxy-4-methyl-thiazole linked to 2-thiazole	2-hydroxy-4-methyl-5-(COCH$_2$Br)-thiazole	HCSNH$_2$	—	187, 107
2-hydroxy-4-methyl-thiazole linked to 4-methylthiazole	2-hydroxy-4-methyl-5-(COCH$_2$Br)-thiazole	CH$_3$CSNH$_2$	—	187, 107
bis-thiazole with CH$_3$, CO$_2$C$_2$H$_5$ and C$_2$H$_5$O$_2$C substituents	ClCH$_2$COCH$_2$CO$_2$C$_2$H$_5$	H$_2$NCSCSNH$_2$	—	187, 107
2-hydroxy-4-methyl-thiazole linked to 2-amino-thiazole	2-hydroxy-4-methyl-5-(COCH$_2$Br)-thiazole	H$_2$NCSNH$_2$	—	187, 107
4-methylthiazole-2-CH$_2$-2-(4-methylthiazole)	CH$_3$COCH$_2$Cl	H$_2$NCSCH$_2$CSNH$_2$	—	188

THE PREPARATION OF THIAZOLES

Structure	Reagents	Amine	Yield	Ref.
H₅C₆–[thiazole]–CH₂–[thiazole]–C₆H₅	C₆H₅COCH₂Br	H₂NCSCH₂CSNH₂	—	188
Polymeric H₂N–[thiazole]–CH₂–[thiazole]–NH₂	BrCH₂COCH₂Br BrCH₂COCH₂COCH₂Br	H₂NCSCH₂CSNH₂ NH₂CSNH₂	— 99	188 188a
H₂N–[thiazole]–(CH₂)₈–[thiazole]–NH₂	ClCH₂CO(CH₂CH₂)₄COCH₂Cl	NH₂CSNH₂	97	176
H₃C–[thiazole]–(CH₂)₄–[thiazole]–CH₃	CH₃COCH₂Cl	H₂NCSCH₂CH₂CH₂CH₂CSNH₂	—	189
H₅C₆–[thiazole]–(CH₂)₄–[thiazole]–C₆H₅	C₆H₅COCH₂Br	H₂NCSCH₂CH₂CH₂CH₂CSNH₂	—	184
Polymeric Polymeric † Polymeric H₃C–[thiazole]–C₆H₄–[thiazole]–CH₃	BrCH₂COCH₂Br ClCH₂COCH₂CH₂CH₂CH₂COCH₂Cl ClCH₂COCH₂Cl CH₃COCH₂Cl	H₂NCSCH₂CH₂CH₂CH₂CSNH₂ H₂NCSCH₂CH₂CH₂CH₂CH₂CSNH₂ p-H₂NSCC₆H₄CSNH₂ p-H₂NSCC₆H₄CSNH₂	— — — —	184 190 191 191
H₅C₆–[thiazole]–C₆H₄–[thiazole]–C₆H₅	C₆H₅COCH₂Br	p-H₂NSCC₆H₄CSNH₂	—	191

* References 83–193 are on pp. 407–409.
† In addition, a macrocyclic compound was obtained in 37% yield.

TABLE XI—Continued

Compounds Containing More Than One Thiazole Ring

Product	Reactants		Yield %	Reference*
Polymeric	BrCH₂COCOCH₂Br	p-H₂NSCC₆H₄CSNH₂	—	191
	BrCH₂CO—C₆H₃—COCH₂Br (1,3)	C₆H₅CSNH₂	—	192
†	BrCH₂CO—C₆H₃—COCH₂Br (1,3) with COCH₂Br at 5	H₂NCSCSNH₂	—	192
Polymeric	BrCH₂CO—C₆H₃—COCH₂Br (1,3) with COCH₂Br at 5	H₂NCSCH₂CH₂CH₂CSNH₂	—	192
Bis(2-hydrazino-thiazole)	ClCH₂CHClOC₂H₅	NH₂CNHNHCNH₂ (=S, =S)	74	193
 Bis(2-hydrazino-4-methylthiazole)	CH₃COCH₂Cl	NH₂CNHNHCNH₂ (=S, =S)	70	193

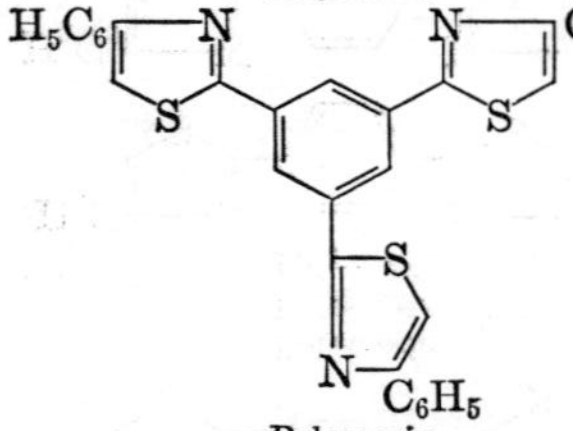

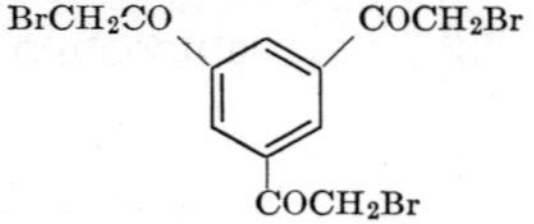

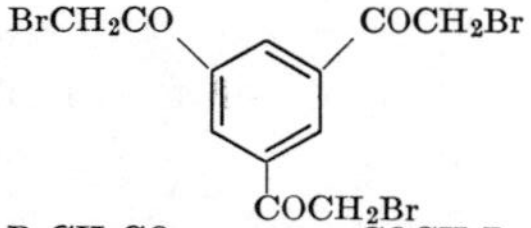

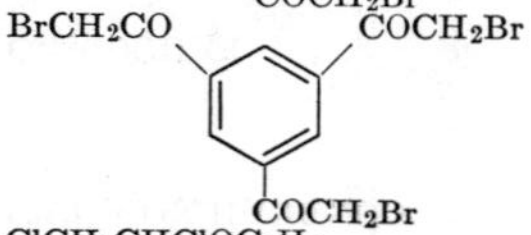

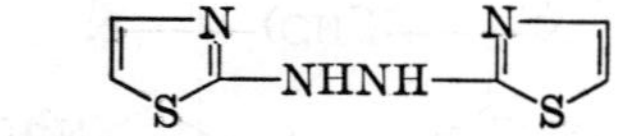

* References 83–193 are on pp. 407–409.
† The structure shown is that mentioned in the original article. By analogy with other thioamide preparations, it appears more likely that the product is 1,3,5-tri(2-phenyl-4-thiazolyl)benzene.

THE PREPARATION OF THIAZOLES

REFERENCES FOR TABLES II–XI

[83] Gabriel and Bachstez, *Ber.*, **47**, 3169 (1914).
[84] Coates, Cook, Heilbron, and Lewis, *J. Chem. Soc.*, **1943**, 419.
[85] Hooper and Johnson, *J. Am. Chem. Soc.*, **56**, 470 (1934).
[86] Wetherill and Hann, *J. Am. Chem. Soc.*, **56**, 970 (1934); **57**, 1752 (1935).
[87] Smith, *J. Chem. Soc.*, **123**, 2288 (1923).
[88] Brody and Bogert, *J. Am. Chem. Soc.*, **65**, 1080 (1943).
[89] Johnson and Gatewood, *J. Am. Chem. Soc.*, **51**, 1815 (1929).
[90] Chi and Tshin, *J. Am. Chem. Soc.*, **64**, 90 (1942).
[91] Olin and Johnson, *J. Am. Chem. Soc.*, **53**, 1475 (1931).
[92] Olin and Johnson, *J. Am. Chem. Soc.*, **53**, 1470 (1931).
[93] Olin and Johnson, *J. Am. Chem. Soc.*, **53**, 1473 (1931).
[94] MacCorquodale and Johnson, *Rec. trav. chim.*, **51**, 483 (1932).
[95] Huntress and Pfister, *J. Am. Chem. Soc.*, **65**, 1667 (1943).
[96] Suter and Johnson, *Rec. trav. chim.*, **49**, 1066 (1930); U. S. pat. 2,014,498 [*C. A.*, **29**, 7344 (1935)].
[96a] Johnson, U. S. pat. 1,743,083 [*C. A.*, **24**, 1126 (1930)].
[97] Hinegardner and Johnson, *J. Am. Chem. Soc.*, **52**, 4139 (1930).
[98] Hinegardner and Johnson, *J. Am. Chem. Soc.*, **52**, 4141 (1930).
[99] Boon, *J. Chem. Soc.*, **1945**, 601.
[100] Smith, Flack, and Inggs, *S. African J. Sci.*, **21**, 227 (1924) [*C. A.*, **19**, 1706 (1925)].
[101] Rüdenburg, *Ber.*, **46**, 3555 (1913).
[102] Gabriel, *Ber.*, **43**, 134 (1910).
[103] Slobodin, Zigel, and Yanishevskaya, *J. Applied Chem. U.S.S.R.*, **16**, 280 (1943) [*C. A.*, **39**, 702 (1945)].
[104] Todd, Bergel, and Jacob, *J. Chem. Soc.*, **1936**, 1555.
[105] Wenz, Ger. pat. 664,789 [*C. A.*, **33**, 177 (1939)].
[106] Hoffmann, La Roche and Co. A.-G., Ger. pat. 663,305 [*C. A.*, **32**, 9099 (1938)].
[107] Ochiai, Tamamushi, and Nagasawa, *Ber.*, **73B**, 28 (1940).
[108] Erlenmeyer and Simon, *Helv. Chim. Acta*, **25**, 528 (1942).
[109] Andersag and Westphal, Ger. pat. 702,831 [*C. A.*, **36**, 784 (1942)].
[110] Beyer, *Ber.*, **74B**, 1100 (1941).
[111] Knott, *J. Chem. Soc.*, **1945**, 455.
[112] Karimullah, *J. Chem. Soc.*, **1937**, 961.
[113] Todd, Bergel, Karimullah, and Keller, *J. Chem. Soc.*, **1937**, 361.
[114] Gotze, *Ber.*, **71B**, 2289 (1938).
[115] Bergel and Todd, *J. Chem. Soc.*, **1937**, 1504.
[116] Todd and Bergel, *J. Chem. Soc.*, **1936**, 1559.
[117] Foldi and Gerecs, U. S. pat. 2,252,921 [*C. A.*, **35**, 7660 (1941)].
[118] Imai and Makino, *Z. physiol. Chem.*, **252**, 76 (1938).
[119] Kofler and Sternbach, *Helv. Chim. Acta*, **24**, 1014 (1941).
[120] Todd and Bergel, *J. Chem. Soc.*, **1937**, 364.
[121] Bergel, Cohen, and Hughes, Brit. pat. 559,106 [*C. A.*, **39**, 4436 (1945)].
[122] Andersag and Westphal, U. S. pat. 2,209,244 [*C. A.*, **35**, 282 (1941)].
[123] I.G. Farbenindustrie A.-G., Fr. pat. 816,432 [*C. A.*, **32**, 1869 (1938)].
[124] Huber, *J. Am. Chem. Soc.*, **65**, 2222 (1943).
[125] Hantzsch, *Ber.*, **25**, 3282 (1892).
[126] Tcherniac, *Ber.*, **25**, 2607 (1892).
[127] Tcherniac, *Ber.*, **25**, 3648 (1892).
[128] Tcherniac and Norton, *Ber.*, **16**, 345 (1883).
[129] Dyckerhoff, *Ber.*, **10**, 119 (1877).
[130] Andersag and Westphal, *Ber.*, **70B**, 2035 (1937).
[131] Bunnett and Tarbell, *J. Am. Chem. Soc.*, **67**, 1944 (1945).
[132] Smith and Sapiro, *Trans. Roy. Soc. South Africa*, **18**, III, 229 (1929) [*C. A.*, **24**, 2130 (1930)].

[133] Erlenmeyer and Meyenburg, *Helv. Chim. Acta*, **20**, 204 (1937).
[134] Erlenmeyer and Marbet, *Helv. Chim. Acta*, **29**, 1946 (1946).
[135] Soc. pour l'ind. chim. à Bâle, Swiss pat. 192,849 [*C. A.*, **32**, 4285 (1938)].
[136] Schoberl and Stock, *Ber.*, **73B**, 1240 (1940).
[137] Erlenmeyer, Buchmann, and Schenkel, *Helv. Chim. Acta*, **27**, 1432 (1944).
[138] Huntress and Pfister, *J. Am. Chem. Soc.*, **65**, 2167 (1943).
[139] Soc. pour l'ind. chim. à Bâle, Swiss pat. 199,647 [*C. A.*, **33**, 3531 (1939)].
[140] Zürcher, *Ann.*, **250**, 281 (1889).
[141] Lakner, *Chem. Folyoirat*, **34**, 129 (1928) [*C. A.*, **23**, 4930 (1929)].
[142] Arapides, *Ann.*, **249**, 27 (1888).
[143] Magidson and Sokolova, U.S.S.R. pat. 66,044 [*C. A.*, **41**, 1713 (1947)].
[144] Britton and Harding, U. S. pat. 2,387,212 [*C. A.*, **40**, 1179 (1946)].
[145] Huapaya, *Farmacia y química* (*Lima, Peru*), **1**, 84 (1944) [*C. A.*, **39**, 2285 (1945)].
[146] Lanfranchi, *Atti reale accad. Italia, Rend. classe sci. fis., mat. e nat.*, [7], **3**, 776 (1942) [*C. A.*, **38**, 5219 (1944)].
[147] Pawlewski, *Ber.*, **21**, 401 (1888).
[148] Jensen and Kjaer, *Dansk Tids. Farm.*, **16**, 110 (1942) [*C. A.*, **38**, 2326 (1944)].
[149] Sprague, Land, and Ziegler, *J. Am. Chem. Soc.*, **68**, 2155 (1946).
[150] Carroll and Smith, *J. Am. Chem. Soc.*, **55**, 370 (1933).
[151] Basu and Das-Gupta, *J. Indian Chem. Soc.*, **15**, 160 (1938) [*C. A.*, **32**, 7039 (1938)].
[152] Horii, *J. Pharm. Soc. Japan*, **55**, 21 (1935) [*C. A.*, **29**, 3338 (1935)].
[153] Conrad, *Ber.*, **29**, 1042 (1896).
[154] Erlenmeyer and Morel, *Helv. Chim. Acta*, **28**, 362 (1945).
[155] Steude, *Ann.*, **261**, 22 (1891).
[156] Ganapathi, Deliwala, and Shirsat, *Proc. Indian Acad. Sci.*, **16A**, 126 (1942) [*C. A.*, **37**, 1404 (1943)].
[157] Land, Sprague, and Ziegler, *J. Am. Chem. Soc.*, **69**, 125 (1947).
[158] Ganapathi, Shirsat, and Deliwala, *Proc. Indian Acad. Sci.*, **14A**, 630 (1941) [*C.A.*, **36**, 4102 (1942)].
[159] Jensen and Thorsteinsson, *Dansk Tids. Farm.*, **15**, 41 (1941) [*C. A.*, **35**, 5109 (1941)].
[160] Todd, Bergel, Fraenkel-Conrat, and Jacob, *J. Chem. Soc.*, **1936**, 1601.
[161] Sprague and Kissinger, *J. Am. Chem. Soc.*, **63**, 578 (1941).
[162] Sprague, Lincoln, and Ziegler, *J. Am. Chem. Soc.*, **68**, 266 (1946).
[163] Nencki and Sieber, *J. prakt. Chem.*, [2], **25**, 72 (1882).
[164] Boettinger, *Arch. Pharm.*, **232**, 349 (1894).
[165] Erlenmeyer and Morel, *Helv. Chim. Acta*, **25**, 1073 (1942).
[166] Backer and de Jonge, *Rec. trav. chim.*, **61**, 463 (1942).
[167] Dann, *Ber.*, **76B**, 419 (1943).
[168] Hirst, Macbeth, and Traill, *Proc. Roy. Irish Acad.*, **37B**, 47 (1925) [*C. A.*, **19**, 2931 (1925)].
[169] Näf, *Ann.*, **265**, 108 (1891).
[170] Burtles, Pyman, and Roylance, *J. Chem. Soc.*, **127**, 581 (1925).
[171] Young and Crookes, *J. Chem. Soc.*, **89**, 59 (1906).
[172] Walther and Roch, *J. prakt. Chem.*, [2], **87**, 27 (1913).
[173] Hunter and Parken, *J. Chem. Soc.*, **1934**, 1175.
[174] Dyson, Hunter, Jones, and Styles, *J. Indian Chem. Soc.*, **8**, 147 (1931) [*C. A.*, **25**, 4880 (1931)].
[175] Ganapathi, *Proc. Indian Acad. Sci.*, **12A**, 274 (1940) [*C. A.*, **35**, 1772 (1941)].
[176] Walker, *J. Chem. Soc.*, **1940**, 1304.
[177] Leitch, Baker, and Brickman, *Can. J. Research*, **23B**, 139 (1945).
[178] Jensen, Falkenberg, Thorsteinsson, and Lauridsen, *Dansk Tids. Farm.*, **16**, 141 (1942) [*C. A.*, **38**, 3263 (1944)].
[179] Bose and Sen, *J. Indian Chem. Soc.*, **5**, 643 (1928) [*C. A.*, **23**, 1409 (1929)].
[180] Das-Gupta and Bose, *J. Indian Chem. Soc.*, **6**, 495 (1929) [*C. A.*, **24**, 1095 (1930)].
[181] Nandi, *J. Indian Chem. Soc.*, **17**, 449 (1940) [*C. A.*, **35**, 2146 (1941)].
[182] Spica and Carrara, *Gazz. chim. ital.*, **21**, 421 (1891).

[183] McLean and Wilson, *J. Chem. Soc.*, **1937**, 556.
[184] Lehr and Erlenmeyer, *Helv. Chim. Acta*, **27**, 489 (1944).
[185] Karrer and Forster, *Helv. Chim. Acta*, **28**, 315 (1945).
[186] Karrer and Sanz, *Helv. Chim. Acta*, **27**, 619 (1944).
[187] Tamamushi and Nagasawa, *J. Pharm. Soc. Japan*, **60**, 127 (1940) [*C. A.*, **34**, 5081 (1940)].
[188] Lehr, Guex, and Erlenmeyer, *Helv. Chim. Acta*, **27**, 970 (1944).
[188a] Ruggli, Wartburg, and Erlenmeyer, *Helv. Chim. Acta*, **30**, 348 (1947).
[189] Erlenmeyer and Bischoff, *Helv. Chim. Acta*, **27**, 412 (1944).
[190] Erlenmeyer and Degen, *Helv. Chim. Acta*, **29**, 1080 (1946); **30**, 592 (1947).
[191] Erlenmeyer, Büchler, and Lehr, *Helv. Chim. Acta*, **27**, 969 (1944).
[192] Bischoff, Weber, and Erlenmeyer, *Helv. Chim. Acta*, **27**, 947 (1944).
[193] Markees, Kellerhals, and Erlenmeyer, *Helv. Chim. Acta*, **30**, 304 (1947).

CHAPTER 9

THE PREPARATION OF THIOPHENES AND TETRAHYDROTHIOPHENES

Donald E. Wolf and Karl Folkers

Merck & Co., Inc.

CONTENTS

	PAGE
Introduction	411
Preparation of Thiophenes	412
Thiophenes by Reaction of 1,4-Difunctional Compounds with Sulfides	412
Syntheses from Succinic Acids	413
Syntheses from γ-Keto Acids	414
Syntheses from 1,4-Diketones	418
Syntheses from Other 1,4-Difunctional Compounds	419
Experimental Conditions	420
Table I. Preparation of 3-Methylthiophene	421
Experimental Procedures	421
3,4-Dimethylthiophene	421
3-*n*-Propylthiophene	421
2-Isopropylthiophene	422
5-Hydroxy-2-methylthiophene	422
2,3-Dimethylthiophene	422
2,3,5-Trimethylthiophene	422
Methyl β-2-(5-Phenylthienyl)propionate	422
Table II. Thiophenes by Reaction of 1,4-Difunctional Compounds with Sulfides	424
Thiophenes by Reaction of Unsaturated Compounds with Sulfides	428
Reaction of Unsaturated Compounds with Metallic Sulfides	428
Reaction of Unsaturated Compounds with Hydrogen Sulfide	429
Reaction of Unsaturated Compounds with Sulfur	430
Table III. Thiophenes by Reaction of Unsaturated Compounds with Sulfides	432
Thiophenes by Reaction of 1,2-Difunctional Compounds with Thiodiacetic Acid Esters	435
Syntheses from α-Diketones	435
Syntheses from α-Keto Esters	436
Syntheses from Oxalic Esters	437
Decarboxylation of 2,5-Thiophenedicarboxylic Acids	437

THIOPHENES AND TETRAHYDROTHIOPHENES

	PAGE
Experimental Procedures	438
Dimethyl 3,4-Dihydroxy-2,5-thiophenedicarboxylate	438
3,4-Diphenyl-2,5-thiophenedicarboxylic Acid	438
Table IV. Thiophenes by Reaction of 1,2-Difunctional Compounds with Thiodiacetic Acid Esters	438
Thiophenes by Reaction of Aryl Methyl Ketones with Sulfides	439
Table V. Thiophenes by Reaction of Aryl Methyl Ketones with Sulfides or Aryl Alkyl Ketone Anils with Sulfur	440
Thiophenes by Miscellaneous Cyclization Reactions	441
Table VI. Thiophenes by Miscellaneous Cyclization Reactions	443
PREPARATION OF TETRAHYDROTHIOPHENES	443
Tetrahydrothiophenes from 1,4-Difunctional Compounds and Sulfides	443
Reaction of 1,4-Dihalides with Sulfides	443
Reaction of a 1,4-Disulfuric Acid Ester with Hydrogen Sulfide	445
Cyclization of δ-Substituted Mercaptobutyl Halides	446
Experimental Conditions	446
Experimental Procedures	447
3,4-Dihydroxytetrahydrothiophene	447
dl-($trans$)-Tetrahydrothiophene-1,5-dicarboxylic Acid	447
Table VII. Tetrahydrothiophenes from 1,4-Difunctional Compounds and Sulfides	448
3,4-Di-n-propyltetrahydrothiophene	449
Tetrahydrothiophenes by the Dieckmann Cyclization Reaction	449
Cyclization of Esters Having Unsubstituted α-Methylene Groups	449
Cyclization of α-Substituted Esters	451
Syntheses from α-Mercapto Esters and Unsaturated Compounds	456
Experimental Conditions	458
Experimental Procedures	459
Methyl dl-4-Benzamido-3-ketotetrahydrothiophene-2-carboxylate (Sodium Salt)	459
Ethyl 3-Keto-2-(4'-methoxybutyl)tetrahydrothiophene-4-carboxylate	459
2-(4'-Methoxybutyl)-3-ketotetrahydrothiophene	459
Table VIII. Tetrahydrothiophenes by the Dieckmann Cyclization Reaction	460
Tetrahydrothiophenes by Catalytic Methods	464
Table IX. Tetrahydrothiophenes by Catalytic Methods.	465
Tetrahydrothiophenes by Miscellaneous Methods	465
Table X. Tetrahydrothiophenes by Miscellaneous Methods	468

INTRODUCTION

Thiophenes and tetrahydrothiophenes are discussed as separate major subdivisions of this chapter because there are significant differences in the general methods by which these two similar types of compounds are prepared. The review is not extended to include reactions that form thiophene or tetrahydrothiophene rings fused to another nucleus, as in benzothiophene, or reactions involving substitutions in the five-membered sulfur-containing ring. The literature on which this chapter is based includes publications reviewed by *Chemical Abstracts* through the 1946 Decennial Index.

The reactions that lead to the formation of thiophenes may be segregated into the following five general classifications:

I. Reaction of 1,4-difunctional compounds with sulfides.
II. Reaction of unsaturated compounds with sulfides.
III. Reaction of 1,2-difunctional compounds with thiodiacetic acid esters.
IV. Reaction of aryl methyl ketones with sulfides.
V. Miscellaneous cyclization reactions.

Similarly, the reactions that form tetrahydrothiophenes may be grouped into the following four general classifications:

I. Reaction of 1,4-difunctional compounds with sulfides.
II. Dieckmann cyclization reaction.
III. Catalytic methods.
IV. Miscellaneous methods.

Discussion of these various types of syntheses follows in the order of their listing above.

PREPARATION OF THIOPHENES

Thiophenes by Reaction of 1,4-Difunctional Compounds with Sulfides

The synthesis of thiophenes from 1,4-difunctional compounds is typified by the classic Volhard and Erdmann synthesis of thiophene itself from sodium succinate (I) and phosphorus trisulfide.[*,1,2] When a mixture of these reactants was heated in a retort over a free flame, a dark brown distillate was formed which contained thiophene.[1,2,3] The crude product was purified by digestion over sodium hydroxide, followed by distillation from sodium, and a 25–30% yield of thiophene (II) was obtained.[3] The method is primarily useful for the synthesis of alkyl- and

---

* There is confusion in the literature as to the exact nature of the sulfides of phosphorus. The commonly mentioned phosphorus trisulfide P_2S_3 does not exist; the product of reaction between red phosphorus and sulfur assigned this formula is probably impure P_4S_7. The phosphorus pentasulfide P_4S_{10} is often written P_2S_5 for convenience. In this review the designations employed in the original literature are used.

The phosphorus sulfides may be prepared in the laboratory (see ref. 3), or they are available from the Oldbury Electro-Chemical Co., Niagara Falls, New York. See Pernert and Brown, *Chem. Eng. News*, **27**, 2143 (1949).

[1] Volhard and Erdmann, *Ber.*, **18**, 454 (1885).
[2] Friedburg, *J. Am. Chem. Soc.*, **12**, 83 (1890); *J. Chem. Soc.*, **58**, 1400 (1890).
[3] Phillips, *Org. Syntheses*, Coll. Vol. **2**, 578 (1943).

aryl-substituted thiophenes; its chief advantage is that it makes possible control of the position of the substituents.

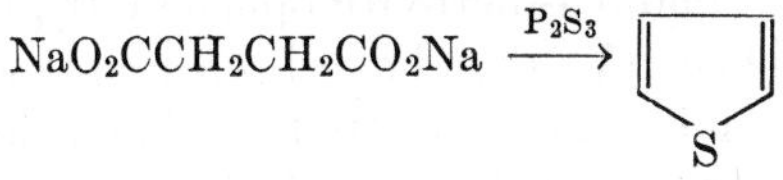

$$NaO_2CCH_2CH_2CO_2Na \xrightarrow{P_2S_3} \underset{S}{[]}$$

I II

The 1,4-difunctional compounds that react with sulfides to form thiophenes are grouped into four subclasses for discussion in the following sections.

Syntheses from Succinic Acids. Thiophene has been prepared by a number of variations of the original method [1] illustrated above. Succinic anhydride reacts with phosphorus pentasulfide to form thiophene; [1] erythritol reacts similarly.[4] When diethyl succinate is heated with 2 parts of phosphorus trisulfide, thiophene together with 2-ethoxythiophene and 2-ethylmercaptothiophene are obtained.[5] 2-Mercaptothiophene has been found as a by-product in the preparation of thiophene from sodium succinate.[6]

Thiophenes with substituents in the 3 or 3 and 4 positions are obtained from salts of substituted succinic acids (III or IV) by reaction with phosphorus sulfides. The 3-alkylthiophenes (V) that have been obtained in this way from alkyl-substituted succinic acids (III) include

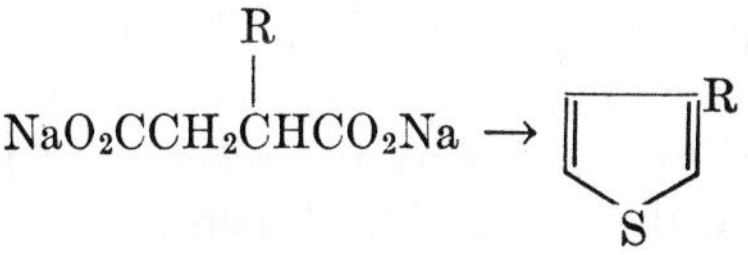

$$\underset{III}{NaO_2CCH_2\overset{R}{\underset{|}{C}}HCO_2Na} \rightarrow \underset{V}{\underset{S}{[]}^R}$$

$$\underset{IV}{NaO_2C\overset{R'}{\underset{|}{C}}H\overset{R}{\underset{|}{C}}HCO_2Na} \rightarrow \underset{VI}{R'\underset{S}{[]}^R}$$

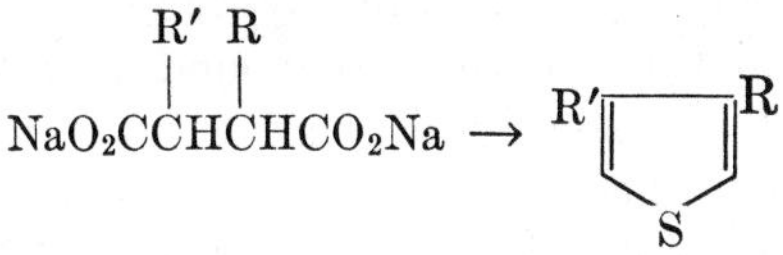

3-methylthiophene (30%),[1,7] 3-ethylthiophene (40-50%),[8,9] 3-isopropyl-thiophene (40%),[10,11] 3-n-propylthiophene (37%),[11] and 3-n-butyl-

[4] Paal and Tafel, *Ber.*, **18**, 688 (1885).
[5] Steinkopf and Leonhardt, *Ann.*, **495**, 166 (1932).
[6] Meyer and Neure, *Ber.*, **20**, 1756 (1887).
[7] Linstead, Noble, and Wright, *J. Chem. Soc.*, **1937**, 911.
[8] Damsky, *Ber.*, **19**, 3282 (1886).
[9] Gerlach, *Ann.*, **267**, 145 (1892).
[10] Thiele, *Ann.*, **267**, 133 (1892).
[11] Scheibler and Schmidt, *Ber.*, **54**, 139 (1921).

thiophene (23%).[12] The 3,4-dialkylthiophenes (VI) obtained from the appropriately disubstituted sodium succinates (IV) include 3,4-dimethylthiophene (43%) [7,13] and 3,4-diethylthiophene (40%).[14]

The 3-arylthiophenes (V) that have been prepared from the sodium salts of the corresponding α-substituted succinic acids (III) with phosphorus trisulfide [15] are 3-phenyl-, 3-*p*-anisyl-, and 3-*p*-tolyl-thiophene.

Syntheses from γ-Keto Acids. The thiophenes that have been prepared from γ-keto acids have substituents in the 2 position, as exemplified by the preparation of 2-methylthiophene (VII) from levulinic acid (VIII) and phosphorus sulfide.[16] Similarly, α,β-disubstituted γ-keto acids (IX) have been converted into 2,3,4-trisubstituted thiophenes (X).

$$CH_3COCH_2CH_2CO_2H \rightarrow$$ [thiophene ring with CH$_3$ at 2-position, S]

VIII   VII

$$\underset{IX}{RCOCHCHCO_2H}\begin{array}{c}R' \; R'' \\ | \; \; | \end{array} \rightarrow$$ [thiophene ring with R'', R', R substituents, S]

IX   X

5-Hydroxy-2-alkylthiophene derivatives are often formed along with the 2-alkylthiophenes from γ-keto acids. These 5-hydroxy derivatives are not formed if the sodium salt of the γ-keto acid is used.[11,15]

The preparation of thiophenes by the reaction of levulinic acid with sulfides has been studied extensively. When mixtures of levulinic acid (VIII) and phosphorus trisulfide or phosphorus pentasulfide are refluxed, there is formed either 5-hydroxy-2-methylthiophene (XI, thiotolenol or thiotenol), or a mixture of this compound and 2-methylthiophene (XII, α-thiotolene), apparently depending upon the amount of the sulfide used.[16] Thus, when a mixture of 3 parts of levulinic acid and 2 parts of phosphorus pentasulfide is heated, only 5-hydroxy-2-methylthiophene is obtained (30%). Two parts of levulinic acid and 3 parts

$$CH_3COCH_2CH_2CO_2H \rightarrow$$ [HO-thiophene-CH$_3$] + [thiophene-CH$_3$]

VIII   XI   XII

---

[12] Scheibler and Rettig, *Ber.*, **59**, 1194 (1926).
[13] Zelinsky, *Ber.*, **21**, 1835 (1888).
[14] Steinkopf, Frömmel, and Leo, *Ann.*, **546**, 199 (1941).
[15] Chrzaszczewska, *Roczniki Chem.*, **5**, 33 (1925) [*C. A.*, **20**, 1078 (1926)].
[16] Kues and Paal, *Ber.*, **19**, 555 (1886).

of phosphorus trisulfide react under similar conditions to give a mixture of 2-methylthiophene (15%) and 5-hydroxy-2-methylthiophene (20–25%);[16] when the mixture of products is treated again with phosphorus trisulfide, the 5-hydroxy-2-methylthiophene is not obtained.[16] Levulinic acid has been found by others[17,18] to react with phosphorus trisulfide to give only 5-hydroxy-2-methylthiophene; sodium levulinate gives only 2-methylthiophene (62%).[15,19]

2-Hydroxythiophene is formed by the reaction of β-formylpropionic acid with phosphorus pentasulfide.[18]

$$OHCCH_2CH_2CO_2H \xrightarrow{P_2S_5} \underset{S}{\boxed{}}OH$$

A number of 2-alkylthiophenes have been prepared from alkyl-substituted levulinic acids by reaction with a sulfide of phosphorus. These derivatives include 2-isopropylthiophene (XIII, 49%) from sodium γ-keto-δ-methylcaproate (XIV);[11] 2,3-dimethylthiophene (XV, 20%) together with some 2,3-dimethyl-5-hydroxythiophene (XVI)[20]

$$(CH_3)_2CHCOCH_2CH_2CO_2Na \rightarrow \underset{S}{\boxed{}}CH(CH_3)_2$$

XIV   XIII

from β-methyllevulinic acid (XVII);[20,21,22] and 3-ethyl-2-methylthio-

$$CH_3COCH(CH_3)CH_2CO_2H \rightarrow \underset{S}{\boxed{}}\begin{array}{l}CH_3\\CH_3\end{array} + HO\underset{S}{\boxed{}}\begin{array}{l}CH_3\\CH_3\end{array}$$

XVII   XV   XVI

phene (XVIII, 23%) from β-ethyllevulinic acid (XIX).[23] 2,4-Dimethyl-

$$CH_3COCH(C_2H_5)CH_2CO_2H \rightarrow \underset{S}{\boxed{}}\begin{array}{l}C_2H_5\\CH_3\end{array}$$

XIX   XVIII

---

[17] Steinkopf and Thormann, *Ann.*, **540**, 1 (1939).
[18] Mentzer and Billet, *Bull. soc. chim. France*, **12**, 292 (1945).
[19] Vlastelitza, *J. Russ. Phys. Chem. Soc.*, **46**, 790 (1914) [*C. A.*, **9**, 1750 (1915)].
[20] Paal and Püschel, *Ber.*, **20**, 2557 (1887).
[21] Grünewald, *Ber.*, **20**, 2585 (1887).
[22] Shepard, *J. Am. Chem. Soc.*, **54**, 2951 (1932).
[23] Steinkopf, Merckoll, and Strauch, *Ann.*, **545**, 45 (1940).

thiophene (XX, 34%) is obtained from α-methyllevulinic acid (XXI),[24, 25] 4-ethyl-2-methylthiophene (XXII) from α-ethyllevulinic acid (XXIII) [22] and 2,3,4-trimethylthiophene (XXIV) from α,β-dimethyllevulinic acid (XXV).[24]

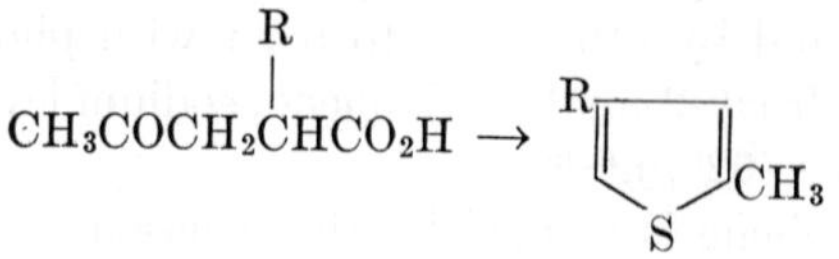

$$CH_3COCH_2\overset{R}{C}HCO_2H \rightarrow \underset{S}{\overset{R}{\boxed{}}}CH_3$$

XXI, R = CH$_3$
XXIII, R = C$_2$H$_5$

XX, R = CH$_3$
XXII, R = C$_2$H$_5$

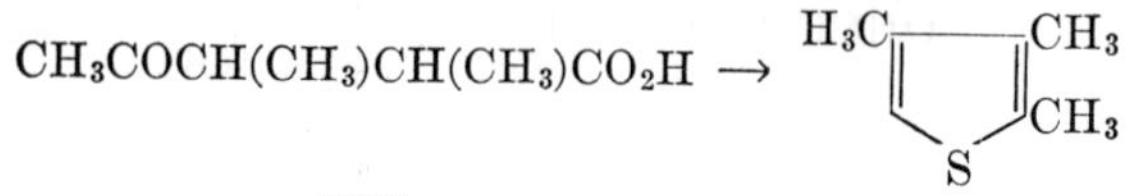

$$CH_3COCH(CH_3)CH(CH_3)CO_2H \rightarrow \underset{S}{\overset{H_3CCH_3}{\boxed{}}}CH_3$$

XXV
XXIV

Of the aryl-substituted thiophenes that may be prepared from γ-keto acids, 2-phenylthiophene (XXVI, 7–10%) is obtained from either phenacylmalonic acid (XXVII) or β-benzoylpropionic acid (XXVIII) [26] by reaction with phosphorus pentasulfide. When the sodium salt of β-benzoylpropionic acid [15] is used, a 30% yield of 2-phenylthiophene is obtained. Similarly, 2-p-tolylthiophene (XXIX) is obtained from β-p-toluylpropionic acid (XXX),[15] and 2-methyl-4-phenylthiophene

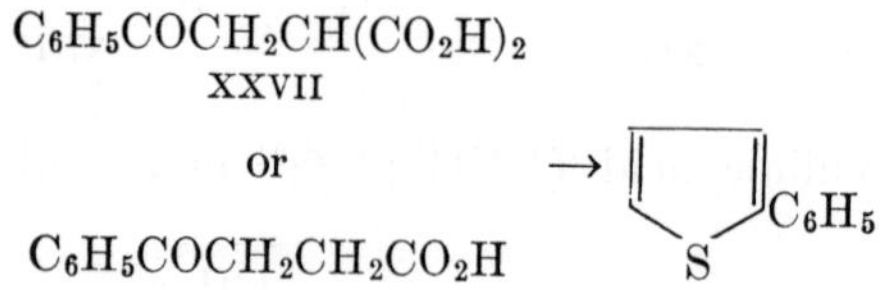

$$C_6H_5COCH_2CH(CO_2H)_2$$
XXVII

or $\rightarrow \underset{S}{\boxed{}}C_6H_5$

$$C_6H_5COCH_2CH_2CO_2H$$
XXVIII

XXVI

(XXXI, 30%) from the sodium salt of α-phenyllevulinic acid (XXXII).[20]

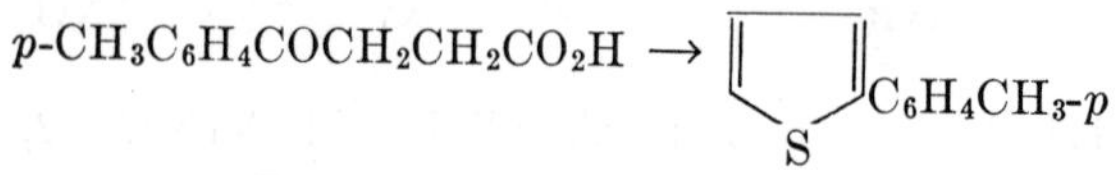

$$p\text{-}CH_3C_6H_4COCH_2CH_2CO_2H \rightarrow \underset{S}{\boxed{}}C_6H_4CH_3\text{-}p$$

XXX
XXIX

$$CH_3COCH_2CH(C_6H_5)CO_2H \rightarrow \underset{S}{\overset{H_5C_6}{\boxed{}}}CH_3$$

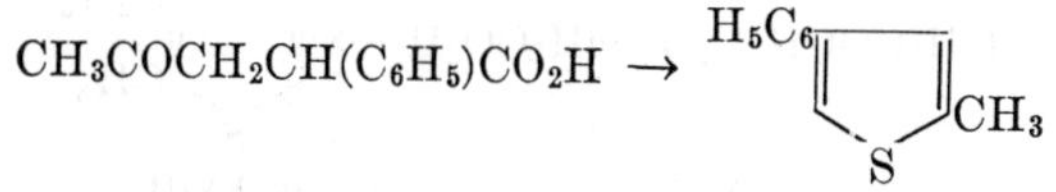

XXXII
XXXI

In contrast with the foregoing syntheses employing phosphorus sulfides, the use of hydrogen sulfide with γ-keto acids leads to alkoxy-substituted thiophenes. Hydrogen sulfide is used in alcoholic solution

[24] Zelinsky, *Ber.*, **20**, 2017 (1887).
[25] Rinkes, *Rec. trav. chim.*, **52**, 1052 (1933).
[26] Kues and Paal, *Ber.*, **19**, 3141 (1886).

saturated with hydrogen chloride; hydroxythiophenes were postulated as intermediates that react further with the alcohol in the reaction medium to give alkoxythiophenes. For example, 5-ethoxy-2-methylthiophene (XXXIII) is prepared from levulinic acid.[27] Similarly,

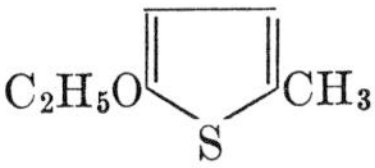

C_2H_5O-[S]-CH_3
XXXIII

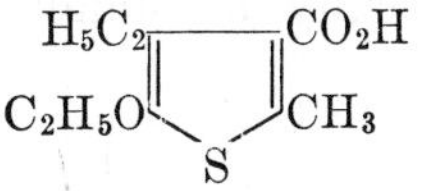

H_5C_2-[CO_2H / CH_3]-C_2H_5O-[S]
XXXIV

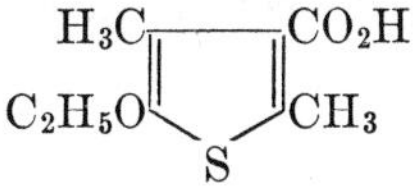

H_3C-[CO_2H / CH_3]-C_2H_5O-[S]
XXXV

5-ethoxy-4-ethyl-2-methyl-3-thiophenecarboxylic acid (XXXIV) is prepared from ethyl β-carbethoxy-α-ethyllevulinate,[27] and 2,4-dimethyl-5-ethoxy-3-thiophenecarboxylic acid (XXXV) from ethyl β-carbethoxy-α-methyllevulinate.[27] The yields of these 5-ethoxythiophene derivatives are 20–25%.

The methyl, ethyl, and n-propyl ethers of ethyl 5-hydroxy-2-methyl-3-thiophenecarboxylate (XXXVI–XXXVIII) are products of the reactions between ethyl β-carbethoxylevulinate (XXXIX) and hydrogen sulfide in the appropriate alcohol-hydrogen chloride mixture.[28]

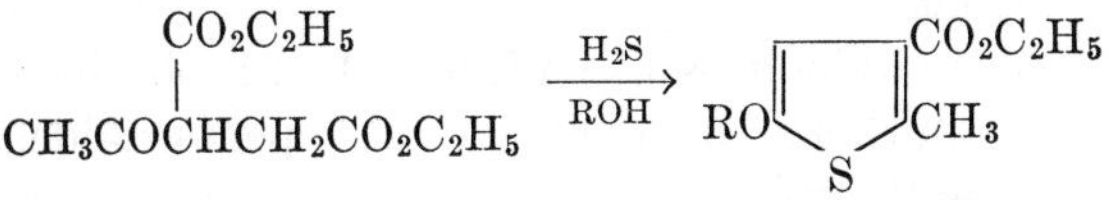

$$\begin{array}{c} CO_2C_2H_5 \\ | \\ CH_3COCHCH_2CO_2C_2H_5 \end{array} \xrightarrow[ROH]{H_2S} RO\text{-}[S]\text{-}CH_3 \text{ with } CO_2C_2H_5$$

XXXIX

XXXVI, R = CH_3
XXXVII, R = C_2H_5
XXXVIII, R = $n\text{-}C_3H_7$

By a variation of the method employing γ-keto acids, 2-thiophenecarboxylic acid (XL, 10–12%) is prepared by reaction of mucic acid (XLI) with barium sulfide.[29] 2-Thiophenealdehyde (XLII) is the product of the reaction between 3-chloro-1,2-cyclopentanedione (XLIII) and hydrogen sulfide in alkaline solution.[30]

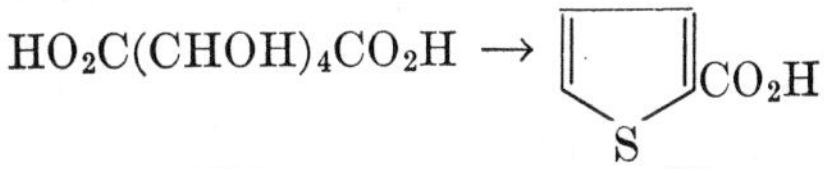

$HO_2C(CHOH)_4CO_2H \rightarrow$ [S]-CO_2H

XLI   XL

$\begin{array}{c} H_2C\text{---}CHCl \\ | \quad\quad | \\ H_2C \quad C=O \\ \backslash\;/ \\ C \\ \| \\ O \end{array} \rightarrow$ [S]-CHO

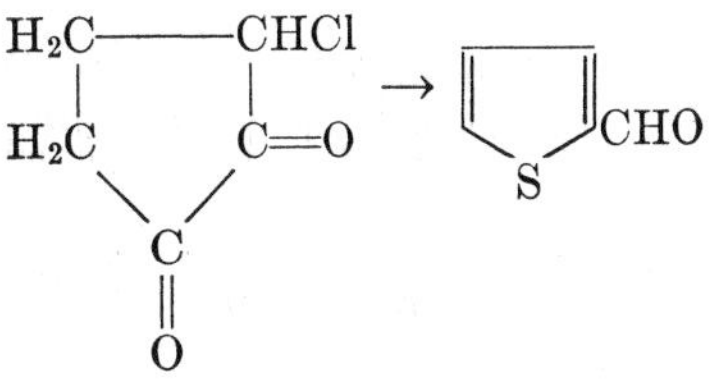

XLIII   XLII

[27] Chakrabarty and Mitra, *J. Chem. Soc.*, **1940**, 1385.
[28] Mitra, Chakrabarty, and Mitra, *J. Chem. Soc.*, **1939**, 1116.
[29] Paal and Tafel, *Ber.*, **18**, 456 (1885).
[30] Hantzsch, *Ber.*, **22**, 2827 (1889).

Syntheses from 1,4-Diketones. 2,5-Disubstituted thiophenes (XLIV) and a few 2,3,4,5-tetrasubstituted thiophenes (XLV) have been prepared by reaction of substituted diketones with sulfides. The application of this method to the preparation of tetrasubstituted derivatives has been limited by the difficulty in obtaining the required diketones.[31]

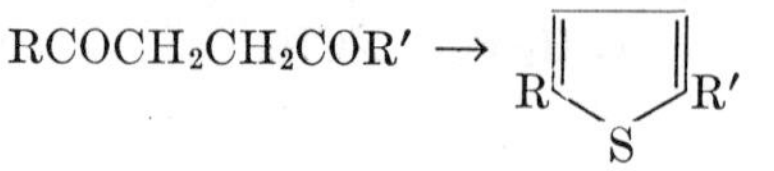

$$RCOCH_2CH_2COR' \rightarrow$$

XLIV

$$RCOCHCHCOR' \text{ (R''R''')} \rightarrow$$

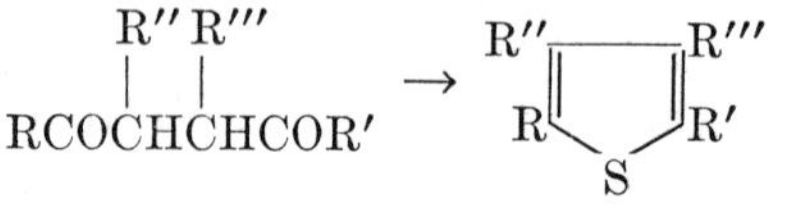

XLV

2,5-Dimethylthiophene (50–60%) results from reaction of 2,5-hexanedione with either phosphorus trisulfide or phosphorus pentasulfide.[32] 2,3,5-Trimethylthiophene (35–40%) and 3-cyano-2,5-dimethylthiophene are prepared from 3-methyl-2,5-hexanedione[31] and 3-cyano-2,5-hexanedione,[33] respectively.

2-Methyl-5-phenylthiophene (XLVI, 60–70%) is obtained by heating 5-phenyl-2,5-pentanedione with phosphorus pentasulfide.[34] Methyl β-2-(5-phenylthienyl)propionate (XLVII, 50%) and methyl β-2-(5-p-methoxyphenylthienyl)propionate (XLVIII) are formed by the reaction of the appropriate methyl 4,7-diketo-7-arylheptanoate with phosphorus pentasulfide.[35]

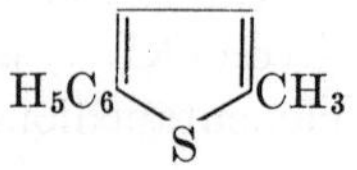

XLVI

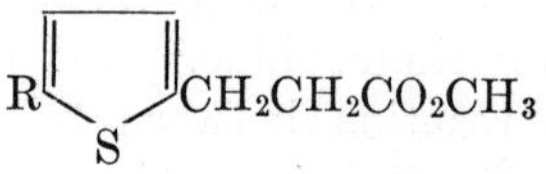

XLVII, R = C₆H₅
XLVIII, R = p-CH₃OC₆H₄

2,5-Diphenylthiophene (XLIX, 60–70%) results from the reaction of either diphenacyl (L)[36] or diphenacyl sulfide (LI)[37] with phosphorus pentasulfide. 2,3,5-Triphenylthiophene is obtained similarly from 1,2-dibenzoyl-1-phenylethane.[38] When diacetylsuccinic acid ester was treated with phosphorus pentasulfide, no thiophene derivative could be isolated.[7]

[31] Youtz and Perkins, *J. Am. Chem. Soc.*, **51**, 3511 (1929).
[32] Paal, *Ber.*, **18**, 2251 (1885).
[33] Justoni, *Gazz. chim. ital.*, **71**, 375 (1941).
[34] Paal, *Ber.*, **18**, 367 (1885).
[35] Robinson and Todd, *J. Chem. Soc.*, **1939**, 1743.
[36] Kapf and Paal, *Ber.*, **21**, 3053 (1888).
[37] Böhme, Pfeifer, and Schneider, *Ber.*, **75**, 900 (1942).
[38] Smith, *J. Chem. Soc.*, **57**, 643 (1890).

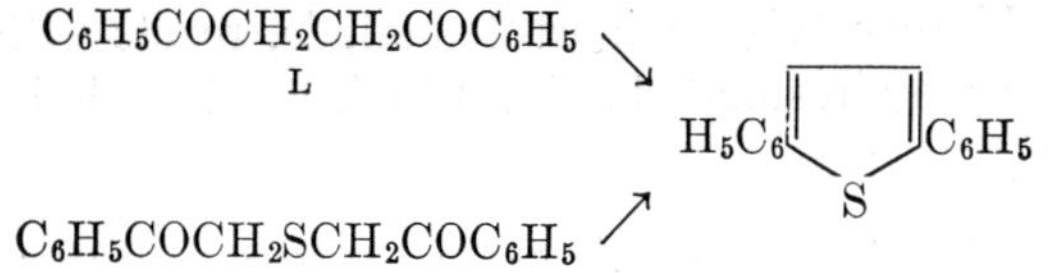

$C_6H_5COCH_2CH_2COC_6H_5$
L

$C_6H_5COCH_2SCH_2COC_6H_5$
LI

[Structure: 2,5-diphenylthiophene with H_5C_6 and C_6H_5 at 2,5-positions]
XLIX

1,2-Dibenzoyl-1-phenylethylene (LII) reacts with hydrogen sulfide in ethanol solution saturated with hydrogen chloride to give 2,3,5-triphenylthiophene (LIII).[39]

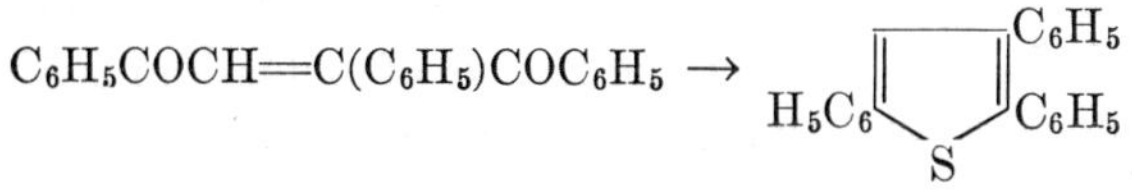

$C_6H_5COCH=C(C_6H_5)COC_6H_5 \rightarrow$ [2,3,5-triphenylthiophene structure]

LII   LIII

Tetraphenylthiophene (LIV) is produced by the reaction of hydriodic acid upon tetraphenyl-2,5-endosulfidothiophene (LV) or its oxygen

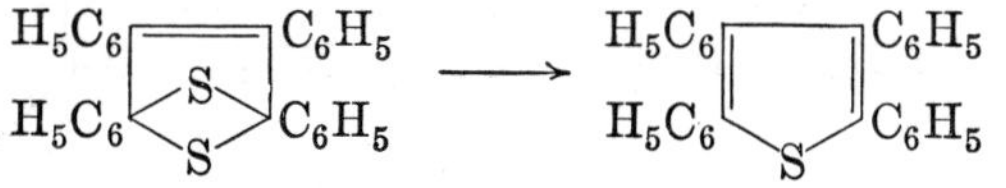

[Structure LV: tetraphenyl-2,5-endosulfidothiophene] → [Structure LIV: tetraphenylthiophene]

LV   LIV

analog.[39] The tetraphenyl-2,5-endosulfidothiophene is formed by passing hydrogen sulfide through a solution of benzoin in either ethanolic hydrogen chloride or a mixture of acetic acid and hydrochloric acid.[39]

Syntheses from Other 1,4-Difunctional Compounds. A limited number of thiophenes have been synthesized from chloroacetyl-substituted esters. Thus ethyl chloroacetylcyanoacetate (LVI) reacts with potassium hydrosulfide to form ethyl 2-amino-4-hydroxy-3-thiophenecarboxylate (LVII, 46%).[40]

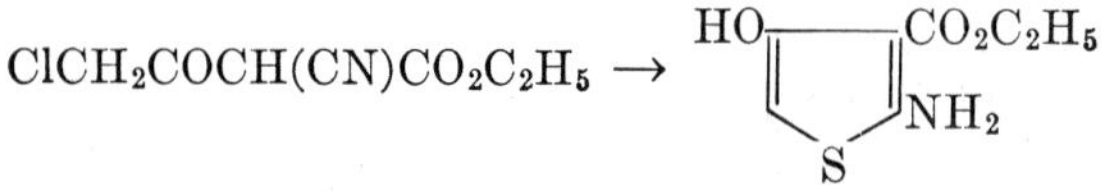

$ClCH_2COCH(CN)CO_2C_2H_5 \rightarrow$ [thiophene with HO, $CO_2C_2H_5$, NH_2 substituents]

LVI   LVII

The methyl and ethyl esters of 4-hydroxy-2-methyl-3-thiophenecarboxylic acid (LVIII, 83%) have been prepared by treating methyl and ethyl α-chloroacetyl-β-aminocrotonate (LIX) with sodium or potassium hydrosulfide in ethanol solution.[41-44] The anilide corresponding to the

---

[39] Mitra, J. Indian Chem. Soc., **15**, 59 (1938) [C. A., **32**, 4982 (1938)].
[40] Benary, Ber., **43**, 1943 (1910).
[41] Benary and Baravian, Ber., **48**, 593 (1915).
[42] Benary and Silberstrom, Ber., **52**, 1605 (1919).
[43] Mentzer, Billet, Molho, and Xuong, Bull. soc. chim. France, **12**, 161 (1945).
[44] Benary, Ger. pat. 282,914 [C. A., **9**, 2568 (1915)].

ester LIX, α-chloroacetyl-β-aminocrotonanilide, reacts with an equivalent of potassium hydrosulfide to give 4-hydroxy-2-methyl-3-thiophenecarbonanilide (LX).[45]

$$\underset{\substack{\text{LIX} \\ \text{R} = \text{CH}_3, \text{C}_2\text{H}_5}}{\underset{\text{H}_2\text{NCCH}_3}{\text{ClCH}_2\text{COCCO}_2\text{R}}} \rightarrow \underset{\substack{\text{LVIII} \\ \text{R} = \text{CH}_3, \text{C}_2\text{H}_5}}{\text{HO}\!\!-\!\!\overset{}{\underset{S}{\bigcirc}}\!\!-\!\!\text{CO}_2\text{R} \atop \text{CH}_3} \quad \underset{\text{LX}}{\text{HO}\!\!-\!\!\overset{}{\underset{S}{\bigcirc}}\!\!-\!\!\text{CONHC}_6\text{H}_5 \atop \text{CH}_3}$$

3-Acetyl-2,4-dihydroxythiophene or 3-acetyl-2,4-diketotetrahydrothiophene (LXI) is prepared by the action of potassium hydrosulfide on ethyl α-chloroacetyl-β-aminocrotonate (LXII).[42] The intermediate amino derivative LXIII is readily hydrolyzed to the ketone LXI.

$$\underset{\text{LXII}}{\begin{array}{c} \text{CH}_3 \\ \text{OC}\!-\!\text{C}\!=\!\text{C} \\ | \quad | \quad \diagdown \text{NH}_2 \\ \text{CH}_2\text{Cl} \; \text{CO}_2\text{C}_2\text{H}_5 \end{array}} \rightarrow \underset{\text{LXIII}}{\begin{array}{c} \text{CH}_3 \\ \text{OC}\!-\!\text{C}\!=\!\text{C} \\ | \quad \quad | \quad \diagdown \text{NH}_2 \\ \text{H}_2\text{C} \quad \text{CO} \\ \diagdown\text{S}\diagup \end{array}} \rightarrow \underset{\text{LXI}}{\begin{array}{c} \text{OC}\!-\!\text{CHCOCH}_3 \\ | \quad \quad | \\ \text{H}_2\text{C} \quad \text{CO} \\ \diagdown\text{S}\diagup \end{array}}$$

Experimental Conditions

There seems to be little difference in the reaction of phosphorus trisulfide or phosphorus pentasulfide with the various difunctional compounds. The yields are about the same from both sulfides, but phosphorus trisulfide is the more common reagent. The proportion of sulfide employed has varied, and an excess up to 1.5 moles is generally used;[3,7] a large excess is reported to have an adverse effect.[15] As summarized in Table II, the yields of products obtained by this method are seldom above 50% except for syntheses involving diketones.

To carry out the reaction, the difunctional compound and the phosphorus sulfide are first mixed intimately. Some investigators advise sand [4,14,15,19,22] as a diluent, the amount used to be either equal to the weight of the sulfide [14,15] or two to ten times the weight of the dicarbonyl compound.[4,22] According to the older literature, the reaction mixture is placed in a retort and heated with a free flame [1] or in a closed tube heated at 160–180°.[36] In more recent procedures, the reaction is carried out in a flask equipped with a condenser for distillation under an atmosphere of carbon dioxide.[3,7,14,17] The carbon dioxide prevents explosions and also carries over the distillate more rapidly. It is not always neces-

---

[45] Benary and Kerckhoff, *Ber.*, **59**, 2548 (1926).

sary to heat the reaction mixture at high temperatures; the two components may be stirred and heated at 90–100° until evolution of hydrogen sulfide ceases.[35] Slow initial heating has been found beneficial,[7] but in general the reaction mixture is finally heated above 150° to complete distillation of the product.

The product is usually distilled from the reaction mixture. However, it has been extracted with ether [35] or steam-distilled.[20] The products are generally purified by washing with strong aqueous alkali and by distilling the dried product over sodium, provided the product does not contain a functional group affected by this treatment.

The effects of minor modification in the procedure on the yield are indicated by a study of the synthesis of 3-methylthiophene from sodium α-methylsuccinate [7] (Table I).

TABLE I

PREPARATION OF 3-METHYLTHIOPHENE

Sodium Salt g.	P_2S_3 g.	Procedure	Yield g.	%
92	140	Rapid initial heating	9.5	18
100	150	Heating in a stream of CO_2	12	22
200	250	Slow initial heating	34	30
200	250	Slow initial heating	22	20
235	295	Slow initial heating	37	28
220	275	Slow initial heating; mixture diluted with sand	22	18

EXPERIMENTAL PROCEDURES

3,4-Dimethylthiophene.[7] A mixture of 195 g. of the sodium salt of α,β-dimethylsuccinic acid (dried at 200°) and 245 g. of phosphorus trisulfide is subjected to dry distillation in a stream of carbon dioxide. The distillate of crude 3,4-dimethylthiophene is allowed to stand in contact with sodium hydroxide for fifteen hours, then refluxed over sodium for six hours, and fractionated. The 3,4-dimethylthiophene boils at 145–148°; yield, 50 g. (43.5%).

3-n-Propylthiophene.[11] A mixture of the dry sodium salt from 26 g. of n-propylsuccinic acid and 60 g. of powdered phosphorus trisulfide is placed in a flask equipped with a condenser for distillation. The mixture is heated until the product distils. The distillate of crude 3-n-propylthiophene is washed with sodium hydroxide solution and with water, and is dried over solid alkali. The product is finally distilled from sodium as a colorless liquid boiling at 160–162° (cor.); yield, 7.6 g. (37%).

2-Isopropylthiophene.[11] A mixture of 34.5 g. of dry sodium δ,δ-dimethyllevulinate, ground to a fine powder, and 80 g. of powdered phosphorus trisulfide is placed in a flask fitted with a condenser for distillation. The flask is heated with a free flame until the reaction starts, when the flame can be removed. The distillate is collected and dissolved in ether; the ethereal solution is washed repeatedly with aqueous sodium hydroxide and then with water and finally dried over solid sodium hydroxide. The ethereal solution is evaporated, and the crude residue is refluxed over sodium and then fractionated. The 2-isopropylthiophene distils at 149–157° as a colorless oil. Refractionation of this distillate over sodium yields 12 g. (49%) of pure 2-isopropylthiophene, b.p. 152–153° (cor.).

5-Hydroxy-2-methylthiophene.[17] A mixture of 60 g. of levulinic acid and 40 g. of finely powdered phosphorus pentasulfide is heated in a 1-l. flask equipped with a condenser for distillation. A stream of carbon dioxide is passed through the flask as it is heated with a free flame. The crude distillate is redistilled under reduced pressure to yield 11 g. (19%) of pure 5-hydroxy-2-methylthiophene, b.p. 94–96°/15 mm., m.p. $-23.5°$ to $-22.5°$.

2,3-Dimethylthiophene.[22] A mixture of 30 g. of β-methyllevulinic acid and 35 g. of powdered phosphorus pentasulfide is heated. A vigorous reaction takes place; as soon as this has subsided the product is distilled from the reaction mixture. The crude distillate is washed with cold sodium hydroxide solution and is then distilled over sodium. The purified 2,3-dimethylthiophene boils at 140.2–141.2°. The yield is 20%.

2,3,5-Trimethylthiophene.[31] To 65–70 g. of powdered phosphorus pentasulfide in a flask fitted with a reflux condenser is added 96 g. of 3-methyl-2,5-hexanedione. The mixture is cooled and allowed to stand for a few minutes to avoid violent reaction, then allowed to warm to room temperature. Finally, it is heated to boiling for three to four hours with the addition of 10 g. of phosphorus pentasulfide after the first hour. The liquid portion of the reaction mixture is decanted from the tarry residue and distilled. The distillate is dried, refluxed over several portions of sodium and then over sodium hydroxide, and finally fractionated. The product is a colorless liquid with a durene-like odor, b.p. 163–165°/746 mm. (cor.). The yield is 35% to 40%.

Methyl β-2-(5-Phenylthienyl)propionate.[35] A mixture of 10 g. of methyl 4,7-diketo-7-phenylheptanoate and 10 g. of phosphorus pentasulfide is heated at 95° and stirred until evolution of hydrogen sulfide has ceased (about one hour). The reaction mixture is a thick brown syrup which solidifies after a few hours. The solid product is extracted with ether; the ethereal solution is filtered, shaken with aqueous sodium bicarbonate, and dried; and the ether is evaporated. The solid residue is dissolved in a small volume of ethanol, and the solution is decolorized with charcoal. The product which crystallizes melts at 75°. The yield is about 50%.

TABLE II

Thiophenes by Reaction of 1,4-Difunctional Compounds with Sulfides

A. Thiophene and Alkylthiophenes

Thiophene	Starting Material	Reagents and Experimental Conditions	Yield %	Reference
Thiophene	NaO$_2$CCH$_2$CH$_2$CO$_2$Na (3 moles)	P$_2$S$_3$ (4.1 moles), free flame, CO$_2$ atmosphere	25–30	1, 2, 3
	COCH$_2$CH$_2$CO (O bridge)	P$_2$S$_5$, 140°	—	1
	HOCH$_2$CHOHCHOHCH$_2$OH (1 part)	P$_2$S$_5$ (1 part), free flame, sand as diluent	—	4
2-Methyl-	C$_2$H$_5$O$_2$CCH$_2$CH$_2$CO$_2$C$_2$H$_5$ (150 g.)	P$_2$S$_3$ (300 g.), 150°	—	5
	CH$_3$COCH$_2$CH$_2$CO$_2$Na	P$_2$S$_3$, sand diluent, distilled	62	15
	CH$_3$COCH$_2$CH$_2$CO$_2$Na	P$_2$S$_3$, sand diluent, distilled	49	19
	CH$_3$COCH$_2$CH$_2$CO$_2$H (1 part)	P$_2$S$_3$ (1.5 parts), distilled	15	16
3-Methyl-	NaO$_2$CCH$_2$CH(CH$_3$)CO$_2$Na (200 g.)	P$_2$S$_3$ (250 g.), slow initial heating	30	1, 7
3-Ethyl-	NaO$_2$CCH$_2$CH(C$_2$H$_5$)CO$_2$Na (100 g.) CH$_2$CH$_2$CH$_3$	P$_2$S$_3$ (150 g.), free flame	40–50	8, 9
3-n-Propyl-	NaO$_2$CCH$_2$CHCO$_2$Na (33 g.)	P$_2$S$_3$ (60 g.), free flame	37	11
2-Isopropyl-	(CH$_3$)$_2$CHCOCH$_2$CH$_2$CO$_2$Na (34.5 g.)	P$_2$S$_3$ (80 g.), free flame	49	11
3-Isopropyl-	NaO$_2$CCH$_2$CH(C$_3$H$_7$-iso)CO$_2$Na	P$_2$S$_3$, free flame	40	11
3-n-Butyl-	NaO$_2$CCH$_2$CH(C$_4$H$_9$-n)CO$_2$Na (120 g.)	P$_2$S$_3$ (120 g.), free flame	30	10
2,3-Dimethyl-	CH$_3$COCH(CH$_3$)CH$_2$CO$_2$H (1 part)	P$_2$S$_3$ (1.5 parts), distilled	23	12
	CH$_3$COCH(CH$_3$)CH$_2$CO$_2$H (30 g.)	P$_2$S$_5$ (35 g.), distilled	25	20
	CH$_3$COCH(CH$_3$)CH$_2$CO$_2$H (10 g.)	P$_2$S$_3$ (17 g.), distilled	20	22
			—	21

2,4-Dimethyl-	CH₃COCH₂CH(CH₃)CO₂H (127 g.)	P₂S₃ (330 g.), distilled	34.5	25
2,5-Dimethyl-	CH₃COCH₂CH(CH₃)CO₂H (20 g.)	P₂S₃ (30–35 g.), distilled	23	24
	CH₃COCH₂CH₂COCH₃ (3 parts)	P₂S₅ or P₂S₃ (2 parts), closed tube at 140–150°	50–60	32
3,4-Dimethyl-	NaO₂CCH(CH₃)CH(CH₃)CO₂Na (195 g.)	P₂S₃ (245 g.), slow initial heating, CO₂ atmosphere	43.5 21–22	7 13
2-Methyl-3-ethyl-	CH₃COCH(C₂H₅)CH₂CO₂H (20 g.)	P₂S₃ (20 g.), 130–140°	23	23
2-Methyl-4-ethyl-	CH₃COCH₂CH(C₂H₅)CO₂Na	P₂S₅, sand diluent, free flame	20	22
3,4-Diethyl-	NaO₂CCH(C₂H₅)CH(C₂H₅)CO₂Na (22 g.)	P₂S₃ (16 g.), free flame, sand diluent, CO₂ atmosphere	40	14
2,3,4-Trimethyl-	CH₃COCH(CH₃)CH(CH₃)CO₂H	P₂S₃, distilled	—	24
2,3,5-Trimethyl-	CH₃COCH(CH₃)CH₂COCH₃ (96 g.)	P₂S₅ (65–70 g.), reflux	35–40	31

B. Arylthiophenes

2-Phenyl-	C₆H₅COCH₂CH₂CO₂Na	P₂S₃, sand diluent, distilled	30	15
	C₆H₅COCH₂CH₂CO₂H or C₆H₅COCH₂CH(CO₂H)₂ (3 parts)	P₂S₅ (2 parts), distilled	7–10	26
3-Phenyl-	NaO₂CCH₂CH(C₆H₅)CO₂Na	P₂S₃, sand diluent, distilled	—	15
2-p-Tolyl-	p-CH₃C₆H₄COCH₂CH₂CO₂H	P₂S₃, sand diluent, distilled	—	15
3-p-Tolyl-	NaO₂CCH₂CH(C₆H₄CH₃-p)CO₂Na	P₂S₃, sand diluent, distilled	—	15
3-p-Anisyl-	NaO₂CCH₂CH(C₆H₄OCH₃-p)CO₂Na	P₂S₃, sand diluent, distilled	—	15
2,5-Diphenyl-	C₆H₅COCH₂CH₂COC₆H₅ (2 parts)	P₂S₅ (3 parts), closed tube 160–180°	60–70	36
	C₆H₅COCH₂SCH₂COC₆H₅ (20 g.)	P₂S₅ (6 g.), 170°	—	37
2,3,5-Triphenyl-	C₆H₅COCH(C₆H₅)CH₂COC₆H₅ (5 g.)	P₂S₅ (2 g.), 150°	—	38
Tetraphenyl-	C₆H₅COC(C₆H₅)=CHCOC₆H₅	H₂S + HCl in absolute ethanol	—	39
	C₆H₅COCHOHC₆H₅	H₂S + HCl in absolute ethanol followed by treatment with HI	—	39

TABLE II—Continued

THIOPHENES BY REACTION OF 1,4-DIFUNCTIONAL COMPOUNDS WITH SULFIDES

C. Alkylarylthiophenes

Thiophene	Starting Material	Reagents and Experimental Conditions	Yield %	Reference
2-Methyl-4-phenyl-	$CH_3COCH_2CH(C_6H_5)CO_2Na$	P_2S_3 or P_2S_5, distilled	30	20
2-Methyl-5-phenyl-	$C_6H_5COCH_2CH_2COCH_3$	P_2S_5, 120–130°	60–70	34

D. Thiophenes Containing Substituents Other Than Hydrocarbon Groups

Thiophene	Starting Material	Reagents and Experimental Conditions	Yield %	Reference
2-Hydroxy-	$OHCCH_2CH_2CO_2H$ (30 g.)	P_2S_5, (20 g.), distilled	—	18
2,5-Dihydroxy-	$HO_2CCH_2CH_2CO_2H$ (1 part)	P_2S_5 (1 part), distilled	—	46
2-Mercapto-—2-carbonal	$NaO_2CCH_2CH_2CO_2Na$	P_2S_3	—	6
	3-Chlorocyclopentane-1,2-dione	H_2S in alkaline buffer solution	—	30
2-Carboxy-	$HO_2C(CHOH)_4CO_2H$ (1 part)	BaS (2 parts), 200–210°	10–12	29
2-Hydroxy-5-methyl-	$CH_3COCH_2CH_2CO_2H$ (3 parts)	P_2S_5 (2 parts), 130–140°	30	16
	$CH_3COCH_2CH_2CO_2H$ (60 g.)	P_2S_5 (40 g.), free flame, CO_2 atmosphere	19	17
2-Methylmercapto-	$CH_3COCH_2CH_2CO_2H$ (25 g.)	P_2S_5 (16.6 g.)	—	18
	$CH_3O_2CCH_2CH_2CO_2CH_3$	P_2S_3	—	5

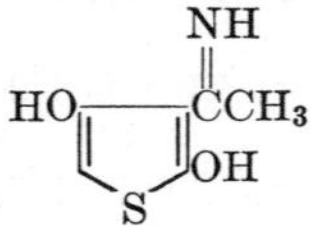 NH ![structure: HO—C(=CCH_3)—... with CH_3, OH, S ring]	$CH_3C{=}CCOCH_2Cl$
	  $\mid$ $\quad\mid$
	  NH_2 $CO_2C_2H_5$
		KSH in ethanol at 0°	—	42

2-Methyl-5-ethoxy-	$CH_3COCH_2CH_2CO_2H$ (25 g.)	H_2S + HCl in absolute ethanol (200 ml.)	—	27

Compound	Starting material	Conditions	Yield	Ref.
2-Ethoxy-	$C_2H_5O_2CCH_2CH_2CO_2C_2H_5$ (150 g.)	P_2S_3 (300 g.), 150°	—	5
2-Hydroxy-4,5-dimethyl-	$CH_3COCH(CH_3)CH_2CO_2H$ (1 part)	P_2S_3 (1.5 parts), distilled	—	20
2-Ethylmercapto-	$C_2H_5O_2CCH_2CH_2CO_2C_2H_5$ (150 g.)	P_2S_3 (300 g.), 150°	—	5
3-Cyano-2,5-dimethyl-	$CH_3COCH(CN)CH_2COCH_3$	P_2S_3 or P_2S_5, 85–90°, violent reaction	—	33
3-Hydroxy-4-carbethoxy-5-amino-	$ClCH_2COCH(CN)CO_2C_2H_5$	KSH in water with warming	46	40
3-Hydroxy-4-carbethoxy-5-methyl-	$CH_3C(NH_2)=C(CO_2C_2H_5)-COCH_2Cl$	KSH in ethanol	83	41, 44
			75	42
		NaSH in ethanol	73–80	43
2-Methoxy-4-carbethoxy-5-methyl-	$CH_3COCH(CO_2C_2H_5)CH_2CO_2C_2H_5$	H_2S + HCl in absolute methanol	—	28
2-Ethoxy-4-carbethoxy-5-methyl-	$CH_3COCH(CO_2C_2H_5)CH_2CO_2C_2H_5$	H_2S + HCl in absolute ethanol	—	28
	$CH_3COCH(CO_2C_2H_5)CH(CO_2C_2H_5)_2$		—	27
2-n-Propoxy-4-carbethoxy-5-methyl-	$CH_3COCH(CO_2C_2H_5)CH_2CO_2C_2H_5$	H_2S + HCl in absolute n-propanol	—	28
2-Ethoxy-4-carbethoxy-3,5-dimethyl-	$CH_3COCH(CO_2C_2H_5)CH(CH_3)CO_2C_2H_5$	H_2S + HCl in absolute ethanol	—	27
HO⎯CONHC6H5 / ⎯CH3 (S)	$CH_3C=CCOCH_2Cl$ / NH_2 $CONHC_6H_5$	KSH in alcohol at 0°	75	45
H_5C_2⎯$CO_2C_2H_5$ / C_2H_5O⎯CH_3 (S)	$CH_3COCH(CO_2C_2H_5)CH(C_2H_5)CO_2C_2H_5$	H_2S + HCl in absolute ethanol	—	27
H_5C_6⎯(CH2)2CO2CH3 (S)	$C_6H_5COCH_2CH_2COCH_2CH_2CO_2CH_3$ (10 g.)	P_2S_5 (10 g.), 95°	50	35
p-$CH_3OH_4C_6$⎯(CH2)2CO2CH3 (S)	p-$CH_3OC_6H_4COCH_2CH_2COCH_2CH_2CO_2CH_3$	P_2S_5, 90–100°	50–60	35

[46] Auger, *Ann. chim. phys.*, (6), **22**, 333 (1891).

Thiophenes by Reaction of Unsaturated Compounds with Sulfides

The second general method for the preparation of thiophenes is typified by the reaction of acetylene with either metallic sulfides, hydrogen sulfide, or sulfur to form thiophene. So many variations upon this general method have been devised that consideration of it has been divided into three parts, which are based upon the three sulfurizing agents mentioned above. Other starting materials that appear to react with the sulfurizing agent through unsaturated intermediates are included.

For the manufacture of thiophene, the method is amenable to large-scale operation. For the preparation of lower alkylthiophenes and some arylthiophenes, particularly tetraphenylthiophene, the method is applicable in the laboratory where the starting materials are readily available. This method has far more limitations than the one involving the reaction of 1,4-difunctional compounds with sulfides, since there is little control of the isomers formed. The preparation of many of the compounds by this method involves apparatus not available in many laboratories. For this reason no experimental procedures are included.

Reaction of Unsaturated Compounds with Metallic Sulfides. The most commonly used metallic sulfide is pyrite, but markasite and synthetic iron sulfide (FeS_2) have also been employed.[47-50] The finely divided pyrite (90-mesh) is generally placed in a heated iron tube equipped with an agitator. The gaseous unsaturated hydrocarbon is then passed through the tube at about 300°.[47,48] Carbon dioxide may be used as a diluent.[49] The exit gases are condensed, and the condensate is fractionated.

The reaction is accompanied by numerous side reactions. For example, in the preparation of thiophene from acetylene,[47,48,49,51,52,53] the crude reaction product contains not only thiophene but also 1,3-butadiene, acetaldehyde, carbon disulfide, acetone, benzene, 2-methylthiophene, 3-methylthiophene, 2,3-dimethylthiophene, 2-ethylthiophene, and 3-ethylthiophene; nevertheless, the crude reaction product yields about 40% of thiophene on fractionation.[49]

Several homologs of thiophene have been prepared by allowing the

---

[47] Steinkopf, *Chem. Ztg.*, **35**, 1098 (1911); *J. Soc. Chem. Ind.*, **30**, 1202 (1911).
[48] Barger and Easson, *J. Chem. Soc.*, **1938**, 2100.
[49] Steinkopf and Kirchhoff, *Ann.*, **403**, 1, 11 (1914).
[50] Steinkopf, *Chem. Ztg.*, **36**, 379 (1912) [*C. A.*, **7**, 1482 (1913)].
[51] Steinkopf and Herold, *Ann.*, **428**, 123 (1922).
[52] Steinkopf and Kirchhoff, Ger. pat. 252,375 [*C. A.*, **7**, 538 (1913)].
[53] Steinkopf and Kirchhoff, Aust. pat. 72,291 [*C. A.*, **11**, 869 (1917)]; Steinkopf and Kirchhoff, Brit. pat. 16,810 [*C. A.*, **8**, 416 (1914)].

appropriate hydrocarbon to react with pyrite, but the yields are low; examples are 3-methylthiophene from isoprene [49,50] and 3,4-dimethylthiophene from 2,3-dimethyl-1,3-butadiene.[49,50]

Reaction of Unsaturated Compounds with Hydrogen Sulfide. When hydrogen sulfide is employed as the sulfurizing agent, the mixture of hydrogen sulfide and the unsaturated compound, which is diluted with carbon dioxide,[54] may be allowed to react directly at high temperature (640–660°). Alternatively, the mixed gases may be passed over a catalyst at 300–600°. The catalysts used include silica gel,[55] a mixture of nickel carbonate with traces of alumina, magnesium carbonate, and manganese dioxide,[56] mixed heavy metal sulfides supported on alumina,[57] bauxite,[58] nickel hydroxide on cement,[58] alumina,[59] and pyrite.[60] Thiophene and several of its homologs have been prepared by this method. A mixture of products results when acetylene reacts with hydrogen sulfide in the presence of a nickel carbonate catalyst containing traces of alumina and magnesium carbonate or bauxite; the crude reaction product contains 40% of thiophene together with small amounts of methylthiophene, dimethylthiophene, and propylthiophene.[56,58] When purified illuminating gas (equivalent to methane) is combined with the acetylene-hydrogen sulfide mixture at 650–670°, a mixture of 1-methylthiophene, 2-methylthiophene, and dimethylthiophene is formed.[54] Experiments with the series of olefinic hydrocarbons, ethylene, propylene, butylene, and isoamylene, have led to the conclusion that the proportion of thiophene derivatives will be smaller as the number of carbon atoms in the olefin becomes larger. It is also found that the proportion of thiophene and carbon disulfide decreases and that of mercaptans and neutral sulfides increases as the number of carbon atoms in the initial hydrocarbons increases.[55]

The reaction temperature influences the yield to a marked extent. For example, when butadiene and hydrogen sulfide were passed over pyrite, the yields of thiophene were 8% at 500°, 22% at 550°, and 32% at 600°.[60]

Furan and pyrrole and their homologs have also been converted to thiophene derivatives. Furan reacts with hydrogen sulfide in the presence of an alumina catalyst at high temperature to give a 31%

[54] Meyer and Wesche, *Ber.*, **50**, 422 (1917).
[55] Mailhe, *Chimie & industrie*, **31**, 255 (1934).
[56] Broun, *J. Applied Chem. U.S.S.R.*, **6**, 262 (1933) [*C. A.*, **28**, 2710 (1934)].
[57] Arnold, U. S. pat. 2,336,916 [*C. A.*, **38**, 3298 (1944)].
[58] Stuer and Grob, U. S. pat. 1,421,743 [*C. A.*, **16**, 3093 (1922)].
[59] Yur'ev, *Ber.*, **69**, 440 (1936); Yur'ev and Tronova, *J. Gen. Chem. U.S.S.R.*, **10**, 31 (1940) [*C. A.*, **34**, 4733 (1940)].
[60] Schneider, Bock, and Häusser, *Ber.*, **70**, 425 (1937).

yield of thiophene.[59] Similarly, 2-methylfuran and hydrogen sulfide react at 350° to form 2-methylthiophene (11%).[61] Pyrrole and hydrogen sulfide react at 450° in the presence of the same catalyst to form thiophene.[61]

Reaction of Unsaturated Compounds with Sulfur. The reaction of hydrocarbons with sulfur at high temperature leads to the synthesis of thiophene and its alkyl and aryl substitution products. Several variations of this method exist that depend upon the nature of the hydrocarbons used. Gaseous or volatile hydrocarbons may be passed into molten sulfur in an iron pot at about 350°, and after condensation of the distillates the recovered hydrocarbons may be recycled.[62] The product is obtained by fractionation of the crude distillates.

When acetylene,[62, 63, 64] ethylene,[63] or butadiene [62] is bubbled through molten sulfur, small yields of thiophene are obtained. The yield of thiophene from acetylene is about 6%.[62]

When isoprene is passed into molten sulfur at 350°, 3-methylthiophene is formed.[62] By diluting (1:1) the isoprene with carbon disulfide and recycling, a 51% yield of 3-methylthiophene is obtained. 3,4-Dimethylthiophene is obtained similarly from dimethylbutadiene and sulfur at 400–420° (31%), and 2,3-dimethylthiophene from 3-methyl-1,3-pentadiene.[62]

A variation of this general method is the reaction of acetylene with carbon disulfide to form thiophene, when a gaseous mixture of the two compounds is passed over broken porous plate at 700°.[65] Since a higher temperature is required in this variation, the thiophene may result from the combination of acetylene with sulfur liberated by decomposition of carbon disulfide.[65] Carbon disulfide is recovered from the reaction at 200°; a trace of thiophene is formed at 350°, and the product contains about 10% thiophene (by volume) after reaction at 700°.[65]

An excellent method has been devised for the large-scale synthesis of thiophene from n-butane.[66] Sulfur and n-butane are allowed to react in the vapor phase at 450–760°; the optimum temperature is about 700°, and the optimum ratio of n-butane to sulfur is 1:1. A mixture of thiophene, butadiene, and butene is formed, and the yield of thiophene can be increased to 50% by recycling unreacted butane, butadiene, and butene. The more unsaturated the hydrocarbon, the lower is the

[61] Yur'ev, *Ber.*, **69**, 1002 (1936); *J. Gen. Chem. U.S.S.R.*, **11**, 1128 (1941) [*C. A.*, **37**, 4071 (1943)].
[62] Shepard, Henne, and Midgley, *J. Am. Chem. Soc.*, **56**, 1355 (1934).
[63] Meyer and Sandmeyer, *Ber.*, **16**, 2176 (1883).
[64] Peel and Robinson, *J. Chem. Soc.*, **1928**, 2068.
[65] Briscoe, Peel, and Robinson, *J. Chem. Soc.*, **1928**, 2857.
[66] Rasmussen, Hansford, and Sachanen, *Ind. Eng. Chem.*, **38**, 376 (1946).

temperature necessary to produce a given yield of thiophene. n-Pentane and isopentane give methylthiophenes, and all the aliphatic hexanes give methylthiophenes or ethylthiophene under the same conditions. Hydrocarbons lower than C_4 do not yield thiophene but are dehydrogenated to olefins.

The reaction of hydrocarbons with sulfur may be carried out in a sealed tube at 270–280°.[67, 68, 69] In this way, 2-octene gives a dimethyldiamylthiophene of unknown structure [67] and octane gives a diethylthiophene of unknown structure [68] in very low yields. Acetylenedicarboxylic acid as its dimethyl or diethyl ester reacts with sulfur at 150–155° in a sealed tube to form the ester of thiophenetetracarboxylic acid.[70]

The starting materials for the synthesis of aryl-substituted thiophenes by this method are relatively non-volatile, and the reaction may be carried out in a flask with a reflux condenser by heating the organic component with sulfur at elevated temperature until evolution of hydrogen sulfide ceases. The product is generally obtained from the residue by recrystallization.

A number of compounds other than hydrocarbons have been found to react with sulfur to give thiophene derivatives. However, unsaturated hydrocarbons may be transitory intermediates since the temperature of the reactions is high. Cinnamic acid reacts with sulfur at 235–240° to give a mixture of 2,5-diphenylthiophene and 2,4-diphenylthiophene;[71,72] styrene reacts with sulfur at 190–195° to give the same products.[71] 2-p-Anisyl-3,4,5-triphenylcyclopentadienone (I) reacts with sulfur at 320° in an atmosphere of carbon dioxide to give about 50% of 2-(4'-methoxyphenyl)-3,4,5-triphenylthiophene (II).[73]

$$\begin{array}{cc}
H_5C_6 \underset{\underset{O}{\overset{\|}{C}}}{\underbrace{}} C_6H_5 & H_5C_6 \underset{S}{\underbrace{}} C_6H_5 \\
H_5C_6 C_6H_4OCH_3\text{-}p & H_5C_6 C_6H_4OCH_3\text{-}p \\
I & II
\end{array}$$

Tetraphenylthiophene has been prepared by the reaction of a number of different compounds with sulfur. Some reactions were carried out in closed vessels, but most were carried out in open flasks at 200–350°.

[67] Friedmann, *Ber.*, **49**, 1551 (1916).
[68] Friedmann, *Ber.*, **49**, 1344 (1916).
[69] Baker and Reid, *J. Am. Chem. Soc.*, **51**, 1566 (1929).
[70] Michael, *Ber.*, **28**, 1633 (1895).
[71] Baumann and Fromm, *Ber.*, **28**, 890 (1895).
[72] Fromm, Fantl, and Leibsohn, *Ann.*, **457**, 267 (1927).
[73] Dilthey, Graef, Dierichs, and Josten, *J. prakt. Chem.*, **151**, 185 (1938).

TABLE III

THIOPHENES BY REACTION OF UNSATURATED COMPOUNDS WITH SULFIDES

Thiophene	Unsaturated Compound	Reagents and Experimental Conditions	Yield %	Reference
Thiophene	CH≡CH	Hydrocarbon passed over pyrite at 300°	40 (of condensate)	47, 48, 49, 51, 52, 53
	CH≡CH	Molten sulfur at 500°	12 (of condensate)	63, 64
	CH≡CH	CS_2, mixed gases passed over porous-plate catalyst at 700°	10	65
	CH≡CH	H_2S and CO_2 passed through glass tube at 640–660°	—	54
	CH≡CH	H_2S over bauxite at 320° or $Ni(OH)_2$ and cement at 300°	—	58
	CH≡CH	H_2S over heavy-metal sulfides on alumina at about 650°	—	57
	CH≡CH	H_2S over catalyst containing $NiCO_3$, Al_2O_3, $MgCO_3$, and MnO_2	40 (of condensate)	56
	CH_2=CHCH=CH_2	H_2S over pyrite at 600°	30–35	60
	CH_2=CHCH=CH_2	Molten sulfur at 320–420°	6	62
	Furan	H_2S over Al_2O_3 at high temperature	31	59
	Pyrrole	H_2S over Al_2O_3 at 450°	Low	61
	CH≡CH	Pyrite at 300°	—	51
	CH≡CH + CH_4 (coal gas)	H_2S passed through glass tube at 650–670°	—	54
2-Methyl-	2-Methylfuran	H_2S over Al_2O_3 at 350°	11	61
	CH≡CH	Pyrite at 300°	—	51
3-Methyl-	CH_2=CHC(CH_3)=CH_2	Pyrite at dull red heat, CO_2 atmosphere	2–3	49, 50
	CH_2=CHC(CH_3)=CH_2	Sulfur at 350°	40	62
2-Ethyl-	CH≡CH	Pyrite at 300°	—	51
3-Ethyl-	CH≡CH	Pyrite at 300°	—	51

THIOPHENES AND TETRAHYDROTHIOPHENES

2,3-Dimethyl-	CH=CH CH$_2$=CHC(CH$_3$)=CHCH$_3$	Pyrite at 300° Sulfur at 400–420°	— Low	51 62
2,5-Dimethyl-	2,5-Dimethylfuran	H$_2$S over Al$_2$O$_3$ at 350°	2	61
3,4-Dimethyl-	CH$_2$=C(CH$_3$)C(CH$_3$)=CH$_2$	Pyrite at dull red heat, CO$_2$ atmosphere	—	49, 50
Mixture of 2,4- and 2,5-diphenyl-	CH$_2$=C(CH$_3$)C(CH$_3$)=CH$_2$ C$_6$H$_5$CH=CHCO$_2$H C$_6$H$_5$CH=CH$_2$	Sulfur at 400–420° Sulfur at 235–240° in open flask Sulfur at 190–195° in open flask	31 — —	62 71, 72 71
2,3,4,5-Tetracarbomethoxy-	CH$_3$O$_2$CC=CCO$_2$CH$_3$	Sulfur at 150–155° in sealed tube	10	70
2,5-Di-n-amyl-3,4-dimethyl- (?)	CH$_3$CH=CH(CH$_2$)$_4$CH$_3$	Sulfur at 270–280° in sealed tube	—	67
2,3,4,5-Tetraphenyl-	(C$_6$H$_5$CH$_2$S)$_2$ (C$_6$H$_5$CH$_2$)$_2$S (C$_6$H$_5$CHS)$_3$ or high polymeric thiobenzaldehyde	Pyrolysis Pyrolysis at 360–460° Pyrolysis	— — —	81, 82, 84 82, 83, 84 88
	(C$_6$H$_5$CO)$_2$S (C$_6$H$_5$COS)$_2$ C$_6$H$_5$COSH	Pyrolysis Pyrolysis Pyrolysis	— — —	85, 86 85, 86, 87 85, 86
	C$_6$H$_5$C=NHC$_6$H$_5$ S C$_6$H$_5$CH$_2$SO$_3$Na C$_2$H$_5$OCS$_2$COC$_6$H$_5$ C$_6$H$_5$CH$_3$ C$_6$H$_5$CH=CHC$_6$H$_5$ C$_6$H$_5$CH=C(C$_6$H$_5$)C(C$_6$H$_5$)=CHC$_6$H$_5$ C$_6$H$_5$CH$_2$CH$_2$C$_6$H$_5$ C$_6$H$_5$CH$_2$OH C$_6$H$_5$CH$_2$OCH$_2$C$_6$H$_5$ C$_6$H$_5$CH$_2$CO$_2$H C$_6$H$_5$CH$_2$COC$_6$H$_5$ 2,3,4,5-Tetraphenylcyclopentadienone 2-p-Anisyl-3,4,5-triphenylcyclopentadienone	Pyrolysis at 270–310° Pyrolysis Pyrolysis in vacuum Sulfur at 250–300° in sealed tube Sulfur at 200° in open flask Sulfur at 250° in open flask Sulfur at 260° Sulfur above 180° Sulfur below 200° Sulfur at 260° Sulfur at elevated temperature Sulfur at 350° Sulfur at 320°	— — — 60–70 56 — — — — — — 70 50	89 90 87 76 77 74 75 75 75 78 78 79, 80 73
2-p-Anisyl-3,4,5-triphenyl-	⬡(CHS)(OCH$_3$)$_n$	Pyrolysis at 250–260°	—	91
2,3,4,5-Tetra-o-anisyl-				
2,3,4,5-Tetra-p-anisyl-	p-CH$_3$OC$_6$H$_4$CH=CHC$_6$H$_4$OCH$_3$-p	Sulfur at 230°	—	71

Those compounds which have been found to react with sulfur to form tetraphenylthiophene, and the yields of this product when reported, are as follows: tetraphenylbutadiene (56%),[74] diphenylethane,[75] benzyl alcohol,[75] benzyl ether,[75] toluene,[76] stilbene (60–70%),[77] phenylacetic acid,[78] desoxybenzoin,[78] and tetraphenylcyclopentadienone (70%).[79, 80] Tetra-*p*-anisylthiophene is obtained similarly from 4,4'-dimethoxystilbene.[71]

Tetraphenylthiophene has also been prepared by the pyrolysis of a number of sulfur-containing compounds. These reactions have not been shown to be generally applicable to the preparation of other thiophenes. When either benzyl sulfide or benzyl disulfide is pyrolyzed at 360–460°, a distillate containing tetraphenylthiophene is obtained.[81–84] It has been suggested that benzyl sulfide first forms stilbene, hydrogen sulfide, and sulfur, which are known pyrolysis products, and that the sulfur and hydrogen sulfide in turn react with stilbene to give tetraphenylthiophene and toluene.[84] The distillation of benzoyl sulfide, benzoyl disulfide, or thiobenzoic acid gives tetraphenylthiophene.[85, 86, 87] Trithiobenzaldehyde or high polymeric thiobenzaldehyde has been pyrolyzed to tetraphenylthiophene.[88]

Pyrolysis of thiobenzanilide at 270–310° gives a small yield of tetraphenylthiophene.[89] Sodium α-toluenesulfonate on dry distillation at high temperatures gives tetraphenylthiophene in addition to benzoic acid, stilbene, and sulfur.[90]

Pyrolysis of polymeric thiosalicylaldehyde methyl ether at 250–260° gives tetra-(2-methoxyphenyl)thiophene.[91]

[74] Smith and Hoehn, *J. Am. Chem. Soc.*, **63**, 1184 (1941).
[75] Szperl and Wierusz-Kowalski, *Chem. Polski*, **15**, 19, 23, 28 (1917) [*J. Chem. Soc.*, **114**(1), 492 (1918)].
[76] Aronstein and Van Nierop, *Rec. trav. chim.*, **21**, 448 (1902).
[77] Baumann and Klett, *Ber.*, **24**, 3307 (1891).
[78] Ziegler, *Ber.*, **23**, 2472 (1890).
[79] Dilthey, Schommer, Höschen, and Dierichs, *Ber.*, **68**, 1159 (1935).
[80] Dilthey, Ger. pat. 628,954 [*C. A.*, **30**, 6009 (1936)].
[81] Laurent, *Ann.*, **52**, 348 (1844).
[82] Märcker, *Ann.*, **136**, 75 (1865).
[83] Forst, *Ann.*, **178**, 370 (1875).
[84] Fromm and Achert, *Ber.*, **36**, 534 (1903).
[85] Fromm and Schmoldt, *Ber.*, **40**, 2861 (1907).
[86] Fromm and Klinger, *Ann.*, **394**, 342 (1912).
[87] Bulmer and Mann, *J. Chem. Soc.*, **1945**, 677.
[88] Baumann and Fromm, *Ber.*, **24**, 1441 (1891).
[89] Chapman, *J. Chem. Soc.*, **1928**, 1894.
[90] Fromm and de Seixas Palma, *Ber.*, **39**, 3308 (1906).
[91] Kopp, *Ber.*, **25**, 600 (1892).

Thiophenes by Reaction of 1,2-Difunctional Compounds with Thiodiacetic Acid Esters

The 1,2-difunctional compounds that have been found to react with esters of thiodiacetic acid to give thiophenes are divided into α-diketones, α-keto esters, and oxalic esters for discussion in this section. This discussion is followed by a description of the formation of thiophene derivatives by decarboxylation of 2,5-thiophenedicarboxylic acids resulting from syntheses with esters of thiodiacetic acid.

In carrying out these reactions, diethyl thiodiacetate and the equivalent weight of the 1,2-difunctional compound are usually mixed and added to an ethanolic solution of a sodium alkoxide. The reaction mixtures are generally allowed to stand at room temperature or in a refrigerator for several days, but they may be heated finally to reflux temperature.[92] When the reactions are complete, the mixtures are poured into water, the ethanol is evaporated, and the 2,5-thiophenedicarboxylic acid esters are saponified. Acidification of the solutions with mineral acid liberates the 2,5-thiophenedicarboxylic acids. If the esters of the 2,5-thiophenedicarboxylic acids are desired, the reaction mixtures are poured into water, and after acidification the esters are extracted immediately with chloroform.[93] Special interest is attached to this method, because many thiophenecarboxylic acid esters may be hydrolyzed to the free acids, which can then be decarboxylated by pyrolysis to give 3,4-disubstituted thiophenes.

Syntheses from α-Diketones. This synthesis of thiophenes from α-diketones, introduced by Hinsberg,[94] is typified by the reaction of α-diketones (I) with thiodiacetic acid esters (II) to give substituted thiophenes (III). Thiophenes with a variety of alkyl and aryl groups in the 3 and 4 positions have been synthesized by this method. There

$$
\begin{array}{c}
\text{R'COCOR''} \\
\text{I} \\
+ \\
RO_2CCH_2 \quad CH_2CO_2R \\
\diagdown \quad \diagup \\
S \\
\text{II}
\end{array}
\quad \longrightarrow \quad
\begin{array}{c}
R' \quad\quad R'' \\
RO_2C \overset{\displaystyle\diagup\!\!\!\diagdown}{\underset{S}{}} CO_2R \\
\text{III}
\end{array}
$$

are few data in the literature on the yields, so that no generalizations can be made about the effect of substituents on the course of the synthesis.

[92] Fager, *J. Am. Chem. Soc.*, **67**, 2217 (1945).
[93] Hinsberg, *Ber.*, **45**, 2413 (1912).
[94] Hinsberg, *Ber.*, **43**, 901 (1910).

α-Diketones react with the methyl or ethyl esters of thiodiacetic acid in presence of sodium alkoxide. Glyoxal (IV), considered here with the α-diketones, and diethyl thiodiacetate (V) react in ethanol in presence of sodium ethoxide at room temperature for five days to give diethyl 2,5-thiophenedicarboxylate (VI), which is saponified and isolated as the free acid.[93] Both alkyl and aryl diketones react similarly with diethyl thiodiacetate in the presence of sodium ethoxide. Diacetyl and ethyl

$$\text{CHOCHO} + \text{C}_2\text{H}_5\text{O}_2\text{CCH}_2\text{SCH}_2\text{CO}_2\text{C}_2\text{H}_5 \rightarrow \underset{\text{S}}{\text{H}_5\text{C}_2\text{O}_2\text{C}}\diagdown\!\!\!\diagup\text{CO}_2\text{C}_2\text{H}_5$$

IV V VI

thiodiacetate yield 3,4-dimethyl-2,5-thiophenedicarboxylic acid diethyl ester, which is hydrolyzed without isolation to give 3,4-dimethyl-2,5-thiophenedicarboxylic acid (VII).[95] 1-Phenyl-1,2-propanedione, benzil, p-tolil, and furil react under similar conditions with diethyl thiodiacetate to yield, after hydrolysis, 3-methyl-4-phenyl-2,5-thiophenedicarboxylic acid (VIII),[96] 3,4-diphenyl-2,5-thiophenedicarboxylic acid (IX) (74%),[94, 97, 98, 99] 3,4-di-(p-tolyl)-2,5-thiophenedicarboxylic acid (X) (74%),[96] and 3,4-di(2-furyl)-2,5-thiophenedicarboxylic acid (XI),[96] respectively.

$$\underset{\text{S}}{\text{HO}_2\text{C}}\diagdown\!\!\!\diagup\text{CO}_2\text{H}\text{ with substituents R', R}$$

VII, R = R' = CH$_3$
VIII, R = CH$_3$, R' = C$_6$H$_5$
IX, R = R' = C$_6$H$_5$

X, R = R' = p-CH$_3$C$_6$H$_4$
XI, R = R' = (2-furyl)

Syntheses from α-Keto Esters. α-Keto esters (XII) react with esters of thiodiacetic acid (II) to give 3-hydroxy-2,5-thiophenedicarboxylic acid esters (XIII). An example of this method is the reaction of ethyl

$$\text{R'COCO}_2\text{R}$$
XII
$$+$$
$$\underset{\text{S}}{\text{RO}_2\text{CCH}_2\diagdown\!\!\!\diagup\text{CH}_2\text{CO}_2\text{R}}$$
II

$$\rightarrow \underset{\text{S}}{\text{RO}_2\text{C}}\diagdown\!\!\!\diagup\overset{\text{R'}\quad\text{OH}}{\text{CO}_2\text{R}}$$
XIII

pyruvate (XIV) with diethyl thiodiacetate (V) to form 2-carbethoxy-3-hydroxy-4-methyl-5-thiophenecarboxylic acid (XV), one of the ester

[95] Seka, *Ber.*, **58**, 1783 (1925).
[96] Backer and Stevens, *Rec. trav. chim.*, **59**, 899 (1940).
[97] Hinsberg, *Ber.*, **48**, 1611 (1915).
[98] Steinkopf, *Ann.*, **424**, 23 (1921).
[99] Backer and Stevens, *Rec. trav. chim.*, **59**, 423 (1940).

groups being hydrolyzed during the reaction.[94] Similarly, ethyl mesoxalate reacts with diethyl thiodiacetate to form 3-hydroxy-2,4,5-thiophenetricarboxylic acid triethyl ester (XVI).[93]

$$CH_3COCO_2C_2H_5 + C_2H_5O_2CCH_2SCH_2CO_2C_2H_5 \rightarrow$$

XIV   V   XV, R = H, R' = CH$_3$
                                       XVI, R = C$_2$H$_5$, R' = CO$_2$C$_2$H$_5$

Syntheses from Oxalic Esters. Diethyl oxalate reacts similarly with dimethyl thiodiacetate to form, after hydrolysis of the ester groups, 3,4-dihydroxy-2,5-thiophenedicarboxylic acid (XVII).[92, 94] When the dihydroxythiophene XVII is treated with dimethyl sulfate, 3,4-dimeth-

XVII

oxy-2,5-thiophenedicarboxylic acid is obtained in 59% yield.[92]

Decarboxylation of 2,5-Thiophenedicarboxylic Acids. 2,5-Thiophenedicarboxylic acid esters are readily hydrolyzed by 10% sodium hydroxide solution. The free acids are stable when the 3 and 4 positions of the thiophene nucleus bear hydrogen atoms or alkyl or aryl groups. Decarboxylation of the acids can be accomplished by pyrolysis at 300° or higher,[94, 97, 99] or by heating the disodium salts of the acids with calcium hydroxide in vacuum.[96]

3,4-Diphenylthiophene (65%) and 3,4-di(p-tolyl)thiophene (83%) are obtained by pyrolysis of the corresponding 2,5-thiophenedicarboxylic acids at 300–360°,[94, 96, 97, 99] 3,4-di(2-furyl)thiophene by pyrolysis of the disodium salt of the dicarboxylic acid,[96] and 3,4-dimethoxythiophene (58%) by heating 3,4-dimethoxy-2,5-thiophenedicarboxylic acid with copper chromite in quinoline solution in a nitrogen atmosphere for thirty minutes at 180°.[92]

When one or both of the 3 and 4 positions of the thiophene nucleus is substituted by a hydroxyl group, hydrolysis of the 2,5-thiophenedicarboxylic acid esters to the dicarboxylic acids is not always possible: 2-carbethoxy-3-hydroxy-4-methyl-5-thiophenecarboxylic acid (XV) on hydrolysis in dilute alkali undergoes partial decarboxylation to form 3-hydroxy-4-methyl-5-thiophenecarboxylic acid (XVIII).[94]

XV   XVIII

Experimental Procedures

Dimethyl 3,4-Dihydroxy-2,5-thiophenedicarboxylate.[94] A mixture of 10 g. of dimethyl thiodiacetate and 8 g. of ethyl oxalate is added to a solution of 4 g. of sodium in 80–100 ml. of methanol. The mixture is shaken during the addition. A yellow precipitate forms immediately. After several days' standing, the reaction mixture is poured into water and the solution is cooled and acidified slowly with hydrochloric acid. The precipitated ester is collected on a filter and washed with water. It is purified by recrystallization from water and melts at 178°.

3,4-Diphenyl-2,5-thiophenedicarboxylic Acid.[99] A solution of 42 g. (0.2 mole) of benzil and 41.2 g. (0.2 mole) of diethyl thiodiacetate in 400 ml. of methanol is added to a solution of 16 g. of sodium in 250 ml. of methanol. After standing for three days, the reaction mixture is diluted with 1 l. of water and the alcohol is distilled at reduced pressure. The residual aqueous solution is acidified with hydrochloric acid. The crystalline precipitate is collected on a filter and washed with water. It is dissolved in ethanol containing 20% of water, and the solution is treated with a small quantity of decolorizing carbon. 3,4-Diphenyl-2,5-thiophenedicarboxylic acid is deposited in small crystals; the yield is about 48 g. (74%). A second recrystallization may be necessary to obtain pure material melting at 341° (dec.).

TABLE IV

Thiophenes by Reaction of 1,2-Difunctional Compounds with Thiodiacetic Acid Esters

Thiophene	Reactants: Diethyl Thiodiacetate and	Reagents and Experimental Conditions	Yield %	Reference
2,5-Dicarboxy-	CHOCHO	$NaOC_2H_5$ in ethanol at 5°	—	93
2,5-Dicarboxy-3,4-dimethyl-	$CH_3COCOCH_3$	$NaOC_2H_5$ in ethanol at 0°	—	95
2,5-Dicarboxy-3,4-dimethoxy-	$(CO_2CH_3)_2$*	$NaOCH_3$ in methanol at 5° followed by methylation with $(CH_3)_2SO_4$	59	92
2,5-Dicarbomethoxy-3,4-dihydroxy-	$(CO_2C_2H_5)_2$*	$NaOCH_3$ in methanol	—	92, 94
2-Carboxy-3-methyl-4-hydroxy-5-carbomethoxy-	$CH_3COCO_2C_2H_5$	$NaOCH_3$ in methanol	—	94
2,5-Dicarbethoxy-3,4-dihydroxy-	$(CO_2C_2H_5)_2$	$NaOC_2H_5$ in ethanol	—	94
2,5-Dicarboxy-3-methyl-4-phenyl-	$CH_3COCOC_6H_5$	$NaOC_2H_5$ in ethanol	—	96
2,3,5-Tricarbethoxy-4-hydroxy-	$CO(CO_2C_2H_5)_2$	$NaOC_2H_5$ in ethanol at 0°	—	93
2,5-Dicarboxy-3,4-di(2'-furyl)-	Furil	$NaOCH_3$ in methanol	—	96
2,5-Dicarboxy-3,4-diphenyl-	$C_6H_5COCOC_6H_5$	$NaOCH_3$ or $NaOC_2H_5$ in alcohol	74 31	99 94, 97, 98
2,5-Dicarboxy-3,4-di-*p*-tolyl-	*p*-$CH_3C_6H_4COCOC_6H_4CH_3$-*p*	$NaOCH_3$ in methanol	74	96

* Dimethyl thiodiacetate was used in this experiment.

Thiophenes by Reaction of Aryl Methyl Ketones with Sulfides

In the Willgerodt reaction,[100] a ketone is heated with ammonium polysulfide; when aryl methyl ketones are employed thiophenes are obtained. The reaction of acetophenone with ammonium sulfide at 215° for six hours in an autoclave gave a mixture containing thiophenes (20%), phenylacetamide (25%), phenylacetic acid (6%), and ethylbenzene (8%). The thiophene fraction was separated by fractional crystallization into 2,4-diphenylthiophene and 2,5-diphenylthiophene.[101,102] In a similar manner, a mixture of 2,4-di-p-tolylthiophene and 2,5-di-p-tolylthiophene was prepared from methyl p-tolyl ketone in about 20% yield.[102,103] This method has been improved and modified for the preparation of 2,4-diphenylthiophene.[104] By heating acetophenone anil and powdered roll sulfur at 220–240° for thirteen hours, 2,4-diphenylthiophene is formed in 28% yield. The anils acetophenone o-tolil and acetophenone p-tolil under the same conditions give 2,4-di-(o-tolyl)- and 2,4-di-(p-tolyl)-thiophene in yields of 24% and 32%, respectively.[104] Extension of the method to the anil of propiophenone gives 3,5-dimethyl-2,4-diphenylthiophene.[105] 3,5-Diethyl-2,4-diphenylthiophene was reported as the product from n-butyrophenone anil, but the identification was incomplete.[105]

In the preparation of thioacetophenone by the reaction of acetophenone with hydrogen sulfide, a disulfide, $C_{24}H_{22}S_2$, was isolated as a by-product.[106] Pyrolysis of this "anhydroacetophenone disulfide" gave 2,4-diphenylthiophene. Of the two formulas suggested for this disulfide, I was considered more probable than II.[106]

$$
\begin{array}{cc}
\text{H}_3\text{C} \quad \text{S}—\text{CC}_6\text{H}_5 \\
\diagdown \; | \quad \| \\
\quad \text{C} \quad \text{CH} \\
\diagup \; | \quad | \\
\text{H}_5\text{C}_6 \quad \text{S}—\text{CC}_6\text{H}_5 \\
\quad \quad \quad | \\
\quad \quad \quad \text{CH}_3 \\
\text{I}
\end{array}
\qquad
\begin{array}{c}
\text{H}_5\text{C}_6 \quad \text{S} \quad \text{C}_6\text{H}_5 \\
\diagdown \; \diagup \diagdown \\
\text{C} \quad \quad \text{C} \\
\diagup \; \diagdown \diagup \diagdown \\
\text{H}_3\text{C} \quad \text{S} \quad \text{CH}=\text{CC}_6\text{H}_5 \\
\quad \quad \quad \quad \quad | \\
\quad \quad \quad \quad \quad \text{CH}_3 \\
\text{II}
\end{array}
$$

Subsequently, a reinvestigation of this work led to the conclusion that the two reactions represented in the accompanying equations are

---

[100] *Organic Reactions*, **3**, 83, John Wiley & Sons, New York, 1946.
[101] Willgerodt and Merk, *J. prakt. Chem.*, (2), **80**, 192 (1909).
[102] Willgerodt and Scholtz, *J. prakt. Chem.*, (2), **81**, 382 (1910).
[103] Willgerodt and Hambrecht, *J. prakt. Chem.*, (2), **81**, 74 (1910).
[104] Bogert and Herrera, *J. Am. Chem. Soc.*, **45**, 238 (1923).
[105] Bogert and Andersen, *J. Am. Chem. Soc.*, **48**, 223 (1926).
[106] Baumann and Fromm, *Ber.*, **28**, 895 (1895).

involved in the formation of 2,4-diphenylthiophene from "anhydroacetophenone disulfide."[107] By placing a hydrogen acceptor, copper chromium oxide catalyst, in the mixture the yield was increased to

$$C_6H_5COCH_3 \xrightarrow{H_2S} C_{24}H_{22}S_2$$

$$\downarrow$$

$$\underset{S}{\overset{\|}{C_6H_5CCH_3}} + \underset{CH_2}{\overset{\|}{C_6H_5C}}\!\!-\!\!\!-\!\!\!-\!\!\underset{CC_6H_5}{\overset{\|}{CH}}$$

$$HS$$

$$\downarrow$$

[2,4-diphenylthiophene structure] + 2H

83%.[107] Since "anhydroacetophenone disulfide" can be prepared in 57% yield by passing hydrogen chloride and hydrogen sulfide into an ethanolic solution of acetophenone, the overall yield of 2,4-diphenylthiophene is 47%. By a similar method, 2,4-bis(p-methoxyphenyl)-3,5-dimethylthiophene can be prepared from p-methoxypropiophenone in 35% yield.[107]

TABLE V

THIOPHENES BY REACTION OF ARYL METHYL KETONES WITH SULFIDES OR ARYL ALKYL KETONE ANILS WITH SULFUR

Thiophene	Starting Material	Reagents and Experimental Conditions	Yield %	Reference
2,4-Diphenyl-	$C_6H_5C(CH_3)\!=\!NC_6H_5$	Sulfur at 220–240°	28	104
	$C_6H_5C(CH_3)\!=\!NC_6H_4CH_3\text{-}o$	Sulfur at 220–240°	23.6	104
	$C_6H_5C(CH_3)\!=\!NC_6H_4CH_3\text{-}p$	Sulfur at 220–240°	32.4	104
	$C_6H_5COCH_3$	H_2S + HCl in absolute ethanol at 0° followed by refluxing with copper chromium oxide catalyst in xylene	47	106, 107
Mixture of 2,4- and 2,5-diphenyl-	$C_6H_5COCH_3$	$(NH_4)_2S$ at 215° in autoclave	20	101, 102
Mixture of 2,4- and 2,5-di-p-tolyl-	$p\text{-}CH_3C_6H_4COCH_3$	$(NH_4)_2S$ at 215° in autoclave	20	102, 103
2,4-Dimethyl-3,5-diphenyl-	$C_6H_5C(C_2H_5)\!=\!NC_6H_5$	Sulfur at 240°	—	105
2,4-Diethyl-3,5-diphenyl-	$C_6H_5C(C_3H_{7}\text{-}n)\!=\!NC_6H_5$	Sulfur at 200–220°	—	105
2,4-Dimethyl-3,5-di-p-anisyl-	$p\text{-}CH_3OC_6H_4COC_2H_5$	H_2S + HCl in absolute ethanol at 0° followed by refluxing with copper chromium oxide catalyst	35	107

[107] Campaigne, *J. Am. Chem. Soc.*, **66**, 684 (1944).

That the reaction actually involves the two steps outlined above is indicated by the results of an experiment in which a solution of "anhydro-p-methoxypropiophenone disulfide" in xylene was refluxed for three hours. The solution, which became deep purple, was evaporated at reduced pressure, and the residual brown oil was dissolved in ethanol. Storage of the cooled solution did not yield a crystalline product. However, when the ethanolic solution was refluxed with added copper chromium oxide catalyst for two hours, 2,4-bis(p-methoxyphenyl)-3,5-dimethylthiophene was obtained.[107]

Thiophenes by Miscellaneous Cyclization Reactions

Hydroxythiophene derivatives have been prepared by cyclization reactions which have not been extensively studied. One method involves the condensation of an α-halogenated fatty ester I with the sodio derivative of a β-mercaptocrotonic ester II, followed by a Dieckmann cyclization of the condensation product III to give the 3-hydroxythiophene IV.[27]

$$\begin{array}{c} CO_2C_2H_5 \\ | \\ RCHCl \end{array} + \begin{array}{c} CH_3 \quad R' \\ \diagdown \diagup \\ C\!=\!C \\ \diagup \diagdown \\ NaS \quad CO_2C_2H_5 \end{array} \rightarrow$$

I, II

$$\begin{array}{c} C_2H_5OCO \quad CH_3 \quad R' \\ | \quad\quad | \quad\diagup \\ RCH \quad\quad C\!=\!C \\ \diagdown \diagup \quad\quad \diagdown \\ S \quad\quad\quad CO_2C_2H_5 \end{array} \rightarrow \begin{array}{c} HO \\ \diagup\!\!\!\diagdown \\ R\diagdown\!\!\!\diagup CHCO_2C_2H_5 \\ S \quad | \\ \quad R' \end{array}$$

III, IV

By this method ethyl 3-hydroxythiophene-5-acetate (V) is obtained from ethyl β-carbethoxymethylthiocrotonate (VI), ethyl 3-hydroxythiophene-5-α-propionate (VII) from ethyl β-carbethoxymethylthio-α-methylcrotonate (VIII), and ethyl 3-hydroxy-2-methylthiophene-5-

$$\begin{array}{c} C_2H_5OCO \quad CH_3 \\ | \quad\quad\quad | \\ CH_2 \quad C\!=\!CCO_2C_2H_5 \\ \diagdown \diagup \quad\quad | \\ S \quad\quad R \end{array} \rightarrow \begin{array}{c} HO \\ \diagup\!\!\!\diagdown \\ \diagdown\!\!\!\diagup CHCO_2C_2H_5 \\ S \quad | \\ \quad R \end{array}$$

VI, R = H
VIII, R = CH₃

V, R = H
VII, R = CH₃

acetate (IX) from either ethyl β-(α'-carbethoxyethylthio)crotonate (X) or ethyl α-(α'-carbethoxyethylthio)ethylidenemalonate (XI).[27]

$$\underset{X}{\underset{\diagdown S \diagup}{CH_3\overset{C_2H_5OCO}{\overset{|}{CH}}\quad \overset{CH_3}{\overset{|}{C}}=CHCO_2C_2H_5}} \qquad \underset{XI}{\underset{\diagdown S \diagup}{CH_3\overset{C_2H_5OCO}{\overset{|}{CH}}\quad \overset{CH_3}{\overset{|}{C}}=C(CO_2C_2H_5)_2}}$$

$$\searrow \qquad \swarrow$$

$$\underset{IX}{\underset{S}{\overset{HO}{\underset{H_3C}{\diagup}}\overset{}{\diagdown}} CH_2CO_2C_2H_5}$$

3-Hydroxy-5-phenylthiophene (XII) has been prepared by heating the carboxymethyl ester of β-phenyl-β-(carboxymethylthio)thioacrylic acid (XIII) with a mixture of sodium acetate and acetic anhydride until the evolution of carbon dioxide was complete. Decomposition of the reaction mixture with water yielded the intermediate 3-acetoxy-5-

$$\underset{XIII}{\underset{\diagdown S \diagup}{C_6H_5\overset{CH—COSCH_2CO_2H}{\overset{\|}{C}}\quad CH_2CO_2H}} \rightarrow \underset{XIV}{\underset{S}{H_5C_6 \diagdown \diagup OCOCH_3}} \rightarrow \underset{XII}{\underset{S}{H_5C_6 \diagdown \diagup OH}}$$

phenylthiophene (XIV) which was hydrolyzed by either acid or alkali to 3-hydroxy-5-phenylthiophene (XII).[108]

2,4-Dihydroxythiophenes (thiotetronic acids) are prepared by reactions somewhat similar to those described above. When α-(acetylthioglycolyl)acetoacetic ester (XV) is treated with alkali, it cyclizes by transesterification to 3-acetyl-2,4-dihydroxythiophene (XVI) or α-acet-

$$\underset{XV}{\underset{\diagdown SCOCH_3}{\overset{CO—CHCOCH_3}{\overset{|}{CH_2}\quad \overset{|}{CO_2C_2H_5}}}} \rightarrow \underset{XVI}{\underset{S}{HO \diagdown \diagup COCH_3 \atop OH}} \quad or \quad \underset{XVII}{\underset{S}{HO \diagdown \diagup COCH_3 \atop =O}}$$

$$\underset{XVIII}{\underset{S}{HO \diagdown \diagup CO_2C_2H_5 \atop =O}} \rightarrow \underset{XIX}{\underset{S}{HO \diagdown \diagup \atop =O}}$$

[108] Friedländer and St. Kielbasinski, Ber., 45, 3389 (1912).

ylthiotetronic acid (XVII).[109] Acetylthioglycolylmalonic ester cyclizes similarly to ethyl 2,4-dihydroxy-3-thiophenecarboxylate or α-carbethoxythiotetronic acid (XVIII), which can be hydrolyzed and decarboxylated to thiotetronic acid (XIX).[109]

TABLE VI

THIOPHENES BY MISCELLANEOUS CYCLIZATION REACTIONS

Thiophene	Starting Material	Experimental Conditions	Yield %	Reference
2,4-Dihydroxy-3-acetyl-	$CH_3COSCH_2COCH(COCH_3)CO_2C_2H_5$	NaOH, dilute solution	—	109
2,4-Dihydroxy-3-carbethoxy-	$CH_3COSCH_2COCH(CO_2C_2H_5)_2$	NaOH, dilute solution	—	109
HO—[ring]—$CH_2CO_2C_2H_5$, S	$CH_3C{=}CHCO_2C_2H_5$ $\,\,\,\|$ $SCH_2CO_2C_2H_5$	Na in dry benzene	—	27
HO—[ring]—$CH(CH_3)CO_2C_2H_5$, S	$CH_3C{=}C(CH_3)CO_2C_2H_5$ $\,\,\,\|$ $SCH_2CO_2C_2H_5$	Na in dry benzene	—	27
HO—[ring]—$CH_2CO_2C_2H_5$, H_3C, S	$CH_3C{=}CHCO_2C_2H_5$ $\,\,\,\|$ $SCH(CH_3)CO_2C_2H_5$ or $CH_3C{=}C(CO_2C_2H_5)_2$ $\,\,\,\|$ $SCH(CH_3)CO_2C_2H_5$	Na in dry benzene	—	27
2-Phenyl-4-acetoxy-	$HO_2CCH_2SC(C_6H_5){=}CHCOSCH_2CO_2H$	$CH_3CO_2Na +$ $(CH_3CO)_2O$, at 100°	—	108

PREPARATION OF TETRAHYDROTHIOPHENES

Tetrahydrothiophenes from 1,4-Difunctional Compounds and Sulfides

Reaction of 1,4-Dihalides with Sulfides. The preparation of tetrahydrothiophenes by the general reaction of 1,4-difunctional compounds with alkali metal sulfides is typified by the preparation of tetrahydrothiophene (I) in nearly quantitative yield by the reaction of either diiodo- or dibromo-butane with potassium sulfide.[110,111,112] The reaction

$$BrCH_2CH_2CH_2CH_2Br \text{ or } ICH_2CH_2CH_2CH_2I \rightarrow \underset{I}{\underset{S}{\underset{|\quad\quad|}{CH_2-CH_2}}\atop{CH_2\quad CH_2}}$$

[109] Benary, *Ber.*, **46**, 2103 (1913).
[110] von Braun and Trümpler, *Ber.*, **43**, 545 (1910).
[111] Bost and Conn, *Oil and Gas J.*, **32**, 17 (1933).
[112] Grishkevich-Trokhimovskii, *J. Russ. Phys. Chem. Soc.*, **48**, 901 (1916) [*C. A.*, **11**, 785 (1917)].

of a 1,4-dihalide with a sulfide is generally carried out in aqueous or alcoholic solution. Tetrahydrothiophenes with a variety of substituent groups, including alkyl, aryl, hydroxyl, keto, and carboxyl, have been prepared by this general reaction.

The alkyl-substituted tetrahydrothiophenes that can be made by this reaction include 2-methyltetrahydrothiophene from 1,4-diiodopentane or 1,4-dibromopentane by reaction with either sodium sulfide or potassium sulfide,[112, 113] 3-methyltetrahydrothiophene from 1,4-dibromo-2-methylbutane,[112] and meso-2,5-dimethyltetrahydrothiophene from 2,5-dibromohexane.[112] The higher alkyl dihalides are also used satisfactorily; both 2,5- and 3,4-di-n-propyltetrahydrothiophene are prepared in 77% yield from 4,7-dibromodecane and 1,4-dibromo-2,3-di-n-propylbutane, respectively.[114]

3,4-Dihydroxytetrahydrothiophene (II) is prepared in 51% yield from 1,4-dichloro-2,3-dihydroxybutane by reaction with sodium sulfide.[115] 3,4-Dichloro- and 3,4-dibromo-tetrahydrothiophene (III) may be made by the action of hydrochloric and hydrobromic acids on the dihydroxy derivative II in yields of 32% and 25%, respectively.[115]

$$\begin{array}{cc} \text{HOCH}\text{—}\text{CHOH} & (\text{Br})\text{ClCH}\text{—}\text{CHCl}(\text{Br}) \\ |\quad\quad | & |\quad\quad | \\ \text{CH}_2\quad \text{CH}_2 \to & \text{CH}_2\quad \text{CH}_2 \\ \diagdown\diagup & \diagdown\diagup \\ \text{S} & \text{S} \\ \text{II} & \text{III} \end{array}$$

3,4-Diethoxytetrahydrothiophene (IV) is prepared by refluxing an ethanol solution of meso-2,3-diethoxy-1,4-diiodobutane and potassium sulfide.[116]

$$\begin{array}{cc} \text{C}_2\text{H}_5\text{OCH}\text{—}\text{CHOC}_2\text{H}_5 & \text{CH}_2\text{—}\text{CO} \\ |\quad\quad | & |\quad\quad | \\ \text{CH}_2\quad \text{CH}_2 & \text{CH}_2\quad \text{CH}_2 \\ \diagdown\diagup & \diagdown\diagup \\ \text{S} & \text{S} \\ \text{IV} & \text{V} \end{array}$$

3-Ketotetrahydrothiophene (V) is made in 22% yield from α-chloromethyl β-iodoethyl ketone.[117]

Both dl- and meso-tetrahydrothiophene-2,5-dicarboxylic acids (VI) are prepared from the corresponding dl- and meso-dibromoadipic acids by reaction with sodium sulfide in about 90% yields.[118]

[113] von Braun, Ber., **43**, 3220 (1910).
[114] Marvel and Williams, J. Am. Chem. Soc., **61**, 2714 (1939).
[115] Kilmer, Armstrong, Brown, and du Vigneaud, J. Biol. Chem., **145**, 495 (1942).
[116] Patterson and Karabinos, U. S. pat. 2,400,436 [C. A., **40**, 4484 (1946)].
[117] Karrer and Schmid, Helv. Chim. Acta, **27**, 116 (1944).
[118] Fredga, J. prakt. Chem., **150**, 124 (1938).

$$\underset{\text{VI}}{\begin{array}{c}\text{CH}_2\text{—CH}_2\\|\quad\quad|\\\text{HO}_2\text{CCH}\quad\text{CHCO}_2\text{H}\\\diagdown\text{S}\diagup\end{array}}\qquad\underset{\text{VII}}{\begin{array}{c}\text{CH——CH}_2\\\diagdown\text{O}\quad\overset{\text{O}}{\underset{\|}{\text{C}}}\\\text{CH}_2\quad\text{CH}\\\diagdown\text{S}\diagup\end{array}}$$

The lactone of 4-hydroxytetrahydrothiophene-2-carboxylic acid (VII) is obtained from α-bromo-δ-chloro-γ-valerolactone by treatment first with potassium iodide to replace the halogens with iodine and then with sodium sulfide.[119]

Ethyl 4-keto-2-phenyltetrahydrothiophene-3-carboxylate (VIII) is the product (67%) of the reaction between ethyl α-benzylidene-γ-chloroacetoacetate (IX) and an ethanolic solution containing sodium ethoxide and saturated with hydrogen sulfide.[120]

$$\underset{\text{IX}}{\begin{array}{c}\text{OC——CCO}_2\text{C}_2\text{H}_5\\|\quad\quad\|\\\text{CH}_2\text{Cl}\quad\text{CHC}_6\text{H}_5\end{array}}\quad\rightarrow\quad\underset{\text{VIII}}{\begin{array}{c}\text{OC——CHCO}_2\text{C}_2\text{H}_5\\|\quad\quad|\\\text{CH}_2\quad\text{CHC}_6\text{H}_5\\\diagdown\text{S}\diagup\end{array}}$$

2,5-Diketotetrahydrothiophene (X), thiosuccinic anhydride, is obtained from succinyl chloride by treatment with sodium sulfide.[46]

$$\underset{\text{X}}{\begin{array}{c}\text{CH}_2\text{—CH}_2\\|\quad\quad|\\\text{OC}\quad\quad\text{CO}\\\diagdown\text{S}\diagup\end{array}}$$

Reaction of a 1,4-Disulfuric Acid Ester with Hydrogen Sulfide. A 1,4-disulfuric acid ester has been used instead of a 1,4-dihalide in one synthesis. This variation is the preparation of 3,4-diaminotetrahydrothiophene (XI) in 25% yield from 2,3-diaminobutane-1,4-disulfuric acid ester (XII).[121]

$$\underset{\text{XII}}{\begin{array}{c}\overset{+}{\text{H}_3}\text{NCH——CHNH}_3^+\\|\quad\quad|\\\text{CH}_2\text{OSO}_3^-\ \text{CH}_2\text{OSO}_3^-\end{array}}\quad\rightarrow\quad\underset{\text{XI}}{\begin{array}{c}\text{H}_2\text{NCH——CHNH}_2\\|\quad\quad|\\\text{CH}_2\quad\text{CH}_2\\\diagdown\text{S}\diagup\end{array}}$$

[119] Karrer and Kehrer, *Helv. Chim. Acta*, **27**, 142 (1944).
[120] Surrey, Hammer, and Suter, *J. Am. Chem. Soc.*, **66**, 1933 (1944).
[121] Kilmer and McKennis, *J. Biol. Chem.*, **152**, 103 (1944).

Cyclization of δ-Substituted Mercaptobutyl Halides. Alkyl and aryl tetramethylenesulfonium halides may be prepared from appropriately δ-substituted mercaptobutyl halides according to the following general reaction.

$$RSCH_2CH_2CH_2CH_2X \rightarrow \begin{array}{c} CH_2\text{---}CH_2 \\ | \quad\quad | \\ CH_2 \quad CH_2 \\ \diagdown \diagup \\ S+ \\ \diagup \\ R \quad X^- \end{array} \quad \begin{array}{l} XIII, R = C_6H_5, X = Br \\ XIV, R = C_2H_5, X = Cl \end{array}$$

Hydroxybutyl sulfides react with fuming hydrobromic acid to give the cyclic sulfonium halides. For example, when phenyl δ-hydroxybutyl sulfide is dissolved in an excess of fuming hydrobromic acid, phenyltetramethylenesulfonium bromide (XIII) is formed.[122] The product is isolated as the bromoaurate (90%). The corresponding chloride, phenyl δ-chlorobutyl sulfide, cyclizes in 50% aqueous acetone solution at 80° to form phenyltetramethylenesulfonium chloride.[123] Similarly, ethyl δ-chlorobutyl sulfide cyclizes to give about 50% of ethyltetramethylenesulfonium chloride (XIV).[123]

Di-δ-benzyloxybutyl sulfide (XV) reacts with 48% hydrobromic acid to form δ-hydroxybutyltetramethylenesulfonium bromide (XVI).[124]

$$(C_6H_5CH_2OCH_2CH_2CH_2CH_2)_2S \rightarrow \begin{array}{c} CH_2\text{---}CH_2 \\ | \quad\quad | \\ CH_2 \quad CH_2 \\ \diagdown \diagup \\ S+ \\ \diagup \\ HO(CH_2)_4 \quad Br^- \\ XVI \end{array}$$
XV

EXPERIMENTAL CONDITIONS

In the preparation of homologs of tetrahydrothiophene from 1,4-diiodobutanes or 1,4-dibromobutanes and sulfides, the dihalide is generally dissolved in ethanol or water and an aqueous or ethanolic solution of sodium or potassium sulfide is added. With diiodides the reaction may proceed satisfactorily at room temperature,[113] but with dibromides higher temperatures are usually necessary.[114] To isolate the product when ethanol has been the solvent, the reaction mixture is diluted with water and the solution is extracted with an immiscible organic solvent.

[122] Bennett and Mosses, *J. Chem. Soc.*, **1930**, 2364.
[123] Bennett, Heathcoat, and Mosses, *J. Chem. Soc.*, **1929**, 2567.
[124] Bennett and Hock, *J. Chem. Soc.*, **1927**, 477.

Special precautions are sometimes necessary, as in the synthesis of 3-ketotetrahydrothiophene.[117] In this reaction the ethanolic solution of α-chloromethyl β-iodoethyl ketone is treated with a saturated aqueous solution of sodium sulfide, and the reaction is allowed to continue in an atmosphere of hydrogen and in the absence of light for about five days or until the color of the mixture disappears. The solution is then neutralized with acetic acid, and the solvent distilled in vacuum. The 3-ketotetrahydrothiophene is isolated as the semicarbazone.[117]

Another technique is used for the preparation of 3,4-diaminotetrahydrothiophene. The aqueous solution of 2,3-diaminobutane-1,4-disulfuric acid ester and sodium sulfide is heated in a sealed tube at 140° for three hours; the solution is then acidified and the 3,4-diaminotetrahydrothiophene isolated as the picrate, the diacetyl derivative, or the dibenzoyl derivative.[121]

Experimental Procedures

3,4-Dihydroxytetrahydrothiophene.[115] To a solution of 4.9 g. of 1,4-dichloro-2,3-dihydroxybutane in 35 ml. of water at 60–70°, about 18 g. of sodium sulfide ($Na_2S \cdot 9H_2O$) in 5 ml. of water is added in portions with stirring, the reaction mixture being kept at 50–60°. The mixture is then heated for two hours on a steam bath. The solution is cooled and acidified to Congo red with 20% hydrochloric acid. The water is evaporated under reduced pressure. The nearly dry residue of organic material and salt is extracted repeatedly with absolute ethanol. The ethanol extract is evaporated in vacuum, leaving a crystalline residue. This residue is dissolved in chloroform, leaving behind extraneous material, and the chloroform is evaporated. The chloroform residue is dried over phosphorus pentoxide and is then sublimed in small portions in a molecular still at 3 mm. to 4 mm. and a bath temperature of 95°. The sublimate weighs about 1.9 g. (51%). After several sublimations, clusters of fine prisms of the product are obtained which melt at 54° to 58°.

dl-(trans)-Tetrahydrothiophene-1,5-dicarboxylic Acid.[118] A solution of 8 g. of sodium hydroxide in 200 ml. of water is cooled in ice, and 30.4 g. (0.1 mole) of dl-α,α'-dibromoadipic acid and a slight excess of crystalline sodium sulfide are added. The reaction mixture is allowed to stand for twenty-four hours and is then acidified with sulfuric acid. The sulfur that precipitates is collected on a filter, and the filtrate is extracted with 400 ml. of ether in eleven portions. By evaporation of the ether extract, 15.9 g. (90%) of crystalline acid is obtained. After recrystallization from a mixture of ethyl acetate and benzene or from

TABLE VII

Tetrahydrothiophenes from 1,4-Difunctional Compounds and Sulfides

In this table the alkyltetrahydrothiophenes are listed first. They are followed by the oxygen-containing derivatives, the nitrogen-containing derivatives, and the sulfonium salts.

Tetrahydrothiophene	1,4-Difunctional Compound	Reagents and Experimental Conditions	Yield %	Reference
Tetrahydrothiophene	$ICH_2CH_2CH_2CH_2I$	K_2S or Na_2S in aqueous alcohol	Quantitative	110, 111, 112
	$BrCH_2CH_2CH_2CH_2Br$	Na_2S in aq. alcohol	—	112
2-Methyl-	$ICH_2CH_2CH_2CHICH_3$	K_2S in aq. alcohol	—	113
	$BrCH_2CH_2CH_2CHBrCH_3$	Na_2S in aq. alcohol	—	112
3-Methyl-	$CH_2BrCH_2CH(CH_3)CH_2Br$	Na_2S in aq. alcohol	—	112
meso-2,5-Dimethyl-	$(CH_3CHBrCH_2)_2$	Na_2S in aq. alcohol	—	112
2,5-Dipropyl-	$(CH_3CH_2CH_2CHBrCH_2)_2$	Na_2S in ethanol at reflux	77	114
3,4-Dipropyl-	$CH_3CH_2CH_2\overset{|}{C}HCH_2Br$ $CH_3CH_2CH_2\overset{|}{C}HCH_2Br$	Na_2S in ethanol at reflux	77.5	114
3-Keto-	$ICH_2CH_2COCH_2Cl$	Na_2S in aq. ethanol in hydrogen atmosphere in absence of light	15	117
2,5-Diketo-	$ClCOCH_2CH_2COCl$	Na_2S in water	—	46
3,4-Dihydroxy-	$XCH_2CHOHCHOHCH_2X(X=Cl, Br)$	Na_2S in water at 100°	51	115
2,5-Dicarboxy- dl (trans) meso (cis)	$NaO_2CCHBrCH_2CH_2CHBrCO_2Na$ (dl or meso)	Na_2S in water, cold	53	118
3,4-Diethoxy-	(meso)$ICH_2CH(OC_2H_5)CH(OC_2H_5)CH_2I$	K_2S in ethanol	—	116
Lactone of 4-hydroxy-2-carboxy-	$CH_2ClCHCH\overset{\|}{C}HBrCO$ $\underset{O\text{———————}}{}$	KI followed by Na_2S	—	119
2-Phenyl-3-carbethoxy-4-keto-	$CH_2ClCOCCO_2C_2H_5$ $\overset{\|}{C}HC_6H_5$	$NaOC_2H_5 + H_2S$ in ethanol	67	120
3,4-Diamino-	$H_3\overset{+}{N}CHCH_2OSO_3^-$ $H_3\overset{+}{N}CHCH_2OSO_3^-$	Na_2S in water at 140° in sealed tube	25	121
![ring]S+ / C2H5 Cl−	$HO(CH_2)_4SC_2H_5$	$SOCl_2$ + dimethylaniline at 40–50°	50	123
![ring]S+ / C6H5 Cl−	$C_6H_5S(CH_2)_4Cl$	50% aq. acetone at 80° in sealed tube	—	123
![ring]S+ / C6H5 Br−	$C_6H_5S(CH_2)_4OH$	HBr (fuming)	90	122
![ring]S+ / HO(CH2)4 Br−	$(C_6H_5CH_2OCH_2CH_2CH_2CH_2)_2S$	HBr (fuming 48%) at room temperature or in sealed tube at 120–150°	—	124

water, the pure product weighs about 9.3 g. and melts at 165–166°. It is soluble in water and ethanol; it is difficultly soluble or insoluble in chloroform, carbon tetrachloride, and the hydrocarbons.

3,4-Di-*n*-propyltetrahydrothiophene.[114] One hundred and twenty-five milliliters of an ethanolic solution of sodium sulfide, prepared according to the method of Bost and Conn,[125] is placed in a 200-ml. three-necked flask equipped with a dropping funnel, a stirrer, and a reflux condenser. The solvent is heated to boiling and the stirrer started. Then 14.7 g. of 1,4-dibromo-2,3-di-*n*-propylbutane in 15 ml. of absolute ethanol is added from the dropping funnel over a period of one hour. Boiling is continued about ten hours; the reaction mixture is cooled and poured into 265 ml. of 25% sodium chloride solution. The organic material is extracted with petroleum ether (b.p. 35–38°), the extract is dried, and the solvent is evaporated. The product is distilled at reduced pressure. 3,4-Di-*n*-propyltetrahydrothiophene is obtained in 77.5% yield; b.p. 65–66°/1 mm.; d_{20}^{20} 0.9129; n_D^{20} 1.4830.

Tetrahydrothiophenes by the Dieckmann Cyclization Reaction

The Dieckmann condensation or cyclization of esters of dibasic acids is a general method of synthesis for 3-ketotetrahydrothiophenes and has

$$\begin{array}{c} \text{CHR'}-\text{C}=\text{O} \\ | \quad\quad | \\ \text{CHR''} \;\; \text{CHR'''} \\ \diagdown \diagup \\ \text{S} \end{array}$$

thus been employed extensively for synthesis of the tetrahydrothiophene nucleus in research on biotin. The primary product of the Dieckmann synthesis is a 3-ketotetrahydrothiophene bearing in the 2 or 4 position a carbalkoxy group, which can be removed by hydrolysis. Problems relating to the nature of R′, R″, and R‴ in this synthesis, and a variant in which the thioether group is formed during the course of the reaction, form the subtopics of the following discussion.

Cyclization of Esters Having Unsubstituted α-Methylene Groups. When neither α-methylene group carries a substituent, as in the ester I,

$$\text{RO}_2\text{CCH}_2\text{CH}_2\text{SCH}_2\text{CO}_2\text{R} \rightarrow \begin{array}{c} \text{RO}_2\text{CCH}-\text{C}=\text{O} \\ | \quad\quad\quad | \\ \text{CH}_2 \quad\; \text{CH}_2 \\ \diagdown \diagup \\ \text{S} \end{array} + \begin{array}{c} \text{CH}_2-\text{C}=\text{O} \\ | \quad\quad | \\ \text{CH}_2 \quad \text{CHCO}_2\text{R} \\ \diagdown \diagup \\ \text{S} \end{array}$$

I II III

[125] Bost and Conn, *Org. Syntheses, Coll. Vol.* **2**, 547 (1943).

cyclization can take place in both directions, giving both products II and III. Work by several investigators has given the following information on the control of the course of the cyclization of the unsubstituted ester I.[126]

When the dimethyl ester I (R = CH_3) is cyclized by the action of sodium methoxide in dry ether or in methanol at room temperature or below, a 75–80% yield of methyl 3-ketotetrahydrothiophene-2-carboxylate (III, R = CH_3) is obtained; there is a small amount of the isomeric product II (R = CH_3).[126, 127, 128] Esters of 3-ketotetrahydrothiophene-2-carboxylic acid (III) are also the predominant isomers when the ring closures are carried out by the action of powdered sodium in benzene [129, 130] on a series of homologous esters (I). However, methyl 3-ketotetrahydrothiophene-4-carboxylate (II, R = CH_3) is the product of the cyclization of the dimethyl ester I in dry toluene solution by the action of sodium methoxide at 80–120°; none of the isomeric ester III is found.[126, 127] An elevated temperature seems to bring about the formation of II when other condensing agents are used also. Thus, II (R = C_2H_5) is produced in about 55% yield by the reaction of the diester (I) and sodium ethoxide in benzene solution at the reflux temperature.[131]

The cyclization of the diethyl ester I (R = C_2H_5) by means of sodium amide in absolute ether or by sodium ethoxide in toluene at 40–50° gave mainly II (R = C_2H_5); [117, 132] the yields of this product were 64% and 72%, respectively, when sodium amide and sodium ethoxide were used. The product, however, was a mixture as shown by the isolation of two phenylhydrazones from the material.[132] On the basis of an analogy drawn from a study of the Dieckmann condensation of nitrogen-containing esters,[133] II (R = C_2H_5) has also been claimed [134] to result from the action of metallic sodium upon the diethyl ester I in benzene solution.

These results are attributed to an electron attracting effect on the attached carbon atom by the sulfur atom in the system —S—CH <. Of IV and V, the two possible intermediary anions, V appears to be the

$$\overset{\ominus}{CH_3O_2CCHCH_2SCH_2CO_2CH_3} \qquad \overset{\ominus}{CH_3O_2CCH_2CH_2SCHCO_2CH_3}$$
$$\text{IV} \qquad\qquad\qquad\qquad \text{V}$$

[126] Woodward and Eastman, *J. Am. Chem. Soc.*, **68**, 2229 (1946).
[127] Woodward and Eastman, *J. Am. Chem. Soc.*, **66**, 849 (1944).
[128] Moore and Moore, *J. Am. Chem. Soc.*, **68**, 910 (1946).
[129] Avison, Bergel, Cohen, and Haworth, *Nature*, **154**, 459 (1944).
[130] Bergel, Haworth, and Avison, Brit. pat. 562,314 [*C. A.*, **40**, 1179 (1946)].
[131] Brown, Baker, Bernstein, and Safir, *J. Org. Chem.*, **12**, 155 (1947).
[132] Hoffmann-LaRoche, Brit. pat. 570,240 [*C. A.*, **40**, 5533 (1946)]; Karrer and Schmid, *Helv. Chim. Acta*, **27**, 124 (1944).
[133] Prill and McElvain, *J. Am. Chem. Soc.*, **55**, 1233 (1933).
[134] Buchman and Cohen, *J. Am. Chem. Soc.*, **66**, 847 (1944).

more probable; it also seems probable that the anion V is formed more rapidly and its cyclization product III is the predominant one at low temperature under non-equilibrium conditions. At higher temperatures, when the reaction is allowed to proceed to equilibrium, a point is finally reached at which the isomer II, formed from the less probable intermediate anion IV, is the sole product.

The condensation of diesters that have a substituent R' as indicated in structure VI usually leads to the expected products, since R' is not on one of the active methylene carbons. When ethyl β-carbethoxymeth-

$$RO_2CCH_2CHR'SCH_2CO_2R$$
$$VI$$

ylmercapto-β-phenylpropionate (VII) in ethereal solution is treated with sodium ethoxide at the temperature of an ice-salt bath for six hours and then at room temperature overnight, condensation takes place to form ethyl 3-keto-5-phenyltetrahydrothiophene-2-carboxylate (VIII).[120]

$$C_2H_5O_2CCH_2CH(C_6H_5)SCH_2CO_2C_2H_5 \rightarrow$$

$$\begin{array}{c} CH_2\!\!-\!\!-\!\!C\!\!=\!\!O \\ |\quad\quad\quad | \\ C_6H_5CH \quad CHCO_2C_2H_5 \\ \diagdown\!\!\diagup \\ S \end{array}$$

VII VIII

Cyclization of α-Substituted Esters. Ordinarily, the monosubstituted structures IX and X are expected to cyclize in only one direction to give the 3-keto derivatives XI and XII. However, the nature of the

$$RO_2CCHR'CH_2SCH_2CO_2R \rightarrow$$

$$\begin{array}{c} R'CH\!\!-\!\!-\!\!C\!\!=\!\!O \\ |\quad\quad | \\ CH_2 \quad CHCO_2R \\ \diagdown\!\!\diagup \\ S \end{array}$$

IX XI

$$RO_2CCH_2CH_2SCHR''CO_2R \rightarrow$$

$$\begin{array}{c} RO_2CCH\!\!-\!\!-\!\!C\!\!=\!\!O \\ |\quad\quad\quad | \\ CH_2 \quad CHR'' \\ \diagdown\!\!\diagup \\ S \end{array}$$

X XII

substituents R' or R" of the diesters IX or X influences the direction of the condensation. When a strongly electronegative group is present, the activity of the adjacent —CH< group is enhanced, and it functions in the condensation to the exclusion of the other available —CH< or CH$_2$< group. For example, the thioanilide of ethyl carbethoxymalonate (XIII) reacts with ethyl chloroacetate in the presence of sodium ethoxide to form diethyl 3-keto-5-phenyliminotetrahydrothiophene-4,4-dicarboxylate (XIV).[135] Similarly, the thioanilide of ethyl

$$C_2H_5O_2CCH_2Cl \quad + \quad \underset{\underset{HS}{\overset{C=NC_6H_5}{\diagup}}}{CH(CO_2C_2H_5)_2}$$

XIII

$$\underset{XIV}{\underset{S}{\overset{OC\!-\!\!-\!\!-\!C(CO_2C_2H_5)_2}{\underset{\diagdown\;\;\diagup}{H_2C\quad C=NC_6H_5}}}} \leftarrow \left[\underset{S}{\underset{\diagdown\;\;\diagup}{\overset{C_2H_5OCO\quad CH(CO_2C_2H_5)_2}{CH_2\quad C=NC_6H_5}}} \right]$$

cyanomalonate (XV) reacts with ethyl chloroacetate to form ethyl 4-cyano-3-keto-5-phenyliminotetrahydrothiophene-4-carboxylate (XVI).[135]

$$C_2H_5O_2CCH_2Cl + \underset{\underset{HS}{\overset{C=NC_6H_5}{\diagup}}}{CH\!\!\!\diagup\!\!\!\overset{CN}{\underset{CO_2C_2H_5}{\diagdown}}} \quad\to\quad \underset{XVI}{\underset{S}{\underset{\diagdown\;\;\diagup}{\overset{O=C\!-\!\!-\!\!-\!C\!\!\!\diagup\!\!\!\overset{CN}{\underset{CO_2C_2H_5}{\diagdown}}}{CH_2\quad C=NC_6H_5}}}}$$

XV

When one of the active methylene groups of the diester I has a substituent such as an alkyl group or an acylamino group, the activity of this substituted methylene group is decreased, and the unsubstituted active methylene group functions in the condensation. Thus, ethyl 3-keto-2-methyltetrahydrothiophene-4-carboxylate (XVII) is obtained by the condensation of ethyl α-(2-carbethoxyethylmercapto)propionate (XVIII) in the presence of either sodium amide [136] at 40–50° or metallic

$$\underset{XVIII}{\underset{S}{\underset{\diagdown\;\;\diagup}{\overset{CO_2C_2H_5}{\underset{CH_2\quad CHCH_3}{\overset{CH_2\quad CO_2C_2H_5}{|\quad\quad\quad|}}}}}} \to \underset{XVII}{\underset{S}{\underset{\diagdown\;\;\diagup}{\overset{CO_2C_2H_5}{\underset{CH_2\quad CHCH_3}{\overset{CH\!-\!\!-\!C=O}{|\quad\quad\quad|}}}}}} \to \underset{XIX}{\underset{S}{\underset{\diagdown\;\;\diagup}{\overset{CH_2\!-\!\!-\!C=O}{\underset{CH_2\quad CHCH_3}{|\quad\quad\quad\;\;|}}}}}$$

[135] Ruhemann, *J. Chem. Soc.*, **93**, 621 (1908); **95**, 117 (1909).
[136] Karrer and Schmid, *Helv. Chim. Acta*, **27**, 124 (1944); Schnider, Bourquin, and Grüssner, *ibid.*, **28**, 510 (1945).

sodium suspended in benzene.[134] The yield from the reaction using sodium amide is 48%.[136] The decarboxylation of the keto ester XVII to 3-keto-2-methyltetrahydrothiophene (XIX, 81%) takes place readily during hydrolysis.[134,136] Similarly, ethyl 3-keto-4-methyltetrahydrothiophene-2-carboxylate (XX) is the product of the reaction between ethyl α-methyl-β-(carbethoxymethylmercapto)propionate (XXI) and sodium ethoxide in toluene solution at the temperature of a hot water bath.[137] The corresponding 4-ethyl derivative, ethyl 4-ethyl-3-ketotetrahydrothiophene-2-carboxylate (XXII) is obtained in a 66% yield from the diester XXIII by reaction with sodium ethoxide in toluene at 40–50°, and in a 30% yield by reaction with sodium ethoxide in ether.[138]

$$C_2H_5O_2CCHRCH_2SCH_2CO_2C_2H_5 \rightarrow$$

```
        RCH────C=O
         |      |
        CH₂    CHCO₂C₂H₅
          \   /
           S
```

XXI, R = CH₃   XX, R = CH₃
XXIII, R = C₂H₅   XXII, R = C₂H₅

Reactants having larger alkyl groups and substituted alkyl groups also cyclize satisfactorily. For example, ethyl α-(2-carbethoxyethylmercapto)-ε-methoxycaproate (XXIV) cyclizes readily in the presence of sodium ethoxide to ethyl 3-keto-2-(4'-methoxybutyl)tetrahydrothiophene-4-carboxylate (XXV, 80%). Hydrolysis and decarboxylation of the latter compound give 3-keto-2-(4'-methoxybutyl)tetrahydrothiophene (XXVI, 77%).[139]

```
    CO₂C₂H₅                         CO₂C₂H₅
     |                               |
    CH₂    CO₂C₂H₅        →         CH────C=O
     |      |                        |     |
    CH₂    CH(CH₂)₄OCH₃             CH₂   CH(CH₂)₄OCH₃
      \   /                           \   /
        S                               S
       XXIV                            XXV
          ↗                              ↓
    CO₂C₂H₅                         CH₂────C=O
     |                               |     |
  CH₃OCH    CO₂C₂H₅                 CH₂   CH(CH₂)₄OCH₃
     |       |                        \   /
    CH₂    CH(CH₂)₄OCH₃                 S
      \   /                            XXVI
        S
      XXVII
```

[137] Larsson, *Svensk Kem. Tid.*, **57**, 24 (1945) [*C. A.*, **40**, 2444 (1946)].
[138] Ghosh, McOmie, and Wilson, *J. Chem. Soc.*, **1945**, 705.
[139] Schmid, *Helv. Chim. Acta*, **27**, 127 (1944).

An alternative synthesis of the ketone XXVI is of interest. Ethyl α-(2-carbethoxy-2-methoxyethylmercapto)-ε-methoxycaproate (XXVII) reacts in the presence of sodium ethoxide in toluene at 40° to form an unidentified substance, apparently the cyclization product XXV. Acid hydrolysis and decarboxylation of this product give the 3-keto-2(4'-methoxybutyl)tetrahydrothiophene (XXVI).[139] Since both α-methylene groups of the ester XXVII are substituted, and a Claisen-type condensation is not expected to take place, it was concluded that the α-methoxyl group was lost before ring closure occurred. The yield of the final ketone XXVI was low.

Ethyl 3-keto-2-(3'-phenoxypropyl)tetrahydrothiophene-4-carboxylate (XXVIII) is the product of the cyclization of ethyl α-(2-carbethoxyethylmercapto)-δ-phenoxyvalerate (XXIX) with sodium ethoxide in benzene (85%).[140] The corresponding benzyloxy derivative, ethyl 2-(3-benzyloxypropyl)-3-ketotetrahydrothiophene-4-carboxylate (XXX, 67%), was prepared similarly from the diester XXXI.[140]

$$
\begin{array}{c}
\text{CO}_2\text{C}_2\text{H}_5 \\
| \\
\text{CH}_2 \quad \text{CO}_2\text{C}_2\text{H}_5 \\
| \quad \quad | \\
\text{CH}_2 \quad \text{CH(CH}_2)_3\text{OR} \\
\diagdown \diagup \\
\text{S}
\end{array}
\longrightarrow
\begin{array}{c}
\text{CO}_2\text{C}_2\text{H}_5 \\
| \\
\text{CH}\text{---}\text{C}\text{=}\text{O} \\
| \quad \quad | \\
\text{CH}_2 \quad \text{CH(CH}_2)_3\text{OR} \\
\diagdown \diagup \\
\text{S}
\end{array}
$$

XXIX, R = C₆H₅
XXXI, R = C₆H₅CH₂

XXVIII, R = C₆H₅
XXX, R = C₆H₅CH₂

By the same general method, the following 3-ketotetrahydrothiophenes have been prepared: ethyl 2-(4'-acetylbutyl)-3-ketotetrahydrothiophene-4-carboxylate (XXXII);[140] ethyl 4-carbethoxy-3-ketotetrahydrothiophene-2-propionate (XXXIII, 67%);[119] ethyl 2-(4'-cyanobutyl)-3-ketotetrahydrothiophene-4-carboxylate (XXXIV, 74%);[141] ethyl 4-carbethoxy-3-ketotetrahydrothiophene-2-valerate (XXXV, 82–89%),[141,142,143] and the corresponding methyl ester XXXVI (80%).[144]

$$
\begin{array}{c}
\text{CO}_2\text{C}_2\text{H}_5 \\
| \\
\text{CH}\text{---}\text{C}\text{=}\text{O} \\
| \quad \quad | \\
\text{CH}_2 \quad \text{CH(CH}_2)_4\text{COCH}_3 \\
\diagdown \diagup \\
\text{S} \\
\text{XXXII}
\end{array}
\qquad
\begin{array}{c}
\text{CO}_2\text{C}_2\text{H}_5 \\
| \\
\text{CH}\text{---}\text{C}\text{=}\text{O} \\
| \quad \quad | \\
\text{CH}_2 \quad \text{CH(CH}_2)_2\text{CO}_2\text{C}_2\text{H}_5 \\
\diagdown \diagup \\
\text{S} \\
\text{XXXIII}
\end{array}
$$

[140] Cheney and Piening, *J. Am. Chem. Soc.*, **67**, 2213 (1945).
[141] Karrer, Keller, and Usteri, *Helv. Chim. Acta*, **27**, 237 (1944).
[142] Cheney and Piening, *J. Am. Chem. Soc.*, **66**, 1040 (1944).
[143] Cheney and Piening, *J. Am. Chem. Soc.*, **67**, 731 (1945).
[144] Baker, Querry, Bernstein, Safir, and Subbarow, *J. Org. Chem.*, **12**, 167 (1947).

$$\begin{array}{c} \text{CO}_2\text{C}_2\text{H}_5 \\ | \\ \text{CH}\!-\!\!-\!\text{C}\!=\!\text{O} \\ | \quad\quad | \\ \text{CH}_2 \quad \text{CH}(\text{CH}_2)_4\text{CN} \\ \diagdown \;\; \diagup \\ \text{S} \end{array}$$

XXXIV

$$\begin{array}{c} \text{CO}_2\text{R} \\ | \\ \text{CH}\!-\!\!-\!\text{C}\!=\!\text{O} \\ | \quad\quad | \\ \text{CH}_2 \quad \text{CH}(\text{CH}_2)_4\text{CO}_2\text{R} \\ \diagdown \;\; \diagup \\ \text{S} \end{array}$$

XXXV, R = C_2H_5
XXXVI, R = CH_3

Use of S-carbalkoxymethyl ethers of N-acylcysteine in the Dieckmann cyclization reaction provides diesters with the α-acylamino substituent. When L-N-benzoyl-β-(carbomethoxymethylmercapto)alanine methyl ester (XXXVII) in methanol solution is treated with sodium methoxide, the sodium salt of enolic methyl 4-benzamido-3-ketotetrahydrothiophene-2-carboxylate (XXXVIII) quickly crystallizes, and an 89% yield is obtained.[145] Similarly, ethyl 4-acetamido-3-ketotetrahydrothiophene-2-carboxylate (XXXIX) is prepared by cyclization of N-acetyl-β-(carbethoxymethylmercapto)alanine ethyl ester (XL) in toluene solution in the presence of either sodium ethoxide or sodium amide.[146]

$$\begin{array}{c} \text{NHCOC}_6\text{H}_5 \\ | \\ \text{CH}\!-\!\!-\!\text{CO}_2\text{CH}_3 \\ | \\ \text{CH}_2 \quad \text{CH}_2\text{CO}_2\text{CH}_3 \\ \diagdown \;\; \diagup \\ \text{S} \end{array} \rightarrow \begin{array}{c} \text{NHCOC}_6\text{H}_5 \\ | \\ \text{CH}\!-\!\!-\!\text{CONa} \\ | \quad\quad \| \\ \text{CH}_2 \quad \text{CCO}_2\text{CH}_3 \\ \diagdown \;\; \diagup \\ \text{S} \end{array}$$

XXXVII → XXXVIII

$$\begin{array}{c} \text{NHCOCH}_3 \\ | \\ \text{CH}\!-\!\!-\!\text{CO}_2\text{C}_2\text{H}_5 \\ | \\ \text{CH}_2 \quad \text{CH}_2\text{CO}_2\text{C}_2\text{H}_5 \\ \diagdown \;\; \diagup \\ \text{S} \end{array} \rightarrow \begin{array}{c} \text{NHCOCH}_3 \\ | \\ \text{CH}\!-\!\!-\!\text{CO} \\ | \quad\quad | \\ \text{CH}_2 \quad \text{CHCO}_2\text{C}_2\text{H}_5 \\ \diagdown \;\; \diagup \\ \text{S} \end{array}$$

XL → XXXIX

More highly substituted tetrahydrothiophene derivatives can also be prepared by this cyclization reaction. Ethyl 4-benzamido-3-keto-5-methyltetrahydrothiophene-2-carboxylate (XLI) was formed from ethyl α-benzamido-β-(carbethoxymethylmercapto)butyrate (XLII) in ethereal solution by the action of sodium ethoxide.[147] Ethyl α-benzamido-β-

[145] Harris, Wolf, Mozingo, Anderson, Arth, Easton, Heyl, Wilson, and Folkers, *J. Am. Chem. Soc.*, **66**, 1756 (1944); Harris, Easton, Heyl, Wilson, and Folkers, *ibid.*, **66**, 1757 (1944).

[146] Karrer and Schmid, *Helv. Chim. Acta*, **27**, 1280 (1944).

[147] Brown, Safir, Baker, Bernstein, and Dorfman, *J. Org. Chem.*, **12**, 483 (1947).

(carbethoxymethylmercapto)suberate (XLIII) under similar conditions

$$\underset{\text{XLII}}{\begin{array}{c}\text{NHCOC}_6\text{H}_5\\|\\\text{CH}-\!\!-\!\!\text{CO}_2\text{C}_2\text{H}_5\\|\qquad\qquad\qquad\\\text{CH}_3\text{CH}\quad\text{CH}_2\text{CO}_2\text{C}_2\text{H}_5\\\diagdown\;\;\diagup\\\text{S}\end{array}}\qquad\underset{\text{XLI}}{\begin{array}{c}\text{NHCOC}_6\text{H}_5\\|\\\text{CH}-\!\!-\!\!\text{C}\!=\!\text{O}\\|\qquad\qquad|\\\text{CH}_3\text{CH}\quad\text{CHCO}_2\text{C}_2\text{H}_5\\\diagdown\;\;\diagup\\\text{S}\end{array}}$$

cyclized to ethyl 4-benzamido-2-carbethoxy-3-ketotetrahydrothiophene-5-valerate (XLIV).[148]

$$\underset{\text{XLIII}}{\begin{array}{c}\text{NHCOC}_6\text{H}_5\\|\\\text{CH}-\!\!-\!\!\text{CO}_2\text{C}_2\text{H}_5\\|\qquad\qquad\qquad\\\text{C}_2\text{H}_5\text{O}_2\text{C}(\text{CH}_2)_4\text{CH}\quad\text{CH}_2\text{CO}_2\text{C}_2\text{H}_5\\\diagdown\;\;\diagup\\\text{S}\end{array}}\quad\underset{\text{XLIV}}{\begin{array}{c}\text{NHCOC}_6\text{H}_5\\|\\\text{CH}-\!\!-\!\!\text{C}\!=\!\text{O}\\|\qquad\qquad|\\\text{C}_2\text{H}_5\text{O}_2\text{C}(\text{CH}_2)_4\text{CH}\quad\text{CHCO}_2\text{C}_2\text{H}_5\\\diagdown\;\;\diagup\\\text{S}\end{array}}$$

The "ketone cleavage" of these 3-ketotetrahydrothiophenes to remove the carbalkoxy groups takes place readily and in good yields. Hydrolyses are carried out in dilute mineral acid, sometimes containing about 50% acetic acid, by refluxing the solution until the decarboxylation is complete. Labile groups such as carbalkoxy and cyano groups may be hydrolyzed during the reaction.[141]

Syntheses from α-Mercapto Esters and Unsaturated Compounds. The formation of the thioether group by the addition of a mercaptan to an olefin can be utilized to carry out a Dieckmann synthesis of a 3-ketotetrahydrothiophene from an α-mercapto ester and an α,β-unsaturated ester (or nitrile) without isolation of the intermediate thioether. Thus, ethyl thioglycolate (XLV) and 2-hexenonitrile (XLVI) in benzene solution condense in the presence of sodium ethoxide at the reflux temperature to form 3-cyano-4-keto-2-n-propyltetrahydrothiophene (XLVII).[149]

$$\underset{\text{XLV}}{\text{C}_2\text{H}_5\text{O}_2\text{CCH}_2\text{SH}}\;+\;\underset{\text{XLVI}}{\begin{array}{c}\text{CHCN}\\\|\\\text{CH}(\text{CH}_2)_2\text{CH}_3\end{array}}\;\rightarrow\;\underset{\text{XLVII}}{\begin{array}{c}\text{O}\!=\!\text{C}-\!\!-\!\!\text{CHCN}\\|\qquad\qquad|\\\text{CH}_2\quad\text{CH}(\text{CH}_2)_2\text{CH}_3\\\diagdown\;\;\diagup\\\text{S}\end{array}}$$

[148] Safir, Bernstein, Baker, McEwen, and Subbarow, *J. Org. Chem.*, **12**, 475 (1947).
[149] Baker, Querry, Safir, and Bernstein, *J. Org. Chem.*, **12**, 138 (1947).

Similarly, ethyl thioglycolate condenses with ethyl 2-hexenoate to form ethyl 4-keto-2-n-propyltetrahydrothiophene-3-carboxylate (XLVIII, 66%), and with methyl 6-phenoxy-2-hexenoate to form methyl 4-keto-2-(γ-phenoxypropyl)tetrahydrothiophene-3-carboxylate (XLIX, 72%).[149]

$$\begin{array}{cc}
O{=}C{\rule{1em}{0.4pt}}CHCO_2C_2H_5 & O{=}C{\rule{1em}{0.4pt}}CHCO_2CH_3 \\
| \quad\quad | & | \quad\quad | \\
CH_2 \quad CH(CH_2)_2CH_3 & CH_2 \quad CH(CH_2)_3OC_6H_5 \\
\diagdown \diagup & \diagdown \diagup \\
S & S \\
\text{XLVIII} & \text{XLIX}
\end{array}$$

These are the products expected by analogy with the reaction of ethyl thioglycolate with the unsaturated nitrile (XLVI). On the other hand, the condensation of methyl β-(carbomethoxymethylmercapto)suberate (L) in toluene solution when treated with sodium methoxide at reflux temperature gave both possible products, LI (67%) and LII (7%).[150] Similarly, ethyl β-(carbethoxymethylmercapto)butyrate (LIII) cyclized in the presence of sodium ethoxide to ethyl 3-keto-5-methyltetrahydrothiophene-2-carboxylate (LIV) when the toluene solution was heated on a hot water bath for five hours.[151]

$$CH_3O_2CCH_2CHSCH_2CO_2CH_3$$
$$|$$
$$CH_3O_2C(CH_2)_4$$
$$\quad\text{L}$$
$$\rightarrow \begin{cases}
CH_2{\rule{1em}{0.4pt}}C{=}O \\
CH_3O_2C(CH_2)_4CH \quad CHCO_2CH_3 \\
\diagdown \diagup \\
S \\
\text{LI} \\
\\
CH_3O_2CCH{\rule{1em}{0.4pt}}C{=}O \\
CH_3O_2C(CH_2)_4CH \quad CH_2 \\
\diagdown \diagup \\
S \\
\text{LII}
\end{cases}$$

$$C_2H_5O_2CCH_2CH(CH_3)SCH_2CO_2C_2H_5 \rightarrow \begin{array}{c}
CH_2{\rule{1em}{0.4pt}}C{=}O \\
| \quad\quad | \\
CH_3CH \quad CHCO_2C_2H_5 \\
\diagdown \diagup \\
S
\end{array}$$

$$\text{LIII} \quad\quad\quad \text{LIV}$$

A number of reactions have been described in which an α-mercapto ester is condensed with methyl acrylate to form an ester of 3-ketotetra-

[150] Brown, Armstrong, Moyer, Anslow, Baker, Querry, Bernstein, and Safir, *J. Org. Chem.*, **12**, 160 (1947).
[151] Larsson and Dahlström, *Svensk Kem. Tid.*, **57**, 248 (1945) [*C. A.*, **40**, 2444 (1946)].

hydrothiophene-4-carboxylic acid in good yield. Methyl 3-keto-2-(3'-phenoxypropyl)tetrahydrothiophene-4-carboxylate (LV, 73%) is prepared by allowing methyl α-mercapto-δ-phenoxyvalerate (LVI) to react with methyl acrylate in the presence of a trace of piperidine and sodium methoxide.[149] In the same way, methyl 2-(3'-chlorophenoxypropyl)-3-ketotetrahydrothiophene-4-carboxylate (LVII)[149] is obtained from methyl α-mercapto-δ-chlorophenoxyvalerate and methyl acrylate, and

$$\begin{array}{c} CH_3O_2CCH \\ \parallel \\ CH_2 \end{array} + \begin{array}{c} CO_2CH_3 \\ | \\ CH(CH_2)_3OC_6H_5 \\ | \\ HS \end{array} \qquad \begin{array}{c} CH_3O_2CCH\!\!-\!\!C\!\!=\!\!O \\ | \quad\quad\quad | \\ CH_2 \quad CH(CH_2)_3OC_6H_5 \\ \diagdown \; \diagup \\ S \end{array}$$

LVI LV

methyl 4-carbomethoxy-3-ketotetrahydrothiophene-2-butyrate (LVIII, 77%)[144] from methyl α-mercaptoadipate and methyl acrylate.[140]

$$\begin{array}{c} CH_3O_2CCH\!\!-\!\!C\!\!=\!\!O \\ | \quad\quad\quad | \\ CH_2 \quad CH(CH_2)_3OC_6H_4Cl \\ \diagdown \; \diagup \\ S \end{array} \qquad \begin{array}{c} CH_3O_2CCH\!\!-\!\!C\!\!=\!\!O \\ | \quad\quad\quad | \\ CH_2 \quad CH(CH_2)_3CO_2CH_3 \\ \diagdown \; \diagup \\ S \end{array}$$

LVII LVIII

Experimental Conditions

In general, yields in the Dieckmann condensations that give ketotetrahydrothiophenes are good, ranging from 50% to 90%. There seems to be no notable variation in yield with the size or nature of the substituent groups. The "ketone cleavage," which brings about decarboxylation, takes place in equally high or higher yields, 80–90%. The procedures employed in these syntheses are the ones commonly used for Claisen-type condensations; effective condensing agents include sodium alkoxide, sodium amide, and metallic sodium. In some cases, sodium alkoxide is used with an inert solvent, such as toluene;[117,126,127,136,139,144,146,149] in others, ethanol is used as a solvent.[135,145] The yields seem to be relatively unaffected by the choice of solvent. When sodium alkoxide or sodium amide with an inert solvent such as ether,[117] toluene, xylene,[119,146] or benzene[140,149] is used, the ester is generally added to a suspension of the condensing agent in the inert solvent at room temperature. The mixture is agitated until the sodium alkoxide or amide is in solution; then the mixture may be heated at slightly elevated temperatures or allowed to stand at room temperature to complete the reaction. The mixture is usually worked up by pouring it into an acidified ice mixture and extracting the product. Copper chelates may be used to purify crude ketotetrahydrothiophenecarboxylic acid esters.[140,143] When ethanol is the solvent, the sodium salt of a ketotetrahydrothiophenecarboxylic acid may crystallize from the reaction mixture.[145]

Experimental Procedures

Methyl *dl*-4-Benzamido-3-ketotetrahydrothiophene-2-carboxylate (Sodium Salt).[145] A solution of sodium methoxide prepared from 57 g. of sodium and 100 ml. of methanol is added to a solution of 770 g. of N-benzoyl-β-(carbomethoxymethylmercapto)alanine methyl ester in 500 ml. of methanol. The sodium salt of enolic methyl *dl*-4-benzamido-3-ketotetrahydrothiophene-2-carboxylate crystallizes quickly. After one hour, the salt is collected on a filter and washed with methanol, then with ether, and air-dried; yield, 663 g. (89%).

Ethyl 3-Keto-2-(4′-methoxybutyl)tetrahydrothiophene-4-carboxylate.[139] A suspension of sodium ethoxide in toluene is prepared by dissolving 1.2 g. of sodium in 3.05 ml. of absolute ethanol and adding 30 ml. of dry toluene. This suspension is covered by a nitrogen atmosphere and protected from moisture while 7.88 g. of ethyl α-(2-carbethoxyethylmercapto)-ε-methoxycaproate is added dropwise. The reaction mixture is heated at 45–50° for six hours, and then allowed to stand at 15° for one hour, during which time the sodium salt of the keto ester crystallizes. The mixture is poured onto ice; then the solution is acidified with 4.5 ml. of acetic acid and extracted with a large volume of ether. The ethereal extract is washed with sodium bicarbonate solution and water and is then dried and evaporated in vacuum. The residue is distilled at reduced pressure. Ethyl 3-keto-2-(4′-methoxybutyl)tetrahydrothiophene-4-carboxylate distils at 115°/0.01 mm. The average yield is 5.49 g. (80%).

2-(4′-Methoxybutyl)-3-ketotetrahydrothiophene.[139] The decarboxylation of ethyl 3-keto-2-(4′-methoxybutyl)tetrahydrothiophene-4-carboxylate is accomplished by refluxing for three hours in a nitrogen atmosphere a mixture containing 20 g. of the ester, 40 ml. of water, 40 ml. of acetic acid, and 8 ml. of concentrated sulfuric acid. The sulfuric acid is neutralized with an equivalent of sodium bicarbonate, and the solution is concentrated in vacuum to remove the acetic acid. The aqueous concentrate is saturated with salt and extracted with ether. The ethereal extract is washed with saturated sodium bicarbonate solution and water, and is then dried and evaporated. The residue is fractionated at 0.05 mm.; 2-(4′-methoxybutyl)-3-ketotetrahydrothiophene distils at 102–103°. The average yield is 11.8 g. (77%).

TABLE VIII
Tetrahydrothiophenes by the Dieckmann Cyclization Reaction

Tetrahydrothiophene	Starting Material	Reagents and Experimental Conditions	Yield %	Reference
2-Carbalkoxy-3-keto-	$RO_2CCH_2CH_2SCH_2CO_2R$	$NaOCH_3$ in ether	72	126, 127
		$NaOCH_3$ in ether or methanol	—	128
		Na powder in benzene	—	129, 130
		$NaNH_2$ in dry ether at reflux	—	117, 132
		$NaOC_2H_5$ in dry toluene at 40°	—	117, 132
3-Carbalkoxy-4-keto-	$RO_2CCH_2CH_2SCH_2CO_2R$	$NaOC_2H_5$ in dry benzene at reflux	55	131
		$NaOCH_3$ in dry toluene 80–120°	30	126, 127
		$NaOC_2H_5$ in dry toluene at 40°	—	117, 132
		$NaNH_2$ in dry ether at reflux	—	117, 132
		Na in benzene	—	134
2-Methyl-3-keto-4-carbethoxy-	$HSCH_2CO_2C_2H_5 + CH_2=CHCO_2C_2H_5$	$NaOC_2H_5$ in benzene at reflux	48	131
	$C_2H_5O_2CCH_2CH_2SCH(CH_3)CO_2C_2H_5$	$NaNH_2$ in dry ether at reflux	48.5	136
		Na in benzene	—	134
2-Carbethoxy-3-keto-4-methyl-	$C_2H_5O_2CCH_2SCH_2CH(CH_3)CO_2C_2H_5$	$NaOC_2H_5$ in toluene at 100°	—	137
2-Carbethoxy-3-keto-5-methyl-	$C_2H_5O_2CCH_2SCH(CH_3)CH_2CO_2C_2H_5$	$NaOC_2H_5$ in toluene at 100°	—	151
2-n-Propyl-3-cyano-4-keto-	$HSCH_2CO_2C_2H_5 + C_3H_7CH=CHCN$	$NaOC_2H_5$ in benzene at reflux	52	149
2-Carbethoxy-3-keto-4-ethyl-	$C_2H_5O_2CCH_2SCH_2CH(C_2H_5)CO_2C_2H_5$	$NaOC_2H_5$ in toluene at 40–50°	66	138
		$NaOC_2H_5$ in ether	30	138

Compound	Reactants	Conditions	Yield	Ref.
2-Carbethoxy-3-keto-4-acetamido-	$CH_2CH(NHCOCH_3)CO_2C_2H_5$ $SCH_2CO_2C_2H_5$	$NaOC_2H_5$ or $NaNH_2$ in toluene at 30–35°	38	146
2-n-Propyl-3-carbethoxy-4-keto- CH_3O_2C O $(CH_2)_3CO_2CH_3$ S	$HSCH_2CO_2C_2H_5 + C_3H_7CH=CHCO_2C_2H_5$ $CH_3O_2C(CH_2)_3CH(SH)CO_2CH_3$ $+ CH_2=CHCO_2CH_3 +$ piperidine	$NaOC_2H_5$ in benzene at reflux $NaOC_2H_5$ in ether	66 77	149 144
$C_2H_5O_2C$ O $CH_2CH_2CO_2C_2H_5$ S	$C_2H_5O_2CCH_2CH_2SCH(CO_2C_2H_5)CH_2CH_2CO_2C_2H_5$	$NaOC_2H_5$ in dry toluene at 55–60°	60	119
CH_3O_2C O $(CH_2)_4CO_2CH_3$ S	$CH(CH_2CO_2CH_3)(CH_2)_4CO_2CH_3$ $CH_2CO_2CH_3$	$NaOCH_3$ in dry toluene at reflux	61	150
O CO_2CH_3 $(CH_2)_4CO_2CH_3$ S	$CH(CH_2CO_2CH_3)(CH_2)_4CO_2CH_3$ $CH_2CO_2CH_3$	$NaOCH_3$ in dry toluene at reflux	7	150
$C_2H_5O_2C$ O $(CH_2)_4OCH_3$ S	$CH(CO_2C_2H_5)(CH_2)_4OCH_3$ $CH_2CH_2CO_2C_2H_5$	$NaOC_2H_5$ in dry toluene at 45–50°	80	139
$C_2H_5O_2C$ O $(CH_2)_4CN$ S	$CH(CO_2CH_3)(CH_2)_4CN$ $CH_2CH_2CO_2C_2H_5$	$NaOC_2H_5$ in dry toluene at 45°	74	141
RO_2C O $(CH_2)_4CO_2R$ S	$CH(CO_2R)(CH_2)_4CO_2R$ $CH_2CH_2CO_2R$	$NaOC_2H_5$ in dry xylene at 35–40° $NaOCH_3$ in dry benzene $NaOC_2H_5$ in dry benzene	82 80 89	141 144 142, 143
2-Phenyl-3-carbethoxy-4-keto-	$ClCH_2COC(CO_2C_2H_5)=CHC_6H_5$	$NaOC_2H_5$ in ethanol saturated with H_2S	67	120
2-Phenyl-4-keto-5-carbethoxy- S	$CH(C_6H_5)CH_2CO_2C_2H_5$ $CH_2CO_2C_2H_5$	$NaOC_2H_5$ in ether	—	120

TABLE VIII—Continued

TETRAHYDROTHIOPHENES BY THE DIECKMANN CYCLIZATION REACTION

Tetrahydrothiophene	Starting Material	Reagents and Experimental Conditions	Yield %	Reference
$C_2H_5O_2C$—(C=O)—$(CH_2)_4COCH_3$ (ring with S)	$CH(CO_2C_2H_5)(CH_2)_4COCH_3$ / S—$CH_2CH_2CO_2C_2H_5$	$NaOC_2H_5$ in dry benzene	—	140
2-Carbomethoxy-3-keto-4-benzamido-	$CH_2CH(NHCOC_6H_5)CO_2CH_3$ / S—$CH_2CO_2CH_3$	$NaOCH_3$ in methanol	89	145
(ring with CN, $CO_2C_2H_5$, NC_6H_5, S)	$C_6H_5NHCSCH(CO_2C_2H_5)CN$ + $ClCH_2CO_2C_2H_5$	$NaOC_2H_5$ in ethanol at reflux	—	135
(ring with CO_2CH_3, $(CH_2)_3OC_6H_5$, S, O)	$HSCH_2CO_2CH_3$ + $C_6H_5O(CH_2)_3CH=CHCO_2CH_3$	$NaOCH_3$ in benzene at reflux	72	149
CH_3O_2C—(C=O)—$(CH_2)_3OC_6H_5$ (ring with S)	$C_6H_5O(CH_2)_3CH(SH)CO_2CH_3$ + $CH_2=CHCO_2CH_3$ + piperidine	$NaOCH_3$ in ether	73	149

Product	Reactants	Conditions	Yield	Ref.
(thiophene with CH$_3$O$_2$C, =O, (CH$_2$)$_3$C$_6$H$_4$Cl)	C$_6$H$_5$O(CH$_2$)$_3$CHCO$_2$C$_2$H$_5$ / S(CH$_2$)$_2$CO$_2$CH$_3$	NaOC$_2$H$_5$ in dry benzene at room temperature then at reflux	85	140
(thiophene with C$_2$H$_5$O$_2$C, =O, NHCOC$_6$H$_5$, CH$_3$)	ClC$_6$H$_4$O(CH$_2$)$_3$CH(SH)CO$_2$CH$_3$ + CH$_2$=CHCO$_2$CH$_3$ + piperidine	NaOCH$_3$ in ether	—	149
(thiophene with (CO$_2$C$_2$H$_5$)$_2$, NC$_6$H$_5$)	CH(CH$_3$)CH(NHCOC$_6$H$_5$)CO$_2$C$_2$H$_5$ / CH$_2$CO$_2$C$_2$H$_5$	NaOC$_2$H$_5$ in dry ether	31	147
(thiophene with C$_2$H$_5$O$_2$C, =O, (CH$_2$)$_3$OCH$_2$C$_6$H$_5$)	C$_6$H$_5$NHCSCH(CO$_2$C$_2$H$_5$)$_2$ + ClCH$_2$CO$_2$C$_2$H$_5$	NaOC$_2$H$_5$ in ethanol at reflux	—	135
(thiophene with NHCOC$_6$H$_5$, (CH$_2$)$_4$CO$_2$C$_2$H$_5$)	CH(CO$_2$C$_2$H$_5$)(CH$_2$)$_3$OCH$_2$C$_6$H$_5$ / CH$_2$CH$_2$CO$_2$C$_2$H$_5$	NaOC$_2$H$_5$ in dry benzene	69	140
(thiophene with C$_2$H$_5$O$_2$C, =O)	CH(NHCOC$_6$H$_5$)CO$_2$C$_2$H$_5$ / (CH$_2$)$_4$CO$_2$C$_2$H$_5$ / CH$_2$CO$_2$C$_2$H$_5$	NaOC$_2$H$_5$ in dry ether in N$_2$ atmosphere	—	148

Tetrahydrothiophenes by Catalytic Methods

Tetrahydrothiophene and a few of its homologs have been prepared from the corresponding tetrahydrofurans by passing a mixture of the tetrahydrofuran and hydrogen sulfide over an aluminum oxide catalyst at an elevated temperature. Sufficient examples of this reaction have not been reported to justify considering the reaction a general one. In the existing examples, the yields of the products are 60–70%.

Tetrahydrothiophene (I) is obtained in a yield of 90% by passing a mixture of tetrahydrofuran (II) and hydrogen sulfide over aluminum oxide, preferably at 400°.[152,153] In like manner, tetrahydrothiophene is

$$\begin{array}{c} CH_2\text{---}CH_2 \\ | \quad\quad | \\ CH_2 \quad CH_2 \\ \diagdown \diagup \\ O \\ II \end{array} \xrightarrow{H_2S} \begin{array}{c} CH_2\text{---}CH_2 \\ | \quad\quad | \\ CH_2 \quad CH_2 \\ \diagdown \diagup \\ S \\ I \end{array}$$

$$\underset{III}{HOCH_2CH_2CH_2CH_2OH} \quad\quad \underset{IV}{HOCH_2CH_2CH_2CH_2Cl}$$

obtained in 62% and 95% yields, respectively, from tetramethylene glycol (III)[154] and tetramethylene chlorohydrin (IV).[152]

Alkyl-substituted tetrahydrofurans have been found to react similarly. 2-Methyltetrahydrofuran (V) is converted to 2-methyltetrahydrothiophene (VI) in 69% yield by reaction with hydrogen sulfide over alumina at 400°.[155] 2-Ethyltetrahydrofuran (VII) and 2,5-dimethyltetrahydrofuran (VIII) react with hydrogen sulfide under similar conditions to give 2-ethyltetrahydrothiophene (IX)[156] and 2,5-dimethyltetrahydrothio-

$$\begin{array}{c} CH_2\text{---}CH_2 \\ | \quad\quad | \\ CH_2 \quad CHR \\ \diagdown \diagup \\ O \end{array} \rightarrow \begin{array}{c} CH_2\text{---}CH_2 \\ | \quad\quad | \\ CH_2 \quad CHR \\ \diagdown \diagup \\ S \end{array}$$

V, R = CH$_3$ VI, R = CH$_3$
VII, R = C$_2$H$_5$ IX, R = C$_2$H$_5$

---

[152] Yur'ev, Minachev, and Samurskaya, *J. Gen. Chem. U.S.S.R.*, **9**, 1710 (1939) [*C. A.*, **34**, 3731 (1940)].

[153] Yur'ev and Tronova, *J. Gen. Chem. U.S.S.R.*, **10**, 31 (1940) [*C. A.*, **34**, 4733 (1940)]; Yur'ev and Prokina, *ibid.*, **7**, 1868 (1937) [*C. A.*, **32**, 548 (1938)].

[154] Yur'ev and Medovshchikov, *J. Gen. Chem. U.S.S.R.*, **9**, 628 (1939) [*C. A.*, **33**, 7779 (1939)].

[155] Yur'ev, *J. Gen. Chem. U.S.S.R.*, **8**, 1934 (1938) [*C. A.*, **33**, 5845 (1939)].

[156] Yur'ev, Gusev, Tronova, and Yurilin, *J. Gen. Chem. U.S.S.R.*, **11**, 344 (1941) [*C. A.*, **35**, 5893 (1941)].

phene (X, 68%).[157] An increase in the number of the carbon atoms in

$$\begin{array}{cc} \mathrm{CH_2\!-\!\!-\!CH_2} & \mathrm{CH_2\!-\!\!-\!CH_2} \\ | \quad\quad | & | \quad\quad | \\ \mathrm{H_3CCH} \quad \mathrm{CHCH_3} & \mathrm{H_3CCH} \quad \mathrm{CHCH_3} \\ \diagdown \,\diagup & \diagdown \,\diagup \\ \mathrm{O} & \mathrm{S} \\ \mathrm{VIII} & \mathrm{X} \end{array}$$

the side chain of the furan is said to result in decreased yields of the tetrahydrothiophene.[156]

TABLE IX

TETRAHYDROTHIOPHENES BY CATALYTIC METHODS

The starting material and hydrogen sulfide were passed over an alumina catalyst at the temperature indicated.

Tetrahydrothiophene	Starting Material	Temperature °C.	Yield %	Reference
Tetrahydrothiophene	Tetrahydrofuran	400	90.5	152
			67	153
	$ClCH_2CH_2CH_2CH_2OH$	400	95	152
	$HOCH_2CH_2CH_2CH_2OH$	400	62.5	154
2-Methyl-	2-Methyltetrahydrofuran	400	69	155
2-Ethyl-	2-Ethyltetrahydrofuran	390	—	156
2,5-Dimethyl-	2,5-Dimethyltetrahydrofuran	400	68	157

Tetrahydrothiophenes by Miscellaneous Methods

Tetraethyl tetrahydrothiophene-3,3,4,4-tetracarboxylate (I) has been characterized as the product of the reaction between tetraethyl ethane-1,1,2,2-tetracarboxylate (II) and bischloromethyl sulfide (III) in the presence of sodium ethoxide.[158,159] No other applications of this reaction have been reported.

Tetrahydrothiophene-3,4-dicarboxylic acid (V) is prepared by hydrolysis of the tetracarboxylic acid ester I and pyrolysis of the intermediate tetracarboxylic acid IV at 140–160°.[158]

Two 2,5-dithionotetrahydrothiophenes have been prepared in about 87% yield by the reaction of bromine in carbon disulfide on ethyl

[157] Yur'ev, Tronova, L'vova, and Bukshpan, *J. Gen. Chem. U.S.S.R.*, **11**, 1128 (1941) [*C. A.*, **37**, 4071 (1943)].
[158] Kilmer, Armstrong, Brown, and du Vigneaud, *J. Biol. Chem.*, **145**, 495 (1942).
[159] Mann and Pope, *J. Chem. Soc.*, **123**, 1172 (1923).

$$(C_2H_5O_2C)_2CHCH(CO_2C_2H_5)_2 + ClCH_2SCH_2Cl \rightarrow$$
$$\text{II} \qquad\qquad\qquad\qquad \text{III}$$

$$\begin{array}{c} (C_2H_5O_2C)_2C\text{------}C(CO_2C_2H_5)_2 \\ |\qquad\qquad\quad| \\ CH_2\qquad\quad CH_2 \\ \diagdown\quad\diagup \\ S \\ \text{I} \end{array}$$

↓

$$\begin{array}{cc} HO_2CCH\text{------}CHCO_2H & (HO_2C)_2C\text{------}C(CO_2H)_2 \\ |\qquad\qquad| & \qquad |\qquad\qquad| \\ CH_2\quad\ CH_2 \quad \leftarrow & CH_2\qquad CH_2 \\ \diagdown\ \diagup & \diagdown\quad\diagup \\ S & S \\ \text{V} & \text{IV} \end{array}$$

sodiomalonate and ethyl sodiocyanoacetate, respectively.[160] The reaction has been postulated to take place as follows: The xanthates VI and VII, believed to be formed first, react in the presence of bromine to give tetraethyl 2,5-dithionotetrahydrothiophene-3,3,4,4-tetracarboxylate (VIII), and diethyl 2,5-dithiono-3,4-dicyanotetrahydrothiophene-3,4-dicarboxylate (IX), respectively.[160]

$$Na[CH(CO_2C_2H_5)_2] \xrightarrow{CS_2} Na\underset{\underset{S}{\|}}{S}CCH(CO_2C_2H_5)_2 \rightarrow$$
$$\text{VI}$$

$$\begin{array}{c} (C_2H_5O_2C)_2C\text{------}C(CO_2C_2H_5)_2 \\ |\qquad\qquad\quad| \\ SC\qquad\quad CS \\ \diagdown\quad\diagup \\ S \\ \text{VIII} \end{array}$$

$$Na(CHCNCO_2C_2H_5) \xrightarrow{CS_2} Na\underset{\underset{S}{\|}}{S}CCH(CN)CO_2C_2H_5 \rightarrow$$
$$\text{VII}$$

$$\begin{array}{c} \quad CN\qquad\quad CN \\ \quad |\qquad\qquad| \\ C_2H_5O_2CC\text{------}CCO_2C_2H_5 \\ |\qquad\qquad\quad| \\ SC\qquad\quad CS \\ \diagdown\quad\diagup \\ S \\ \text{IX} \end{array}$$

[160] Wenzel, *Ber.*, **33**, 2041 (1900); **34**, 1043 (1901).

3,4-Dichloro-3,4-dimethyltetrahydrothiophene (X) has been prepared in about 1% yield by the action of sulfur dichloride on 2,3-dimethyl-1,3-butadiene (XI).[161] 3,4-Dichloro-3-methyltetrahydrothiophene (XII) was obtained similarly from isoprene.[161]

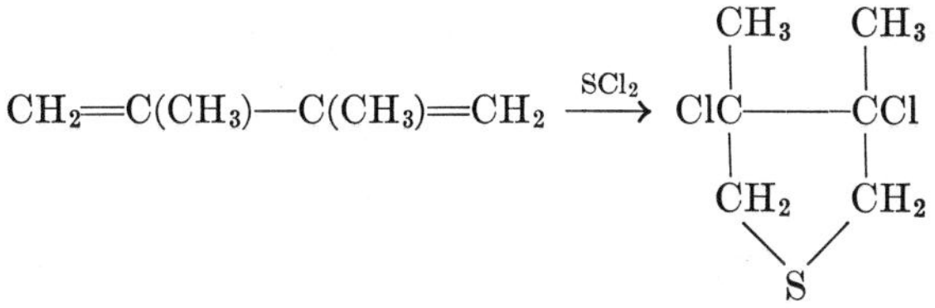

$$CH_2=C(CH_3)-C(CH_3)=CH_2 \xrightarrow{SCl_2}$$

<pre>
              CH₃   CH₃
               | |
         ClC———CCl
               | |
              CH₂   CH₂
                \\ /
                 S
</pre>

XI   X

$$CH_2=CHC(CH_3)=CH_2 \xrightarrow{SCl_2}$$

<pre>
                    CH₃
                     |
         ClCH———CCl
               | |
              CH₂   CH₂
                \\ /
                 S
</pre>

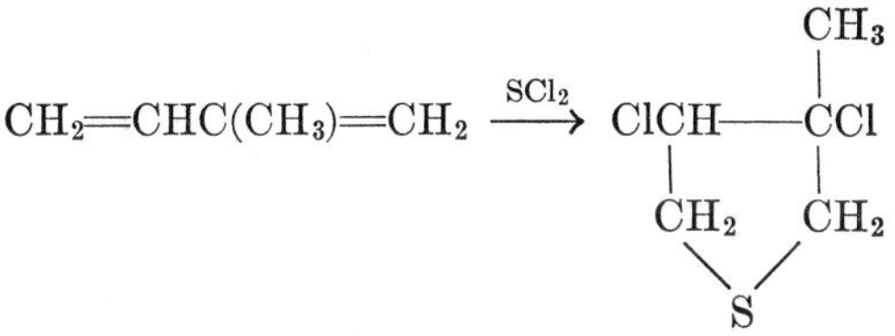

XII

2-Ketotetrahydrothiophene (XIII) or γ-thiobutyrolactone has been made by the slow distillation of γ-mercaptobutyric acid (XIV).[162]

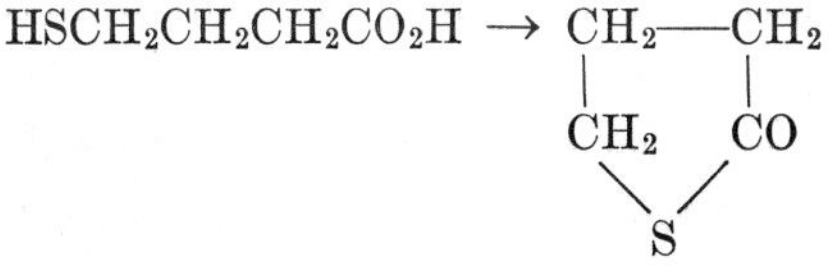

$$HSCH_2CH_2CH_2CO_2H \rightarrow$$

<pre>
         CH₂———CH₂
          | |
         CH₂   CO
           \\ /
            S
</pre>

XIV   XIII

2,5-Diketotetrahydrothiophene (XV), thiosuccinic anhydride, is formed readily when an aqueous solution of potassium thiosuccinate (XVI) is acidified with sulfuric acid.[163]

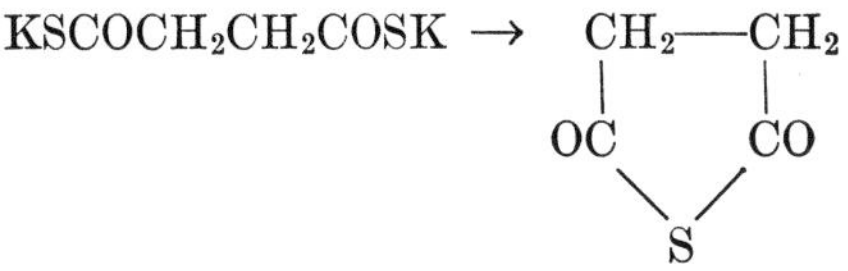

$$KSCOCH_2CH_2COSK \rightarrow$$

<pre>
         CH₂———CH₂
          | |
         OC   CO
           \\ /
            S
</pre>

XVI   XV

---

[161] Backer and Strating, *Rec. trav. chim.*, **54**, 52 (1935).
[162] Holmberg and Schjanberg, *Arkiv Kemi, Mineral. Geol.*, **14A**, No. 7, 22 pp. (1940) [*C. A.*, **35**, 2113 (1941)].
[163] Weselsky, *Ber.*, **2**, 518 (1869).

TABLE X
Tetrahydrothiophenes by Miscellaneous Methods

Tetrahydrothiophene	Starting Material	Reagents and Experimental Conditions	Yield %	Reference
3,3,4,4-Tetracarbethoxy-	$(C_2H_5O_2C)_2CHCH(CO_2C_2H_5)_2$ + $(ClCH_2)_2S$	$NaOC_2H_5$ in ethanol at reflux	27	158, 159
2,5-Dithiono-3,3,4,4-tetracarbethoxy-	$Na[CH(CO_2C_2H_5)_2]$ + CS_2	Br_2 in CS_2	5 to 20	160
2,5-Dithiono-3,4-dicyano-3,4-dicarbethoxy-	$Na(CHCNCO_2C_2H_5)$ + CS_2	Br_2 in CS_2	87	160
3,4-Dichloro-3-methyl-	CH_2=$C(CH_3)CH$=CH_2	SCl_2 in petroleum ether	1	161
3,4-Dichloro-3,4-dimethyl-	CH_2=$C(CH_3)C(CH_3)$=CH_2	SCl_2 in petroleum ether	1	161
2-Keto-	$HSCH_2CH_2CH_2CO_2H$	Slow distillation	—	162
2,5-Diketo-	$KSCOCH_2CH_2COSK$	H_2SO_4 in aqueous solution	—	163

One method of preparing tetrahydrothiophenes, which does not involve formation of the heterocyclic ring and is therefore beyond the scope of this chapter, requires mention. Tetrahydrothiophene and a number of substituted tetrahydrothiophenes have been prepared by catalytic hydrogenation of thiophene and substituted thiophenes over palladium-carbon or palladium-barium sulfate.[164] The tetrahydrothiophenes prepared in this way have not been included in Table X.

[164] Mozingo, Harris, Wolf, Hoffhine, Jr., Easton, and Folkers, *J. Am. Chem. Soc.*, **67**, 2092 (1945).

CHAPTER 10

REDUCTIONS BY LITHIUM ALUMINUM HYDRIDE

WELDON G. BROWN

University of Chicago

CONTENTS

	PAGE
INTRODUCTION	470
Table I. Functional Groups Reduced by Lithium Aluminum Hydride	471
MECHANISM	471
SCOPE AND LIMITATIONS	473
Compounds Containing Active Hydrogen	473
Reduction of Aldehydes and Ketones	474
Reduction of Epoxides	476
Reduction of Esters	477
Reduction of Carboxylic Acids	478
Reduction of Amides	479
Reduction of Nitriles	480
Reduction of Halogen Compounds	480
Reduction of Double Bonds	481
Reduction of Heterocyclic Nitrogen Compounds	483
THE LITHIUM ALUMINUM HYDRIDE REAGENT	483
Formation and Properties of Lithium Aluminum Hydride	483
Preparation and Analysis of Solutions of Lithium Aluminum Hydride	484
EXPERIMENTAL CONDITIONS	486
Solvents	486
Hydride Solution vs. Slurry	487
Alternative Methods of Introducing Reactants	487
Alternative Methods of Decomposing Excess Hydride	487
Alternative Methods of Isolating Products	488
Fire Hazard	489
EXPERIMENTAL PROCEDURES	489
2,2,2-Trichloroethanol (Reduction of Chloral Hydrate)	489
Cinnamyl Alcohol (Reduction of Cinnamaldehyde)	490
Vitamin A Alcohol (Reduction of the Ethyl Ester of Vitamin A Acid)	490
o-Aminobenzyl Alcohol (Reduction of Anthranilic Acid)	491
3,5-Dimethoxybenzyl Alcohol (Reduction of 3,5-Dimethoxybenzoic Acid)	491
N-Phenylpyrrolidine (Reduction of N-Phenylsuccinimide)	492

	PAGE
TABULAR SURVEY OF REDUCTIONS WITH LITHIUM ALUMINUM HYDRIDE	493
Table II. Aldehydes, Ketones, Quinones	494
Table III. Epoxides	497
Table IV. Esters and Lactones	498
Table V. Carboxylic Acids and Anhydrides	504
Table VI. Amides and Nitriles	505
Table VII. Miscellaneous Nitrogen Compounds	506
Table VIII. Halogen Compounds	507
Table IX. Sulfur Compounds	508

INTRODUCTION

Lithium aluminum hydride,* one of a group of recently discovered complex metal hydrides, is a useful and convenient reagent for the selective reduction of various polar functional groups. It is used in diethyl ether solution, less commonly in higher-boiling ethers, following the conventional procedures for syntheses employing Grignard reagents which the hydride closely resembles in its general pattern of behavior. Normally, the reactions proceed with extraordinary rapidity and are relatively free from side reactions. The principal limitation on yield is the loss entailed in isolation of the product. As in Grignard syntheses, the reactions usually give rise to intermediate metal alkoxides from which the desired products are liberated by hydrolysis.

The types of organic compounds reduced by lithium aluminum hydride, and the nature of the reduction products, are set forth in Table I. Certain of the reactions indicated in the table are known to be quite general; others are known to be subject to definite limitations, as the later discussion will show. Still others can be substantiated as yet by such a limited number of observations that generalizations would be premature; the data pertaining to these will be presented in the tabular survey without comment.

It is perhaps equally important to define the functional groups that are not reduced by lithium aluminum hydride, but this cannot be done without qualification as to experimental conditions or without recognizing that there may be exceptions. Under normal operating conditions the following types are reduced either slowly or not at all: alcohols, ethers, ketals, carbon-carbon double and triple bonds, diaryl sulfones,

* The first account of the reactions of lithium aluminum hydride was presented in a joint paper by Finholt, Nystrom, Brown, and Schlesinger before the Symposium on Hydrides and Related Compounds at the Chicago meeting of the American Chemical Society, September 10, 1946. The subject matter of this paper was later published in a paper by Finholt, Bond, and Schlesinger (ref. 56) dealing with the discovery of the reagent and certain inorganic applications, and in a series of three papers by Nystrom and Brown (refs. 10, 27, and 36) dealing with organic applications.

and dialkyl peroxides. Some of the exceptions, particularly those involving the reduction of double bonds, will be specifically noted later.

TABLE I

FUNCTIONAL GROUPS REDUCED BY LITHIUM ALUMINUM HYDRIDE

Functional Group	Product	Moles LiAlH$_4$ Required (Theoretical)
Aldehyde	Primary alcohol	0.25
Ketone	Secondary alcohol	0.25
Quinone	Hydroquinone	0.25
Epoxide	Alcohol	0.25
Ester	Primary alcohol	0.5
Lactone	Diol	0.5
Carboxylic acid	Primary alcohol	0.75
Anhydride	Primary alcohol	1
Amide, —CONH$_2$	Primary amine	1
Amide, —CONHR	Secondary amine	0.75
Amide, —CONR$_2$	Tertiary amine	0.5
	Aldehyde	0.25
Nitrile	Primary amine	0.5 *
	Imine (aldehyde)	0.25
Nitro (aryl)	Azo compound	1
Nitro (aliphatic)	Amine	1.5
Azoxy	Azo compound	0.5
Anil	Amine	0.25
Nitroso	Azo compound	0.5
Acid chloride	Primary alcohol	0.5
Alkyl halide	Hydrocarbon	0.25
Disulfide	Thiol	0.5
Sulfoxide	Thioether	0.5
Sulfonyl chloride	Thiol	0.5
Sulfonic ester	Various products	—

* It has been reported, reference 42a, that 1 mole of hydride is required.

MECHANISM

The constitution of lithium aluminum hydride can only be inferred, reasoning by analogy with lithium borohydride which it closely resembles in properties and reactions. X-ray observations on the crystalline borohydride point toward a polar structure consisting of lithium ions and tetrahedral borohydride ions.[1] Lithium aluminum hydride is possibly somewhat less polar than the borohydride, but it is reasonable to suppose

[1] Harris and Meibohm, *J. Am. Chem. Soc.*, **69**, 1231 (1947).

that in ether solutions it exists largely as ionic aggregates of strongly solvated lithium ions and aluminohydride anions (AlH_4^-).

Nearly all the normal reduction reactions involve the displacement of a strongly electronegative atom (O, N, halogen, etc.) and the accession of a hydrogen atom to the electron deficient center, usually a carbon atom. Assuming the reactive species to be the aluminohydride ion, the most plausible mechanism would appear to be one in which hydrogen is transferred as hydride in a bimolecular nucleophilic displacement.[2] Illustrated with reference to the reduction of an epoxide, the initial step would occur as shown in the equation. It is probable that the neutral

$$AlH_4^- + \underset{\diagdown}{\overset{\diagup}{C}}\!\!\!\!\underset{\diagdown}{\overset{\diagup}{\underset{C}{\overset{}{|}}}}\!\!\!\!O \longrightarrow AlH_3 + \underset{\diagdown}{\overset{\diagup}{C}}\!\!-\!\!O^- \atop H\!-\!\underset{\diagdown}{\overset{\diagup}{C}}$$

aluminum hydride immediately coordinates with the alkoxide anion, forming a new ion of the form AlH_3OR^-, which, by successive bimolecular reactions of a similar kind with additional molecules of the reactant, is eventually converted to $Al(OR)_4^-$. In the general case it is by no means certain that the aluminum hydride formed in the first step must necessarily coordinate with the available anions and thereafter continue the sequence of nucleophilic displacements. In the reduction of certain alkyl halides, the reaction comes virtually to a halt after one of the four hydrogens of the original lithium aluminum hydride has reacted.

The assumed mechanism is supported experimentally by the demonstration of inversion of configuration in the reduction of epoxides, by the observation that the mode of ring opening in unsymmetrical epoxides is the same as in known bimolecular nucleophilic displacements, and by a comparison of reactivities of alkyl halides.[2] The prediction that reduction of an optically active secondary alkyl halide by lithium aluminum deuteride would lead to an optically active hydrocarbon has also been verified.[3] Further evidence for the interpretation of reduction by lithium aluminum hydride as a nucleophilic displacement reaction is to be found in the mode of reaction with toluenesulfonic esters.[4]

A more complicated sequence of reactions is involved in the reduction of nitro groups, sulfoxides, etc., where the reactions are accompanied by the evolution of hydrogen gas. It is apparent that an initial transfer of hydrogen to a nitrogen or sulfur atom creates an active hydrogen

[2] Trevoy and Brown, *J. Am. Chem. Soc.*, **71**, 1675 (1949).
[3] Eliel, *J. Am. Chem. Soc.*, **71**, 3970 (1949).
[4] Kenner and Murray, *J. Chem. Soc.*, **1950**, 406.

atom, which must subsequently be removed by further reaction with the metal hydride.

The reduction of double bonds, which occurs with cinnamyl alcohol, is known not to proceed by the addition of two hydrogen atoms supplied by the hydride. Instead an aluminum atom becomes bonded to the ethylenic carbon atom nearer the benzene ring and a hydrogen atom supplied by the hydride adds to the other carbon atom of the ethylenic center. On hydrolysis the aluminum atom is replaced by hydrogen supplied by the hydrolyzing agent.[5]

SCOPE AND LIMITATIONS

Compounds Containing Active Hydrogen

The use of lithium aluminum hydride to determine quantitatively the active hydrogen in organic compounds will not be reviewed here in detail.[6,7,8] From the standpoint of syntheses employing the hydride it is important, however, to consider the reactions, the extent to which they interfere with concurrent reductions, and means of avoiding such interference when it arises.

In broad terms, any and all hydrogen atoms attached to nitrogen, oxygen, or sulfur are active hydrogens with respect to lithium aluminum hydride and will react with the liberation of one mole of hydrogen gas and the consumption of one-quarter mole of the hydride per active hydrogen. So far as is known, all such reactions are fast and complete, provided the compound can be brought into solution in ether. The reactions parallel the well-known reactions of methylmagnesium iodide (Zerewitinoff procedure for active hydrogen), but there are notable differences in degree and in the response of enolizable compounds. For example, primary amines ordinarily generate only one mole of methane from the Grignard reagent, but two moles of hydrogen are formed with the hydride.[8]

It is probable that the hydrogen liberated by enolizable substances corresponds very closely to the true enol content. This is a consequence of the rapid reaction with both tautomeric forms, with one by replacement of active hydrogen and with the other by reduction, thus effectively freezing the interconversion. Acetomesitylene, although it reacts with methylmagnesium iodide to form methane, reacts normally with the hydride and shows a negligible enolic content.[8] However, some nitriles

---

[5] Hochstein and Brown, *J. Am. Chem. Soc.*, **70**, 3484 (1948).
[6] Krynitsky, Johnson, and Carhart, *J. Am. Chem. Soc.*, **70**, 486 (1948).
[7] Zaugg and Horrom, *Anal. Chem.*, **20**, 1026 (1948).
[8] Hochstein, *J. Am. Chem. Soc.*, **71**, 305 (1949).

are reduced slowly by lithium aluminum hydride and some hydrogen is evolved as a consequence of the greater opportunity for enolization.[7]

The reaction of an enol with the hydride presumably forms the lithium aluminum enolate, which upon hydrolysis will regenerate the original functional group. It is perhaps for this reason that the yields reported in reductions of malonic esters are not invariably good. A lithium aluminum enolate is probably formed during the reduction of α-angelica lactone which furnishes γ-acetopropanol as the product.[8] The nonreduction, or partial reduction, of enol forms thus constitutes a limitation on the hydride process.

Two aspects of the presence of ordinary active hydrogens (hydroxyl groups, amino groups, etc.) are to be considered. First, and incidentally, the wasteful consumption of reagent by such groups is undesirable. More important, if several such groups are present in a molecule the complex formed in the rapid reaction may throw the material out of solution before the reduction of other functional groups is complete. This difficulty frequently arises in the reduction of hydroxy acids and of amino acids.

It is frequently necessary to convert hydroxyl groups to acetoxy groups in order to achieve ether solubility. During the course of the hydride reduction the acetyl groups are eliminated and the formation of highly insoluble intermediate products is not avoided, but it may be sufficiently delayed to achieve the desired result.

Acylation of amino groups is effective in improving the ether solubility of amino acids but may lead to undesired products because the acylamino group is normally reduced to an alkylamino group by lithium aluminum hydride. However, the attack on the acylamino group may be relatively slow, making possible a selective reduction such as that reported for the methyl ester of dibenzoylhistidine, which was converted to monobenzoylhistidinol by selective reduction of the ester group.[9] It is not clear in this example whether the removal of one benzoyl group occurred by reaction with lithium aluminum hydride or during the subsequent operations.

Reduction of Aldehydes and Ketones (Table II)

The reduction of carbonyl groups seldom presents any great difficulty, and the alcohols are obtained in uniformly good yields. Ketones, such as acetomesitylene [10] and hexamethylacetone,[11] that show steric hindrance in their reactions with Grignard reagents and other nucleophilic

---

[9] Karrer, Suter, and Waser, *Helv. Chim. Acta*, **32**, 1936 (1949).
[10] Nystrom and Brown, *J. Am. Chem. Soc.*, **69**, 1197 (1947).
[11] Cook and Percival, *J. Am. Chem. Soc.*, **71**, 4141 (1949).

reagents behave normally toward the hydride. Cyclopentanone is converted to cyclopentanol in relatively poor yield (60%) by the normal procedure,[10] evidently because of the formation of a highly insoluble intermediate product that removes active hydride from the solution. If the mixture is refluxed for one hour an 85% yield is obtained,[12] and in boiling tetrahydrofuran the formation of cyclopentanol takes place in nearly quantitative yield.[8,13]

Unsymmetrical ketones introduce the problem of stereochemical specificity, owing to the appearance of a new asymmetric carbon atom on conversion to a secondary alcohol. In the reduction of several keto steroids, both epimeric alcohols are formed,[14,15,16] but in connection with the reduction of 7-ketocholesteryl acetate it has been noted [16] that the reduction proceeds "more efficiently and more predominately in one steric sense" than does the Meerwein-Ponndorf-Verley reduction.* A similar comment could be made with reference to camphor, which, in the hydride reduction, is converted almost exclusively to isoborneol,[2] but which, in the Meerwein-Ponndorf-Verley reduction, forms comparable amounts of borneol and isoborneol.[17] It is stated that the reduction of amidone forms one of the two possible products to the extent of 98%; the same product is formed by catalytic hydrogenation.[18] The stereochemical specificity shown in the reduction of benzil (81% mesohydrobenzoin) is augmented somewhat by conducting the reduction at −80° (90% mesohydrobenzoin).[2] Both *cis* and *trans* glycols are formed from acenaphthenequinone, and the composition of the mixture is not markedly influenced by the reaction temperature.[2]

Although the hydride method lacks the specificity for carbonyl groups that is characteristic of the Meerwein-Ponndorf-Verley method, the reduction by lithium aluminum hydride is advantageous with respect to the time required and the freedom from side reactions, and generally but not always with respect to yield. In no reduction yet reported is the yield in the Meerwein-Ponndorf-Verley process significantly higher. With respect to selectivity, sodium borohydride, a milder reducing agent than lithium aluminum hydride, is comparable to the Meerwein-Ponndorf-Verley method.[19]

* The Meerwein-Ponndorf-Verley reduction has been reviewed by Wilds, *Organic Reactions*, Vol. II, Chapter 5, John Wiley & Sons, New York, 1944.

[12] Roberts and Sauer, *J. Am. Chem. Soc.*, **71**, 3925 (1949).
[13] Nystrom, unpublished work.
[14] Plattner, Heusser, and Feurer, *Helv. Chim. Acta*, **31**, 2210 (1948).
[15] Plattner, Heusser, and Kulkarni, *Helv. Chim. Acta*, **32**, 265 (1949).
[16] Fieser, Fieser, and Chakravarti, *J. Am. Chem. Soc.*, **71**, 2226 (1949).
[17] Lund, *Ber.*, **70**, 1520 (1937).
[18] Speeter, Byrd, Cheney, and Binkley, *J. Am. Chem. Soc.*, **71**, 57 (1949).
[19] Chaikin and Brown, *J. Am. Chem. Soc.*, **71**, 122 (1949).

Certain ketones that are resistant to catalytic hydrogenation, e.g., isoamidone [20] and the morpholinyl analogs of both amidone and isoamidone,[18] have been successfully reduced by lithium aluminum hydride.

Where it is desired to effect the reduction of other functional groups without at the same time reducing carbonyl groups, blocking of the latter may be accomplished in various ways. The use of acetal derivatives is illustrated by the reduction of a sugar epoxide.[21] A somewhat similar treatment of the problem is involved in a reported synthesis of 17-α-hydroxypregnenolone, wherein the carbonyl group was protected by conversion to a ketal with ethylene glycol.[22] An alternative device, used in different forms by different workers, is the conversion of the carbonyl compound to a derivative of the enol form. Enol ethyl ethers,[23,24] benzyl thio-enol ethers, and β-hydroxyethyl thio-enol ethers [25] have been employed. The use of the unsaturated bromo derivative,[26] which upon hydrolysis generates a carbonyl group, falls in the same category.

Reduction of Epoxides (Table III)

The reductive cleavage of epoxide rings has proved to be a useful synthetic procedure in the steroid field for introducing a hydroxyl group at the former site of a double bond. Catalytic hydrogenolysis of the epoxides frequently fails either because the epoxide is unaffected or, at the other extreme, the oxygen may be completely removed. Numerous applications of the hydride to the reduction of steroidal epoxides will be found in the tables. No failures have been reported.

Unsymmetrical epoxides containing a primary and a secondary oxide linkage undergo mainly rupture of the primary linkage, forming secondary alcohols.[2] Styrene oxide is converted almost entirely to α-phenylethanol;[27] 3,4-epoxy-1-butene furnishes a mixture of 3-buten-1-ol and 3-buten-2-ol, the latter predominating.[2] A secondary oxide linkage is attacked in preference to a tertiary, and the normal product from such a combination is a tertiary alcohol.[28] An exception to this rule has been reported; β-cholesteryloxide acetate (I) yielded 20% of the expected product, 3β,5-dihydroxycoprostane (II), and 60% of the "abnormal" product, 3β,6β-dihydroxycholestane (III).[28] The occurrence of inversion of configuration in the formation of III will be noted; inversion also

[20] May and Mosettig, *J. Org. Chem.*, **13**, 663 (1948).
[21] Prins, *J. Am. Chem. Soc.*, **70**, 3955 (1948).
[22] Julian, Meyer, and Ryden, *J. Am. Chem. Soc.*, **71**, 756 (1949).
[23] Meystre and Miescher, *Helv. Chim. Acta*, **32**, 1758 (1949).
[24] Meystre and Wettstein, *Helv. Chim. Acta*, **32**, 1978 (1949).
[25] Rosenkranz, St. Kaufmann, and Romo, *J. Am. Chem. Soc.*, **71**, 3689 (1949).
[26] Wagner and Moore, *J. Am. Chem. Soc.*, **71**, 4160 (1949).
[27] Nystrom and Brown, *J. Am. Chem. Soc.*, **70**, 3738 (1948).
[28] Plattner, Heusser, and Feurer, *Helv. Chim. Acta*, **32**, 587 (1949).

accompanies the reduction of 1,2-epoxy-1,2-dimethylcyclohexane, the products in each case being *trans* alcohols.[2]

Reduction of Esters (Table IV)

The reduction of esters to primary alcohols is perhaps the most widely exploited reaction of lithium aluminum hydride. The examples reported thus far cover a wide range of types, the yields of alcohols are uniformly good, and relatively few reports of anomalous behavior have been recorded.

Under forcing conditions (elevated temperatures for long periods) reduction may be carried beyond the primary alcohol stage to the hydrocarbon,[13] but this behavior has not been encountered under normal conditions of operation.

An interesting anomaly appears in the behavior of 3-carbethoxy-4-ketoquinolizidine (IV), from which the only product, isolated in very small yield, was 4-ketoquinolizidine (V).[29]

The selective reduction of one ester group in esters of dicarboxylic acids is evidently not possible if the two ester groups are of comparable reactivity. Diethyl sebacate, treated with sufficient hydride to reduce one ester group, furnished only the diol and unchanged ester. A successful selective reduction of the primary carbomethoxyl group in dimethyl *cis*-2-methyl-2-carboxycyclohexaneacetate (VI) is reported.[30]

The reduction of optically active esters in which the α-carbon atom is asymmetric, as in the esters of the natural amino acids, occurs without

[29] Boekelheide and Rothchild, *J. Am. Chem. Soc.*, **71**, 879 (1949).
[30] Bachmann and Dreiding, *J. Am. Chem. Soc.*, **71**, 3222 (1949).

racemization.[31] Likewise, epimerizations due to labile α-hydrogen atoms, such as are known to occur in the reduction of esters of lysergic and of isolysergic acids by sodium, do not occur in the hydride reduction.[32]

The reduction of esters has been utilized as a means of recovering the alkoxy component where ordinary hydrolytic procedures might cause undesired racemization of the alcohol.[33,34]

Reduction of Carboxylic Acids (Table V)

The reduction of the free carboxylic acid is generally somewhat less satisfactory than the reduction of the corresponding ester or acid chloride. The acidic hydrogen consumes one-quarter mole of hydride in the initial reaction, and there is frequently formed an insoluble derivative which is slowly and sometimes incompletely reduced. A further disadvantage is that the acid itself is often of very limited solubility in ether, necessitating long periods of extraction in order to introduce the compound. Some acids, e.g., aliphatic amino acids, are so slightly soluble in ether that even this technique fails.

VII

Podocarpic acid (VII) was reduced to podocarpinol in 4.6% yield in two hours, and in 56% yield when the mixture was allowed to stand four days.[35] The ester and acid chloride of the O-methyl ether were readily reduced in 92% and 93% yield, respectively. Triphenylacetic acid is not reduced under ordinary conditions [36] but can be converted to the carbinol in good yield either by carrying out the reduction at a higher temperature, in tetrahydrofuran solution,[37] or by first converting to the acid chloride which is readily reduced under the usual conditions.[13] Pivalic acid is readily reduced to neopentyl alcohol;[36] slowness of reaction is therefore not invariably characteristic of tertiary acids.

[31] Karrer, Portmann, and Suter, *Helv. Chim. Acta*, **31**, 1617 (1948).
[32] Stoll, Hofmann, and Schlientz, *Helv. Chim. Acta*, **32**, 1947 (1949).
[33] Doering and Zeiss, *J. Am. Chem. Soc.*, **72**, 147 (1950).
[34] Cram, *J. Am. Chem. Soc.*, **71**, 3863 (1949).
[35] Zeiss, Slimowicz, and Pasternak, *J. Am. Chem. Soc.*, **70**, 1981 (1948).
[36] Nystrom and Brown, *J. Am. Chem. Soc.*, **69**, 2548 (1947).
[37] Hochstein, unpublished work.

Reduction of Amides (Table VI)

The normal reduction product of an amide, when excess lithium aluminum hydride is employed, appears to be the amine resulting from conversion of $RCONH_2$ to RCH_2NH_2. Exceptions have been reported in the formation of benzyl alcohol from diethylbenzamide,[27] and of 2-aminobutane-1,4-diol from ethyl asparaginate.[31] Since benzamide is converted to benzylamine in good yield,[38] the behavior of the diethyl derivative should perhaps be re-investigated; the earlier reduction of the diethyl derivative was carried out in the hope of obtaining benzaldehyde as an intermediate reduction product and not under conditions favoring reduction to the amine.

Certain cyclic amines not previously obtainable by any convenient methods are now easily prepared from the more readily available cyclic amides, e.g., phenylpyrrolidine from N-phenylsuccinimide [39] and cyclic polymethyleneimines from the lactams.[40]

An interesting reductive cyclization is shown in the synthesis of the yohimbine skeleton, X from VIII or IX.[41]

VIII → X ← IX

The amido ester XI, treated with a quantity of lithium aluminum hydride (0.3 mole) which would be insufficient for the complete reduction of either the amide or the ester group, furnished the amido alcohol XII in unstated yield.[42]

XI XII

[38] Matlow, unpublished work.
[39] Spitzmueller, unpublished work.
[40] Ruzicka, Kobelt, Häfliger, and Prelog, *Helv. Chim. Acta*, **32**, 544 (1949).
[41] Julian and Magnani, *J. Am. Chem. Soc.*, **71**, 3207 (1949).
[42] Swan, *J. Chem. Soc.*, **1949**, 1720.

The anomalous behavior of the methyl ester of dibenzoylhistidine, which loses one benzoyl group entirely while the other is unchanged, has been mentioned earlier.

Reduction of Nitriles (Table VI)

Benzonitrile and o-tolunitrile have been reduced to the corresponding amines in 72% and 88% yields, respectively.[27] Mandelonitrile and sebaconitrile gave lower yields (48% and 40%, respectively) while lauryl cyanide gave a 90% yield of amine.[27] The lower yields are believed to be due to the precipitation of intermediate products rendered highly insoluble through the bifunctionality of these substances. A more recent procedure describes the reduction of five aliphatic and aromatic nitriles to the corresponding primary amines in high yields.[42a]

The discovery [43] that the reduction of nitriles can be so conducted as to furnish aldehydes is certain to extend very greatly the utility of hydride reduction procedures. It also demonstrates quite clearly that the steps involved in the reduction of a nitrile are the following, where $M = \dfrac{LiAl}{4}$.

$$RC\equiv N \xrightarrow{MH} RCH=NM \xrightarrow{MH} RCH_2NM_2$$
$$\downarrow H_2O \qquad\qquad \downarrow H_2O$$
$$RCHO \qquad\qquad RCH_2NH_2$$

The complete reduction of a nitrile, i.e., reduction to the amine, may be slow or may require elevated temperature if no more than the calculated quantity of hydride is employed, and a substantial excess is usually advisable. It is also advisable to conduct the reduction of nitriles under nitrogen as there is evidence that the intermediate products are oxygen-sensitive.[27] The same is true also of the reduction of nitro compounds.

Reduction of Halogen Compounds (Table VIII)

Replacement of the halogen atom of alkyl halides by hydrogen by the action of lithium aluminum hydride shows the general characteristics of nucleophilic displacement reactions, and the wide variation in the ease and completeness of reaction can be regarded as normal. For practical purposes, the reaction is limited to primary and secondary halides of the aliphatic type, and, among the halogens, the usual order of reactivity holds, i.e., iodides > bromides > chlorides.

[42a] Amundsen and Nelson, *J. Am. Chem. Soc.*, **73**, 242 (1951).
[43] Friedman, Abstracts of Papers, 116th meeting American Chemical Society, September 18–23, 1949, p. 5M.

Deviations from the normal replacement of halogen by hydrogen have been observed in the formation of olefins from 1,2-dibromides and from tertiary alkyl halides. The normal reduction of diphenylbromomethane, and of 9-bromofluorene, is accompanied by the formation of dimeric reduction products. Color phenomena and other evidence point toward intermediate organometallic compounds in these reductions.[2] Triphenylchloromethane, with excess hydride, is largely converted to a colored organometallic derivative.[13]

Lithium aluminum hydride may act as a catalyst for the reduction of alkyl halides by lithium hydride.[44] Aluminum hydride formed in the initial reaction of lithium aluminum hydride with the alkyl halide re-forms lithium aluminum hydride by reaction with lithium hydride.

$$LiAlH_4 + RCl \rightarrow LiCl + AlH_3$$

$$AlH_3 + LiH \rightarrow LiAlH_4$$

The catalysis is essentially similar to the catalysis by lithium aluminum hydride of its own formation from lithium hydride and aluminum chloride.

Reduction of Double Bonds

There are several compounds in which the reduction of a polar functional group is accompanied by the complete or partial reduction of a carbon-carbon double bond in the α,β position. With few exceptions, this behavior is confined to aromatic systems containing the structural grouping ArC=CCO, or ArC=CN$\langle$.

Among purely aliphatic compounds, reduction of the double bond has been observed with allyl alcohol under forcing conditions,[5] and with α-ethylcrotonamide,[45] which is reported to yield α-ethylbutylamine * on prolonged treatment (twenty-four hours' refluxing). One instance of carbon-carbon triple bond reduction has been reported, namely, that of 1-(1'-cyclohexenyl)-1-butyn-3-ol (XIII) to the diene, XIV.[46]

    C≡CCHOHCH$_3$ →   CH=CHCHOHCH$_3$

    XIII   XIV

---

* The product is thus designated by the authors. If the starting material is given correctly as α-ethylcrotonamide, reduction of the amide group and of the double bond should have given β-ethylbutylamine.

[44] Johnson, Blizzard, and Carhart, *J. Am. Chem. Soc.*, **70**, 3664 (1948).
[45] Uffer and Schlittler, *Helv. Chim. Acta*, **31**, 1397 (1948).
[46] Chanley and Sobotka, *J. Am. Chem. Soc.*, **71**, 4140 (1949).

The reduction of the double bond of cinnamyl alcohol occurs by way of an oxygen-sensitive intermediate organometallic addition compound, believed to contain a carbon-aluminum bond that upon hydrolysis is replaced by a hydrogen atom derived from the solvent.[5] This addition to the double bond occurs at a moderate rate at room temperature. Consequently, it is possible to direct the reduction of the aldehyde, ester, etc., to give either cinnamyl alcohol or β-phenethyl alcohol in satisfactory yields by appropriate choice of conditions. Likewise, the reduction of benzalacetophenone can be controlled so as to provide either the saturated or unsaturated alcohol.

In some other substances of the cinnamyl type, double-bond reduction appears to proceed less readily. *p*-Methylcinnamic acid furnishes mainly the unsaturated alcohol,[47] and even upon prolonged refluxing in diethyl ether with excess hydride conversion to the saturated alcohol is incomplete. Coumarin is reported by one investigator [8] to be reduced mainly to 3-(*o*-hydroxyphenyl)propanol, together with some of the normal product, *o*-hydroxycinnamyl alcohol, but another group [48] obtained only the normal product under all conditions tried. However, the same group observed double-bond reduction with ethyl coumarate, and in fact the abnormal product was obtained exclusively under all conditions tried. Ethyl acetoferulate (XV) was observed to form the normal product XVI, but the relatively low yield (43%, or 67% when isolated as the benzoate) does not exclude the possibility of some double-bond reduction.[49]

$$CH_3CO_2\text{-Ar}(CH_3O)\text{-}CH=CHCO_2C_2H_5 \rightarrow HO\text{-Ar}(CH_3O)\text{-}CH=CHCH_2OH$$

$$\text{XV} \qquad\qquad \text{XVI}$$

Double-bond reduction is involved in the action of the hydride upon perinaphthenone, benzanthrone, and β-angelica lactone.[8]

In systems containing the grouping $\text{ArC}=\text{CN}\langle$, double-bond reduction is represented by the formation of saturated amines from ω-nitrostyrenes,[27,50] and by the partial reduction of the indole ring that occurs as a side reaction with methyl-substituted oxindoles.[51] Indole itself is not reduced by lithium aluminum hydride, but 1-methylindole and 1,3-dimethylindole are converted to the corresponding indolines to the extent of 25–30%.[51] Several other compounds containing the indole structure listed in the tables are reported to furnish the normal products.

---

[47] Collins, unpublished work.
[48] Karrer and Banerjea, *Helv. Chim. Acta*, **32**, 1692 (1949).
[49] Allen and Byers, *J. Am. Chem. Soc.*, **71**, 2683 (1949).
[50] Hamlin and Weston, *J. Am. Chem. Soc.*, **71**, 2210 (1949).
[51] Julian and Printy, *J. Am. Chem. Soc.*, **71**, 3206 (1949).

Reduction of Heterocyclic Nitrogen Compounds

As far as the limited data at present permit any conclusion, it may be inferred that the pyrazole [52] and the imidazole [53] rings are stable toward lithium aluminum hydride. In several successful reductions of functional groups in pyridine derivatives the pyridine ring remains intact. However, pyridine itself is slowly attacked with the formation of dihydropyridine,[13] and phenanthridine is converted to 5,6-dihydrophenanthridine.[54]

Quaternary iodides in the quinoline and isoquinoline series are readily reduced, the products being N-alkyldihydroquinolines or the analogous dihydroisoquinolines.[55]

THE LITHIUM ALUMINUM HYDRIDE REAGENT

Formation and Properties of Lithium Aluminum Hydride

Lithium aluminum hydride is formed by the reaction of lithium hydride with anhydrous aluminum chloride in ether solution.[56] To a slurry of finely powdered lithium hydride in ether containing some previously formed lithium aluminum hydride, a solution of aluminum chloride is added at a rate sufficient to maintain refluxing conditions. Stirring is continued for a considerable period after the addition is complete. Lithium chloride precipitates during the reaction, and this, together with the excess lithium hydride, is separated by filtration under nitrogen pressure. The yield, based upon aluminum chloride, is practically quantitative under favorable conditions.

If aluminum chloride is present in excess or if the reaction is terminated before completion, aluminum hydride is formed. It is probably an intermediate in the autocatalytic formation of lithium aluminum hydride in accordance with the scheme shown below.

$$3LiAlH_4 + AlCl_3 \rightarrow 3LiCl + 4AlH_3$$

$$LiH + AlH_3 \rightarrow LiAlH_4$$

Aluminum hydride remains dissolved in ether for a time, but it is eventually transformed to an insoluble, non-volatile form containing firmly bound ether. The soluble form is an active reducing agent toward aldehydes, ketones, and esters.[13] The insoluble form is possibly a polymer of saltlike structure.

---

[52] Jones, *J. Am. Chem. Soc.*, **71**, 3994 (1949).
[53] Jones, *J. Am. Chem. Soc.*, **71**, 383 (1949).
[54] Wooten and McKee, *J. Am. Chem. Soc.*, **71**, 2946 (1949).
[55] Schmid and Karrer, *Helv. Chim. Acta*, **32**, 960 (1949).
[56] Finholt, Bond, and Schlesinger, *J. Am. Chem. Soc.*, **69**, 1199 (1947).

Lithium aluminum hydride likewise retains ether tenaciously. In order to obtain a product substantially free of ether, it is necessary to heat the residue left after evaporation of the bulk of the ether under high vacuum at 70°. There has been no conclusive evidence that the intensively dried solid is not a mixture of lithium hydride and aluminum hydride from which lithium aluminum hydride slowly re-forms when the material is suspended in ether.

Thermal decomposition of lithium aluminum hydride sets in at about 120°, is rapid at 150°, and complete at 220° in accordance with the equation.[56]

$$LiAlH_4 \rightarrow LiH + Al + 1.5H_2$$

The approximate solubilities of the hydride, in grams per hundred grams of solvent at 25°, are as follows: [56]

Diethyl ether	25–30
Tetrahydrofuran	13
Di-n-butyl ether	2
Dioxane	0.1

The solid reacts superficially with atmospheric moisture and carbon dioxide. With water in large amounts it reacts in accordance with the following equation.

$$LiAlH_4 + 4H_2O \rightarrow LiOH + Al(OH)_3 + 4H_2$$

When the hydride is in excess the reaction takes the course: [8]

$$LiAlH_4 + 2H_2O \rightarrow LiAlO_2 + 4H_2$$

In ether solution, the hydride reacts slowly with atmospheric oxygen, liberating hydrogen.[8]

Preparation and Analysis of Solutions of Lithium Aluminum Hydride

Stock solutions of the hydride are most conveniently prepared by the following procedure. If the reagent is available only in lump form, it is crushed to a powder in a dry atmosphere. Grinding in a mortar should not be attempted except with care and in an atmosphere of nitrogen. Avoiding as far as possible exposure to atmospheric moisture, the powder is transferred to a dry two-necked flask and covered at once with dry ether. The quantity of reagent and the volume of ether may be conveniently taken so as to make up a 1 M solution, i.e., 38 g./l., a 5–10% excess of the hydride being added to allow for insoluble material and other impurities.

The flask is equipped with a sealed stirrer, driven preferably by an explosion-proof motor, and a reflux condenser provided with a soda-lime drying tube at its open end. With moderately vigorous stirring, the mixture is maintained under gentle reflux for several hours, the time required being somewhat variable, depending upon the degree of subdivision of the reagent, the condition of the surface, and the grade of hydride. The technical grade will leave a substantial amount of gray residue of undissolved material, and the stirring may be discontinued when it is judged that the residue is no longer diminishing.*

The procedure from this point may vary with the preferences of the operator. If the solution is to be clarified by sedimentation, the contents of the flask are transferred rapidly, without cooling, to a tall cylinder. Some gas is liberated by moisture on the surface of the cylinder and moisture picked up during the transfer, but this soon subsides and the cylinder may be loosely stoppered or capped. Alternatively, the cap may be provided with an opening to a soda-lime drying tube. After a day or two, sedimentation will have progressed to the point where supernatant liquid may be withdrawn either by decantation or by means of a fitting which carries a delivery tube extending into the liquid and through which the liquid is forced by a slight pressure of nitrogen gas.

If the solution is to be clarified by filtration, a suitable procedure is the following. The filter is constructed from a large sintered-glass funnel of the Büchner type having at its lower end a male joint fitting to the receiver and having the upper part sealed to a reservoir large enough to take the entire charge at one filling. By means of a connection through a stopper in the opening, pressure (nitrogen gas) is applied cautiously. The pressure should be no more than a few centimeters of mercury if a high frequency of breakage of filter disks is to be avoided.

The sludge collected on the filter and remaining on the flask is disposed of by covering with dry dioxane and then cautiously adding wet dioxane or a mixture of ethanol and dioxane. When all the active hydride contained therein has been destroyed, the apparatus may be safely cleaned with aqueous acid.

Hydride hydrogen may be determined by measurement of the hydrogen gas evolved upon hydrolysis.[56] In the analysis of ether solutions it is necessary to correct the measured gas volume for ether vapor carried over; this correction becomes small, and the uncertainty becomes less, if the reaction vessel is immersed in an ice bath throughout the determination. Alternatively the hydrolysis may be carried out in an appa-

* The procedures described in this section are applicable to the preparation of solutions from lithium aluminum hydride of the grade hitherto available commercially. The currently available grade is said to be freely soluble in ether with little or no residue.

ratus so designed that the gas volume remains constant and the increase in pressure is measured.[57] Here also, in order to avoid the change in volatility of ether with temperature, the reaction vessel is maintained at ice-bath temperature, but in this method no correction for the partial pressure of ether is necessary.

EXPERIMENTAL CONDITIONS

Solvents. Although the great majority of hydride reductions have been carried out in diethyl ether solution, other solvents have been employed to permit operations at temperatures above the boiling point of diethyl ether, or for other reasons. Of the common solvents, tetrahydrofuran has been a frequent and di-n-butyl ether a somewhat less frequent choice.

Bis(β-ethoxyethyl)ether (Diethyl Carbitol) was chosen as the solvent for the reduction of radioactive carbon dioxide to methanol;[58] here the problem of isolating a volatile reduction product necessitated the use of a non-volatile solvent.

Where the reduction is impeded by the formation of highly insoluble precipitates, an alternative to operating at a higher temperature is the use of N-ethylmorpholine,[8] which has good solvent characteristics not only for lithium aluminum hydride but also for the intermediate reduction products. Unfortunately this solvent is not readily available in pure form, and the purification is somewhat troublesome.

Pyridine is unsuitable because it is attacked by the reagent. The ethers, tetrahydrofuran and di-n-butyl ether, are also attacked by the reagent at elevated temperatures over a long period of time, but apart from the small loss of reagent this reaction causes no serious interference.

Dioxane has been used rarely. It is not a particularly good solvent for lithium aluminum hydride, and moreover the isolation of products is complicated by its miscibility with water.

Solutions of lithium aluminum hydride in solvents other than diethyl ether may be prepared by the direct method, which is slow, or by addition of the solvent in question to a diethyl ether solution followed by evaporation of the diethyl ether under reduced pressure. The latter procedure permits the preparation of more concentrated solutions, and the hydride is probably present in such solutions as the diethyl etherate.

The purification of solvents for use in hydride reductions requires much the same care as would ordinarily be taken in work with Grignard reagents. Freedom not only from water, but also from alcohols, alde-

[57] Krynitsky, Johnson, and Carhart, *Anal. Chem.*, **20**, 311 (1948).
[58] Nystrom, Yanko, and Brown, *J. Am. Chem. Soc.*, **70**, 441 (1948).

hydes, ketones, esters, etc., is desirable. Treatment with sodium does not completely remove these impurities but is a useful preliminary to a final treatment with lithium aluminum hydride. In the recovery of higher-boiling ethers after treatment with the hydride, vacuum distillation should be used to avoid as much as possible the ether cleavage reaction which occurs on prolonged heating. The purification of commercial tetrahydrofuran may require several prolonged treatments with sodium to arrive at a product that will not discolor when subjected to further treatment.

Hydride Solution vs. Slurry. Most workers have used the hydride in the form of a clarified solution, but it is becoming increasingly common practice to use directly the slurry that is obtained upon stirring the solid hydride with ether. This avoids the troublesome filtration, the transfers of material, and the sludge disposal. It is without doubt the most economical procedure when hydride reductions are to be carried out only occasionally. If such reductions are being done routinely it is advantageous to have a stock solution of known hydride content. In those rare instances requiring inverse addition of reagents, it is essential to have, if not a clear solution, one that will flow freely through the stopcock of a dropping funnel.

Alternative Methods of Introducing Reactants. In the normal procedure the substance to be reduced is added to a solution or slurry of the hydride. If the substance to be reduced is a liquid or solid, soluble in ether, an ether solution is added in order that the reaction, usually vigorously exothermic, may be moderated. For solids of limited solubility in ether, it is convenient to place the material in the thimble of an extractor inserted between the reaction flask and the reflux condenser; then, with the application of external heat, the substance is eventually carried into the reaction flask. Acids of moderate ether solubility, when handled in this way with a Soxhlet extractor, produce an undesirably large surge of gas each time the extractor reservoir discharges its contents; for such compounds a continuous-return type of extractor is preferable.

Some workers have introduced solid reactants by means of a mechanically operated hopper;[59] others have introduced the solid manually, in small portions, through the opening in a wide-bore reflux condenser.[60]

Alternative Methods of Decomposing Excess Hydride. It is usually, but not invariably, true that hydride reductions are best accomplished by having the hydride in excess of that consumed in the reduction, and occasionally quite a large excess (2- to 4-fold) is used. The destruction

[59] Ehrlich, *J. Am. Chem. Soc.*, **70**, 2286 (1948).
[60] Neville, unpublished work.

of this excess presents no problem on a small scale and may be accomplished by the cautious addition of wet ether, an ethanol-ether mixture, or (with extra caution) water. When water is used, it is desirable to employ a large flask on account of the frothing that takes place. If the amount of hydride to be destroyed is considerable, the hazard may be greatly reduced by the employment of a reactant which does not generate hydrogen gas. Ethyl acetate is suitable for this purpose, as its reduction product, ethanol, does not interfere in the subsequent isolation; it is used routinely by some workers.

Alternative Methods of Isolating Products. Isolation presents a variety of problems differing according to the solubility and the stability of the product. If the product is ether-soluble and stable to acid, the reaction mixture, after destruction of excess hydride, may be poured into a mixture of ice and dilute acid; the procedure thenceforth is the same as in a Grignard synthesis. If the product is an ether-soluble amine, the isolation will usually be accomplished more directly by treatment of the mixture, after hydrolysis of excess hydride, with strong sodium hydroxide solution, which will dissolve the precipitated alumina and allow a clean-cut separation of phases. If the basic compound will not tolerate contact with concentrated alkali, the precipitated alumina may be dissolved by sodium potassium tartrate.

It is not always essential, however, to dissolve the alumina to permit a satisfactory isolation by means of extraction procedures. If the amount of water added to the reaction mixture is limited to a small excess over that required for hydrolysis of both excess hydride and the product complex, a granular mass, consisting essentially of lithium aluminate, is obtained. The ether solution can then be separated without difficulty by filtration or decantation, and the solid mass can be triturated with further quantities of solvent to effect substantially complete product recovery in favorable cases.

Another method, applicable to the isolation of substances that will undergo the Schotten-Baumann reaction, consists of treatment of the mixture resulting from hydrolysis with an excess of an acid chloride, e.g., benzoyl chloride, thereby converting the product to an acyl derivative. This procedure is advantageous in furnishing more readily crystallizable products, in furnishing the product in a form less sensitive to decomposition, or in furnishing the product in a form more readily extractable by ether.

The isolation of water-soluble products (glycols, polyamines, amino alcohols, etc.) presents problems that cannot invariably be solved adequately by the above-mentioned procedure employing the Schotten-Baumann reaction. In a limited way, these problems have been resolved by the application of devices providing automatic continuous extraction. Ion-exchange resins should provide an elegant method for the solution of some of these problems; however, no procedures employing resins in this connection have been reported.

Fire Hazard. The hazard involved in the use of lithium aluminum hydride is probably less than with most other metal hydrides and, except for the fact that hydrogen gas is evolved during some reactions, is not significantly greater than with Grignard reagents. It may not be amiss, however, to direct attention to the potential fire hazard in large-scale operations. Adequate provision should be made to discharge hydrogen gas from the reactor to the atmosphere without risk from nearby flames, hot plates, brush-type motors, etc. Carbon dioxide-filled fire extinguishers are not ideal because of the rapid exothermic reaction between the hydride and carbon dioxide, but perhaps they are less objectionable than other available types.

There is evidence for the formation of an intermediate product in the carbon dioxide reaction that is explosive when dry.[61]

EXPERIMENTAL PROCEDURES

2,2,2-Trichloroethanol (Reduction of Chloral Hydrate).[13] The apparatus consists essentially of a 2-l. three-necked flask provided with a mercury-sealed mechanical stirrer, a dropping funnel, and a reflux condenser. Normal precautions are taken to ensure that the apparatus is dry, and the opening of the reflux condenser is fitted with a drying tube. The operation is conducted in a hood with good draft, and an induction-type motor is used to drive the stirrer.

Six hundred milliliters of a 0.5 M stock solution of lithium aluminum hydride in ether is transferred to the reaction flask. A solution of 35 g. (0.2 mole) of chloral hydrate in 100 ml. of dry ether is added dropwise from the dropping funnel at such a rate that the capacity of the reflux condenser is not exceeded. The addition will require thirty to sixty minutes, and spontaneous refluxing of the ether solution will continue for a short time thereafter. The mixture is allowed to stand with continued stirring for two hours after the addition has been completed. Water is then placed in the dropping funnel, and, with an ice bath surrounding the reaction vessel, it is added cautiously, one drop at a time, until there is no further evidence of hydrogen gas evolution. This is followed by 250 ml. of 10% sulfuric acid, which will cause the precipitated alumina to dissolve. The contents of the flask are then transferred to a separatory funnel, and the aqueous phase is extracted twice with

---

[61] Barbaras, Barbaras, Finholt, and Schlesinger, *J. Am. Chem. Soc.*, **70**, 877 (1948).

200-ml. portions of ether. The combined ether solutions are dried over potassium carbonate and distilled, first at atmospheric pressure to remove most of the ether, and then under reduced pressure using a 24-in. helical-wire packed column. The product, 2,2,2-trichloroethanol, is collected at 61°/20 mm.; the yield is 26 g. (50%). With p-nitrobenzoyl chloride, it reacts to form a p-nitrobenzoate, m.p. 71°.

Trichloroethanol has also been prepared in 65%, 64%, and 31% yields by the reduction of ethyl trichloroacetate, trichloroacetyl chloride, and trichloroacetic acid, respectively.[62]

Cinnamyl Alcohol (Reduction of Cinnamaldehyde).[5] This procedure illustrates the conditions under which reduction of a double bond may be avoided: inverse order of addition, low temperature, and minimum quantity of hydride. The normal procedure results in the formation of hydrocinnamyl alcohol.

A solution of 31 g. (0.23 mole) of cinnamaldehyde in 80 ml. of dry ether is placed in a 300-ml. three-necked flask to which are fitted a stirrer, a dropping funnel, and a thermometer reaching into the liquid. A side arm below the tip of the dropping funnel is open to the atmosphere through a drying tube. The solution is cooled to $-10°$ by means of an ice-salt bath, and there is added from the dropping funnel 40 ml. of a solution of lithium aluminum hydride in diethyl ether containing 0.065 mole of hydride, which is 10% in excess of the theoretical requirement. During the addition, which lasts about thirty minutes, the temperature is not allowed to rise above 10°. An additional ten minutes is allowed for completion of the reaction, and water is then added, cautiously at first, to decompose excess hydride. This is followed by 80 ml. of 10% sulfuric acid, and the product is taken up in ether in the usual way. Upon evaporation of the ether the residue solidifies to a mass of crystals, and after vacuum distillation, there is obtained 28 g. (90%) of cinnamyl alcohol, m.p. 33–34°.

Vitamin A Alcohol (Reduction of the Ethyl Ester of Vitamin A Acid).[63] A 3-l. three-necked flask is equipped with a stirrer, a dropping funnel, and a thermometer. In the flask is placed a solution of 15.9 g. (0.42 mole) of lithium aluminum hydride in 1280 ml. of diethyl ether. The solution is cooled to $-65°$, and a solution of 115 g. (0.5 mole) of the ethyl ester of vitamin A acid in 400 ml. of ether is added dropwise at a rate such that the temperature does not exceed $-60°$. Upon completion of the addition, the solution is held at $-30°$ for one hour. Decomposition of excess hydride is effected by the rapid addition of 12.4 g. (0.141

---

[62] Sroog, Chih, Short, and Woodburn, *J. Am. Chem. Soc.*, **71**, 1710 (1949).

[63] Schwarzkopf, Cahnmann, Lewis, Swidinsky, and Wuest, *Helv. Chim. Acta*, **32**, 443 (1949).

mole) of ethyl acetate, which causes the solution to become viscous. Hydrolysis is then brought about by the addition of 88 ml. of saturated ammonium chloride solution, and the mixture is allowed to reach 20°. The fine precipitate that has formed is separated by filtration and washed with ether. After evaporation of the ether at 50°, the remaining volatile impurities are removed by the application of high vacuum, leaving a residue of orange-colored viscous oil. The crude product, obtained in quantitative yield, may be purified by conversion to the acetate.

The same authors report the reduction of the methyl ester and of the acid to vitamin A. The synthesis of vitamin A, one of the obvious industrial applications of lithium aluminum hydride from the outset, has been accomplished by other investigators also.[64,65] See also reference 66.

o-Aminobenzyl Alcohol (Reduction of Anthranilic Acid).[36] In this procedure, a compound of low solubility in ether is placed in the thimble of an extractor and is carried into the reaction vessel by refluxing ether.

A 3-l. three-necked flask is arranged with a sealed stirrer and a Soxhlet extractor surmounted by an efficient reflux condenser, and the third neck is stoppered. A wide-bore drying tube is attached to the upper opening of the reflux condenser. A solution of 9.1 g. (0.24 mole) of lithium aluminum hydride in 600 ml. of ether is placed in the flask, and 13.7 g. (0.1 mole) of anthranilic acid is placed in the extractor thimble. By means of a heating mantle, the hydride solution is maintained at a moderate rate of boiling until all the acid in the thimble has been dissolved. The flask is then cooled; the Soxhlet extractor is removed, and the condenser, without the drying tube, is connected directly to the flask; finally, a dropping funnel is placed in the opening previously stoppered. Sufficient water is then added, cautiously at first, to decompose excess hydride. This is followed by 250 ml. of 10% sodium hydroxide solution. The ether layer is separated and combined with two further ether washings of 200 ml. each and dried, first over sodium sulfate, then over Drierite. Evaporation of the ether leaves a solid residue that is further dried over calcium hydride in vacuum for five hours. The product without further purification melts at 82°; the yield is 97%.

3,5-Dimethoxybenzyl Alcohol (Reduction of 3,5-Dimethoxybenzoic Acid).[67] In this example an ether-insoluble compound is added to the hydride solution as a suspension in ether. The authors state that the use of a Soxhlet extractor offers no advantage in this reaction.

[64] Cawley, Robeson, Weisler, Shantz, Embree, and Baxter, Abstracts of Papers, 112th meeting American Chemical Society, September, 1947, p. 26C.
[65] Wendler, Rosenblum, and Tishler, *J. Am. Chem. Soc.*, **72**, 234 (1950).
[66] Milas and Harrington, *J. Am. Chem. Soc.*, **69**, 2247 (1947).
[67] Adams, Harfenist, and Loewe, *J. Am. Chem. Soc.*, **71**, 1624 (1949).

A suspension of 91 g. of 3,5-dimethoxybenzoic acid in 1.5 l. of ether is added, as rapidly as the vigorous boiling of the solution will allow, to a solution of 24 g. of lithium aluminum hydride (94% purity) in 1.5 l. of anhydrous ether in a flask equipped with an efficient Hershberg stirrer,[67a] an addition funnel with a wide-bore stopcock, and a condenser. The solution is refluxed for fifty minutes after the addition. The flask is then cooled by the external application of ice while 150 ml. of water is added, the first few milliliters being added with extreme caution. An iced solution of 100 ml. of concentrated sulfuric acid in 2 l. of water is then added slowly. The ethereal layer is separated, washed with dilute acid, aqueous sodium bicarbonate, and water, and is then dried over magnesium sulfate. Distillation of the tan-colored oil obtained by removal of the ether, all the material that distils up to 170°/0.6 mm. being collected, furnishes 76 g. of product, m.p. 46°. The yield, corrected for 2.5 g. of acid recovered from the bicarbonate extract, is 93%.

N-Phenylpyrrolidine (Reduction of N-Phenylsuccinimide).[39] A 1-l. three-necked flask is equipped with a sealed stirrer, a Soxhlet extractor connected to a reflux condenser, and a dropping funnel. Four hundred milliliters of a solution containing 2.0 g. of lithium aluminum hydride, prepared by diluting a stock solution with dry ether, is placed in the flask, and 4.0 g. of N-phenylsuccinimide is placed in the extractor thimble. The flask is warmed until all the compound has been carried into the reaction flask by the refluxing ether (thirty hours). Upon each discharge of the extractor, a precipitate appears which slowly redissolves. At the end of the reduction period alcohol is slowly added from the dropping funnel and then sufficient 10% sodium hydroxide solution to dissolve the precipitated alumina. The apparatus is then arranged to permit steam distillation of the contents of the flask. The aqueous layer of the distillate is saturated with sodium chloride and further extracted with ether. After the ether solution has been dried over potassium hydroxide pellets, the ether is evaporated, leaving an oily residue which is transferred to a Hickman alembic * and distilled at a pressure below 2 mm. There is obtained 2.9 g. (69%) of product, a colorless liquid when freshly distilled. It readily forms a methiodide, m.p. 149°.

* The alembic was essentially of the form described by Hickman, *J. Phys. Chem.*, **34**, 643 (1930), Fig. 7.

[67a] Hershberg, *Ind. Eng. Chem., Anal. Ed.*, **8**, 313 (1936); see also *Org. Syntheses*, **17**, 31 (1937).

TABULAR SURVEY OF REDUCTIONS WITH LITHIUM ALUMINUM HYDRIDE

In the following survey the compounds that have been reported to be reduced by lithium aluminum hydride are arranged in tables according to the type of functional group that is reduced and within each table in order of empirical formulas. Tables II to V list compounds with functional groups containing oxygen, in the order aldehydes and ketones, epoxides, esters, carboxylic acids, and anhydrides. Compounds containing more than one reducible functional group are listed somewhat arbitrarily according to the group deemed to be of principal interest. Thus the reductive elimination of acetoxy groups in the reactions of epoxysterol acetates with lithium aluminum hydride is incidental to the reduction of the epoxide groups, and such compounds are therefore listed in the table of epoxide reductions.

Tables VI and VII list reductions in which the functional group reduced contains nitrogen; Tables VIII and IX deal with reductions of halogen compounds and of sulfur compounds, respectively, in which these elements, or functional groups containing these elements, undergo reduction.

The survey covers the literature available to the author up to January, 1950.

TABLE II
Aldehydes, Ketones, Quinones

Compound Reduced		Product		Yield %	Reference *
CH_2O	Formaldehyde	CH_4O	Methanol	Quantitative	13
$C_2H_3Cl_3O_2$	Chloral hydrate	$C_2H_3Cl_3O$	Trichloroethanol	50	13
C_4H_6O	Cyclobutanone	C_4H_8O	Cyclobutanol	90	12
C_4H_8O	2-Butanone	$C_4H_{10}O$	sec-Butyl alcohol	80	10
C_5H_8O	Methyl cyclopropyl ketone	$C_5H_{10}O$	Methylcyclopropylcarbinol	76, 80	68, 69
	Cyclopentanone		Cyclopentanol	60, 85	10, 12
$C_6H_4O_2$	p-Benzoquinone	$C_6H_6O_2$	Hydroquinone	70	27
$C_6H_8O_2$	Cyclohexane-1,2-dione	$C_6H_{10}O_2$	Cyclohexanol-2-one	41	2
C_7H_6O	Benzaldehyde	C_7H_8O	Benzyl alcohol	86	10
$C_7H_{14}O$	n-Heptaldehyde	$C_7H_{16}O$	n-Heptyl alcohol	86	10
C_8H_8O	Acetophenone	$C_8H_{10}O$	α-Phenylethanol	90	12
$C_8H_{12}O_2$	4-Octene-2,7-dione	$C_8H_{16}O_2$	4-Octene-2,7-diol	79	70
C_9H_8O	Cinnamaldehyde	$C_9H_{10}O$	Cinnamyl alcohol	90	5
			Hydrocinnamyl alcohol	93	5
$C_9H_{18}O$	Hexamethylacetone	$C_9H_{20}O$	Di-t-butylcarbinol	—	11
$C_{10}H_{10}O$	Benzylideneacetone	$C_{10}H_{12}O$	Styrylmethylcarbinol	Quantitative	71
$C_{10}H_{19}NO$	N-Methyl-3,5-diethyl-4-piperidone	$C_{10}H_{21}NO$	N-Methyl-3,5-diethyl-4-piperidinol	95	72
$C_{11}H_{11}NO$	![structure]	$C_{11}H_{13}NO$	![structure with OH]	75	73
$C_{11}H_{14}O$	Acetomesitylene	$C_{11}H_{16}O$	Mesitylmethylcarbinol	Quantitative	10
$C_{10}H_{14}O_2$	(+)-2,3-Camphorquinone	$C_{10}H_{18}O_2$	(+)-2,3-Camphaneglycol	97	2
$C_{11}H_{14}O_3$	3,5-Dihydroxyphenyl butyl ketone	$C_{11}H_{16}O_3$	3,5-Dihydroxyphenylbutylcarbinol	90	67
$C_{12}H_6O_2$	Acenaphthenequinone	$C_{12}H_{10}O_2$	cis-Acenaphthyleneglycol, trans-acenaphthyleneglycol	15, 45	2

REDUCTIONS BY LITHIUM ALUMINUM HYDRIDE

$C_{12}H_9ClO$	6-Chloro-2-acetonaphthalene		
	7-Chloro-2-acetonaphthalene		
$C_{13}H_{10}O$	Fluorenone		
$C_{13}H_{20}O$	β-Ionone		
$C_{14}H_8O_2$	Anthraquinone		
	Phenanthraquinone		
$C_{14}H_{10}O_2$	Benzil		
$C_{14}H_{22}O$	![structure: H3C, CH3, CH2CH=CCHO, CH3 on cyclohexene]		
$C_{18}H_{26}O$	![structure: H3C, CH3, CH=CHC=CHCH=CHCOCH3, CH3 on cyclohexene]		
$C_{19}H_{16}O_2$	![structure: tetracyclic diketone with CH3]		
$C_{20}H_{28}O$	Vitamin A aldehyde		
$C_{21}H_{27}NO$	Isoamidone		
	Amidone		
$C_{12}H_{11}ClO$	6-Chloro-2-naphthylmethylcarbinol	75–80	74
	7-Chloro-2-naphthylmethylcarbinol	90–92	75
$C_{13}H_{10}O$	9-Fluorenol	99	8
$C_{13}H_{22}O$	β-Ionol	85	76
$C_{14}H_{10}O_2$	Anthrahydroquinone	95	27
	Phenanthrahydroquinone	98	27
$C_{14}H_{14}O_2$	*meso*-Hydrobenzoin, isohydrobenzoin	81, 5	2
$C_{14}H_{24}O$	![structure: H3C, CH3, CH2CH=CCH2OH, CH3 on cyclohexene]	89	76
$C_{18}H_{28}O$	![structure: H3C, CH3, CH=CHC=CHCH=CHCHOHCH3, CH3 on cyclohexene]	—	76
$C_{19}H_{20}O_2$	![structure: tetracyclic diol with CH3, OH, OH]	Quantitative crude	77 †
$C_{20}H_{30}O$	Vitamin A alcohol	70–75	65
$C_{21}H_{29}NO$	6-Dimethylamino-4,4-diphenyl-5-methyl-3-hexanol	75, Quantitative, crude	20, 18 ‡
	6-Dimethylamino-4,4-diphenyl-3-heptanol	90, —	20, 18

* References 68–114 are on pp. 508–509.
† These authors also reported the reduction of two other isomeric diketones.
‡ These authors also reported the reduction of various morpholinyl analogs of amidone and isoamidone.

496   ORGANIC REACTIONS

TABLE II—*Continued*

ALDEHYDES, KETONES, QUINONES

	Compound Reduced		Product	Yield %	Reference*
$C_{21}H_{28}O_2$	$\Delta^{4:11}$-Androstadiene-3,17-dione, 3-enol ethyl ether	$C_{21}H_{30}O_2$	11-Dehydrotestosterone, 3-enol ethyl ether	—	24
$C_{21}H_{30}O_2S$	Δ^4-Androstene-3,17-dione, 3-(β-hydroxyethyl)-thioenol ether	$C_{21}H_{32}O_2S$	Testosterone, 3-(β-hydroxyethyl)thioenol ether	66	25
$C_{26}H_{32}OS$	Δ^4-Androstene-3,17-dione, 3-benzylthioenol ether	$C_{26}H_{34}OS$	Testosterone, 3-benzylthioenol ether	66	25
$C_{27}H_{44}O$	3-Δ^4-Cholestenone	$C_{27}H_{46}O$	Allocholesterol, epiallocholesterol	—, —	78
$C_{29}H_{46}O_3$	7-Ketocholesteryl acetate	$C_{29}H_{48}O_3$	7β-Hydroxycholesterol, 7α-hydroxycholesterol	59, 5	16
$C_{29}H_{48}O_3$	7-Ketocholestanyl acetate	$C_{29}H_{50}O_3$	7α-Hydroxycholestanol, 7β-hydroxycholestanol	—, —	16
$C_{31}H_{48}O_3$	7-Ketostigmasterol acetate		7α-Hydroxystigmasterol, 7β-hydroxystigmasterol	7, 76	16
$C_{32}H_{44}O_3$	3β-Acetoxy-24-keto-24-phenyl-5-cholene	$C_{30}H_{44}O_2$	24-Phenyl-5-cholene-3β,24-diol §	88	79

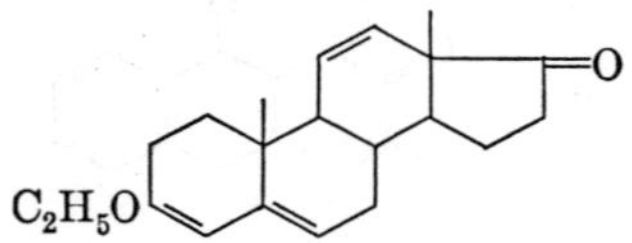

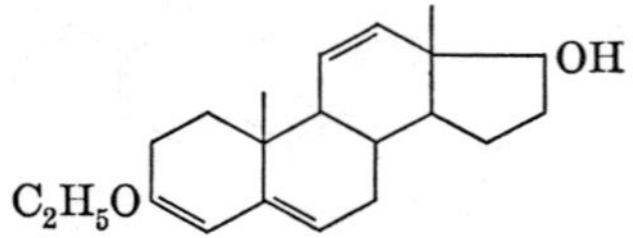

* References 68–114 are on pp. 508–509.
§ The product was a mixture of epimers.

TABLE III

Epoxides

Compound Reduced		Product		Yield %	Reference *
C_3H_5ClO	Epichlorohydrin	C_3H_8O	Isopropyl alcohol	88	2
C_4H_6O	3,4-Epoxy-1-butene	C_4H_8O	1-Buten-3-ol	58	2
			1-Buten-4-ol	13	
$C_6H_{10}O$	Epoxycyclohexane	$C_6H_{12}O$	Cyclohexanol	91	2
$C_7H_{12}O$	1,2-Dimethyl-1,2-epoxy-cyclopentane	$C_7H_{14}O$	trans-1,2-Dimethylcyclopentan-1-ol	40	2
C_8H_8O	Styrene oxide	$C_8H_{10}O$	α-Phenylethanol	94, 75	10, 2
$C_8H_{14}O$	1,2-Dimethyl-1,2-epoxy-cyclohexane	$C_8H_{16}O$	trans-1,2-Dimethylcyclohexane-1-ol	74	2
$C_{14}H_{16}O_5$	Methyl 2,3-anhydro-4,6-benzylidene-α-D-allo-pyranoside	$C_{14}H_{18}O_5$	Methyl 4,6-benzylidene-2-desoxy-α-D-allo-pyranoside	56	21
	HCOCH₃ / HC—O—CH / HCO / HCO—CHC₆H₅ / CH₂O		HCOCH₃ / CH₂ / HCOH / HCO / HCO—CHC₆H₅ / CH₂O		
$C_{15}H_{12}O_2$	Benzalacetophenone oxide	$C_{15}H_{16}O_2$	1,3-Diphenylpropane-1,2-diol	79	2
$C_{23}H_{32}O_4$	3β-Acetoxy-16,17-epoxy-5-pregnen-20-one	$C_{21}H_{34}O_3$	3β,17,20-Trihydroxy-5-pregnene	—	22
		$C_{21}H_{34}O_3$	3β,20β-Dioxy-16α,17α-epoxy-5-allopregnane	20	14
		$C_{21}H_{36}O_3$	3β,17α,20α-Trihydroxy-5-allopregnane (substance "O")	20	
$C_{23}H_{34}O_4$	3β-Acetoxy-16α,17α-epoxy-20-keto-5-allopregnane		3β,17α,20β-Trihydroxy-5-allopregnane (substance "J")	40	
$C_{25}H_{36}O_5$	3β-Acetoxy-16α,17α-epoxy-5-pregnene-20-one, ethylene ketal	$C_{23}H_{36}O_4$	3β,17α-Dihydroxy-5-pregnene-20-one, ethylene ketal	—	22, 80
$C_{27}H_{46}O$	2α,3α-Epoxycholestane	$C_{27}H_{48}O$	Epicholestanol	59	81
	2β,3β-Epoxycholestane		2β-Hydroxycholestane	87	81
$C_{27}H_{48}O_2$	3-Keto-4β,5-epoxycoprostane	$C_{27}H_{48}O_2$	3α,5-Dihydroxycoprostane	18	15
			3β,5-Dihydroxycoprostane	23	
$C_{29}H_{48}O_3$	3β-Acetoxy-5,6α-epoxycholestane		3β,5-Dihydroxycholestane	95	28
	3β-Acetoxy-5,6β-epoxycholestane		3β,6β-Dihydroxycholestane	60	28
			3β,5-Dihydroxycoprostane	20	
	3β-Acetoxy-4β,5-epoxycoprostane		3β,5-Dihydroxycoprostane	90 crude	82
	3α, Acetoxy-4β,5-epoxycoprostane		3α,5-Dihydroxycoprostane	94 crude	82
	3α-Acetoxy-4α,5-epoxycholestane		3α,5-Dihydroxycholestane	22	83
	3β-Acetoxy-4α,5-epoxycholestane		3β,5-Dihydroxycholestane	Quant.	83

* References 68–114 are on pp. 508–509.

TABLE IV
Esters and Lactones

	Compound Reduced		Product	Yield %	Reference*
$C_4H_5Cl_3O_2$	Ethyl trichloroacetate	$C_2H_3Cl_3O$	Trichloroethanol	65	62
$C_4H_6Cl_2O_2$	Ethyl dichloroacetate	$C_2H_4Cl_2O$	Dichloroethanol	65	62
$C_4H_7ClO_2$	Ethyl chloroacetate	C_2H_5ClO	Ethylene chlorohydrin	37	62
$C_4H_9NO_3$	dl-Serine, methyl ester	$C_3H_9NO_2$	2-Amino-1,3-propanediol	30	31
$C_5H_6O_2$	α-Angelica lactone	$C_5H_{10}O_2$	γ-Acetopropanol	65	8
	β-Angelica lactone	$C_5H_{12}O_2$	2,4-Pentanediol	10	8
$C_5H_{11}NO_2$	L(+)Alanine, ethyl ester	C_3H_9NO	L(+)2-Aminopropanol	50	31
$C_6H_8N_2O_2$	Ethyl 3-pyrazolecarboxylate	$C_4H_6N_2O$	3-Hydroxymethylpyrazole	84	52
	Ethyl 4-pyrazolecarboxylate		4-Hydroxymethylpyrazole	86	52
$C_6H_{10}O_2$	Methyl 3-pentenoate	$C_5H_{10}O$	3-Pentenol	75	84
$C_6H_{12}N_2O_2$	L-Asparagine, ethyl ester	$C_4H_{11}NO_2$	L(+)2-Amino-1,4-butanediol	70	31 †
$C_7H_{12}O_5$	Dimethyl L-methoxysuccinate	$C_5H_{12}O_3$	L-2-Methoxybutane-1,4-diol	69	85
$C_7H_{13}NO_2$	L-Proline, ethyl ester	$C_5H_{11}NO$	L(+)-2-Hydroxymethylpyrrolidine	73	31
$C_8H_{13}NO_2$	Arecolin	$C_7H_{13}NO$	1-Methyl-3-hydroxymethyl-1,2,5,6-tetrahydropyridine	80	108
	![structure: CO₂CH₃ on N-CH₃ tetrahydropyridine] Guvacin, ethyl ester	$C_6H_{11}NO$	3-Hydroxymethyl-1,2,5,6-tetrahydropyridine	60	108
$C_8H_{17}NO_2$	L-Leucine, ethyl ester	$C_6H_{15}NO$	L(+)-4-Methyl-2-aminopentanol	85	31
$C_9H_8O_2$	Coumarin	$C_9H_{12}O_2$	3-(o-Hydroxyphenyl)propanol	50	8; cf. 48
		$C_9H_{10}O_2$	o-Hydroxycinnamyl alcohol	10	
$C_9H_{10}O_2$	Ethyl benzoate	C_7H_8O	Benzyl alcohol	90	10
$C_9H_{12}O_4$	Methyl anhydrocrotalate	$C_8H_{16}O_3$	2,3,4-Trimethyl-2-pentene-1,4,5-triol	86	86
$C_9H_{14}O_4$	Methyl dihydroanhydrocrotalate	$C_8H_{18}O_3$	2,3,4-Trimethylpentane-1,4,5-triol	93	86

REDUCTIONS BY LITHIUM ALUMINUM HYDRIDE

Formula	Structure/Starting Material	Product	Yield 1	Yield 2
$C_9H_{14}O_5$	Methyl monocrotalate (OH, CH₃, H₃CC—C—CO₂CH₃, H₃CCH—C=O)	2,3,4-Trimethylpentane-1,3,4,5-tetrol	92	86
$C_9H_{16}O_2$	Ethyl 2-heptenoate	2-Heptenol	79 crude	87
$C_9H_{17}NO_4$	Diethyl L-glutamate	L-2-Aminopentane-1,5-diol	58	108
$C_{10}H_{16}O_4$	Diethyl allylmalonate	2-Hydroxymethyl-4-pentenol	—	88
$C_{10}H_{18}O_4$	Diethyl adipate	1,6-Hexanediol	52	10
$C_{11}H_{12}O_3$	Ethyl o-coumarate	2-(o-Hydroxyphenyl)propanol	83	48
$C_{11}H_{15}NO_2$	L-Phenylalanine, ethyl ester	L(−)-2-Amino-3-phenylpropanol	—	31
$C_{11}H_{15}NO_3$	L-Tyrosine, ethyl ester	L(−)-2-Amino-3-(p-hydroxyphenyl)propanol	75	8
			60	
$C_9H_{18}O$	Ethyl 2-nonenoate	2-Nonenol	65	89
$C_7H_{16}O_2$	Diethyl n-butylmalonate	2-n-Butyl-1,3-propanediol	98	87
	Diethyl isopropylsuccinate	2-Isopropyl-1,4-butanediol	94	13
$C_{11}H_{14}N_2O$	L-Tryptophane, methyl ester	"Tryptophanol"	96	90
$C_9H_{15}NO_3$	3-Carbethoxy-4-ketoquinolizidine	4-Ketoquinolizidine	90	91
$C_{11}H_{20}O_3$	Methyl cis-2-methyl-2-carbomethoxycyclohexaneacetate	cis-β-2-Methyl-2-carbomethoxycyclohexaneëthanol ‡	20	29
			53	30
$C_{10}H_{20}O_2$		cis-β-2-Methyl-2-hydroxymethylcyclohexaneethanol §	80	30
$C_8H_{18}O_2$	Diethyl sec-butylsuccinate	2-sec-Butyl-1,4-butanediol	96	90
	Diethyl isobutylsuccinate	2-Isobutyl-1,4-butanediol	88	90
$C_{11}H_{12}N_2O_2$	Ethyl 1-benzyl-4-pyrazolecarboxylate	1-Benzyl-4-hydroxymethylpyrazole	92–96	52
$C_{13}H_{16}O_2$	Ethyl 1,2,3,4-tetrahydro-2-naphthoate	1,2,3,4-Tetrahydro-2-naphthylcarbinol	95	94
$C_{13}H_{16}O_5$	Diethyl phenoxymalonate	2-Phenoxy-1,3-propanediol	95	93
$C_{11}H_{21}NO$	3-Carbethoxy-4-methylquinolizidine	3-Hydroxymethyl-4-methylquinolizidine	50	29
$C_9H_{20}O_2$	Ethyl 4,5-dimethyl-3-carbethoxyhexanoate	4,5-Dimethyl-3-hydroxymethylhexanol	85	90
$C_{12}H_{26}O$	Methyl laurate	1-Dodecanol	94	10

* References 68–114 are on pp. 508–509.
† The same authors also reported reduction of the racemic ester.
‡ This reduction was run at −10°.
§ This reduction was run under normal conditions.

TABLE IV—Continued

ESTERS AND LACTONES

	Compound Reduced	Product		Yield %	Reference*
$C_{14}H_{16}O_5$	Ethyl acetoferulate	Coniferyl alcohol	$C_{10}H_{12}O_3$	43 ‖	49
$C_{14}H_{19}NO_4$	Diethyl β-(2-pyridyl)ethylmalonate	2-Hydroxymethyl-4-(2'-pyridyl)-1-butanol	$C_{10}H_{15}NO_2$	24	29
$C_{15}H_{12}O_2$	Methyl 9-fluorenecarboxylate	9-Fluorenylcarbinol	$C_{14}H_{12}O$	87 crude	95
$C_{15}H_{28}O_4$	Diethyl n-heptylsuccinate	2-n-Heptyl-1,4-butanediol	$C_{11}H_{24}O_2$	93	90
$C_{16}H_{16}O_3$	Ethyl α-phenoxyphenylacetate	2-Phenyl-2-phenoxyethanol	$C_{14}H_{14}O_2$	89	96
$C_{16}H_{22}O_4$	(+)-Hydrogen 2,4-dimethylhexyl-4-phthalate ¶	(−)-2,4-Dimethylhexan-4-ol	$C_8H_{18}O$	80–85	33
		Phthalyl alcohol	$C_8H_{10}O_2$	—	
$C_{17}H_{18}N_2O_2$	Methyl lysergate	Lysolysergol	$C_{16}H_{18}N_2O$	90	32
	Methyl isolysergate	Isolysergol		90	32
$C_{17}H_{20}N_2O_2$	Methyl dihydrolysergate	α-Dihydrolysergol	$C_{16}H_{20}N_2O$	74	32
	Methyl dihydroisolysergate "I"	β-Dihydrolysergol		75	32
	Methyl dihydroisolysergate	γ-Dihydrolysergol		80	32
$C_{17}H_{26}O_2$	Ethyl β-ionylideneacetate	β-Ionylideneëthyl alcohol	$C_{15}H_{24}O$	Quantitative, 82, 85	66 76, 97
	Ethyl α-ionylideneacetate	α-Ionylideneëthyl alcohol		95	98
$C_{18}H_{14}O_3$	Pseudo ethyl 5-formyl-4-phenanthrenecarboxylate	4,5-Dihydroxymethylphenanthrene	$C_{16}H_{14}O_2$	90	99

[structure: pseudo ethyl 5-formyl-4-phenanthrenecarboxylate showing CO–O–CHOC$_2$H$_5$ bridge on phenanthrene]

[structure: 4,5-dihydroxymethylphenanthrene showing two CH$_2$OH groups on phenanthrene]

	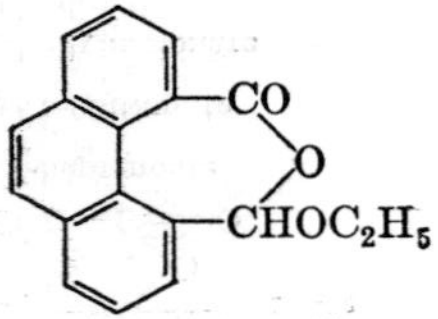		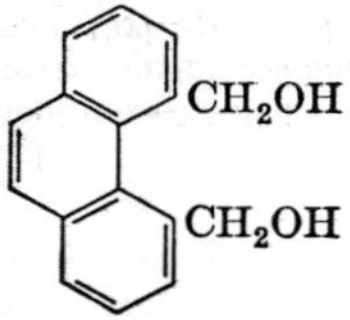		
$C_{18}H_{26}O_2$	C_{17} acid, methyl ester	C_{17} alcohol	$C_{17}H_{26}O$	85	76

REDUCTIONS BY LITHIUM ALUMINUM HYDRIDE

$C_{19}H_{30}O_2$	H₃C–C(CH₃)(CH₃)–cyclohexenyl–CH=CHC(CH₃)=CHCH=CHCO₂CH₃		$C_{17}H_{30}O$	H₃C–C(CH₃)(CH₃)–cyclohexenyl–CH=CHC(CH₃)=CHCH=CHCH₂OH 68 100
	decalin–CH₂CH₂–CO₂CH₃			decalin–CH₂CH₂–CH₂OH
$C_{18}H_{34}O_4$	Diethyl n-decylsuccinate		$C_{14}H_{30}O_2$	2-n-Decyl-1,4-butanediol 94 90
$C_{18}H_{36}O_2$	Ethyl palmitate		$C_{16}H_{34}O$	1-Hexadecanol 98 10
$C_{20}H_{24}O_3$	Methyl α,α-dimethyl-β-(6-methoxy-2-naphthyl)-valerate		$C_{19}H_{24}O_2$	2,2-Dimethyl-3-(6-methoxy-2-naphthyl)pentan-1-ol — 101
$C_{19}H_{26}O_3$	Methyl O-methylpodocarpate H₃C⋯CO₂CH₃ / OCH₃		$C_{18}H_{26}O_2$	O-Methylpodocarpinol H₃C⋯CH₂OH / OCH₃ 93 35
$C_{19}H_{36}O_2$	Methyl oleate		$C_{18}H_{36}O$	Oleyl alcohol 86 10
$C_{20}H_{14}O_2$	Diphenylphthalide		$C_{20}H_{18}O_2$	o-Hydroxymethyltriphenylcarbinol 94 crude 102

* References 68–114 are on pp. 508–509.

‖ The yield was 67% when the product was isolated as the benzoate.

¶ The same authors reported reduction of the levo ester in 83% yield.

TABLE IV—Continued
Esters and Lactones

Compound Reduced	Product	Yield %	Reference*
$C_{20}H_{20}N_2O_3$ (indole-CH$_2$-NH-C(=O)-O-CH$_2$-C$_6$H$_5$ with CO$_2$CH$_3$)	$C_{19}H_{20}N_2O_2$ (indole-CH$_2$-NH-C(=O)-O-CH$_2$-C$_6$H$_5$ with CH$_2$OH)	—	42
$C_{20}H_{38}O_4$ Diethyl n-dodecylsuccinate	$C_{16}H_{34}O_2$ 2-n-Dodecyl-1,4-butanediol	95	90
$C_{21}H_{16}O_2$ 9-Carbomethoxy-9-phenylfluorene	$C_{20}H_{16}O$ 9-Phenylfluorenylcarbinol	93 crude	102
$C_{21}H_{19}N_3O_4$ Dibenzoyl-L-histidine, methyl ester	$C_{13}H_{15}N_3O_2$ Monobenzoyl-L-histidinol	48	9
$C_{22}H_{32}O_2$ C_{20} acid, ethyl ester	$C_{20}H_{30}O$ Vitamin A alcohol	95	63

REDUCTIONS BY LITHIUM ALUMINUM HYDRIDE

$C_{22}H_{33}BrO_2$	Methyl 17-pregnen-3-β-ol-20-bromo-21-oate	—	$C_{21}H_{33}O_2Br$	3(β),21-Dihydroxy-20-bromo-17-pregnene	26
$C_{22}H_{36}O_2$	Methyl allopregnane-21-carboxylate	95	$C_{21}H_{36}O$	21-Hydroxyallopregnane	103
	Methyl 17-isoallopregnane-21-carboxylate	86		21-Hydroxy-17-iso-allopregnane	103
$C_{25}H_{38}O_3$	Methyl $\Delta^{3:5}$-3-ethoxybisnorcholestadienoate	—	$C_{24}H_{38}O_2$	$\Delta^{3:5}$-3-Ethoxy-22-hydroxybisnorcholestadien	23

* References 68–114 are on pp. 508–509.

TABLE V

Carboxylic Acids and Anhydrides

Compound Reduced		Product		Yield %	Reference [*]
$C_2HCl_3O_2$	Trichloroacetic acid	$C_2H_3Cl_3O$	Trichloroethanol	31	62
$C_2H_2Cl_2O_2$	Dichloroacetic acid	$C_2H_4Cl_2O$	Dichloroethanol	65	62
$C_2H_3ClO_2$	Chloroacetic acid	C_2H_5ClO	Ethylene chlorohydrin	13	62
$C_2H_4O_2$	Acetic acid	C_2H_6O	Ethanol	Quantitative	13
$C_4H_6O_2$	Cyclopropanecarboxylic acid	C_4H_8O	Cyclopropylcarbinol	95	13
$C_5H_4O_3$	Furoic acid	$C_5H_6C_2$	Furfuryl alcohol	85	36
$C_5H_{10}O_2$	Trimethylacetic acid	$C_5H_{12}O$	Neopentyl alcohol	92	36
$C_6H_8O_2$	Sorbic acid	$C_6H_{10}O$	Sorbyl alcohol	92	36
$C_7H_5ClO_2$	p-Chlorobenzoic acid	C_7H_7ClO	p-Chlorobenzyl alcohol	85	60
	m-Chlorobenzoic acid		m-Chlorobenzyl alcohol	85	60
	o-Chlorobenzoic acid		o-Chlorobenzyl alcohol	97	36
$C_7H_6O_2$	Benzoic acid	C_7H_8O	Benzyl alcohol	81	36
$C_7H_6O_3$	Salicylic acid	$C_7H_8O_2$	o-Hydroxybenzyl alcohol	99	36
$C_7H_7NO_2$	Anthranilic acid	C_7H_9NO	o-Aminobenzyl alcohol	97	36
$C_8H_4O_3$	Phthalic anhydride	$C_8H_{10}O_2$	Phthalyl alcohol	87	10
$C_8H_6O_3$	Phenylglyoxylic acid	$C_8H_{10}O_2$	Phenylethyleneglycol	80	36
$C_8H_8O_2$	Phenylacetic acid	$C_8H_{10}O$	β-Phenylethanol	92	36
$C_8H_8O_2$	p-Toluic acid	$C_8H_{10}O$	p-Tolylcarbinol	85	60
	m-Toluic acid		m-Tolylcarbinol	85	60
$C_8H_8O_3$	p-Anisic acid	$C_8H_{10}O_2$	p-Anisylcarbinol	85	60
$C_9H_8O_2$	Cinnamic acid	$C_9H_{12}O$	Hydrocinnamyl alcohol	85	36
$C_9H_{10}O_3$	m-Methoxyphenylacetic acid	$C_9H_{12}O_2$	β-(m-Methoxyphenyl)ethanol	90	104
$C_9H_{10}O_4$	3,5-Dimethoxybenzoic acid	$C_9H_{12}O_3$	3,5-Dimethoxybenzyl alcohol	93	67
$C_{10}H_9NO_2$	3-Indoleacetic acid	$C_{10}H_{11}NO$	Tryptophol	65	105
$C_{10}H_{10}O_2$	p-Methylcinnamic acid	$C_{10}H_{12}O$	p-Methylcinnamyl alcohol	90	47
$C_{10}H_{12}O_2$	p-Methylhydrocinnamic acid	$C_{10}H_{14}O$	p-Methylhydrocinnamyl alcohol	96	47
$C_{10}H_{18}O_4$	Sebacic acid	$C_{10}H_{22}O_2$	Decane-1,10-diol	97	36
$C_{12}H_9ClO_2$	7-Chloro-1-naphthylacetic acid	$C_{12}H_{11}ClO$	2-(7-Chloro-1-naphthyl)ethanol	75	75
$C_{12}H_{22}O_4$	Ethyl hydrogen sebacate	$C_{10}H_{22}O_2$	Decane-1,10-diol	91	36
$C_{14}H_{10}O_3$	Benzoic anhydride	C_7H_8O	Benzyl alcohol	87	10
$C_{16}H_{16}O_2$	(+)-2,4-Diphenylbutanoic acid	$C_{16}H_{18}O$	(+)-2,4-Diphenylbutanol	96	106
	(−)-2,4-Diphenylbutanoic acid		(−)-2,4-Diphenylbutanol	77	106
$C_{17}H_{22}O_3$	Podocarpic acid	$C_{17}H_{24}O_2$	Podocarpinol	56	35
$C_{18}H_{36}O_2$	Stearic acid	$C_{18}H_{38}O$	1-Octadecanol	91	36
$C_{20}H_{16}O_2$	o-Carboxytriphenylmethane	$C_{20}H_{18}O$	o-Hydroxymethyltriphenylmethane	95	102
$C_{20}H_{28}O_2$	Vitamin A acid	$C_{20}H_{30}O$	Vitamin A alcohol		63
$C_{21}H_{20}O_2$	2-Benzyl-2,3-diphenylpropanoic acid	$C_{21}H_{22}O$	2-Benzyl-2,3-diphenylpropanol	50	107
$C_{22}H_{22}O_2$	2-Benzyl-2,4-diphenylbutanoic acid	$C_{22}H_{24}O$	2-Benzyl-2,4-diphenylbutanol	63	107

[*] References 68–114 are on pp. 508–509.

TABLE VI

Amides and Nitriles

Compound Reduced		Product		Yield %	Reference *
C_4H_7N	n-Butyronitrile	$C_4H_{11}N$	n-Butylamine	57	42a
$C_5H_5N_3$	3-Cyanomethylpyrazole	$C_5H_9N_3$	3-(β-Aminoethyl)pyrazole	53	52
$C_5H_{11}NO$	N-Ethylpropionamide	$C_5H_{13}N$	Ethylpropylamine	53	45
$C_6H_{11}NO$	α-Ethylcrotonamide	$C_6H_{15}NO$	α-Ethylbutylamine (?)	—	45
$C_6H_{11}NO$	Cyclohexanone isoöxime	$C_6H_{13}N$	Hexamethyleneimine	—	40 †
$C_6H_{13}NO$	N,N-Diethylacetamide	$C_6H_{15}N$	Triethylamine	50	45
C_7H_4ClN	p-Chlorobenzonitrile	C_7H_8ClN	p-Chlorobenzylamine	81	42a
C_7H_5N	Benzonitrile	C_7H_9N	Benzylamine	72	27
C_7H_5N	Benzonitrile	C_7H_9N	Benzylamine	83	42a
C_7H_7NO	Benzamide	C_7H_9N	Benzylamine	85	38
$C_7H_7NO_2$	p-Hydroxyformanilide	C_7H_9NO	N-Methyl-p-aminophenol	92	59
$C_8H_5NO_2$	Phthalimide	C_8H_9N	Isoindoline	—	45
C_8H_7N	o-Tolunitrile	$C_8H_{11}N$	o-Xylylamine	88	27
C_8H_7NO	Mandelonitrile	$C_8H_{11}NO$	β-Hydroxy-β-phenethylamine	48	27
C_8H_9NO	Acetanilide	$C_8H_{11}N$	N-Ethylaniline	60	27
	Phenylacetamide		β-Phenethylamine	—	45
$C_8H_9NO_2$	Phenoxyacetamide	$C_8H_{11}NO$	β-Phenoxyethylamine	80	45
$C_8H_{15}N$	Caprylonitrile	$C_8H_{19}N$	n-Octylamine	90	42a
C_9H_9NO	1-Methyloxindole	C_9H_9N	1-Methylindole	62	51
		$C_9H_{11}N$	1-Methylindoline	12	
$C_9H_{11}NO$	N-Methylacetanilide	$C_9H_{13}N$	N-Methyl-N-ethylaniline	91	27
$C_{10}H_9NO_2$	N-Phenylsuccinimide	$C_{10}H_{13}N$	N-Phenylpyrrolidine	69	39
$C_{10}H_{11}NO$	1,3-Dimethyloxindole	$C_{10}H_{11}N$	1,3-Dimethylindole	86	51
		$C_{10}H_{13}N$	1,3-Dimethylindoline	13	
$C_{10}H_{14}N_2O$	N-Diethylnicotinamide	$C_{10}H_{16}N_2$	β-Pyridylmethyldiethylamine	55	45
$C_{10}H_{15}NO$	N-Diethylbenzamide	C_7H_8O	Benzyl alcohol	—	27
$C_{10}H_{16}N_2$	Sebaconitrile	$C_{10}H_{24}N_2$	1,10-Diaminodecane	40	27
$C_{10}H_{19}N$	Caprinonitrile	$C_{10}H_{23}N$	n-Decylamine	92	42a
$C_{11}H_9N_3$	1-Benzyl-4-cyanopyrazole	$C_{11}H_{13}N_3$	1-Benzyl-4-aminomethylpyrazole	72	52
$C_{11}H_{11}NO_2$	N-Phenylglutarimide	$C_{11}H_{15}N$	N-Phenylpiperidine	52	39
$C_{11}H_{13}NO_2$	1-Methyl-5-ethoxyoxindole	$C_{11}H_{13}NO$	1-Methyl-5-ethoxyindole	60	51
		$C_{11}H_{15}NO$	1-Methyl-5-ethoxyindoline	—	
$C_{11}H_{19}NO$	N-Acetyldecahydroisoquinoline	$C_{11}H_{21}N$	N-Ethyldecahydroisoquinoline	84	45
$C_{12}H_{11}N_3$	1-Benzyl-2-cyanomethylimidazole	$C_{12}H_{15}N_3$	1-Benzyl-2-(β-aminoethyl)-imidazole	88	53
$C_{12}H_{17}NO_4$	N,N-Dimethyl-3,4,5-trimethoxybenzamide	$C_{12}H_{19}NO_3$	N,N-Dimethyl-3,4,5-trimethoxybenzylamine	54	45
$C_{13}H_{25}N$	Lauryl cyanide	$C_{13}H_{29}N$	Tridecylamine	90	27
$C_{19}H_{14}N_2O$	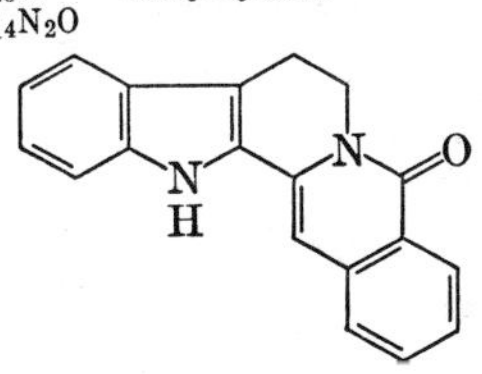 (structure shown)	$C_{19}H_{16}N_2$	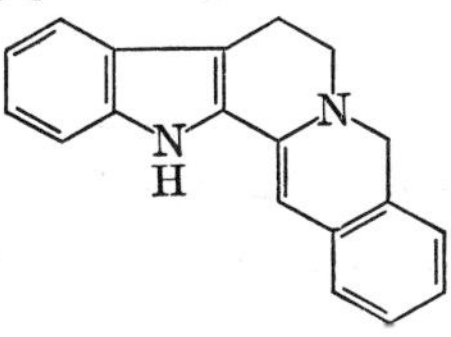 (structure shown)	Quantitative	42
$C_{19}H_{21}NO_5$	N-Formyl-N-(3-methoxybenzyl)-3-methoxy-4,5-methylenedioxyphenethylamine	$C_{19}H_{23}NO_4$	N-(3-Methoxybenzyl)-N-methyl-3-methoxy-4,5-methylenedioxyphenethylamine	87	50
$C_{20}H_{20}N_2O$	1-Methyl-3-{2-N-(1,2-dihydroisoquinolylethyl)]oxindole (VIII, p. 479)	$C_{20}H_{20}N_2$	Compound X, p. 479	70	41
$C_{20}H_{18}N_2O_2$	1-Methyl-3-{2-N-(1-oxo-1,2-dihydroisoquinolylethyl)}-oxindole (IX, p. 479)	$C_{20}H_{20}N_2$	Compound X, p. 479	75	41
$C_{21}H_{22}N_2O_2$	Strychnine	$C_{21}H_{24}N_2O$	Strychnidine	91	109

* References 68–114 are on pp. 508–509.
† The same authors reported the preparation of all the polymethyleneimines from C_6 to C_{20} in yields of 60–95%.

TABLE VII

Miscellaneous Nitrogen Compounds

Compound Reduced		Product		Yield %	Reference *
$C_4H_9NO_2$	2-Nitrobutane	$C_4H_{11}N$	2-Aminobutane	85	27
$C_6H_4BrNO_2$	p-Bromonitrobenzene	$C_{12}H_8Br_2N_2$	4,4'-Dibromoazobenzene	88	27
$C_6H_5NO_2$	Nitrobenzene	$C_{12}H_{10}N_2$	Azobenzene	84	27
$C_8H_7NO_2$	ω-Nitrostyrene	$C_8H_{11}N$	β-Phenethylamine	60	27
$C_8H_{10}N_2O$	p-Nitrosodimethylaniline	$C_{16}H_{20}N_4$	4,4'-Bisdimethylaminoazobenzene	80	13
$C_9H_{11}NO_2$	Nitromesitylene	$C_{18}H_{22}N_2$	Azomesitylene	71	27
$C_{10}H_9NO_5$	ω-Nitro-3-methoxy-4,5-methylenedioxystyrene	$C_{10}H_{13}NO_3$	3-Methoxy-4,5-methylenedioxyphenethylamine	49	50
$C_{10}H_{10}IN$	Isoquinoline methiodide	$C_{10}H_{11}N$	2-Methyl-1,2-dihydroisoquinoline	70	55
	Quinoline methiodide		1-Methyl-1,2-dihydroquinoline	37	55
$C_{12}H_8N_2O_4$	2,2'-Dinitrobiphenyl	$C_{12}H_8N_2$	Azobiphenyl	90	13
$C_{12}H_{10}N_2O$	Azoxybenzene	$C_{12}H_{10}N_2$	Azobenzene	99	27
$C_{13}H_9N$	Phenanthridine	$C_{13}H_{11}N$	5,6-Dihydrophenanthridine	74	54
$C_{13}H_{11}N$	Benzalaniline	$C_{13}H_{13}N$	N-Benzylaniline	93	27
$C_{13}H_{11}NO$	Benzophenone oxime	$C_{13}H_{13}N$	Benzhydrylamine	60	8
$C_{13}H_{16}IN$	Isoquinoline butiodide	$C_{13}H_{17}N$	2-n-Butyl-1,2-dihydroisoquinoline	76	55
	Quinoline butiodide		1-n-Butyl-1,2-dihydroquinoline	42	55
$C_{16}H_{14}IN$	1-Phenylisoquinoline methiodide	$C_{16}H_{15}N$	1-Phenyl-2-methyl-1,2-dihydroisoquinoline	66	55
	2-Phenylquinoline methiodide		1-Methyl-2-phenyl-1,2-dihydroquinoline	—	55
$C_{20}H_{18}NO_4HSO_4$	Berberin sulfate	$C_{20}H_{19}NO_4$	Dihydroanhydroberberine	—	55
$C_{21}H_{22}N_2O_2(CH_3)_2SO_4$	Strychnine methosulfate	$C_{21}H_{24}N_2O$	Strychnidine	63	4
$C_{21}H_{24}INO_4$	Papaverin methiodide	$C_{21}H_{25}NO_4$	N-Methyl-1,2-dihydropapaverine	—	55

* References 68–114 are on pp. 508–509.

TABLE VIII
Halogen Compounds

Compound Reduced		Product		Yield %	Reference *
CH_3I	Methyl iodide	CH_4	Methane	100	27
C_2F_3ClO	Trifluoroacetyl chloride	$C_2H_3F_3O$	Trifluoroethanol	85	110
C_2Cl_4O	Trichloroacetyl chloride	$C_2H_3Cl_3O$	Trichloroethanol	64	62
C_2HCl_3O	Dichloroacetyl chloride	$C_2H_4Cl_2O$	Dichloroethanol	63	62
$C_2H_2Cl_2O$	Chloroacetyl chloride	C_2H_5ClO	Ethylene chlorohydrin	62	62
$C_3H_4Cl_2$	cis-1,3-Dichloropropene	C_3H_5Cl	cis-1-Chloropropene	46	111
C_3H_5Br	Allyl bromide	C_3H_6	Propene	85	27
$C_4H_6Br_2$	trans-1,4-Dibromo-2-butene	C_4H_8	trans-2-Butene	72	2
$C_4H_8Cl_2O$	Ethyl α,β-dichloroethyl ether	C_4H_9ClO	Ethyl β-chloroethyl ether	53	2
C_4H_9Cl	n-Butyl chloride		No reduction at 25°	—	27
C_4H_9I	t-Butyl iodide	C_4H_8	Isobutylene	—	2
		C_4H_{10}	Isobutane	—	
$C_5H_8Br_4$	Pentaerythrityl bromide		No reaction at 65°		2
C_5H_9ClO	Trimethylacetyl chloride	$C_5H_{12}O$	Neopentyl alcohol	86	10
C_6H_4ClI	1-Chloro-2-iodobenzene	C_6H_5Cl	Chlorobenzene	40	2
C_6H_7ClO	Sorboyl chloride	$C_6H_{10}O$	Sorbyl alcohol	98	10
$C_6H_{11}Cl$	Chlorocyclohexane		No reaction		44
$C_6H_{11}ClO$	Isocaproyl chloride	$C_6H_{14}O$	Isohexyl alcohol	95	10
$C_6H_{11}Br$	Bromocyclohexane	C_6H_{12}	Cyclohexane	10	44
C_7H_5ClO	Benzoyl chloride	C_7H_8O	Benzyl alcohol	72	10
C_7H_7Cl	Benzyl chloride	C_7H_8	Toluene	72	2
C_7H_7Br	Benzyl bromide	C_7H_8	Toluene	78	2
	p-Bromotoluene	C_7H_8	Toluene	4–14	44
C_7H_7I	Benzyl iodide	C_7H_8	Toluene	86	2
$C_7H_{11}BrO_4$	Diethyl bromomalonate	$C_3H_8O_2$	Trimethyleneglycol	5	2
$C_7H_{15}Br$	2-Bromoheptane	C_7H_{16}	Heptane	76–92	44
$C_8H_4Cl_2O_2$	sym-o-Phthalyl chloride	$C_8H_{10}O_2$	Phthalyl alcohol	95	10
$C_8H_6Br_2O$	p-Bromophenacyl bromide	C_8H_9BrO	α-(p-Bromophenyl)ethanol	85	2
C_8H_7Br	ω-Bromostyrene	C_8H_8	Styrene	49	2
$C_8H_8Br_2$	Styrene dibromide	C_8H_8	Styrene	71	2
$C_8H_{16}Br_2$	1,2-Dibromoöctane	C_8H_{16}	1-Octene	17	2
		$C_8H_{17}Br$	2-Bromoöctane	26	
	1,2-Dibromoöctane	C_8H_{18}	n-Octane	80	44
$C_8H_{17}Cl$	3-(Chloromethyl)heptane	C_8H_{18}	3-Methylheptane	52–96	44
$C_8H_{17}Br$	3-(Bromomethyl)heptane	C_8H_{18}	3-Methylheptane	98	44
	2-Bromoöctane		n-Octane	30	2
	1-Bromoöctane		n-Octane	40–96	44
$C_9H_{19}Br$	2-Bromo-2-methyloctane	C_9H_{18}	2-Methyloctene	76	2
$C_{10}H_{21}Br$	1-Bromodecane	$C_{10}H_{22}$	n-Decane	72	2
$C_{12}H_{25}Cl$	1-Chlorododecane	$C_{12}H_{26}$	n-Dodecane	80–98	44
$C_{11}H_{12}Br_2O_2$	Ethyl 2,3-dibromo-3-phenylpropionate	$C_9H_{12}O$	Hydrocinnamyl alcohol	59	2
$C_{13}H_9Br$	9-Bromofluorene	$C_{13}H_{10}$	Fluorene	30	2
		$C_{26}H_{18}$	Dibiphenyleneëthane	34	
$C_{13}H_{11}Br$	Diphenylbromomethane	$C_{13}H_{12}$	Diphenylmethane	38	2
		$C_{26}H_{22}$	Tetraphenylethane	25	
$C_{13}H_{27}Cl$	5-Chloro-5-n-butylnonane		No reaction		44
$C_{14}H_{12}Br_2$	meso-1,2-Diphenyl-1,2-dibromoethane	$C_{14}H_{12}$	trans-Stilbene	98	2
$C_{16}H_{31}ClO$	Palmitoyl chloride	$C_{16}H_{34}O$	1-Hexadecanol	98	10
$C_{16}H_{33}I$	Cetyl iodide	$C_{16}H_{34}$	n-Hexadecane	95	27
$C_{18}H_{29}ClO_2$	O-Methylpodocarpoyl chloride	$C_{18}H_{32}O_2$	O-Methylpodocarpinol	92	35

* References 68–114 are on pp. 508–509.

TABLE IX

Sulfur Compounds

Compound Reduced		Product		Yield %	Reference [*]
$C_4H_9ClO_2S$	1-Butanesulfonyl chloride	$C_4H_{10}S$	n-Butyl mercaptan [†]	45	113
$C_6H_5ClO_2S$	Benzenesulfonyl chloride	C_6H_6S	Thiophenol	60	13
		$C_{12}H_{10}S_2$	Diphenyl disulfide [‡]	32	
$C_7H_7ClO_2S$	p-Toluenesulfonyl chloride	C_7H_8S	p-Thiocresol	50	113
$C_8H_{18}S_2$	Di-n-butyl disulfide	$C_4H_{10}S$	n-Butyl mercaptan	96	114
	n-Butyl t-butyl disulfide	$C_4H_{10}S$	n-Butyl mercaptan	96	114
			t-Butyl mercaptan		
	Di-t-butyl disulfide		No reaction		114
$C_{10}H_{22}S_2$	Di-isoamyl disulfide	$C_5H_{12}S$	Isoamyl mercaptan	—	114
$C_{12}H_{10}OS$	Diphenyl sulfoxide	$C_{12}H_{10}S$	Diphenyl sulfide	—	13
$C_{12}H_{10}O_2S$	Diphenyl sulfone		No reaction		13
$C_{12}H_{10}S_2$	Diphenyl disulfide	C_6H_6S	Thiophenol	95	114
$C_{12}H_{22}F_2OS$	Ethyl difluorothiodecanoate [§]	$C_{10}H_{20}F_2O$	Difluorodecanol	76	92
$C_{13}H_{12}O_3S$	Phenyl p-toluenesulfonate [‖]		Phenol	Small	112
			p,p'-Ditolyl disulfoxide	Small	
$C_{14}H_{14}S$	Dibenzyl disulfide	C_7H_8S	Benzyl mercaptan	95	114
$C_{16}H_{34}S_2$	Di-n-octyl disulfide	$C_8H_{18}S$	n-Octyl mercaptan	98	114
$C_{17}H_{26}O_3S$	(−)-Menthyl p-toluenesulfonate	$C_{10}H_{20}$	p-Menthane	—	112
$C_{19}H_{26}O_8S$	6-p-Toluenesulfo-diacetone-D-galactose <1,5>	$C_{12}H_{20}O_5$	Diacetone-D-fucose	61	112
	3-p-Toluenesulfo-diacetone-D-glucose <1,4>	$C_{12}H_{20}O_6$	Diacetone-D-glucose	—	112
		$C_{14}H_{14}S_2$	Ditolyl disulfide	—	
	1-p-Toluenesulfo-β-diacetone-D-fructose <2,6>	$C_{12}H_{20}O_6$	β-Diacetone-D-fructose	—	112
		$C_7H_8O_2S$	p-Toluenesulfinic acid	—	
		$C_{14}H_{14}S_2$	Di-p-tolyl disulfide	—	
$C_{24}H_{50}S_2$	Di-t-dodecyl disulfide	$C_{12}H_{26}S$	t-Dodecyl mercaptan [¶]	—	114
$C_{24}H_{50}S_3$	Di-t-dodecyl trisulfide	$C_{12}H_{26}S$	t-Dodecyl mercaptan [**]	—	114
$C_{34}H_{52}O_3S$	Cholesteryl p-toluenesulfonate	$C_{27}H_{46}$	Cholestene	—	112
			i-Cholestene	—	

[*] References 68–114 are on pp. 508–509.
[†] The product was isolated as mercury n-butyl mercaptide.
[‡] This product presumably resulted from atmospheric oxidation of the alkaline solution resulting after hydrolysis of the reaction mixture.
[§] The starting material was a mixture of the ethylthiol and the n-butylthiol esters of 5,5- and 6,6-difluorodecanoic acid.
[‖] The compound was recovered largely unchanged after two days' boiling.
[¶] Product not isolated. The yield was 67% based on the hydrogen evolved.
[**] Product not isolated. The yield was 100% based on the hydrogen evolved.

REFERENCES FOR TABLES II–IX

[68] Slabey and Wise, *J. Am. Chem. Soc.*, **71**, 3252 (1949).
[69] van Volkenburgh, Greenlee, Derfer, and Boord, *J. Am. Chem. Soc.*, **71**, 3595 (1949).
[70] Karrer and Eugster, *Helv. Chim. Acta*, **32**, 1934 (1949).
[71] Meek, Lorenzi, and Cristol, *J. Am. Chem. Soc.*, **71**, 1830 (1949).
[72] Witkop, *J. Am. Chem. Soc.*, **70**, 3716 (1948).
[73] Uhle, *J. Am. Chem. Soc.*, **71**, 765 (1949).
[74] Price and Schilling, *J. Am. Chem. Soc.*, **70**, 4265 (1948).
[75] Price and Voong, *J. Org. Chem.*, **14**, 111 (1949).

[76] Inhoffen, Bohlmann, and Bohlmann, *Ann.*, **565**, 35 (1949).
[77] Newman and Gaertner, *J. Am. Chem. Soc.*, **72**, 264 (1950).
[78] McKennis and Gaffney, *J. Biol. Chem.*, **175**, 217 (1948).
[79] Levin, Spero, McIntosh, and Rayman, *J. Am. Chem. Soc.*, **70**, 2958 (1948).
[80] Julian, Meyer, and Ryden, *J. Am. Chem. Soc.*, **72**, 367 (1950).
[81] Fürst and Plattner, *Helv. Chim. Acta*, **32**, 275 (1949).
[82] Plattner, Heusser, and Kulkarni, *Helv. Chim. Acta*, **31**, 1885 (1948).
[83] Plattner, Heusser, and Kulkarni, *Helv. Chim. Acta*, **32**, 1070 (1949).
[84] Goering, Cristol, and Dittmer, *J. Am. Chem. Soc.*, **70**, 3314 (1948).
[85] Lardon and Reichstein, *Helv. Chim. Acta*, **32**, 2003 (1949).
[86] Adams and Govindachari, *J. Am. Chem. Soc.*, **72**, 158 (1950).
[87] Martin, Schepartz, and Daubert, *J. Am. Chem. Soc.*, **70**, 2601 (1948).
[88] Kharasch and Büchi, *J. Org. Chem.*, **14**, 84 (1949).
[89] Karrer, Portmann, and Suter, *Helv. Chim. Acta*, **32**, 1156 (1949).
[90] Overberger and Roberts, *J. Am. Chem. Soc.*, **71**, 3618 (1949).
[91] Karrer and Portmann, *Helv. Chim. Acta*, **32**, 1034 (1949).
[92] Newman, Renoll, and Auerbach, *J. Am. Chem. Soc.*, **70**, 1023 (1948).
[93] Chaikin, *J. Am. Chem. Soc.*, **70**, 3522 (1948).
[94] Newman and Mangham, *J. Am. Chem. Soc.*, **71**, 3342 (1949).
[95] Collins, *J. Am. Chem. Soc.*, **70**, 2418 (1948).
[96] Guss, *J. Am. Chem. Soc.*, **71**, 3460 (1949).
[97] Wendler, Slates, and Tishler, *J. Am. Chem. Soc.*, **71**, 3267 (1949).
[98] Karrer, Karanth, and Benz, *Helv. Chim. Acta*, **32**, 436 (1949).
[99] Newman and Whitehouse, *J. Am. Chem. Soc.*, **71**, 3664 (1949).
[100] Dürst, Jeger, and Ruzicka, *Helv. Chim. Acta*, **32**, 46 (1949).
[101] Wieland and Miescher, *Helv. Chim. Acta*, **31**, 1844 (1948).
[102] van Dyken, unpublished work.
[103] Casanova and Reichstein, *Helv. Chim. Acta*, **32**, 647 (1949).
[104] Hunter and Hogg, *J. Am. Chem. Soc.*, **71**, 1922 (1949).
[105] Blicke and Sheets, *J. Am. Chem. Soc.*, **70**, 3768 (1948).
[106] Baker and Jenkins, *J. Am. Chem. Soc.*, **71**, 3969 (1949).
[107] Baker, *J. Am. Chem. Soc.*, **70**, 3857 (1948).
[108] Karrer and Portmann, *Helv. Chim. Acta*, **31**, 2088 (1948).
[109] Karrer, Eugster, and Waser, *Helv. Chim. Acta*, **32**, 2381 (1949).
[110] Henne, Alm, and Smook, *J. Am. Chem. Soc.*, **70**, 1968 (1948).
[111] Hatch and Perry, *J. Am. Chem. Soc.*, **71**, 3262 (1949).
[112] Schmid and Karrer, *Helv. Chim. Acta*, **32**, 1371 (1949).
[113] Marvel and Caesar, *J. Am. Chem. Soc.*, **72**, 1033 (1950).
[114] Arnold, Lien, and Alm, *J. Am. Chem. Soc.*, **72**, 731 (1950).

INDEX

Numbers in **bold-face** type refer to experimental procedures.

Acetoacetic ester condensation, *Vol. I*
Acetylenes, *Vol. V*
Acyloins, *Vol. IV*
Adamkiewicz test, 155
Aldehydes, preparation by lithium aluminum hydride reduction, 480
  preparation by Oppenauer oxidation, 222–223
Alkenylsuccinic acids, 48
Alkylation of aromatic compounds by Friedel-Crafts reaction, *Vol. III*
Alkylideneparaconic acids, 48
Alkylidenesuccinic acids, 48
Amination of heterocyclic bases by alkali amides, *Vol. I*
Aminoacetal, **199**
o-Aminobenzyl alcohol, **491**
2-Amino-4-methylthiazole, **380**
2-Amino-4-phenylthiazole, **380**
2-Aminothiazole, **379**
5-Amino-2-thiocarbonylthiazole, 369
Ammonium dithiocarbamate, use in synthesis of thiazoles, 374–376
Anhydroacetophenone disulfide, 439–440
Antipyrine, phosphonation of, 326–327
Arbuzov transformation, 276
Arndt-Eistert reaction, *Vol. I*
Arsinic and arsonic acids, synthesis by Bart, Bechamp, and Rosenmund reactions, *Vol. II*
Azlactones, *Vol. III*

Benzalaminoacetal, 192
Benzenephosphonic acid, **302**
Benzoins, *Vol. IV*
Benzoquinolizines, 79
Benzoquinones, preparation by oxidation, *Vol. IV*
1-Benzyl-1-carboxy-1,2,3,4-tetrahydro-2-carboline, **174**
1-Benzyl-3,4-dihydro-2-carboline, **103**

1-Benzyl-1,2,3,4-tetrahydro-2-carboline, **174**
Biaryls, *Vol. II*
Biogenesis of alkaloids, 154–155
Bischler-Napieralski reaction, 74–150
  condensing agents, 98–100
  direction of ring closure, 80–83
  ease of cyclization, 90–98
  experimental conditions, 98–100
  experimental procedures, 100–103
  location of double bond formed, 83–85
  side reactions, 85–90
  tables, 91, 94, 95, 97, 103–150
Bis(4-dimethylaminophenyl)phosphinic acid, **324–325**
Bis(4-dimethylaminophenyl)phosphonic acid, **324–325**
8-Bromoisoquinoline, **200**
p-Bromophenyltriphenylsilane, **355**
3-Bromopropanephosphonic acid, **287**
Bucherer reaction, *Vol. I*
t-Butyl alcohol, anhydrous, **44**
n-Butyldichlorophosphine, **319**
n-Butyllithium, **352**

Cannizzaro reaction, *Vol. II*
β-Carbethoxy-γ,γ-diphenylvinylacetic acid, **42**, 47
3-Carbethoxy-4-phenyl-3-pentenoic acid, **46**
2-Carbolines, 79, 94, **103**, 152–153, **173–174**
  tables, 142–143, 185–186
  tryptophan test, 155
β-Carbomethoxy-β-(3-methyl-1,2,3,4-tetrahydro-1-phenanthrylidene)propionic acid, **44**
3-Carboxy-1,2,3,4-tetrahydro-2-carboline, 155
Chloroacetyl-substituted esters, use in synthesis of thiophenes, 419–420

Chloromethylation of aromatic compounds, *Vol. I*
α-Chloro-α-phenylethanephosphonic acid, **313**
Cholestenone, 234
Chrysean, 369
Cinnamyl alcohol, **490**
Claisen rearrangement, *Vol. II*
Clemmensen reduction, *Vol. I*
Curtius reaction, *Vol. III*
Cyanoethylation, *Vol. V*
Cyclic ketones, preparation by intramolecular acylation, *Vol. II*
α-Cyclocitral, **236**

Darzens glycidic ester condensation, *Vol. V*
cis-α-Decalone, **235**
Desoxycorticosterone acetate, **235**
Dialkylanilines, conversion to phosphinic and phosphonic acids, 323–325
Dialkylidenesuccinic acids, 48
2,3-Diaminobutane-1,4-disulfuric acid ester, 445
Dibenzophosphazinic acid, **325**
Dibenzylphosphinic acid, **306**
Dibutyl alkanephosphonates, **289**
Di-*t*-butyl succinate, use in Stobbe condensation, 39
Di-β-chloroethyl β-chloroethylphosphonate, **287**
Dichlorophosphines, conversion to phosphinic and phosphonic acids, 297–298
preparation, 297–303, 316–319
Dieckmann condensation, use in synthesis of tetrahydrothiophenes, 449–463
Diels-Alder reaction, with cyclenones, *Vol. V*
with ethylenic and acetylenic dienophiles, *Vol. IV*
with maleic anhydride, *Vol. IV*
Diene synthesis, *see* Diels-Alder reaction
Diethyl benzenephosphonate, 302–303
Diethyl benzoylphosphonate, **286**
Diethyl cyclohexane-1,4-dione-2,5-dicarboxylate, 38
Diethyl ethanephosphonate, **286**
Diethyl 2-hydroxyethanephosphonate, **290**
Diethyl succinate, self-condensation, 38

Diethyl thiodiglycolate, use in Stobbe condensation, 20
1,4-Dihalides, use in preparation of tetrahydrothiophenes, 443–445
3,4-Dihydroisoquinolines, preparation by Bischler-Naperialski reaction, 74–150
disproportionation, 90
oxidation by atmospheric oxygen, 88
Di(α-hydroxyisopropyl)phosphinic acid, **306**
3,4-Dihydroxytetrahydrothiophene, **447**
1,2-Diketones, use in preparation of thiophenes, 435–436
1,4-Diketones, use in preparation of thiophenes, 418–419
Dilactones from Stobbe condensation, 21, 26
p-Dilithiumaminophenyllithium, 347
3,5-Dimethoxybenzyl alcohol, **491**
1-(2,3-Dimethoxybenzyl)-6,7-dimethoxy-3,4-dihydroisoquinoline, **100**
3,11-Dimethoxy-5,6-dihydro-8H-dibenzo[*a,g*]quinolizine, **102**
Dimethyl 3,4-dihydroxy-2,5-thiophenedicarboxylate, **438**
1,3-Dimethyl-6,7-dimethoxyisoquinoline, **101**
Dimethyl succinate, **45**
2,4-Dimethylthiazole, **379**
2,3-Dimethylthiophene, **422**
3,4-Dimethylthiophene, **421**
Diphenylchlorophosphine, **321**
$\Delta^{20,23}$-24,24-Diphenylcholadiene-3,11-dione, **235**
Diphenyl-2,4-dimethoxy-5-bromophenylcarbinol, **356**
α,α-Diphenyl-α-hydroxymethanephosphonic acid, **314**
Diphenylphosphinic acid, **296**
3,4-Diphenyl-2,5-thiophenedicarboxylic acid, **438**
Di-*n*-propylchlorophosphine, **321**
3,4-Di-*n*-propyltetrahydrothiophene, **449**

Elbs reaction, *Vol. I*
Equilenones, preparation by Stobbe condensation, 25, 36
Eschweiler reaction with homopiperonylamine, 162
See also Leuckart reaction, *Vol. V*

INDEX

Ethanephosphinic acid, **297**
Ethyl diphenylphosphinate, **302–303**
Ethyl 3-keto-2-(4'-methoxybutyl)tetrahydrothiophene-4-carboxylate, **459**
9-Ethylphenanthridine, **102**

Fluorine compounds, aliphatic, *Vol. II*
  aromatic, *Vol. V*
Friedel-Crafts reaction with aliphatic dibasic acid anhydrides, *Vol. V*
Fries reaction, *Vol. I*
Fulgenic acids, **7**, 17–18
Fulgides, **7**, 18
Fused-ring ketones, preparation by Diels-Alder reaction, *Vol. V*

Gattermann-Koch reaction, *Vol. V*
Glycidic esters, preparation by Darzens reaction, *Vol. V*
Glyoxal semiacetal, 193

Halogen-metal interconversions with organolithium compounds, 339–366
  experimental conditions, 351–352
  experimental procedures, 352–356
  mechanism, 341–342
  scope, 342–348
  side reactions, 348–351
  tables, 356–366
Harman, 155
Hoesch synthesis, *Vol. V*
Hofmann reaction, *Vol. III*
1-Homoveratryl-6,7-dimethoxy-3,4-dihydroisoquinoline, **100**
Hopkins-Cole test, 155
Hydrohydrastinine, 162
7-Hydroxy-8-chloroisoquinoline, **200**
α-Hydroxyethanephosphonic acid, **307**
Hydroxymethanephosphonic acid, 313
2-Hydroxy-4-methylthiazole, **382**
5-Hydroxy-2-methylthiophene, **422**
α-Hydroxy-α-phenylethanephosphonic acid, 314
α-Hydroxy-α-toluenephosphonic acid, 313, 314

Indones, preparation by Stobbe condensation, 23, 29–30
ψ-Ionone, 223, **237**

Irone, preparation by Oppenaurer oxidation, 223
Isoparaconic acids, 48
Isopropyl isopropylphenylphosphinate, 288
2-Isopropylthiophene, **422**
Isoquinoline, **201**
Isoquinolines, preparation by Pictet-Gams reaction, 76, 125
  preparation by Pomeranz-Fritzsch reaction, 191–206
  1-substituted, 196–197
  See also Bischler-Napieralski reaction; Pictet-Spengler reaction

Jacobsen reaction, *Vol. I*

Ketenes and ketene dimers, preparation, *Vol. III*
γ-Keto acids, use in preparation of thiophenes, 414–417
β-Ketobutyraldehyde 2-methylpentane-2,4-diol acetal, 237
α-Keto esters, use in preparation of thiophenes, 436–437

γ-Lactones from Stobbe condensation, 22, 26–27
Lactonic acids from Stobbe condensation, 21, 25–26
Leuckart reaction, *Vol. V*
Lithium aluminum hydride, conditions for use, 486–489
  fire hazard, 489
  formation, 483
  properties, 483–484
  solutions, preparation and analysis, 484–486
  See also Reduction with lithium aluminum hydride

Mannich reaction, *Vol. I*
Meerwein, Ponndorf, Verley reduction, *Vol. II*
2-Mercapto-4-methylthiazole, **381**
2-(4'-Methoxybutyl)-3-ketotetrahydrothiophene, **459**
1-(4'-Methoxyphenyl)-2-benzoylethane-1-phosphonic acid, **315**
6-Methoxy-1,2,3,4-tetrahydroisoquinoline, **172**

514 INDEX

Methyl *dl*-4-benzamido-3-ketotetrahydrothiophene-2-carboxylate (sodium salt), **459**
1-Methyl-1-carboxy-6,7-dihydroxy-1,2,3,4-tetrahydroisoquinoline, **173**
1-Methyl-3,4-dihydroisoquinoline, **100**
1-Methyl-6,7-dihydroxy-1,2,3,4-tetrahydroisoquinoline, **172**
1-Methyl-6,7-dimethoxy-8-hydroxyisoquinoline, **201**
2,3-Methylenedioxy-10,11-dimethoxy-5,6,13,13a-tetrahydro-8H-dibenzo-[*a,g*]quinolizine, **173**
2,3-Methylenedioxy-11,12-dimethoxy-5,6,8,9-tetrahydrodibenzo[*a,h*]quinolizinium iodide, 102
4-Methyl-5-(β-hydroxyethyl)thiazole, **379**
Methyl $\Delta^{4,6}$-3-ketoetiocholadienate, **236**
1-Methyl-7-methoxyisoquinoline, **202**
Methyl β-2-(5-phenylthienyl)propionate, **422**
1-Methyl-1,2,3,4-tetrahydro-2-carboline, **173**

Naphthalic anhydrides from Stobbe condensation, 25, 35
Naphthols from Stobbe condensation, 22–23, 28–29
1-Naphthylmethanephosphonic acid, **286**
1-(*o*-Nitrobenzyl)-6,7-dimethoxy-3,4-dihydroisoquinoline, **101**
7-Nitro-9-phenylphenanthridine, 102
Norcoralydine, 165
Norharman, 89
Norhydrohydrastinine, 156

Oppenauer oxidation, 207–272
    catalysts, 225–228
    condensation of aldehydes formed, 223–224
    experimental conditions, 225–234
    experimental procedures, 234–238
    hydrogen acceptors, 228–231
    isolation of products, 233–234
    mechanism, 209
    migration of double bonds, 212, 215
    of nitrogen-containing alcohols, 219–224
    of polyhydroxyl compounds, 216–219

Oppenauer oxidation, of polyhydroxyl compounds, preferential oxidation, 216–219
    of primary alcohols, 222–224
    of saturated alchols, 210–211
    of steroids, 210–211, 212–214, 216–219, 221
    of unsaturated alcohols, 212–216
    scope, 210–224
    side reactions, 224–225
    solvents, 231–232
    tables, 238–272
Organolithium compounds, color test for, 343
    p-dilithiumaminophenyllithium, 347
    halogen-metal interconversion, 339–366
    metal-metal interconversion, 349
Organomercury compounds, use in preparation of dichlorophosphines, 317–319
Oxalic esters, use in preparation of thiophenes, 437
Oxazoles from N-acylphenacylamines, 86
Oxazolines from hydroxyphenethylamines, 87

Papaverine, 76
Paraconic acids, 48
Periodic acid oxidation, *Vol. II*
Perkin reaction and related reactions, *Vol. I*
Phenanthridines, 79, 81, 91, 93, **103**
    tables, 131–133
Phenyl-*p*-bromophenylchlorophosphine, **319**
Phenyl-di-(*p*-aminophenyl)arsenic, **354**
Phenyldichlorophosphine, **318**
2-Phenyl-4,5-dimethylthiazole, **379**
Phenylethynephosphonic acid, **293**
1-Phenylisoquinoline, **101**
Phenyllithium, **353**
1-Phenyl-3-methyl-5-chloropyrazole-4-phosphonic acid, **327**
N-Phenylpyrrolidine, **492**
5-Phenylthiazole, **381**
γ-Phenylvalerolactone, **47**
Phosphines, oxidation to phosphinic and phosphonic acids, 322
Phosphinic acids, preparation, 273–338
    by addition phosphorous acids to carbonyl compounds, 304–308

INDEX

Phosphinic acids, preparation, by addition phosphorus chlorides to carbonyl compounds, 308–315
  by miscellaneous methods, 322–327
  by pyrolysis, 316–317
  from Grignard reagents, 293–297
  from organomercury compounds, 317–319
  from phosphonites, 279, 283, 288
  from phosphonium compounds, 319–321
  tables, 327–338
Phosphonic acids, preparation, 273–338
  by addition phosphorous acids to carbonyl compounds, 304–308
  by Friedel-Crafts reaction, 297–303
  by miscellaneous methods, 322–327
  by pyrolysis, 316–317
  β-keto, 309–310, 315
  from alkyl phosphites, 276–291
    experimental conditions, 284–286
    experimental procedures, 286–291
    mechanism, 278–279
    scope and limitations, 279–283
  from Grignard reagents, 293–297
  from organomercury compounds, 317–319
  from phosphonium compounds, 319–321
  from phosphonous acids, 321–322
  from phosphorus pentachloride and unsaturated compounds, 291–293
  α-hydroxy, 306–307, 313–315
  tables, 327–338
  α,β-unsaturated, 291–293
Phosphonitrilic chloride, 300
Phosphonium compounds, thermal decomposition, 319–321
α-Phosphono-α-hydroxypropionic acid, **315**
Phosphonous acids, disproportionation, 321–322
Phosphorus pentachloride, addition to unsaturated compounds, 291–293
Phosphorus pentasulfide, use in preparation of phosphonic acids, 295–297
Phosphorus sulfides, 412
  in synthesis of phosphonic acids, 295–297
  in synthesis of thiazoles, 376–377

Phosphorus sulfides, in synthesis of thiophenes, 412–419
Phthalazines, 79
Pictet-Gams synthesis of isoquinolines, 76
Pictet-Spengler reaction, 151–190
  condensing agents, 168–169, 172
  direction of ring closure, 157–162
  ease of cyclization, 164–168
  experimental conditions, 168–172
  experimental procedures, 172–174
  mechanism of cyclization, 156–157
  side reactions, 162–163
  tables, 166, 174–190
  under physiologically possible conditions, 153–155, 165, 168–172
N-Piperidylphosphoryldichloride, use in preparation of phosphinic acids, 294–296
Pomeranz–Fritsch reaction, 191–206
  application, 199
  experimental procedures, 199–202
  mechanism, 193
  scope and limitations, 193–199
  tables, 202–206
Potassium, handling, 42–43
Potassium t-butoxide, **42–44**
3-n-Propylthiophene, **421**

Quininone from Oppenauer oxidation, 220
Quinolizines, 102, 173
  tables, 137–140, 181–183
  See also Bischler-Napieralski reaction; Pictet-Spengler reaction

Rearrangement, allylic, in alkylation of phosphate esters, 281–282
  of double bonds in Oppenauer oxidation, 212, 215
Reduction with aluminum alkoxides, *Vol. II*
Reduction with lithium aluminum hydride, 469–509
  aldehydes, 474–476
  amides, 479–480
  carboxylic acids, 478
  double bonds, 481–483
  epoxides, 476–477
  esters, 477–478
  experimental conditions, 486–489

Reduction with lithium aluminum hydride, experimental procedures, 489–492
  fire hazard, 489
  functional groups reduced, 471
  halogen compounds, 480–481
  heterocyclic nitrogen compounds, 483
  interference by active hydrogen, 473–474
  ketones, 474–476
  mechanism, 471–473
  of nitriles, 480
  stereochemical specificity, 475, 477–478
  tables, 493–509
  See also Lithium aluminum hydride
Reductive alkylation, *Vol. IV*
  See also Leuckart reaction, *Vol. V*
Reformatsky reaction, *Vol. I*
Replacement of aromatic primary amino groups by hydrogen, *Vol. II*
Resolution of alcohols, *Vol. II*
Retronecanol, 219–220
Rosenheim test, 155
Rosenmund reduction, *Vol. IV*

Schiemann reaction, *Vol. V*
Schiff bases in Pomeranz-Fritsch reaction, 192–194
Schlittler-Müller reaction, 193, 197–198
Schmidt reaction, *Vol. III*
Selenium dioxide oxidation, *Vol. V*
Sodium hydride, handling, 46
Stearamidomethanephosphonic acid, **290**
Stobbe condensation, 1–73
  applications, 21–36
  condensing agents, 36–40
  experimental conditions, 36–41
  experimental procedures, 41–47
  isolation of products, 14–15
  mechanism, 2
  proof of configuration of products, 16–17
  proof of structure of products, 15–16
  related condensations, 19–21
  scope and limitations, 5–21
  side reactions, 36–38, 40–41
  tables, 48–73
  with aldehydes, 5–7
  with cyano ketones, 13
  with diketones, 12

Stobbe condensation, with hindered ketones, 10
  with keto esters, 12
  with ketones, 7–10
  with substituted succinic esters, 17–19
  with α,β-unsaturated ketones, 11
α-Styrenephosphonic acid, **314**
β-Styrenephosphonic acid, **292**
Succinic acids, use in preparing thiophenes, 413–414
Succinoylation, 40
Sulfonation of aromatic hydrocarbons and their halogen derivatives, *Vol. III*

Tetrahydro-ψ-berberine, 158–159
Tetrahydroharman, 154
Tetrahydroindanones from Stobbe condensation, 23, 31–32
Tetrahydroisoquinolones, 151–190
dl-($trans$)-Tetrahydrothiophene-1,5-dicarboxylic acid, **447**
Tetrahydrothiophenes, 443–468
  by catalytic methods, 464–465
  by Dieckmann cyclization, 449–463
    experimental conditions, 458–459
    experimental procedures, 459
  by miscellaneous methods, 465–468
  from 1,4-difunctional compounds and sulfides, 443–448
    experimental conditions, 446–447
    experimental procedures, 447, 449
  from α-mercapto esters and unsaturated compounds, 456–458
  tables, 448, 460–463, 465, 468
Tetralonecarboxylic acids from Stobbe condensation, 24, 32–33
Tetralones from Stobbe condensation, 24, 34
2,3,9,10-Tetramethoxy-7,12,12a,13-tetrahydro-5H-dibenzo[b,g]quinolizine, **173**
Tetraphenyl ethane-1,2-diphosphonate, **288**
Tetraphenylthiophene, 419, 431–432
Thiazoles, 367–409
  2-alkoxy, 372
  2-amino, 372–373
    experimental procedures, 378–382
  from α-acylaminocarbonyl compounds and phosphorus pentasulfide, 376–377, 381

Thiazoles, from ammonium dithiocarbamate and α-halo ketones, 374–376, 380
  from thioamides and α-halo carbonyl compounds, 369–374, 378
  from α-thiocyanoketones, 377–378, 381–382
  2-mercapto, 374
  tables, 370, 382–409
Thiazolium salts, 372
Thioamides, use in preparation of thiazoles, 369–374
γ-Thiobutyrolactone, 467
Thiocyanogen, addition and substitution reactions, *Vol. III*
Thiophene, 430
2,5-Thiophenedicarboxylic acid, 355
Thiophenes, 410–443
  alkoxyl, 416–417
  alkyl, 413–415
  aryl, 414, 416
  by decarboxylation of thiophene-2,5-dicarboxylic acids, 437
  by miscellaneous cyclizations, 441–443
  2,5-disubstituted, 418–419
  from aryl methyl ketones, 439–441
  from 1,4-difunctional compounds and sulfides, 412–427
    experimental conditions, 420–421
    experimental procedures, 421–423
  from thiodiacetic acid esters, 435–438
  from unsaturated compounds, 428–434
  tables, 421, 424–427, 432–433, 438, 440, 443
Thiosuccinic anhydride, 461

Thiourea, 372
α-Toluenephosphonic acid, **306**
p-Toluenephosphonic acid, **301**
p-Tolylphenylchlorophosphine, **318**
Trialkyl phosphites, **286**
  conversion to phosphonic acids, 276–291
Triarylcarbinols, reaction with phosphorus trichloride, 277, 281–282
Triarylmethanephosphonic acids, 277, 281–282
2,2,2-Trichloroethanol, 489
Triethyl β-phosphonopropionate, **289**
2-(*m*-Trifluoromethylphenyl)quinoline, **354**
2,3,5-Trimethylthiophene, **422**
2,4,5-Triphenyl-3-furancarboxylic acid, **354**
Triphenyl-*o*-hydroxymethylphenyllead, **355**
Triphenylmethanephosphonic acid, 290
Tryptophan tests, 155

Unsaturated acids from Stobbe condensation, 22, 26–27

Vitamin A alcohol, **490**

Willgerodt reaction, *Vol. III*
Wolff-Kishner reduction, *Vol. IV*
Wurtz reaction, use in preparation of phosphinic acids, 326

Yobyrone, 89
Yohimbone, 220